System Modeling and Identification

Rolf Johansson

Department of Automatic Control
Lund Institute of Technology

PRENTICE HALL, Englewood Cliffs, NJ 07632

Library of Congress Cataloging-in-Publication Data

Johansson, Rolf.
System modeling and identification / by Rolf Johansson
p. cm.
Includes bibliographical references and index.
ISBN 0-13-482308-7
1. System identification--Mathematical models I. Title
QA402.J62 1993
003'.1--dc20
92-32044
CIP

Acquisitions editor: **PETER JANZOW**
Editorial/production supervision and
interior design: **RICHARD DeLORENZO**
Cover design: **KAREN MARSILIO**
Prepress buyer: **LINDA BEHRENS**
Manufacturing buyer: **DAVID DICKEY**
Editorial assistant: **PHYLLIS MORGAN**
Supplements editor: **ALICE DWORKIN**

To the memory of Nils and Olga Johansson

Printed in the United States of America

10 9 8 7 6 5 4 3 2 1

ISBN 0-13-482308-7

Prentice-Hall International (UK) Limited, London
Prentice-Hall of Australia Pty. Limited, Sydney
Prentice-Hall Canada Inc., Toronto
Prentice-Hall Hispanoamericana, S.A., Mexico
Prentice-Hall of India Private Limited, New Delhi
Prentice-Hall of Japan, Inc., Tokyo
Simon & Schuster Asia Pte. Ltd., Singapore
Editora Prentice-Hall do Brasil, Ltda., Rio de Janeiro

Contents

Preface

In many scientific problems an essential step toward their solution is to accomplish modeling and identification of some object or system under investigation. As defined here, system identification is the process of deriving a mathematical system model from observed data in accordance with some predetermined criterion. The increasing expansion in the use of system identification is the result of demands imposed by advances in other scientific and technological areas such as biomedicine, physics, electrical engineering, and computer science.

Modeling and identification as a methodology dates back to Galileo (1564-1642), who also is important as the founder of dynamics. Galileo was the first to establish the law of falling bodies, a law which states, *i.e.*, that if a body is falling freely in vacuum, its velocity increases at a constant rate. The key to Galileo's success was his combination of theoretical and experimental work, with patience in observation and boldness in framing hypotheses. Unfortunately, his ideas brought him into conflict with Aristotelian physics and the Church, and in 1633 Galileo was forced to abjure his "heresies" by the Inquisition, which was successful in putting an end to science in Italy.

Hence, the important role of modeling and identification in science and technology is to establish empirical relationships between observed variables. The standard view is that mathematical models are computational devices that should be distinguished from theories about physical structure. In a mature form, modeling can even be used in a theory when attempting to explain empirical laws by incorporating them into a deductive system. Although analogies might certainly be of great value to guide further research, it is important to state that an approach solely based on appeal to a model or an analogy is insufficient for the purposes of scientific explanation. For this reason modeling and identification are important in the early phases of scientific work where hypotheses are formulated and tested, refuted, or confirmed. It is hoped that the text will prove useful in such work.

As modeling and identification are omitted or neglected in many graduate curricula, the knowledge aquired by students is largely confined to techniques whose applicability they cannot ascertain. This volume represents an attempt to remedy this by providing an integrated collection of laboratory experiments that illustrate the variety of situations to which quantitative identification and modeling methods may be applied. The book has grown out of lecture notes for a course on system identification held at the Lund Institute of Technology. The text is intended for a senior-level or graduate-level course in system identification for students with some background in applied mathematics (control theory) and statistics (stochastic processes). As a basic course text, it is intended to furnish a broad perspective of this area of research and prepare the student for various forms of further study and research. In the book some sections and exercises are marked $*$ to indicate a more advanced or specialised subject matter. I have tried to establish some generalizations for this field, which to some extent is little more than an empirical art. Accordingly, I try to develop an appreciation of limitations and capabilities by discussing representative examples in major areas of application.

A necessary complement to this book is some software for basic numerical computation such as Ctrl-C, Mathematica, Matlab, Matrix-X, X-math, etc. Implementation of identification algorithms without such support may easily lead to numerically inaccurate results. Basic prerequisites include numerical implementations of linear algebra, and preferably some numerical optimization. Other valuable tools include software for the simulation of dynamical systems, such as the Omola, Simnon, or Simulink programs, and some graphics interface. Several software houses also offer supplementary software for solving problems of signal processing and identification.

Among those who corrected errors and suggested improvements are: Leif Andersson, Bo Bernhardsson, Ola Dahl, Per A. Fransson, Kjell Gustafsson, Anders Hansson, Ulf Holmberg, Ulf Jönsson, Mats Lilja, and Henrik Olsson, to all of whom I owe sincere thanks. I would also like to thank Professor Måns Magnusson and our staff at the Lund University Hospital, with whom I have collaborated in the practical application of identification.

I would also like to thank my esteemed colleagues Leif Andersson, Karl J. Åström, Per Hagander, Jan Holst, Ulla Holst, Tore Hägglund, Georg Lindgren, Holger Rootzén, and Björn Wittenmark at the Departments of Automatic Control and Mathematical Statistics at Lund Institute of Technology for creating the atmosphere in which I teach and work. I am also grateful for the research semester granted by the Lund Institute of Technology, which has been helpful during preparation of the manuscript. Compiling a textbook is naturally a time-consuming undertaking, and I would like to thank my family for generous support of an occasionally preoccupied family member. Acknowledgement is also extended to Mr. Peter Janzow, Editor, and to Professor Thomas Kailath, Series Editor, and their staff at Prentice Hall for their enthusiasm and patience during preparation of the manuscript. Aside from these personal acknowledgements the book owes much to the numerous authors of original articles and textbooks.

Lund, Scandinavia, Midsummer Eve 1992

Rolf Johansson

1

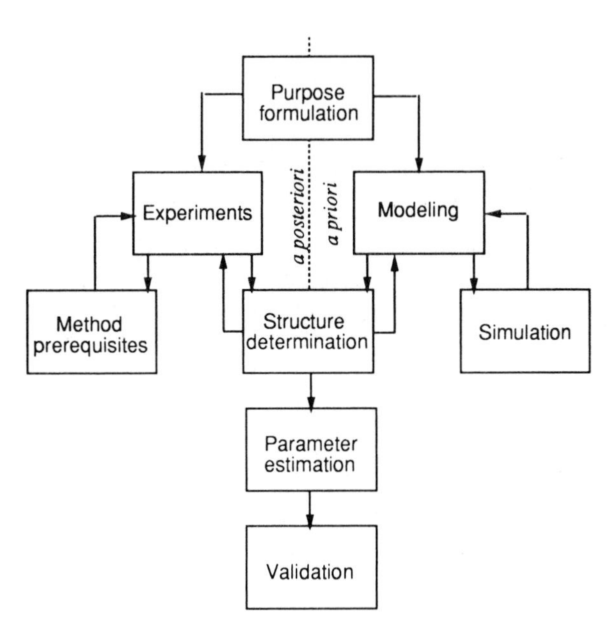

Introduction

1.1 WHY MODELS?

Decision making and problem solving are dependent upon access to adequate information about the problem to be solved. Often the available information is originally in the form of data or observations that require interpretation before further analysis (and decisions) can be made. The derivation of a relevant system description from observed data is termed *system identification*, and the resultant system description a *model*.

Why are models needed? A general answer is that modeling and identification methods are needed for the interpretation of—often indirect—observations and measurements obtained from some system of study. As models constitute the necessary link between experiments and decision making, modeling and identification are manifestly important for all applied science.

A *model* represents essential aspects of a system with respect to certain purposes and may take on several different forms such as

- Cognitive models (human concepts)
- Normative models (purpose oriented)
- Descriptive models (behavior oriented)
- Functional models (action and control oriented)

Cognitive models are the conceptual models underlying human reasoning and perception, inductive learning, decision making, and planning—*i.e.*, human effort to understand and control the ambient world. Another category of models is that of *normative models*, which define the specified or desired function, goal, or purpose of a system or process. Such models are often found in engineering design and government regulations.

Other classes of models arise from the need for *descriptive* and *functional models* for scientific and technological purposes. Such models are often subdivided into *quantitative models* (described by numbers or parameters) and *qualitative models* (described by categorical data). A necessary scientific background for the development of more precise normative models as well as of cognitive models includes an understanding of descriptive and functional quantitative models based on empirical data as used in science, technology, and economics. The value of quantitative models also derives from their ability to predict, and therefore to act upon, phenomena. For this reason, models for scientific and technological use are often quantitative models, and a central problem is to fit such models to data. From an empirical point of view, it is natural to start considering the collecting of input-output data from a system in operation where experiments are performed by manipulating the input. Such models serve to determine criteria of lawful change and thus fulfill at least three purposes:

- *Prediction:* Given a description of the system over some period of time and the set of rules governing the change, predict the way the system will behave in future time.
- *Learning new rules:* Given a description of the world at different times, produce a set of rules which accounts for regularities in the system.

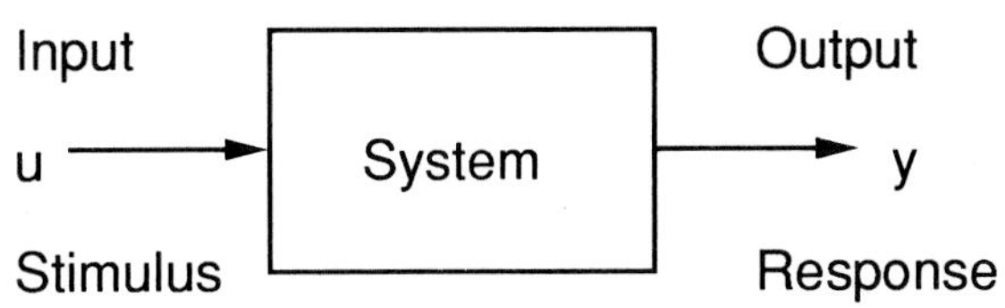

Figure 1.1 An input-output relationship or a stimulus-response relationship.

– *Data compression:* Produce a model that represents data on a compact form and with low complexity.

Quantitative models may, of course, be formulated with different degrees of complexity, detail, and internal structure. A purely behavioristic model devoid of assumptions regarding internal structure is the *black box model*, which simply models a causal relationship between *input* and *output* (*cf.* Fig. 1.1). The model may be *static* or *dynamic*, where a dynamical system is understood to mean one where output is determined not only by its input but also by some internal *state* that, in turn, may depend on previous input. If the state variables vary with time, then the dynamical system is said to undergo a *process*.

Recourse to the lack of internal structure in the black box model may be motivated by a desire to avoid irrelevant detail or simply by inability to connect into the system under consideration. Such approaches are common in control systems analysis and in biological and biomedical research.

1.2 MODELING

The search for relevant models may also start from another point. Provided that the structure or design of a system is largely known, it is often possible to produce a block diagram or some network sketch of the system and its functional components. Such practice is called *modeling*, which usually proceeds by starting from a set of (ideal) model components and gives rise to a physical model with some network structure. The resultant network behavior can be determined from the network structure and from the properties of the balance equations between all interacting components. The manner in which individual subsystems interconnect and act upon each other provides the overall system with an organizational pattern.

The behavior of a system obtained from balance equations derived from physics may be described in detail by a set of algebraic and differential equations, which in turn may be solved analytically or by simulation.

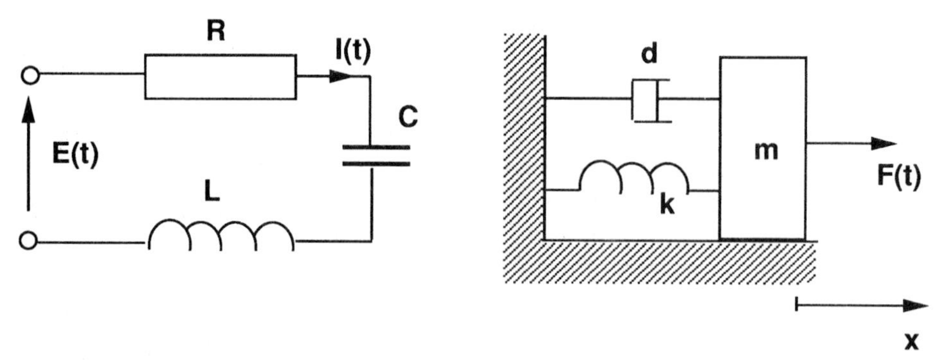

Figure 1.2 Physical models of an electrical model and a mechanical model with components (capacitor, resistor, and a coil) and (mass, damping action, spring) with coefficients C, R, L and m, d, k.

Example 1.1—Simple electrical and mechanical models
Consider Fig. 1.2, which shows simple physical models, one an electrical circuit with a resistor, a capacitor, and an inductor, and the other a mechanical model with a spring, a damper, and mass.

$$\begin{aligned} E(t) &= \mathbf{R}I(t) + \frac{1}{\mathbf{C}}\int I(t)dt + \mathbf{L}\frac{dI(t)}{dt} && \text{Kirchhoff voltage law} \\ \mathbf{m}\ddot{x}(t) &= -\mathbf{d}\dot{x}(t) - \mathbf{k}x(t) + F(t) && \text{Newton's second law} \end{aligned} \tag{1.1}$$

It is often natural to distinguish between *parameters* and *variables*, where parameters denote the constant (or slowly varying) coefficients, *e.g.*, **R, C, L** and **m, d, k**, and variables (or *signals*) denote time-varying quantities, *e.g.*, voltage $E(t)$, current $I(t)$, force $F(t)$, velocity $\dot{x}(t)$, and position $x(t)$. A parameter can also be regarded as a variable that has a constant value for a specific purpose or process.

The causal relationship between different variables can be emphasized by adopting the terminology of *input* (*e.g.*, voltage $E(t)$ or force $F(t)$) for the forcing variable, and *output* (*e.g.*, current $I(t)$ or momentum $p = \mathbf{m}\dot{x}$) for the dependent variable. If we apply the Laplace transform $\mathcal{L}\{\cdot\}$ and try to solve for the relationship between input and output, then we have the *transfer function*

$$\begin{aligned} G_1(s) &= \frac{\mathcal{L}\{I(t)\}}{\mathcal{L}\{E(t)\}} = \frac{s\mathbf{C}}{s^2\mathbf{LC} + s\mathbf{RC} + 1} \\ G_2(s) &= \frac{\mathcal{L}\{p(t)\}}{\mathcal{L}\{F(t)\}} = \frac{\mathbf{m}s}{\mathbf{m}s^2 + \mathbf{d}s + \mathbf{k}} \end{aligned} \tag{1.2}$$

The transfer functions obtained with *physical parametrizations* in terms of the parameters $\mathbf{R}, \mathbf{L}, \mathbf{C}$ (or $\mathbf{m}, \mathbf{d}, \mathbf{k}$) are special cases of input-output models. In general, the models thus obtained are called *parametric models* (or *parametrized models*) and represent a given structure where the parameters are sometimes not known and must be estimated. During modeling and experimentation, the engineer or scientist drafts a block diagram or a sketch of the system and its functional components. Then he disturbs the system and traces response through the system. This procedure requires continuous measurement of responses as well as quantitative control over inputs. The subsequent mathematical problem consists in determining which differential equations govern the behavior of the system from which data have been obtained. This procedure differs from the task of solving differential equations, and the parameter estimation problem is therefore called an *inverse problem*. Inverse problems typically involve models of known physical structure and physical parameterization and interconnections but with unknown parameters. Such models are often designated *grey box models* and are commonly found in physics, technology, and control systems analysis. An example from physics is the inverse scattering problem. Here the scientific objective is to determine the physical structure that generates a given scattering pattern, which, of course, is valuable for the interpretation of data.

Many modeling problems describe a condition of equilibrium associated with a minimum of energy. Such approaches often suffice to describe static properties and are natural first steps in many cases of modeling. A second step is to describe oscillations or transients around that equilibrium when the energy is conserved, released, or dissipated. Such approaches are useful for modeling resonances and vibrations in mechanical structures, monitoring variations in the neutron flow dynamics of nuclear reactors, or modeling vocal tract dynamics in speech processing.

Modeling of oscillating resonant behavior is important in order to predict the stability, amplitude, and frequency of resonant or limit cycle oscillations. Commonly used models are autoregressive linear models (*cf.* Chapter 5), or nonlinear models such as *describing function analysis* (*cf.* Chapter 12). More recently, the study of nonlinear oscillations has been focused to some extent on models of aperiodic oscillative behavior ("chaos"), for instance, in hydrodynamics.

The nature of some sustained oscillating behavior is difficult to explain from conditions of equilibrium and conservation of energy. Consider, for instance, such periodic biological phenomena as circadian, monthly, and annual rhythms or periodic phenomena observed in glycolytic metabolism and respiratory mechanics. Some of these systems exhibit *limit cycle* behavior where the system evolves toward a well-defined oscillation, no matter what initial conditions are imposed.

The modeling complexity required obviously depends upon the purpose of the modeling and identification. Modeling is in this respect an art based on the ability to visualize physical and other interconnections where all basic and applied knowledge contributes to modeling expertise.

1.3 THE PURPOSE OF IDENTIFICATION

As the modeling complexity required depends upon the purpose of the modeling and identification, it is desirable to distinguish a number of important application areas. For instance, it is necessary to stress the basic scientific need of quantitative models, which in turn presupposes an ability to predict new phenomena. *Prediction* or *forecasting* are areas closely related to modeling and identification, as the attempts to predict future states of a system are limited by the accuracy of the model used and the range of correlations of random processes affecting the system. In this context, it is important to represent external actions and external perturbations, extracting and using knowledge of statistical characteristics of random variables, as there is usually little theoretical or practical possibility of determining such characteristics in advance.

Control systems analysis and design provide a rich field for the application of modeling and identification. A *control* mechanism is one that senses the control error, *i.e.*, the difference between desired and actual states, and then initiates a series of processes and actions, which in turn produce counteractive effects to minimize the control error. This control principle introduces the important concept of *feedback*, which for linear systems can be illustrated in terms of a transfer function. This case also allows quantitative predictions to be made concerning crucial features of control systems such as stability conditions and the development of oscillatory behavior.

Another established application area is signal processing for state estimation of variables not available to measurement. An example is how to estimate the velocity from data records of position in Example 1.1. This can be viewed as a type of indirect measurement where modeling and identification are necessary for correct calibration—*e.g.*, by comparison with some other standardized method. A calibration can often be reduced to a linear regression problem, which provides a linear fit between the outputs of two different measurement devices.

Simulation based on mathematical models is widely used for the assessment of model complexity, for engineering design, or for operator training, all of which require adequate modeling and adequate input. Examples from electric power engineering are abundant, and such simulation models often serve as the basis of monitoring or supervision, *error detection*, and *process diagnosis* in large systems.

For their process economy and maintenance, industrial processes in continuous operation require *system optimization*, which in turn requires very accurate modeling. The result of optimization is often given as a function of system parameters, which are contingent upon reliable and accurate modeling and identification. A special case is autonomous systems automation (adaptation), which for its implementation presupposes some learning capacity or parameter estimation.

1.4 SYSTEM AND MODEL COMPLEXITY

The modeling complexity needed depends on the purpose of the modeling and identification, and it is natural to distinguish among the following classes of models depending upon the detail and precision of modeling:

Qualitative and categorical models (often used for fault detection and process diagnosis) are often easy to derive from physical principles. Although the basic idea of qualitative modeling is simple and straightforward, reasoning based on qualititative models may lead to ambiguous results. The models are defined by causal relationships or categorical data (*e.g.*, models based on Boolean algebra, such as logical circuits), and they often serve as a complement or a user interface to more detailed quantitative models.

A step toward quantitative models can be taken by adopting semi-quantitative models, where variables take on such categorical values as hot-normal-cold or high-normal-low. Such models are popular in statistics and in various branches of applied computer science.

Static quantitative models are steady-state models described by algebraic equations involving such relationships as those between stresses and strains in mechanical structures or those among pressure, volume, and temperature in thermodynamics. Moreover, dynamic modeling expresses some proportional dependence, and it is natural to distinguish between the *a priori* models derived from physical principles, balance equations, and structural interconnections, and *a posteriori* models derived from experimental data. Both kinds of models may take on several different forms, although *a priori* models tend to be organized as structured systems or networks with various subsystems, components, and internal structure, whereas *a posteriori* models tend to be formulated as behavioral models. As *a posteriori* models are derived from experimental data, they often use abstract or experiment-dependent parameterizations such as black box models (Chapter 2), linear regression models (Chapter 5) or time series models (Chapter 6). Another difference is that the *a priori* models often manifest clear physical and causal relationships, whereas the data-oriented *a posteriori* models express relationships, such as covariance between variables formulated in statistical notions.

Linear systems have a particular significance in this comparison because both modeling and identification often presume linear, proportional relationships to exist between variables. The powerful *principle of superposition*, then, simplifies the solution of the mathematical problems involved. In addition, the linear model is often useful as a *small signal model* around some equilibrium point. Linear systems comprise a number of highly different classes of models, and it is natural to distinguish, on the basis of their complexity, between certain alternative types of models:

i. Composite models versus Submodels

Obviously, this dichotomy is often a trivial one; for in composite models structure is determined by the model's components and detail is determined by the purpose of modeling. Methodologies to structure composite models from interconnected submodels or components are closely related to the analytic approach adopted in science and technology (see Chapter 7 for some examples).

Dynamical systems with outputs that change over time are often subdivided into

ii. Time domain models versus Frequency domain models,

where the Laplace transform, the Fourier transform, and the z-transform are the major analytic instruments used.

An objective of modeling is to reduce uncertainties about the behavior of a system, although the remaining sources of uncertainty also require some modeling. In all cases it is customary to use probability theory and stochastic processes to model uncertainty, interference of other subsystems, or input variablity according to some statistical distribution. Stochastic properties can be evaluated in the framework of linear systems with additive and multiplicative characteristics, ensemble, and temporal statistics (Chapter 6). The differences in analytic approaches thus motivate a distinction between

iii. Deterministic versus Stochastic systems

Subjective uncertainty (or ignorance) is often poorly modeled by stochastic processes and thus requires other modeling. Recently there has been an increase of interest in the domain of fuzzy set theory to model subjective uncertainty, although this trend has hitherto had little impact on system identification.

Worst-case uncertainties and modeling of hostile interference or damage require modeling by other means—*e.g.*, differential games where uncertain interference is modeled and solved as an optimization problem.

Again, on the basis of complexity, it is natural to distinguish between

iv. Single-input single-output (SISO) versus Multi-input multi-output systems (MIMO)

This difference in modeling complexity reflects dependencies not only between input and output but also dependencies and interaction between the output variables.

The mathematical methods of data analysis motivate the distinctions of

v. Continuous-time versus Discrete-time models

The continuous-time models are closer to physical considerations, whereas with discrete-time models system behavior is considered to be defined at a sequence of time-instants related to measurement. The discrete-time models are closely related to implementation problems of digital signal processing.

It is possible to describe the model obtained with many very different parameter sets, and it is sometimes a controversial and purpose-related problem to evaluate the significance of the parameters finally chosen to describe the model. For this reason it is standard to distinguish between

vi. Parametric versus Non-parametric models

Many models have been defined by a given form but are dependent on a finite number of real parameters. Such models are said to be *parameterized* or *parametric models*, although there is no clear-cut distinction from other models such as spectra or impulse responses, which are sometimes referred to as *non-parametric* models. (A strict use of "non-parametric" would, however, refer to situations and experiments in which the outcomes are assigned to categories rather than measured on numerical scales.) Other distinctions such as that between structured and unstructured models would appear to be no more successful.

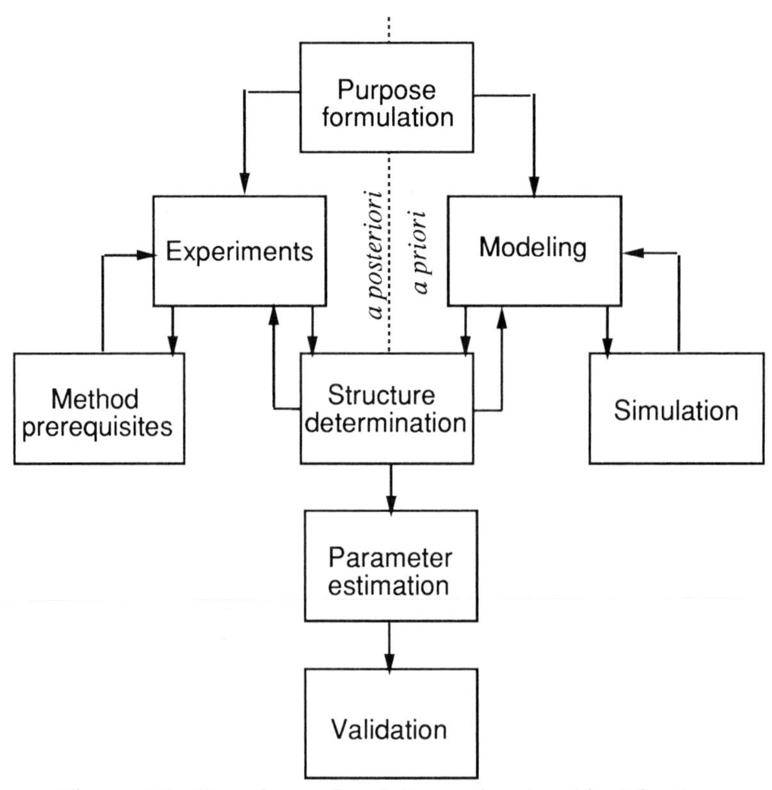

Figure 1.3 Procedures of modeling and system identification.

Methods of introducing state variables and associated parameters, to describe a certain external behavior and various equivalence classes thereof, are commonly referred to as *realization theory*. As there are usually several possible models to describe a given external behavior, it is desirable to distinguish a *minimal realization* according to some complexity criterion.

1.5 THE PROCEDURE OF IDENTIFICATION

Identification has many aspects and phases, and it is customary to organize identification by considering a certain number of steps. We start with the object of identification designated

$\mathcal{S}$: A system to study

It is also necessary to consider the application context—*i.e.*, whether the model derived is intended for scientific use or engineering design (*e.g.*, simulation models, fault detection models, control systems analysis, and/or human intervention) or for some other use. We designate this context

$\mathcal{P}$: Purpose and problem formulation

The experimental conditions of experiment and the reliability of data must be reported in support of an estimated model. This is designated

$\mathcal{X}$: Experimental planning and operation to ensure that the prerequisites of the identification methods are used in the experimental procedure.

Experimental design is often used to circumvent instrumental difficulties or to compensate for other inherent difficulties or interaction from other control loops.

It is often natural to restrict the complexity of modeling to a certain model structure. The class of models thus adopted often belong to some standard category of models such as linear systems or ARMAX models associated with certain mathematical properties. Another standard approach often used is to make assumptions as to the physical nature of the system or other restrictions that define physical parameterizations. This we designate:

$\mathcal{M}$: A model set

The algorithms and the software used (including filtering, estimation methods, numerical optimization) are often designated

$\mathcal{I}$: Identification and parameter estimation methods

The quality and the limitations of an estimated model in the form of statistical tests, simulations, etc., should be stated in support of the model and are necessary in order to make modeling statements precise. Such tests are referred to as

$\mathcal{V}$: Model validation

The task of identification is to choose that element in a model class which explains a given set of observations and as little else as possible. According to Popper, this is called the *most powerful unfalsified model.* The contexts $\mathcal{S}, \mathcal{P}, \mathcal{X}, \mathcal{M}, \mathcal{I}, \mathcal{V}$ thus represent stages or aspects of interest during the identification procedure, and Fig. 1.3 illustrates some of these relationships. The book is organized to cover these aspects beginning with the identification of simple linear models (Chapters 2–6) and proceeding with modeling issues (Chapters 7–10), followed by a review of topics in structured modeling and identification in Chapters 11–15, as outlined below.

Chapter 2 contains a presentation of some standard-type identification methods for black box models such as transient analysis with test signals in the form of impulse and step signals. In addition, frequency response analysis with sinusoidal inputs and correlation analysis with white noise input are treated.

Chapter 3 provides a systematic description of the use of Laplace transforms, the Fourier transform, and the z-transform in the context of signals and systems. The effects of discretization and finite measurement time are also treated at this early stage, as they are fundamental properties in the context of measurements. Such spectra as autospectrum, cross spectrum, power spectrum, coherence spectrum and correlation, and covariance are defined, and their dependence on input-output properties is treated.

Chapter 4 covers both spectral estimation techniques and some standard modifications of these techniques intended to compensate for distortion due to discretization and finite measurement times. In addition, covariance analysis and frequency response analysis are examined.

Linear regression is explored in Chapter 5 with emphasis on the least-squares problem as an optimization problem and statistical properties of linear regression estimates. The problems of bias in the context of least-squares identification are observed. Finally, the discrete Fourier transform as a least-squares estimation problem is also treated.

Identification of time-series models in the form of autoregressive models, moving average models, and their generalizations are examined in Chapter 6. The problems of estimating a transfer function by means of time-series models in terms of prediction error problems or output error problems are examined. Lattice algorithms are also presented. Finally, some aspects of application such as bias reduction, assessment of periodic phenomena, and numerical optimization methods are discussed.

Physical modeling principles for the determination of structured models such as mechanical models, thermodynamic models, and compartment models are presented in Chapter 7 with such common concepts as potentials, gradients, and flows. Identifiability problems of parameters in physical models are also discussed.

Principles and methods of experimental procedure are to be found in Chapter 8. Problems of experimental conditions with respect to the choice of input are discussed, and some special topics such as identification of systems in closed-loop operation are analyzed. The conclusion of Chapter 8 offers practical hints on the planning of experiments.

Model validation techniques are considered in Chapter 9, which covers the role of fulfilled method prerequisites, coherence test, model order determination, residual correlation tests, and other statistical tests.

Model approximation and model reduction methods are considered in Chapter 10, with some emphasis on the balanced realizations applicable to linear systems.

Real-time identification and recursive algorithms are considered in Chapter 11 and provide a necessary basis for adaptive and learning systems that operate in real-time conditions.

Continuous-time linear models are covered in Chapter 12, with further discusssion of structural identifiability as motivated by physical modeling.

Multidimensional modeling and identification methods are treated in Chapter 13, including some methods for certain complex systems described by delay-differential systems or partial differential equations.

Examples of nonlinear system modeling and identification are discussed in Chapter 14.

Adaptive systems in Chapter 15 are related to identification in that they incorporate real-time identification mechanisms permitting adaptatation to parameter changes. Some important classes of adaptive systems are reviewed from the perspective of identification.

The appendices comprise basic linear algebra, time-series analysis, statistical inference, numerical optimization, and a case study.

1.6 HISTORICAL REMARKS AND BIBLIOGRAPHY

As early as the fourth century before our present era, Aristotle recognized the importance of numerical and geometrical relations within science, and he singled out astronomy, optics, harmonics, and mechanics as sciences whose subject matter is mathematical relationships among physical objects. A few centuries later, Ptolemy (c.A.D. 85-165),who described planetary motion, emphasized that more than one mathematical model can be constructed to describe the astronomical observations. As two models may be mathematically equivalent, the scientist is at liberty to employ whichever model is the more convenient.

In the fourteenth century, William of Occam used simplicity as a criterion of concept formation and modeling. According to his view, it is desirable to eliminate superfluous concepts so that the simpler of two theories that account for a type of phenomenon is to be preferred. This methodological principle favoring low model complexity has often been referred to as "Occam's razor."

Dynamic modeling and identification as a methodology dates back to Galileo (1564-1642), who also is important as the founder of dynamics. In addition, Galileo made contributions to the design and application of scientific instruments such as the telescope, the thermometer, and the clock.

Hence, system identification and modeling have several different roots and belong with equal right to mathematics, statistics, computer science, control theory, systems analysis, and signal processing, with applications in technology, natural sciences, and econometrics. Recent years have witnessed continuing parallel development in statistics, econometrics, speech processing, geophysics, and structural mechanics.

The standard treatment of spectral analysis is to be found in the following reference:

– G.M. JENKINS AND D.G. WATTS, *Spectral Analysis and Its Applications*. San Francisco: Holden-Day, 1968.

System identification with some focus on time series analysis is presented in

– G.E.P. BOX AND G.M. JENKINS. *Time Series Analysis, Forecasting and Control*. San Francisco: Holden-Day, 1970.

– P.E. CAINES, *Linear Stochastic Systems*. New York: John Wiley, 1988.

– P. EYKHOFF, *System Identification: Parameter and State Estimation*. London: John Wiley, 1974.

– G.C. GOODWIN AND R.L. PAYNE, *Dynamic System Identification: Experiment Design and Data Analysis*. New York: Academic Press, 1977.

– L. LJUNG, *System Identification: Theory for the User*. Englewood Cliffs: Prentice-Hall, 1987.

– T. SÖDERSTRÖM AND P. STOICA. *System Identification*. London: Prentice-Hall Int., 1989.

Although purposes of system modeling and identification are similar throughout the world, philosophical attitudes differ perceptibly. The American and British cultural spheres and Scandinavia, with their tradition of analytic and empiricist philosophy, sometimes tend to emphasize statistical aspects, whereas scientists in continental Europe with a strong background of idealistic philosophy more often emphasize modeling and approximation approaches.

Some philosophical background to scientific modeling is to be found in the following works:

- K.R. POPPER, *Conjectures and Refutations.* London: Harper and Row, 1963.
- P. FEYERABEND, *Against Method,* 2d ed. London: Verso, 1988.

Some of the philosophy of K.R. Popper adapted to the context of identification is to be found in the following essays:

- J.C. WILLEMS, "From time series to linear systems," *Automatica.* "Part I: Finite dimensional linear time invariant systems." Vol. 22., 1986, pp. 561–80, 1986. "Part II: Exact modelling." Vol. 22, 1986, pp. 675–694. "Part III: Approximate modelling." Vol. 23, 1987, pp. 87–115,

Suitable background material is provided in either of the three works below

- K.J. ÅSTRÖM AND B. WITTENMARK, *Computer-Controlled Systems,* 2d ed. Englewood Cliffs, NJ: Prentice-Hall, 1990.
- R. ISERMANN, *Digital Control Systems, Vols. I–II.* Berlin and Heidelberg: Springer-Verlag, 1991.
- R.H. MIDDLETON AND G.C. GOODWIN, *Digital Control and Estimation: A Unified Approach.* Englewood Cliffs, NJ: Prentice-Hall, 1990.

Important scientific journals covering topics in identification include *Advances in Applied Probability, Annals of Statistics, Automatica, Econometrica, Econometric Theory, IEEE Transactions Automatic Control, IEEE Transactions Signal Processing, Journal of Econometrics,* and *Time Series Analysis.* Further references are provided in each chapter. ■

2

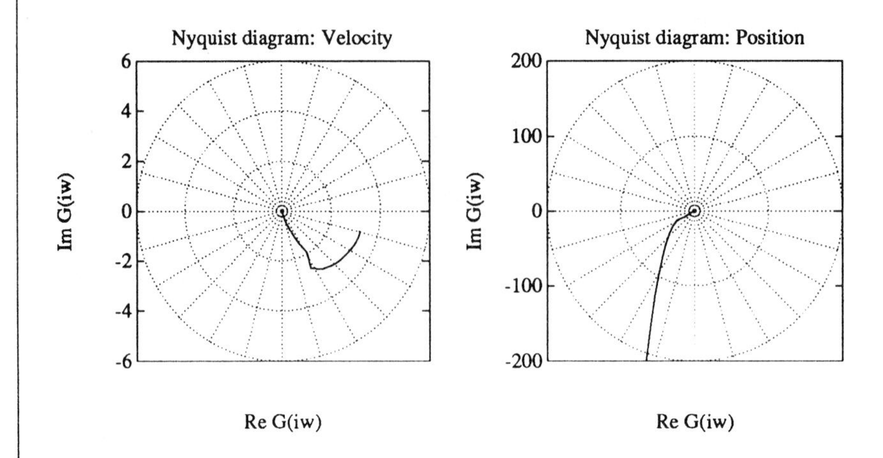

Black Box Models

2.1 INTRODUCTION

This chapter reviews some representative methods for simple analysis of dynamical systems such as black box models fitted with data from transient response analysis. Frequency response analysis with sinusoidal inputs and correlation analysis with white noise input are also covered.

The *impulse response* or the *weighting function* is fundamental in the description of a linear system because it gives a complete characterization of of the input-output map a linear time invariant system. Let the impulse response be designated $g(t)$ and consider the integral

$$y(t) = \int_0^t g(\tau)u(t-\tau)d\tau \tag{2.1}$$

The function g determines a causal relationship between input u and output y if $g(t) \equiv 0$ for $t < 0$. Assuming the system to be at rest at time $t = 0$, it follows that application of a test signal u in the form of an impulse

$$u(t) = \delta(t) \tag{2.2}$$

gives the response

$$y(t) = \int_0^t g(\tau)\delta(t-\tau)d\tau = g(t) \tag{2.3}$$

which, in turn, justifies the use of the term impulse response. Moreover, if the input is chosen as $u(t) = \alpha u_1(t) + \beta u_2(t)$ for constants α, β, we find that

$$y(t) = \int_0^t g(\tau)u(t-\tau)d\tau = \alpha \int_0^t g(\tau)u_1(t-\tau)d\tau + \beta \int_0^t g(\tau)u_2(t-\tau)d\tau \tag{2.4}$$

from which it is clear that the system is *linear* in its response. The linear dynamic properties of (2.1–2.4) are also of interest for the purpose of approximating nonlinear systems with linear ones. Such approximations may be valid at least as *small signal models.*

2.2 TRANSIENT RESPONSE ANALYSIS

An identification methodology based on the application of an impulse on the input of a system is known as *impulse response analysis*. A similar method is to observe the transient output behavior produced by the system proceeding from a certain given initial condition (*initial condition response*). Several other identification methods such as *step response analysis* are based on similar special choices of inputs, and together constitute a class of methods known as *transient response analysis*. Such methods are often simple to apply and understand and often provide information good enough for estimates of input-output gain, dominating time constants, and time delays. These properties

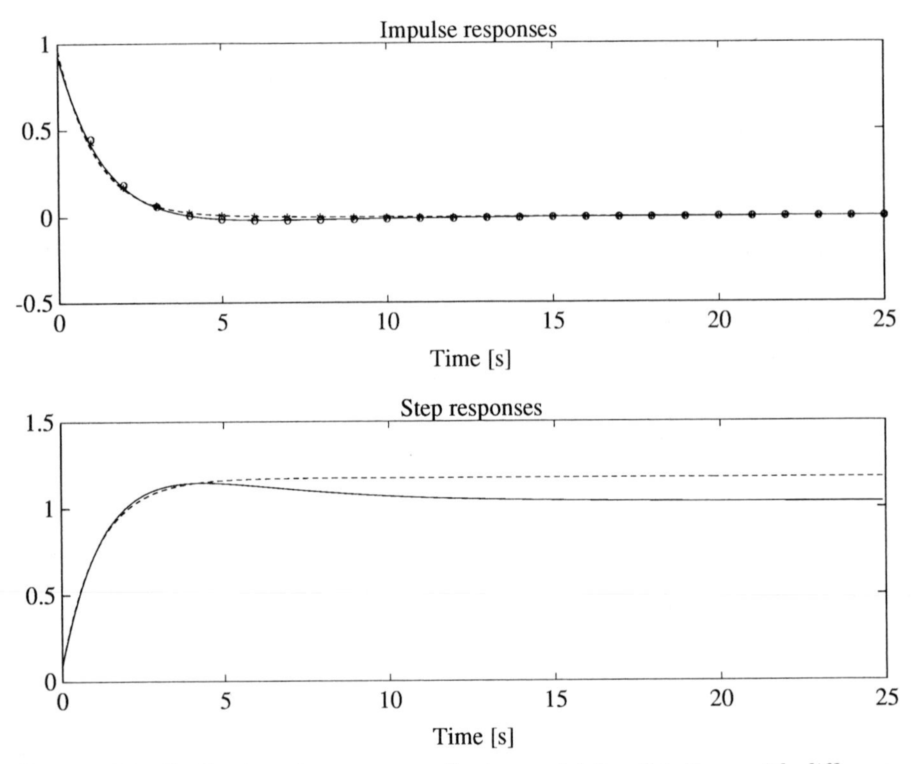

Figure 2.1 Similar impulse responses for two weighting functions with different step responses. Samples collected every second are denoted by '*' and 'o'. The example demonstrates the difficulty in obtaining correct low-frequency properties from impulse response tests.

make the methods suitable for first-stage experiments to prepare for other experiments in system identification.

In order to demonstrate properties of impulse response analysis we may begin with two examples:

Example 2.1—Impulse response analysis
Consider the two weighting functions

$$
\begin{aligned}
g_1(t) &= 1.05 \cdot 0.41^t \\
g_2(t) &= 1.2 \cdot 0.5^t - 0.2 \cdot 0.75^t
\end{aligned}
\tag{2.5}
$$

which are shown in Fig. 2.1 together with samples collected every second and their corresponding step responses. The two data sequences obtained from the impulse responses are virtually indistinguishable. Nevertheless, they result in considerable differences in the corresponding step responses. This example

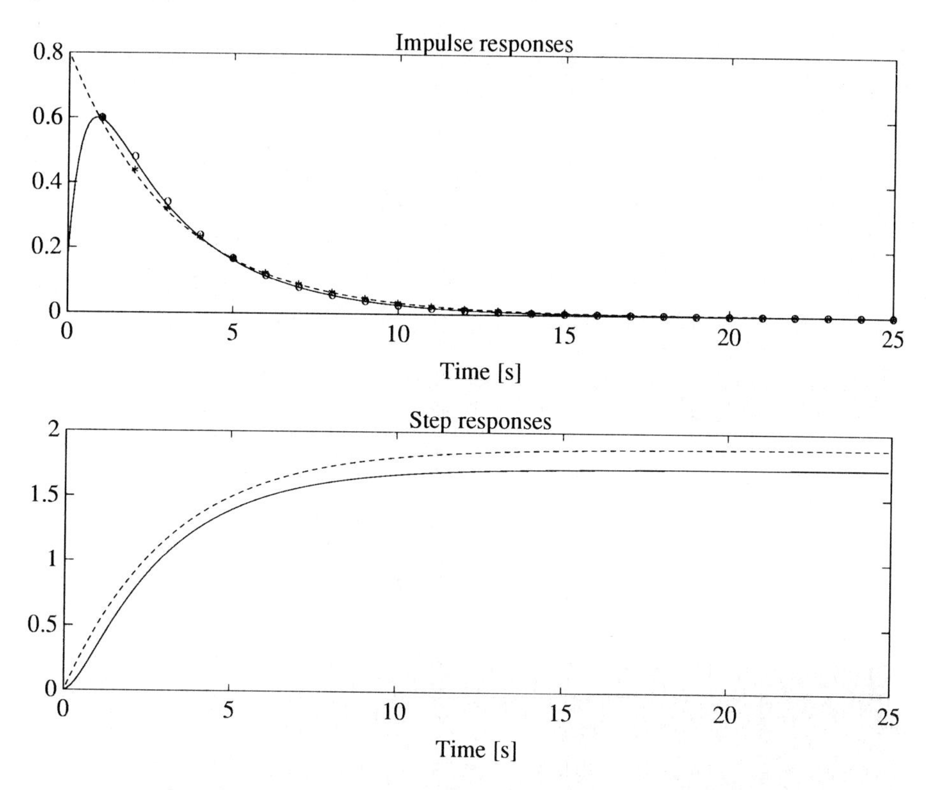

Figure 2.2 Similar impulse responses for two weighting functions of different initial responses and samples of the impulse responses with period $h = 1$. The example demonstrates the difficulty in obtaining correct high-frequency properties from impulse response tests. In particular, sampled impulse responses exhibit this weakness.

thus illustrates the problem of finding accurate estimates of the static (and low frequency) properties from an impulse response. ■

The next example demonstrates the difficulties in determination of high-frequency properties

Example 2.2—Impulse response analysis
Consider the two impulse responses

$$\begin{aligned} g_3(t) &= 0.7^t - 0.1^t \\ g_4(t) &= 0.8 \cdot 0.73^t \end{aligned} \tag{2.6}$$

These signals and their corresponding step responses are shown in Fig. 2.2 along with samples of the two impulse responses collected every second. Notice that the two sequences of samples are almost indistinguishable despite the marked initial difference between the two impulse responses. An adequate assessment of the initial response would require much more frequent sampling than was used in this example. Also, the synchronization between the generation of the input, *i.e.*, the impulse, and the recording of the response requires attention for correct estimation of the initial properties. ■

As a conclusion it might be stated that it is difficult to handle both low- and high-frequency properties and synchronization effects in impulse response analysis even under noise-free conditions. Another problem is how to perform impulse response analysis in the presence of saturations and nonlinearities. In fact, the following practical problems of impulse response analysis can be listed:

- Restriction to stable systems
- Difficulties in generating impulses
- Dynamics of sample and hold circuits
- Synchronization between impulse and sampling
- Difficulties for the system in managing inputs of large magnitude
- Saturations
- Nonlinearities
- Difficulties in handling the "tails" of responses due to their long duration and low amplitudes, with consequent problems of quantization and numerical accuracy
- Sensitivity to noise

Obviously, several of these problems are dependent upon experimental conditions.

Step response analysis

A complement to impulse reponse analysis is *step response analysis* which is often easier to perform than impulse response analysis. The test signal is

$$u(t) = \begin{cases} 1, & t > 0 \\ 0, & t \le 0 \end{cases} \tag{2.7}$$

Under noise-free conditions, a step input to a system described by Eq. (2.1) generates the response

$$y(t) = \int_0^t g(\tau)d\tau \tag{2.8}$$

The step response test has an advantage in that it provides a good estimate of any static gain. Also, it is obviously possible to obtain an estimate of the

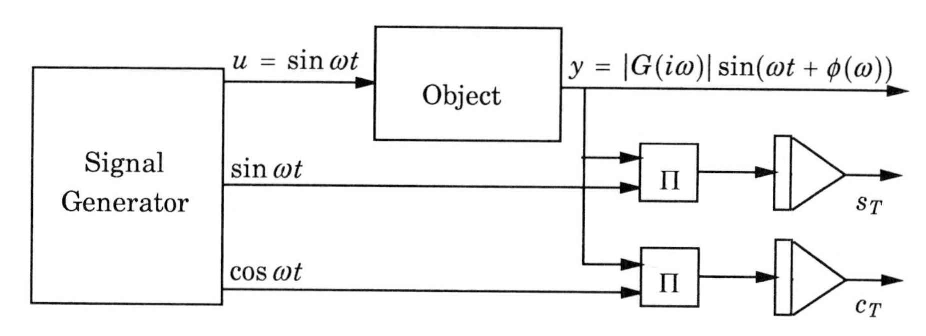

Figure 2.3 Frequency response analysis.

weight function $g(t)$ *via* differentiation of the step response

$$\hat{g}(t) = \frac{d}{dt} y(t) \tag{2.9}$$

which according to Eq. (2.8) is equal to $g(t)$. Such a calculation involves differentiation of the recorded output, where the accuracy of calculation may be modified or improved by bandpass filtering.

The above examples illustrate some of the fundamental problems of system identification and such specific questions of how to choose experimental conditions (experiment duration, sampling frequency, test signals, data filtering, computation, *etc.*). Subsequently, it is necessary to select an appropriate level of model complexity. The answers to such questions are dependent on the purpose of system identification.

2.3 FREQUENCY RESPONSE ANALYSIS

The basis of frequency response analysis as an identification principle may be outlined as follows: Assuming the identification object to be a linear time invariant dynamic system, it can be described by some weighting function $g(t)$. The Laplace transform $\mathcal{L}\{\cdot\}$ of the weighting function $g(t)$ provides a transfer function $G(s) = \mathcal{L}\{g(t)\}$ that relates the Laplace-transformed input $U(s) = \mathcal{L}\{u(t)\}$ to the output $Y(s) = \mathcal{L}\{y(t)\}$.

$$\begin{aligned} y(t) &= \int_0^\infty g(\tau)u(t-\tau)d\tau \\ Y(s) &= G(s)U(s) \end{aligned} \tag{2.10}$$

Assuming that we can ignore the transient from initial conditions and only consider the input-output response, the steady-state response of a stable system to a sinusoidal input

$$u(t) = u_1 \sin \omega t \tag{2.11}$$

is then characterized by the *gain* $|G(i\omega)|$ and the *phase shift* $\phi(\omega)$

$$y(t) = |G(i\omega)| u_1 \sin(\omega t + \phi(\omega)); \qquad \phi(\omega) = \arg G(i\omega) \tag{2.12}$$

Simple observation of a sinusoidal input plotted against the corresponding output response allows direct computation of the gain and phase shifts from Lissajou contours. Results are presented in the form of a Bode diagram where gain and phase shift are plotted against frequency in a specified range. However, all Lissajou-type methods are sensitive to disturbances and transients, and do not yield any statistical averaging over the measurement interval.

A better approach to reduce disturbance is the following averaging method (sometimes also called the *correlation method of frequency response analysis*; see Fig. 2.3): Let the output of the system subject to frequency analysis be multiplied with sinusoids of the frequency of the input with a subsequent integration during a specified measurement interval T. To minimize the effect of disturbance, the measurement duration T is always chosen as a number k of full periods of the periodic test signal $u(t) = u_1 \sin \omega t$, where $T = k \cdot (2\pi/\omega)$. The outputs from these computations are obtained as follows: The "sine channel" provides

$$s_T(\omega) = \int_0^T y(t) \sin \omega t dt = \frac{1}{2} T |G(i\omega)| u_1 \cos \phi(\omega); \qquad T = kh = \frac{2\pi}{\omega} k \tag{2.13}$$

whereas the "cosine channel" provides the signal

$$c_T(\omega) = \int_0^T y(t) \cos \omega t dt = \frac{1}{2} T |G(i\omega)| u_1 \sin \phi(\omega) \tag{2.14}$$

Estimates of the gain $|G(i\omega)|$ and the phase shift $\phi(\omega)$ are obtained as

$$|\widehat{G}(i\omega)| = \frac{2}{T u_1} \sqrt{s_T^2(\omega) + c_T^2(\omega)} \quad \text{and} \quad \widehat{\phi}(\omega) = \arctan \frac{c_T(\omega)}{s_T(\omega)} + k\pi \tag{2.15}$$

Hence, frequency response analysis can be viewed as an instrumentation of the Fourier transform (or as the computation of coefficients of a Fourier series expansion). The measurement intervals are usually specified in absolute time or as a number of periods of the frequency investigated. Another standard

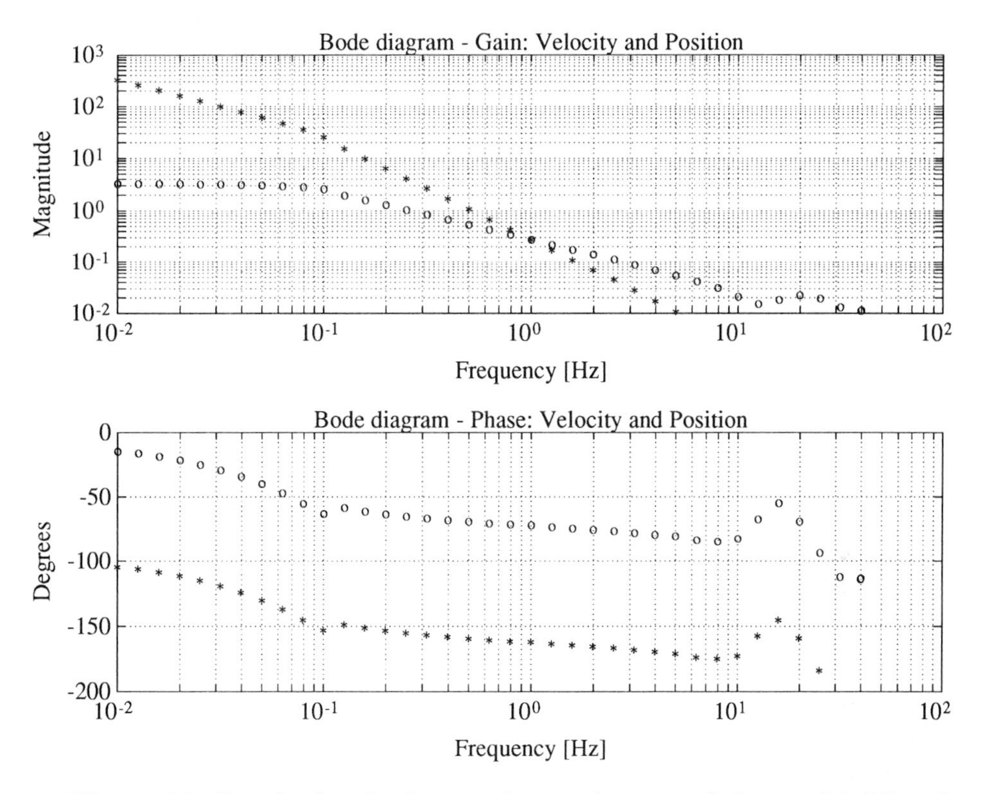

Figure 2.4 Transfer function between input voltage u and the speed $\dot{q}$ ('o') and the position q ('*') of a DC motor as obtained from frequency response analysis. To avoid inaccuracy due to transients from initial conditions, each input of a new test frequency is started 5 s before measurement is started.

feature is to avoid effects of non-zero initial conditions by introducing a time delay between the application of a new test frequency and the start of the measurement interval.

Presentation of results

Experimental results with estimates of $|G(i\omega)|$, $\phi(\omega)$ over some frequency range are generally presented graphically in the form of *Bode, Nyquist,* or *Nichols diagrams*. The *Bode diagram* contains the gain and phase response versus frequency, with the gain represented in a log-log-diagram and the phase delay in a log-lin-diagram, and is the standard representation of frequency response analysis (Fig. 2.4).

The *Nyquist diagram* is a polar diagram containing the gain as the radial coordinate and the phase shift as the phase coordinate (Fig. 2.5). This repre-

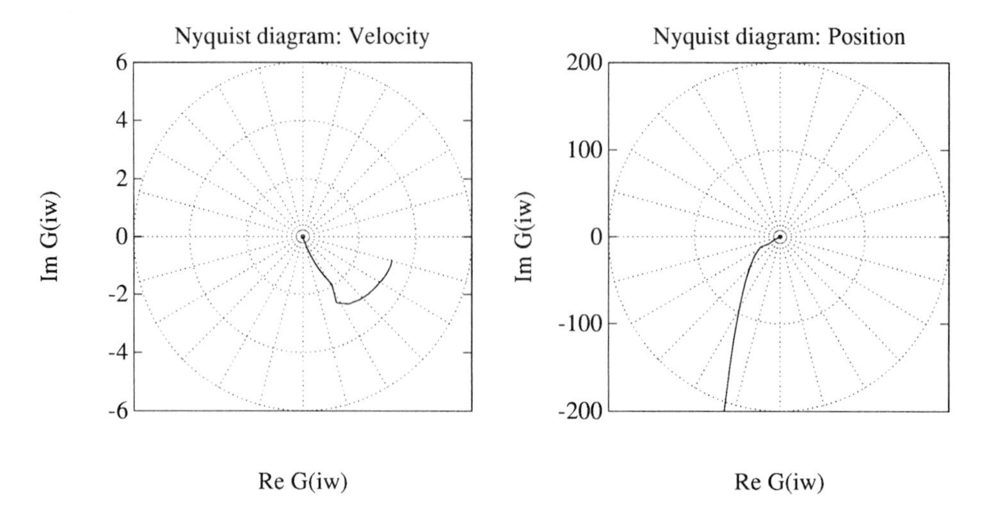

Figure 2.5 Nyquist diagrams representing the transfer function of a DC-motor velocity and position as obtained from frequency response analysis (*cf.* Fig. 2.4). To avoid inaccuracy due to transients from initial conditions, each input of a new test frequency is started 5 s before measurement is started.

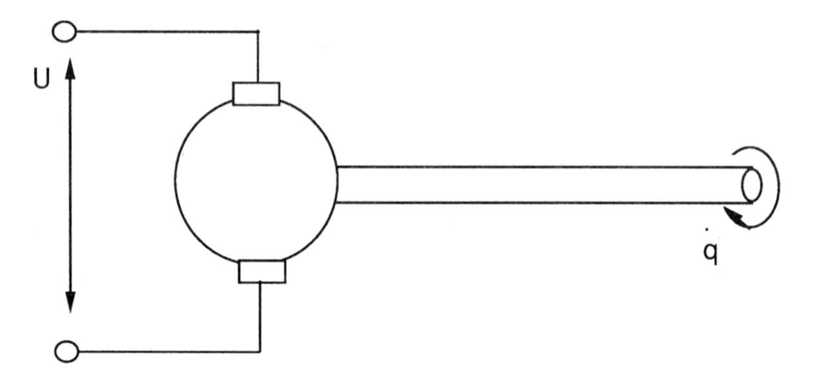

Figure 2.6 A DC motor with input voltage u and the speed $\dot{q}$ as the output.

sentation is specifically used in control systems analysis.

$$\begin{aligned} \text{Re} \quad \widehat{G}(i\omega) &= \frac{2}{u_1 T} s_T(\omega) \\ \text{Im} \quad \widehat{G}(i\omega) &= \frac{2}{u_1 T} c_T(\omega) \end{aligned} \tag{2.16}$$

Example 2.3—Frequency response analysis of a DC motor

Consider frequency response analysis of a DC motor from the input voltage u to the outputs measured as angular position q and angular velocity $v = \dot{q}$

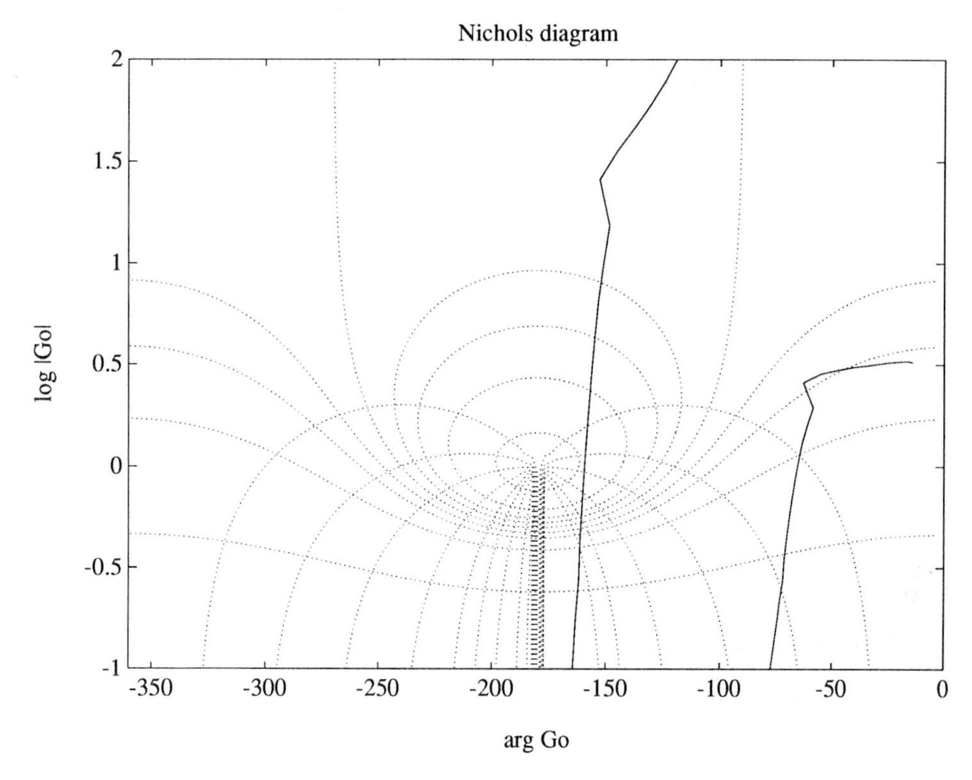

Figure 2.7 Nichols diagram of the transfer function between input voltage u and the speed $\dot{q}$ and the position q of a DC motor as obtained from frequency response analysis in Example 2.3.

(Fig. 2.6) of the shaft. The Bode diagram of Fig. 2.4 displays the result of frequency response analysis of the DC-motor velocity over a frequency range 0.01–40 Hz.

To avoid inaccuracy due to transients from initial conditions, the DC motor input was exposed to each new test frequency for 5 s before measurement was started. The result is also presented in the form of a Nyquist diagram (Fig. 2.5). ■

The *Nichols diagram* (Fig. 2.7) depicts the frequency response as a diagram with the gain $\log |G(i\omega)|$ on the vertical axis versus the phase shift $\phi(\omega)$ on the horizontal axis. The Nichols diagram contains level surfaces for $|G/(1+G)|$ and $\arg G/(1+G)$, which is valuable in control systems analysis for quantification of stability margins.

Sensitivity to disturbances

There are some obvious sources of disturbances:

- Disturbances (constant; white noise) acting on the output
- Disturbances (trends) acting on the output
- Sampling interference
- Unmodeled nonlinearities
- The presence of higher-order harmonics

Assume the output y to be corrupted by a disturbance v so that

$$y(t) = |G(i\omega)|u_1 \sin(\omega t + \phi(\omega)) + v(t); \qquad \phi(\omega) = \arg G(i\omega) \tag{2.17}$$

Sensitivity to this type of disturbance gives rise to errors of the form

$$\begin{cases} \Delta s_T = \int_0^T \sin(\omega t)v(t)dt \\ \Delta c_T = \int_0^T \cos(\omega t)v(t)dt \end{cases} \tag{2.18}$$

The error due to a constant disturbance is thus zero. Another case of low disturbance sensitivity is white noise and bandwidth limited noise. The relative error of the transfer function estimate can be estimated as

$$\frac{|\Delta G(i\omega)|}{|G(i\omega)|} = \sigma_v \sqrt{\frac{\omega}{2k\omega_c}} \tag{2.19}$$

where σ_v^2 is the variance of the output disturbance v, ω_c the bandwidth of the disturbance, and k the number of full periods of the sinusoid completed during the measurement time. A weak performance may be expected in the presence of ramp disturbances and periodic disturbances of the same frequency as the test frequency. Hence, the correlation method of frequency response analysis effectively eliminates the effect of constant disturbances but unfortunately does not effectively eliminate that of trend disturbances. To reduce estimation error due to trend disturbances, more complicated versions of the correlation are needed.

Example 2.4—Sensitivity to sinusoidal disturbances
Consider a first-order system

$$Y(s) = G(s)U(s) + V(s) \tag{2.20}$$

with output $Y(s)$, input $U(s)$, a disturbance $V(s)$, and the transfer function

$$G(s) = \frac{1}{s+1} \tag{2.21}$$

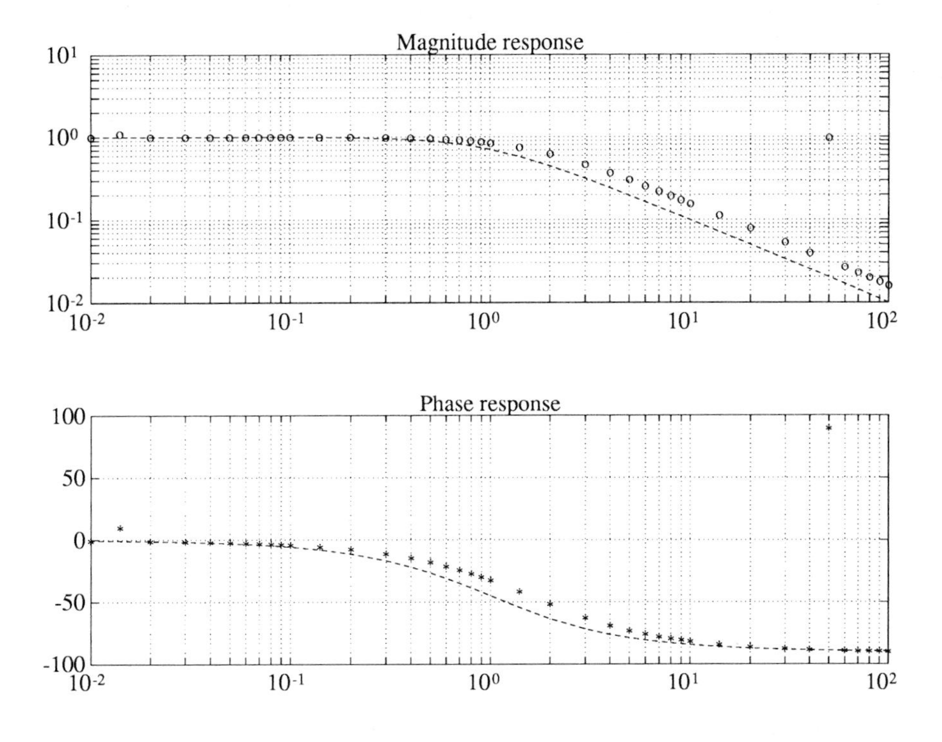

Figure 2.8 Frequency response analysis of a transfer function $G(s) = 1/(s+1)$ when subject to a sinusoidal disturbance at 10 Hz = 62.8 [rad/s] with signal-to-noise ratio $S/N = 1$. Frequency axis in [rad/s].

The disturbance $v(t)$ is assumed to be sinusoidal with the period 10 Hz and with the same magnitude as that of the input test frequencies $u(t)$. The result of frequency response analysis is seen in Fig. 2.8 where both the experimental values and the theoretically calculated transfer functions are shown. The disturbance has strong impact on the experimental result at the frequency of the disturbance, *i.e.*, at 10 Hz or 62.8 rad/s. The disturbance is, however, noticeable over a large frequency interval and corrupts both gain and phase measurements. A longer measurement time T tends to reduce the corrupted frequency interval and result in a narrow-band error.

■

Example 2.5—Sampling interference in frequency response analysis

It is, of course, often natural to implement frequency response analysis by discrete-time methods where the sinusoidal inputs and outputs are sampled. In Fig. 2.9 is shown the result of frequency response analysis applied to identification of the transfer function $G(s) = 1/(s+1)$ sampled at a sampling

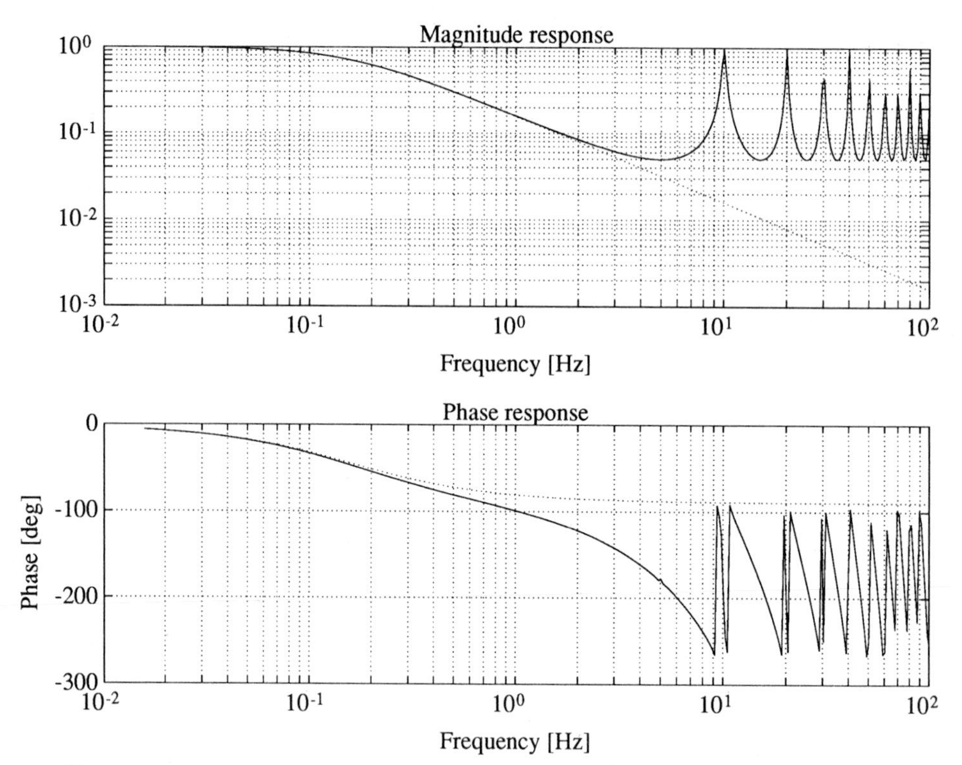

Figure 2.9 Frequency response analysis of a transfer function $G(s) = 1/(s+1)$ when implemented by discrete-time methods at 10 Hz sampling rate (*solid line*) as compared to true transfer function (*dotted line*). Notice that the estimate is poor for large frequency due to sampling interference. Frequency axis in Hz.

frequency of 10 Hz. It is clear that there is a significant distortion of the transfer function estimate in the frequency range 1–10 Hz. This style of implementation thus requires a sufficiently high sampling frequency, *vis-à-vis* the frequency range to be investigated, in order to avoid detrimental sampling interference. ■

Frequency response analysis is applicable to linear time-invariant systems but may have difficulty in providing meaningful results when applied to nonlinear systems. A difficulty associated with nonlinearities is the ambiguity problem of *jump resonances*, which may appear in certain saturation nonlinearities. Systems with jump resonances may give irreproducible and different results depending upon the sequence of frequency response tests, *i.e.*, results obtained when the experiments are performed with increasing test frequency may differ from those obtained with decreasing test frequency. Another problem is the dependence of the frequency response analysis on the input magnitude, an issue discussed in greater detail in conjunction with the *describing function method* (see Chapter 12 on model approximation and model reduction).

Problems of another type arise in applications to time-varying systems where distorting interference may appear. In particular, all frequency response analyses applied to periodically sampled and controlled systems are error-prone due to interference between the test frequency used and the frequency of periodic time-variant behavior. For instance, periodic systems such as combustion engines, thyristor controlled systems, and other pulse-width modulated systems may suffer from this type of problem. Synchronization of the periodic system and the phase of the frequency response test signal often provides the means of solving these problems in the experimental procedure.

2.4 APPLICATION OF FREQUENCY RESPONSE ANALYSIS

The application of sinusoidal inputs to a dynamic system with recording of gain and phase lag is widely used in such contexts as electronic circuits, electromechanical devices, acoustics, audiometry, and data communication. This methodology instruments the Fourier transform and is the principle of many commercial transfer function analyzers.

The *experimental procedure* must fulfill the methodological prerequisites, and it is essential to ensure that the system is linear in response to the input magnitude used.

Another concern is that frequency response analysis of devices that contain friction and similar nonlinearities may produce poor results for tests with zero mean velocity. A possible alternative is to choose an input signal

$$u(t) = u_0 + u_1 \sin(\omega t) \tag{2.22}$$

in order to avoid some of these problems.

Transients from non-zero initial conditions may result in problems of accuracy. As the method is dependent upon steady-state conditions, the introduction of a delay between the application of a new test frequency and the start of the measurement interval is necessary to minimize the detrimental effects of initial transients.

The method normally yields the gain without ambiguity. The phase delay is obtained uniquely except for a multiple of 2π [rad] (or 360°). This ambiguity may be resolved by choosing a certain experimental procedure: For low-pass type transfer functions it may be known that the phase delay is, say, zero for the low frequencies. Let the frequency analysis be made for frequency points in increasing order with the phase delay at zero for the low frequencies. The phase lag for higher frequencies may then be obtained without ambiguity.

There are several advantages in using this procedure:

- A Bode diagram is obtained directly from data.
- Phase lag and delay time are easy to measure.
- Improved accuracy can be obtained if the measurement time is increased.
- Good results may be expected, even with poor signal-to-noise ratios
- Averaging of nonlinearities provides approximate linear models.

However, a number of restrictions can also be noticed, such as

- A special test signal is needed (only sinusoidal inputs).
- One experiment is needed for each test frequency.
- Long measurement time is required.
- To avoid biased results, the effects of initial conditions must be allowed to disappear before measurements can be made.
- Frequency sweeping may not be possible.
- Experiments are only possible on stable systems.
- Frequency response analysis presupposes application to time-invariant linear systems.

Obviously these restrictions determine the range of possible applications of frequency response analysis.

2.5 SUMMARY

The examples given in this chapter illustrate some common approaches to system identification in both the time domain and frequency domain. Given an object to identify, all approaches are predicated upon certain choices of experiment durations, test signals, and sampling frequency. The identification methods have been used to fit a model in some model set—*e.g.*, a weighting function, a transfer function, or some parametric model.

As in all experimental work it is necessary to consider carefully the reproducibility of results, the statistical properties of data, and the possibility of inaccurate measurements. These concerns affect the choice of experimental conditions and identification method and will be discussed in greater detail in the following chapters.

2.6 EXERCISES

2.1 Consider the single-input single-output state space system

$$\begin{aligned} \dot{x}(t) &= Ax(t) + Bu(t), \qquad x \in R^n \\ y(t) &= Cx(t) \end{aligned} \tag{2.23}$$

Show that the impulse response $g(t) = Ce^{At}B$. What is the effect on the measured impulse response if there is a non-zero initial condition $x(0) = x_0$?

2.2 Assume that the ideal impulse cannot be implemented, and that it is replaced by the input

$$u(t) = \begin{cases} 1/T, & 0 < t \le T \\ 0, & t > T \end{cases} \tag{2.24}$$

How should the impulse response be determined from data obtained when using the input (2.24)?

2.3 Consider the setup for frequency analysis in Fig. 2.10.

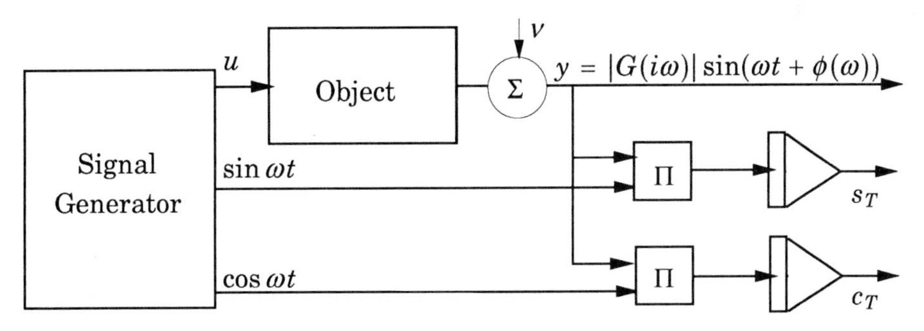

Figure 2.10 Noise-corrupted frequency response analysis.

Assume that the disturbance v affects the output of the system so that the resultant identification will be compromised. Also assume that the measurement time is n periods for each test signal with frequency ω_i.

a. How will the Bode plot be affected if the disturbance is a sinusoid of frequency ω_v, *i.e.*, $v(t) = A_v \sin(\omega_v t)$?

b. Assume that frequency response analysis has been performed with the measurement duration T and assume that v is high bandwidth noise with mean 0 and variance σ^2. How much is it necessary to increase the measurement duration in order to reduce the variance of $\widehat{G}$ by a factor of two?

c. Assume that the frequency analysis and the Bode plots will be used as a basis for regulator design. Discuss how to choose a finite measurement time at each frequency to obtain optimal closed-loop control. Is it possible to choose a measurement strategy for this purpose?

2.4 The frequency response method allows estimation of the transfer function $G_0(i\omega)$. The estimate may be written

$$\widehat{G}(i\omega) = G_0(i\omega) + \Delta G(i\omega)$$

where $G_0(i\omega)$ is the "true" transfer function and $\Delta G(i\omega)$ is the contribution from measurement noise. Assume that the measurement noise is "white" with mean value 0 and variance σ^2. Then it can be shown that

$$\mathcal{E}\{\Delta G(i\omega)\} = 0, \qquad \text{Var}\,\{\Delta G(i\omega)\} = \frac{4\sigma^2}{T}.$$

Moreover, $\arg \Delta G(i\omega)$ has a square distribution in the interval $[0, 2\pi)$.

The transfer function was estimated for 40 different frequencies. The result, $\widehat{G}(i\omega)$, is shown in Fig. 2.11. The correlation time T was chosen as 50 s and the noise variance σ^2 was estimated to be 0.1 s.

The system will be controlled by a proportional regulator with a gain of 1. In order to decide the closed-loop system stability properties, it is helpful to ascertain whether the Nyquist contour encloses the point -1. Is it possible to extract this information from Fig. 2.11?

2.5 Assume that there is a nonlinearity at the output of the analyzed system (see Fig. 2.12). Then the response to a sinusoidal input will also contain frequencies other than the input. What is the impact on the Bode plot? What does this mean for the choice of input signal?

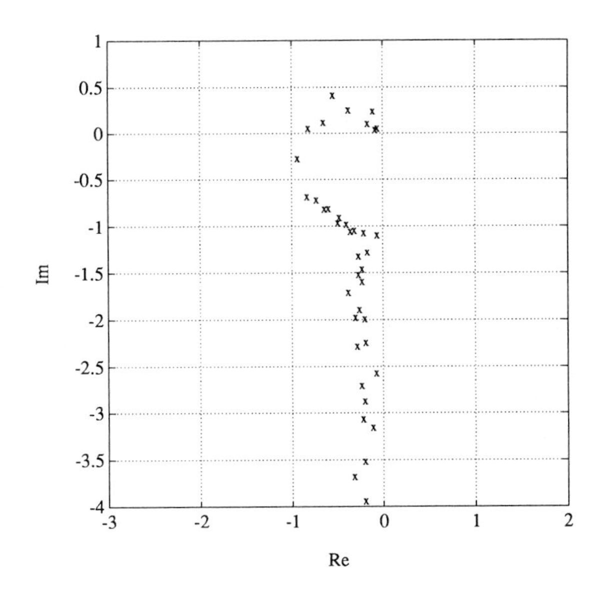

Figure 2.11 The transfer function for $\widehat{G}(i\omega)$ in Exercise 2.4 measured at 40 different frequencies.

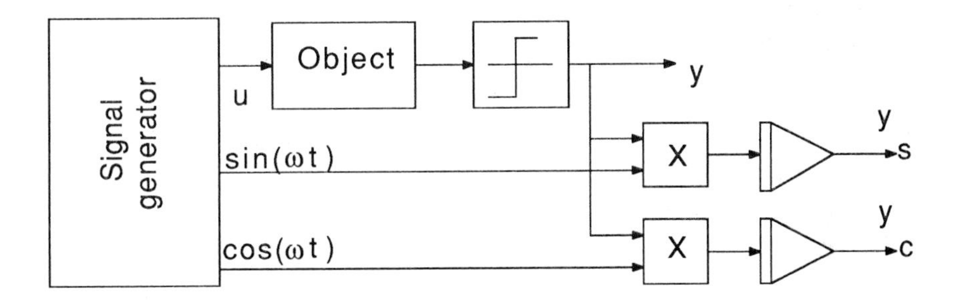

Figure 2.12 Frequency response analysis with nonlinearity.

2.6 Discuss the requirements on sampling frequency of a digital implementation of frequency analysis for a given desired bandwidth.

2.7 Show that it is possible to perform frequency response analysis according to Eq. (2.13–2.14), *i.e.*, the correlation method, with a selected measurement interval of $T = (\pi/\omega)k$. In addition, show that it is possible to eliminate constant perturbations in the output by choosing $T = (2\pi/\omega)k$ according to (2.13). ■

3 Signals and Systems

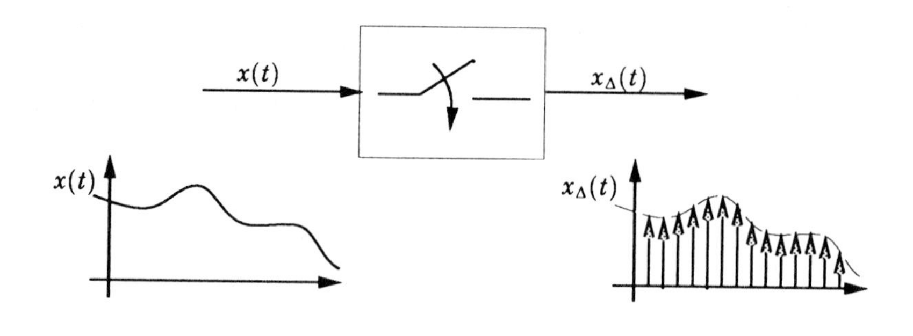

3.1 INTRODUCTION

This chapter provides a brief systematic description of the use of the Laplace transforms, the Fourier transform, and the z-transform in the context of signals and systems. Also, effects of discretization and finite measurement time are treated at this early stage, as they are fundamental properties in the context of measurements. Definitions are given of such spectra as autospectra, cross spectra, power spectra, correlation, covariance, and coherence spectra, and their dependence on input-output properties is treated. Spectrum analysis and other signal processing based on these spectra follow in Chapter 4.

3.2 TIME-DOMAIN AND FREQUENCY-DOMAIN TRANSFORMS

Consider a function $x(t)$ which represents a measured or observed variable and which is assumed to be a function defined for all time t. Let the Laplace transform be defined as

$$\begin{aligned} X(s) = \mathcal{L}\{x(t)\} &= \int_{-\infty}^{\infty} x(t)e^{-st}dt \\ x(t) = \mathcal{L}^{-1}\{X(s)\} &= \frac{1}{2\pi i}\int_{\sigma-i\infty}^{\sigma+i\infty} X(s)e^{st}ds \end{aligned} \tag{3.1}$$

where the argument $s = \sigma + i\omega$ is called *complex frequency*. Let the *spectrum* be defined as the Fourier transform of $x(t)$

$$\begin{aligned} X(i\omega) = \mathcal{F}\{x(t)\} &= \int_{-\infty}^{\infty} x(t)e^{-i\omega t}dt \\ x(t) = \mathcal{F}^{-1}\{X(i\omega)\} &= \frac{1}{2\pi}\int_{-\infty}^{+\infty} X(i\omega)e^{i\omega t}d\omega \end{aligned} \tag{3.2}$$

Clearly, the Fourier transform and the Laplace transform coincide for the choice $s = i\omega$ in cases when the Fourier integral (3.2) exists. The Fourier transform provides a result in the form of a function $X(i\omega)$ with the argument ω interpreted as angular frequency $[rad/s]$. The functions $x(t)$ and $X(s)$ are called *time-domain* and *frequency-domain* representations, respectively. From the definitions (3.1) and (3.2), we notice that the Fourier transform and Laplace transforms exist if the integrals take finite values (*i.e.*, remain bounded). A major difference between the Laplace and Fourier transforms is that the Laplace transform is valuable for analysis of transient behavior, whereas the Fourier transform is mainly applicable to periodic signals.

Most problems can be formulated in a manner which permits all signals to be zero for $t \le 0$, and it is customary to restrict the Laplace transform to the one-sided Laplace transform

$$X(s) = \mathcal{L}\{x(t)\} = \int_{0}^{\infty} x(t)e^{-st}dt \tag{3.3}$$

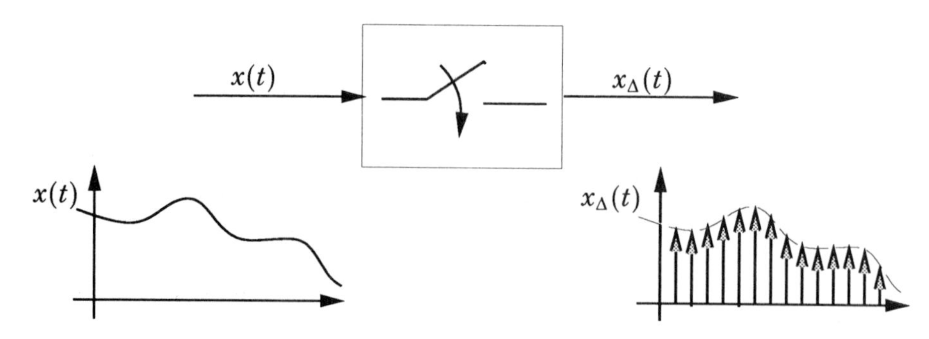

Figure 3.1 Discretized data x_Δ obtained from periodic sampling of a continuous-time variable $x(t)$.

This is identical to the two-sided Laplace transform if and only if $x(t) = 0$ for $t \leq 0$. An important property is uniqueness, *i.e.*, if $f(t)$ and $g(t)$ both have the same Laplace transform, then $f(t)$ and $g(t)$ can differ only at a countable number of points. It is also straightforward to show that the Laplace transformation is a linear operation, *i.e.*,

$$\mathcal{L}\{ax_1(t) + bx_2(t)\} = a\mathcal{L}\{x_1(t)\} + b\mathcal{L}\{x_2(t)\} \tag{3.4}$$

for arbitrary numbers a and b.

3.3 DISCRETIZED DATA

A measured variable $x(t)$ is in many cases available only as periodic observations of $x(t)$ sampled with a time interval h (the sampling period). Let the sampled values of $x(t)$ be represented by the sequence

$$\{x_k\}_{-\infty}^{\infty}; \quad x_k = x(kh) \qquad \text{for} \quad k = \ldots, -1, 0, 1, 2, \ldots \tag{3.5}$$

For ideal sampling, it is required that the duration of each sampling be very short and the sampled function may be represented by a sequence of infinitely short impulses; see Fig. 3.1. Let the sampled function of time be expressed thus

$$x_\Delta(t) = x(t) \cdot h \sum_{k=-\infty}^{\infty} \delta(t - kh) = x(t) \cdot Ш_h(t) \tag{3.6}$$

where

$$Ш_h(t) \stackrel{\Delta}{=} h \sum_{k=-\infty}^{\infty} \delta(t - kh) \tag{3.7}$$

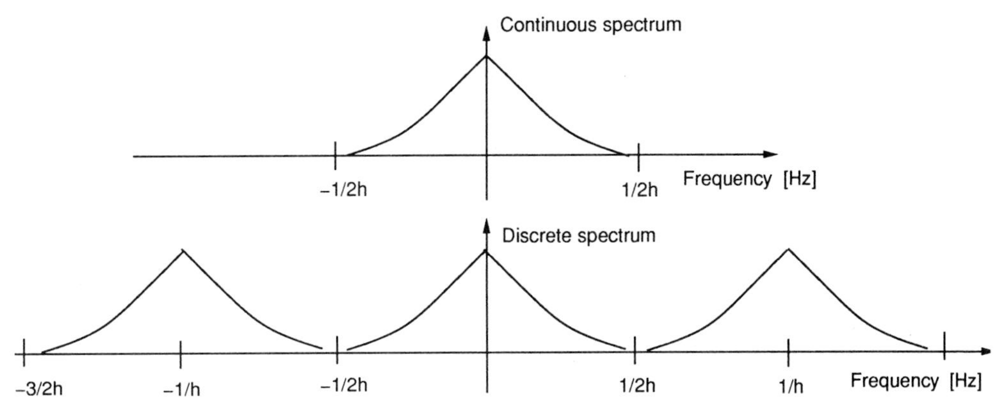

Figure 3.2 Continuous and sampled spectra.

and where the sampling period h is multiplied to assure that the averages over a sampling period of the original variable x and the sampled signal x_Δ, respectively, are of the same magnitude.

Obviously, the original variable $x(t)$ and the sampled data are not identical, and thus it is necessary to consider the distortive effects of discretization. Consider the spectrum of the sampled signal $x_\Delta(t)$ obtained as the Fourier transform

$$X_\Delta(i\omega) = \mathcal{F}\{x_\Delta(t)\} = \mathcal{F}\{x(t)\} * \mathcal{F}\{Ш_h(t)\} \tag{3.8}$$

where

$$\mathcal{F}\{Ш_h(t)\} = \sum_{k=-\infty}^{\infty} \delta(\omega - \frac{2\pi}{h}k) = \frac{h}{2\pi} Ш_{2\pi/h}(\omega) \tag{3.9}$$

so that

$$X_\Delta(i\omega) = \mathcal{F}\{x(t)\} * \mathcal{F}\{Ш_h(t)\} = \sum_{k=-\infty}^{\infty} X\left(i(\omega - \frac{2\pi}{h}k)\right) \tag{3.10}$$

The Fourier transform X_Δ of the sampled variable is thus a periodic function of the original spectrum $X(i\omega)$ along the frequency axis with a period equal to the sampling frequency $\omega_s = 2\pi/h$; see Fig. 3.2.

Theorem 3.1—The sampling theorem (SHANNON)

The continuous-time variable $x(t)$ may be reconstructed from the samples $\{x_k\}_{-\infty}^{+\infty}$ if and only if the sampling frequency is at least twice that of the highest frequency for which $X(i\omega)$ is non-zero.

Proof: Let $W_a(\omega)$ denote the *spectral window*

$$W_a(\omega) = \begin{cases} 1, & |\omega| \le a \\ 0, & |\omega| > a \end{cases} \tag{3.11}$$

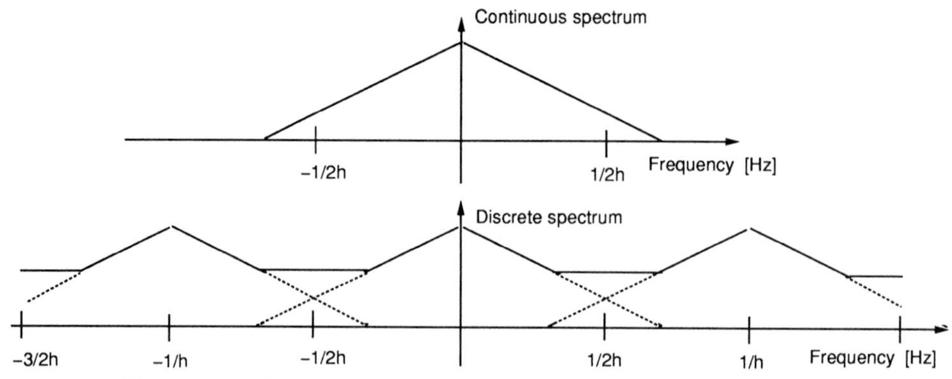

Figure 3.3 Continuous and sampled spectra in case of aliasing.

Subject to the assumption that $|X(i\omega)| = 0$ for $|\omega| > \omega_s/2 = \pi/h$, it holds that

$$X_\Delta(i\omega) \cdot W_{\pi/h}(\omega) = X(i\omega) \tag{3.12}$$

where the original spectrum $X(i\omega)$ thus may be recovered. The original variable $x(t)$ may thus be recovered from (3.12) as

$$\begin{aligned} x(t) &= \mathcal{F}^{-1}\{X_\Delta(i\omega) \cdot W_{\pi/h}(\omega)\} \\ &= \mathcal{F}^{-1}\{X_\Delta(i\omega)\} * \mathcal{F}^{-1}\{W_{\pi/h}(\omega)\} = x_\Delta(t) * (\frac{1}{h}\frac{\sin \pi t/h}{\pi t/h}) \\ &= \sum_{k=-\infty}^{\infty} x_k \frac{\sin \frac{\pi}{h}(t-kh)}{\frac{\pi}{h}(t-kh)} \end{aligned} \tag{3.13}$$

■

The formula (3.13) is called *Shannon interpolation* and is valid only for infinitely long data sequences. Notice also that it would require a noncausal filter to reconstruct the continuous-time signal $x(t)$ in real-time operation. The frequency $\omega_n = \omega_s/2 = \pi/h$ is called the *Nyquist frequency* and indicates the upper limit of distortion-free sampling. Failure to respect this limit leads to interference between the sampling frequency and the sampled signal (*aliasing*); see Fig. 3.3.

3.4 THE Z-TRANSFORM

The z-transform of a signal $x(t)$ discretized as the sequence $\{x_k\}_{-\infty}^{+\infty}$ is defined as

$$X_z(z) = \mathcal{Z}\{x\} = \sum_{k=-\infty}^{\infty} x_k z^{-k} \tag{3.14}$$

with the inverse transform

$$x_k = \frac{1}{2\pi i} \oint X_z(z) z^{k-1} dz \tag{3.15}$$

The z-transform is an infinite power series in the complex variable z^{-1} where $\{x_k\}$ constitutes a sequence of coefficents. As the z-transform is an infinite power series, it exists only for those values of z for which this series converges. A sufficient condition for existence of the z-transform is convergence of the power series

$$\sum_{k=-\infty}^{\infty} |x_k| \cdot |z^{-k}| < \infty \tag{3.16}$$

The region of convergence for a finite-duration signal is the entire z-plane except $z = 0$ and $z = \infty$. For a one-sided infinite-duration signal $\{x_k\}_{k=0}^{\infty}$, a number r can usually be found so that the power series converges for $|z| > r$.

A direct application of the discretized variable $x_\Delta(t)$ in (3.6) verifies that the spectrum of x_Δ is related to the z-transform $X_z(z)$ as

$$X_\Delta(i\omega) = \mathcal{F}\{x(t) \cdot Ш_h(t)\} = h \sum_{k=-\infty}^{\infty} x_k \exp(-i\omega kh) = hX_z(e^{i\omega h}) \tag{3.17}$$

3.5 FINITE MEASUREMENT TIME

Sampling with interpolation according to the Shannon theorem presupposes assumptions on infinitely long data records, which is clearly not possible in the context of identification. In order to represent finite measurement time $T = Nh$ with all measurements starting at time $t = 0$ and ending after N measurements, we introduce the functions

$$\sqcap_T(t) = \begin{cases} 0, & t < 0 \\ 1, & 0 \le t \le T \\ 0, & t > T \end{cases} \quad \rightleftharpoons \quad \mathcal{F}\{\sqcap_T(t)\} = 2T \frac{\sin(\omega T/2)}{\omega T} e^{-i\omega T/2} \tag{3.18}$$

and in discrete time

$$\sqcap_N(k) = \begin{cases} 0, & k < 0 \\ 1, & 0 \le k \le N-1 \\ 0, & k \ge N \end{cases} \quad \rightleftharpoons \quad \mathcal{Z}\{\sqcap_N\} = \frac{1 - z^{-N}}{1 - z^{-1}} \tag{3.19}$$

Table 3.1 Properties of the Fourier transform

	Time function	$\rightleftharpoons$	Fourier transform
Linearity	$f(t-\tau)$	$\rightleftharpoons$	$F(i\omega)e^{-i\omega\tau}$
	$f(at)$	$\rightleftharpoons$	$\frac{1}{\lvert a\rvert}F(i\frac{\omega}{a})$
	$af(t)+bg(t)$	$\rightleftharpoons$	$a\mathcal{F}\{f\}+b\mathcal{F}\{g\}$
Plancherel theorem	$f*g$	$\rightleftharpoons$	$\mathcal{F}\{f\}\cdot\mathcal{F}\{g\}$
	$f(t)\cdot g(t)$	$\rightleftharpoons$	$\mathcal{F}\{f\}*\mathcal{F}\{g\}$
Dirac impulse	$\delta(t)$	$\rightleftharpoons$	1
	$\delta(t-\tau)$	$\rightleftharpoons$	$e^{-i\omega\tau}$
Poisson's formula			
	$Ш_h(t)=h\sum_{-\infty}^{+\infty}\delta(t-kh)$	$\rightleftharpoons$	$\frac{h}{2\pi}Ш_{2\pi/h}(\omega)=\sum_{-\infty}^{+\infty}\delta(\omega-\frac{2\pi}{h}k)$
	$w_T(t)=\begin{cases}1, & \lvert t\rvert\le T\\ 0, & \lvert t\rvert>T\end{cases}$	$\rightleftharpoons$	$\mathcal{F}\{w_T\}=2T\frac{\sin\omega T}{\omega T}$
	$\sqcap_T(t)=\begin{cases}1, & 0\le t\le T\\ 0, & \text{otherwise}\end{cases}$	$\rightleftharpoons$	$\mathcal{F}\{\sqcap_T\}=2T\frac{\sin(\omega T/2)}{\omega T}e^{-i\omega T/2}$

The spectrum of any signal is distorted by finite measurement time, as for any variable $x(t)$ measured during a finite measurement interval T it holds that the spectrum is

$$\mathcal{F}\{x(t)\cdot\sqcap_T(t)\}=\mathcal{F}\{x(t)\}*\mathcal{F}\{\sqcap_T(t)\} \tag{3.20}$$

All spectrum estimation based on finite-time measurement is thus distorted by the convolution with $\mathcal{F}\{\sqcap_T\}$; see Fig. 3.4. The original spectrum is thus dispersed over large frequency ranges and provides a non-zero spectrum beyond all choices of the Nyquist frequency. Hence, the conditions of Shannon's sampling theorem are clearly not satisfied. *A rigorous attitude toward sampling is, therefore, that a variable cannot be sampled in a finite measurement interval without spectral distortion arising.*

Table 3.2 Properties of the Laplace transform

	Time function	$\rightleftharpoons$	Laplace transform
Linearity	$af(t) + bg(t)$	$\rightleftharpoons$	$a\mathcal{L}\{f\} + b\mathcal{L}\{g\}$
Time derivative	$\dfrac{df}{dt}$	$\rightleftharpoons$	$sF(s) - f(0)$
Time translation	$f(t-\tau)$	$\rightleftharpoons$	$F(s)e^{-s\tau}$
Multiplication	$f(t)e^{-at}$	$\rightleftharpoons$	$F(s+a)$
Multiplication by t	$tf(t)$	$\rightleftharpoons$	$-\dfrac{dF(s)}{ds}$

Table 3.3 Properties of the z-transform

Convolution	$\mathcal{Z}\{f * g\}$	$=$	$\mathcal{Z}\{f\} \cdot \mathcal{Z}\{g\}$
	$\mathcal{Z}\{f \cdot g\}$	$=$	$\mathcal{Z}\{f\} * \mathcal{Z}\{g\}$
Time translation	$\mathcal{Z}\{f((k-d)h)\}$	$=$	$z^{-d}\mathcal{Z}\{f(kh)\}$
Linearity	$\mathcal{Z}\{af + bg\}$	$=$	$a\mathcal{Z}\{f\} + b\mathcal{Z}\{g\}$
Multiplication	$\mathcal{Z}\{a^k f(k)\}$	$=$	$F_z(a^{-1}z)$
Final value	$f(\infty)$	$=$	$\lim_{z\to 1}(1-z^{-1})F(z)$
Initial value	$f(0)$	$=$	$\lim_{z\to\infty} F_z(z)$

Example 3.1—Spectral effects of finite measurement time

Consider a signal

$$x(t) = \sin \omega t \cdot \sqcap_T(t) \tag{3.21}$$

consisting of a sinusoid that is truncated after some time T. The Fourier

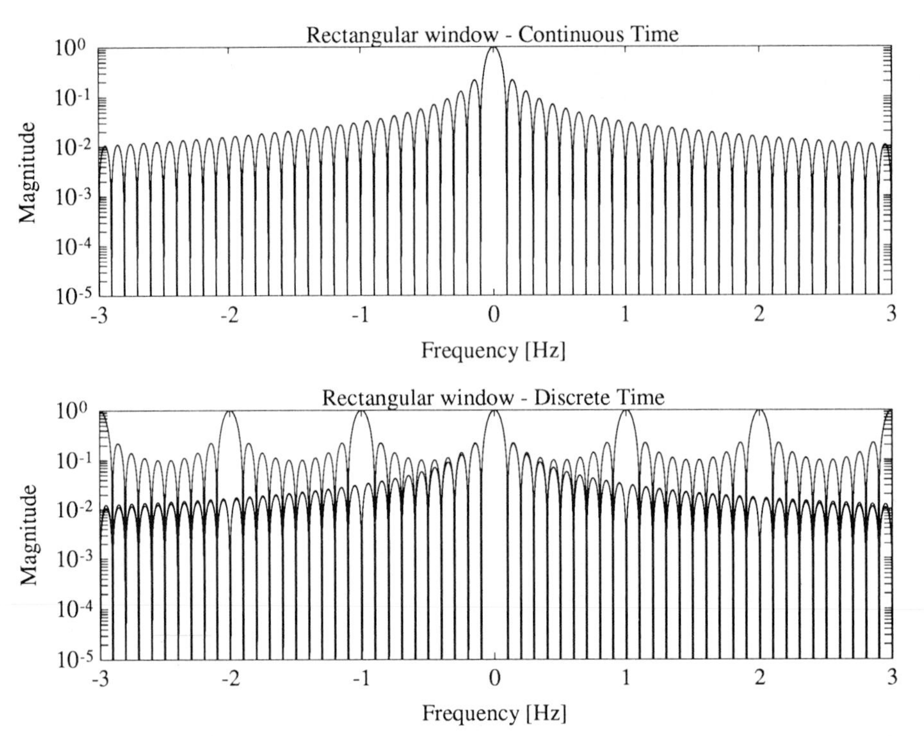

Figure 3.4 The spectral properties of $\mathcal{F}\{\sqcap_T(t)\}$ shown for a measurement duration $T = 10$. The zero crossings appear at frequencies that are multiples of $1/T$. The lower graph shows a discrete-time counterpart for $T = Nh = 10$ for $N = 10, 100, 1000$, respectively. The maximum magnitude of all graphs has been normalized to one.

transform of the signal is

$$X(i\omega) = \mathcal{F}\{\sin \omega t \cdot \sqcap_T(t)\} = \mathcal{F}\{\sin \omega t\} * \mathcal{F}\{\sqcap_T(t)\} \tag{3.22}$$

where $\mathcal{F}\{\sqcap_T(t)\}$ is shown in Fig. 3.4. A remarkable conclusion is that the spectrum of the pure sinusoid does not appear within any finite measurement time. Thus, the effect of finite measurement time on the spectrum is considerable and the resultant distortion of the spectrum *(the spectral leakage)* cannot be ignored. ■

The Discrete Fourier transform

Consider a finite length sequence $\{x_k\}_{k=0}^{N-1}$ that is zero outside the interval $0 \le k \le N-1$. Evaluation of the z-transform $X_z(z)$ at N equally spaced points on the unit circle $z = \exp(i\omega_k h) = \exp(i(2\pi/Nh)kh)$ for $k = 0, 1, \ldots, N-1$ defines the discrete Fourier transform (DFT) of a signal x with a sampling

Figure 3.5 Input-output models with transfer functions $H(z)$ and $G(s)$.

period h and N measurements

$$X_k = \mathcal{F}_{\Delta(h,N)}\{x(kh)\} = h\sum_{\ell=0}^{N-1} x_\ell \exp(-i\omega_k \ell h) = hX_z(e^{i\omega_k h}) \tag{3.23}$$

Notice that the discrete Fourier transform $\{X_k\}_{k=0}^{N-1}$ is only defined at the discrete frequency points

$$\omega_k = \frac{2\pi}{Nh}k, \qquad \text{for} \qquad k = 0, 1, \ldots, N-1 \tag{3.24}$$

For a finite-time measurement (N samples with a sampling period h) of a variable x, the relationship between the Fourier transforms $\mathcal{F}$ of the continuous-time variables, the discrete Fourier transform $\mathcal{F}_\Delta$, and the z-transform are as follows:

$$\begin{aligned} X_k = \mathcal{F}_{\Delta(h,N)}\{x(kh)\} &= \mathcal{F}\{x(t)\cdot ш_h(t)\cdot \sqcap_{Nh}(t)\} \\ &= hZ\{x(kh)\cdot \sqcap_N(k)\}\,|_{z=exp(i\omega_k h)} \\ &= hZ\{x\} * Z\{\sqcap_N(k)\}\,|_{z=exp(i\omega_k h)} \end{aligned} \tag{3.25}$$

which follows from (3.6) and (3.19). We conclude that

$$X_k = \mathcal{F}_{\Delta(h,N)}\{x(kh)\} = X_\Delta(i\omega_k h) * \frac{1-e^{-i\omega_k hN}}{1-e^{-i\omega_k h}} \tag{3.26}$$

In fact, the discrete Fourier transform adapts the Fourier transform and the z-transform to the practical requirements of finite measurements.

3.6 THE TRANSFER FUNCTION

An alternative to the input-output representations is the state-space representation. Consider the following system with input u (stimulus) and output y (response). The dependency of the output of a linear system is characterized by the convolution equation

$$y(t) = \int_0^\infty g(\tau)u(t-\tau)d\tau + v(t) \tag{3.27}$$

where $v(t)$ is some external input that represents errors and disturbances and $g(t)$ is the *impulse response* or *weighting function*. Application of the Laplace transform to Eq. (3.27) gives

$$Y(s) = G(s)U(s) + V(s) \tag{3.28}$$

and provides the frequency domain input-output relation with $G(s) = \mathcal{L}\{g\}$ as the *transfer function*. A similar relationship holds for discretized input-output data. Consider the model

$$y_k = \sum_{\ell=0}^{\infty} h_\ell u_{k-\ell} + v_k = \sum_{\ell=-\infty}^{k} h_{k-\ell} u_\ell + v_k, \qquad k = \ldots, -1, 0, 1, 2, \ldots \tag{3.29}$$

with the *pulse response* $h(kh) = \{h_k\}_{k=0}^{\infty}$ and its z-transform, the *pulse transfer function*

$$H(z) = \mathcal{Z}\{h(kh)\} = \sum_{k=0}^{\infty} h_k z^{-k} \tag{3.30}$$

The pulse transfer function $H(z)$ is obtained as the ratio

$$H(z) = \frac{Y_z(z)}{U_z(z)} \tag{3.31}$$

when the disturbance V is zero.

State-space systems

Alternatives to the input-output representations by means of transfer functions are the state-space representations. Consider the following finite dimensional discrete state-space equation with a state vector $x_k \in R^n$, input $u_k \in R^p$, and observations $y_n \in R^m$.

$$\begin{cases} x_{k+1} = \Phi x_k + \Gamma u_k \\ y_k = C x_k + D u_k \end{cases} \qquad k = 0, 1, \ldots \tag{3.32}$$

with the pulse transfer function

$$H(z) = C(zI - \Phi)^{-1}\Gamma + D \tag{3.33}$$

and the output variable

$$Y(z) = C \sum_{k=0}^{\infty} \Phi^k z^{-k} x_0 + H(z)U(z) \tag{3.34}$$

where possible effects of initial conditions x_0 appear as the first term. Notice that the initial conditions x_0 can be viewed as the net effects of the input in the time interval $(-\infty, 0)$ which can be verified by comparison with Eq. (3.29).

3.7 SIGNAL POWER AND ENERGY

Assuming that for a signal $x(t)$ at a fixed time t the *instantaneous power* at time t is

$$p_{xx}(t) = x(t) \cdot x^*(t) \tag{3.35}$$

where the asterisk denotes complex conjugate and transpose. For example, for a scalar signal $x(t) = a + ib$ the instantaneous power is $p_{xx}(t) = x(t) \cdot x^*(t) = a^2 + b^2$. The instantaneous power of the interaction between two signals x and y is

$$p_{xy}(t) = x(t) \cdot y^*(t) = p^*_{yx}(t) \tag{3.36}$$

The average power over an interval $[t_0, t_0 + T]$ is

$$\begin{aligned} \bar{p}_{xx}(t_0, T) &= \frac{1}{T}\int_{t_0}^{t_0+T} x(t)x^*(t)dt \\ \bar{p}_{xy}(t_0, T) &= \frac{1}{T}\int_{t_0}^{t_0+T} x(t)y^*(t)dt \end{aligned} \tag{3.37}$$

The energy of a signal is the integral of the power over time

$$e_{xx} = \int_{-\infty}^{+\infty} x(t)x^*(t)dt \tag{3.38}$$

and the interaction energy between two signals x and y is defined as

$$\begin{aligned} e_{xy}(\tau) &= \int_{-\infty}^{+\infty} x(t)y^*(t-\tau)dt \\ e_{yx}(\tau) &= \int_{-\infty}^{+\infty} y(t)x^*(t-\tau)dt = e^*_{xy}(-\tau) \end{aligned} \tag{3.39}$$

The signals x and y are said to be *uncorrelated* if the interaction energy $e_{xy}(\tau) = 0$.

Remark: The energy definitions (3.35–3.39) should not be confused with the energy definitions used in physics.

3.8 SPECTRA AND COVARIANCE FUNCTIONS

The *spectral density of energy* or the *energy spectrum* is defined as

$$E_{xx}(i\omega) = X(i\omega)X^*(i\omega) \tag{3.40}$$

whereas the *cross energy spectrum* between two signals x and y with Fourier transforms X and Y is defined as

$$E_{xy}(i\omega) = X(i\omega)Y^*(i\omega) \tag{3.41}$$

According to the Parseval relations it is known that the total signal energy

$$\int_{-\infty}^{+\infty} x(t)y^*(t)dt = \frac{1}{2\pi}\int_{-\infty}^{+\infty} X(i\omega)Y^*(i\omega)d\omega \tag{3.42}$$

which verifies that the signal energy is independent of the choice of representation in time or frequency. According to the *Plancherel theorem*, the product of two Fourier transforms equals the Fourier transform of the convolution of the two time-domain signals; thus it follows that the energy cross spectrum $E_{xy}(i\omega)$ is

$$\begin{aligned} E_{xy}(i\omega) &= X(i\omega)Y^*(i\omega) = \mathcal{F}\{x\} \cdot \mathcal{F}\{y^*\} \\ &= \mathcal{F}\{\int_{-\infty}^{+\infty} x(t)y^*(t-\tau)dt\} = \mathcal{F}\{e_{xy}(\tau)\} \end{aligned} \tag{3.43}$$

The relationship (3.43) is known as the *Wiener-Khintchine theorem*. For signals with infinite energy it makes better sense to consider the *cross covariance*

$$C_{xy}(\tau) = \lim_{T\to\infty}\frac{1}{2T}\int_{-T}^{T} x(t)y^*(t-\tau)dt \tag{3.44}$$

and the *power cross spectrum* (or *cross spectrum*) between the signals x and y. The power cross spectrum S_{xy} and the cross-covariance function are related as

$$S_{xy}(i\omega) = \mathcal{F}\{C_{xy}(\tau)\} \tag{3.45}$$

Similar relationships exist between the autospectrum and the *autocovariance* function as

$$S_{xx}(i\omega) = \mathcal{F}\{C_{xx}(\tau)\} = \mathcal{F}\{\lim_{T\to\infty}\frac{1}{2T}\int_{-T}^{T} x(t)x^*(t-\tau)dt\} \tag{3.46}$$

Table 3.4 Some properties of spectra S and covariances C obtained from a linear process $y(t) = g(t) * u(t) + v(t)$

Cross-covariance function

$$C_{yu}(\tau) = \lim_{T\to\infty} \frac{1}{2T} \int_{-T}^{T} y(t)u^*(t-\tau)dt$$

$$C_{uy}(\tau) = \lim_{T\to\infty} \frac{1}{2T} \int_{-T}^{T} u(t)y^*(t-\tau)dt$$

Autocovariance function

$$C_{uu}(\tau) = \lim_{T\to\infty} \frac{1}{2T} \int_{-T}^{T} u(t)u^*(t-\tau)dt$$

$$C_{yu}(\tau) = g(\tau) * C_{uu}(\tau)$$

Autospectrum

$$S_{uu}(i\omega) = \mathcal{F}\{C_{uu}(\tau)\}$$

Power cross spectra (Wiener-Khintchine theorem)

$$S_{uy}(i\omega) = \mathcal{F}\{C_{uy}(\tau)\}$$

$$S_{yu}(i\omega) = \mathcal{F}\{C_{yu}(\tau)\}$$

Power cross spectra from linear systems

$$S_{yu}(i\omega) = G(i\omega)S_{uu}(i\omega)$$

$$S_{uy}(i\omega) = S_{uu}(i\omega)G^*(i\omega)$$

$$S_{yy}(i\omega) = G(i\omega)S_{uu}(i\omega)G^*(i\omega) + S_{vv}(i\omega)$$

The power spectra and covariance functions are thus related according to the Wiener-Khintchine theorem (3.45–3.46).

3.9 CORRELATION AND COHERENCE

Assume that

$$y(t) = x(t) + v(t) = g(t) * u(t) + v(t) \tag{3.47}$$

where $y(t)$ is an observation of a variable $x(t)$ corrupted by a variable $v(t)$ which is some external input that represents disturbances. A crude measure to indicate the relative magnitudes of x and the noise v to the power of the observed output y is the signal-to-noise ratio (SNR)

$$\mathrm{SNR} = \frac{e_{xx}}{e_{vv}} = \frac{e_{yy}}{e_{vv}} - 1 \tag{3.48}$$

where the second equality holds if the signal x and the noise v are uncorrelated, *i.e.*, if the interaction energy $e_{xv} = 0$. The correlation coefficient ρ between two signals x and y is defined as the ratio

$$\rho(\tau) = \frac{C_{xy}(\tau)}{\sqrt{|C_{xx}(\tau)|}\sqrt{|C_{yy}(\tau)|}} \tag{3.49}$$

The quadratic coherence spectrum between the two signals x and y is defined as the ratio

$$\gamma_{xy}^2(\omega) = \frac{|S_{xy}(i\omega)|^2}{S_{xx}(i\omega)S_{yy}(i\omega)} \tag{3.50}$$

where the quadratic coherence always takes on a value in the interval $0 \leq \gamma^2(\omega) \leq 1$ with a value close to one if the noise level is low ($S_{vv} \ll S_{uu}$). The coherence spectrum is particularly interesting as a test of linearity in an input-output relationship. For instance, given a linear model (3.47) with observations y, input u, and with x, v not available to measurement, it holds that

$$\begin{aligned}\gamma_{uy}^2(\omega) &= \frac{|S_{uy}(i\omega)|^2}{S_{uu}(i\omega)S_{yy}(i\omega)} = \frac{|G(i\omega)|^2 S_{uu}^2(i\omega)}{S_{uu}(i\omega)(|G(i\omega)|^2 S_{uu}(i\omega) + S_{vv}(i\omega))} = \\ &= \frac{1}{1 + \dfrac{S_{vv}(i\omega)}{S_{uu}(i\omega)|G(i\omega)|^2}}\end{aligned} \tag{3.51}$$

Conversely, if in examining an input-output relationship a coherence value close to one is obtained, it may be inferred that the noise level is low and that there is a linear response of the type (3.47) between input and output.

The coherence function thus expresses the degree of linear correlation in the frequency domain between the input u and the output y. The coherence function may be viewed as a type of correlation function in the frequency domain. There is no immediate counterpart in the time domain—*i.e.*, the inverse Laplace transform of γ_{yu} has no interpretation.

3.10 STATISTICAL CHARACTERIZATION OF DISTURBANCES

Thus far we have discussed disturbances in quite general terms except for disturbances in the form of an initial condition or special cases of inputs in the form of impulses, steps, and sinusoids. Given the transfer function model, it is quite straightforward to predict the response to such inputs as they are completely determined by their initial behavior. In order to consider more complicated disturbances in the form of consecutive events, we thus introduce some notions from the theory of stochastic processes, which, as formulated by Kolmogorov and Wiener, has emerged from attempts to solve prediction problems; see Appendix D.

A function $x(t) = x(t,\omega)$ whose values depend on a random variable ω is called a random or *stochastic process*. For each fixed ω, $x(t,\omega)$ is a function called a *trajectory* or *realization* or *sample function*. In discrete time we find a stochastic process in the form of a sequence $\{x_k\}_{k=0}^{N}$ over some interval of time. It is standard practice to use a statistical model to express the behavior of x_k in terms of a probability distribution function

$$F_k(x) = \mathcal{P}\{x_k \leq x\}, \qquad 0 \leq F_k(x) \leq 1, \quad \forall x \in R \tag{3.52}$$

where $\mathcal{P}\{x_k \leq x\}$ denotes the probability that $x_k \leq x$. Given the distribution function, the *mathematical expectation* can be determined as follows

$$\mathcal{E}\{g(x_k)\} = \int_{-\infty}^{\infty} g(x_k)dF(x_k) \tag{3.53}$$

for various functions $g(x_k)$. Several of the concepts introduced earlier, such as covariance functions and spectra, can be applied to stochastic processes by taking the mathematical expectation of the corresponding function.

An important special case of stochastic processes is the white-noise process defined below.

Definition 3.1—White noise
A sequence of N uncorrelated, identically distributed stochastic variables $\{x_k\}_{k=1}^{N}$ with $\mathcal{E}\{x_k\} = \mu_x$, $\mathcal{E}\{(x_i - \mu_x)(x_j - \mu_x)^T\} = \delta_{ij}\Sigma_x$ for all i,j is known as *white noise* in the domain of time-series analysis. ■

It should be borne in mind that there is little possibility—both theoretical and practical—of determining such probabilistic perturbation characteristics in advance, and it may also be difficult to justify any assumptions as to the stochastic nature of a disturbance. Instead, it is a part of the identification procedure to determine such characteristics. In fact, the mean value and the covariance function are much easier to obtain than the probability distribution of the process. In particular, this is helpful in determining the normally distributed white-noise process. Let $\{x_k\}$ be a white-noise sequence with normally distributed components $x_k \in \mathcal{N}(\mu_x, \Sigma_x)$. Because the sequence is normally distributed, it is described by its mean

$$\mu_x = \mathcal{E}\{x_k\} \tag{3.54}$$

and its covariance matrix

$$C_{xx}(i,j) = \mathcal{E}\{(x_i - \mathcal{E}\{x_i\})(x_j - \mathcal{E}\{x_j\})^T\} \tag{3.55}$$

Because the sequence is white with mutually independent x_k values

$$C_{xx}(i,j) = \mathcal{E}\{(x_i-\mathcal{E}\{x_i\})(x_j-\mathcal{E}\{x_j\})^T\} = \Sigma_x \delta_{ij}, \qquad \delta_{ij} = \begin{cases} 1, & i = j \\ 0, & i \neq j \end{cases} \tag{3.56}$$

For normally distributed processes it is thus necessary to estimate μ_x and Σ_x as parameters along with other parameters. A standard approach to justifying assumptions on normal distributions is as follows: Experience suggests that noise encountered in physical systems is often characterized by normal distribution. In addition, according to the central limit theorem of statistics (see Appendix B), when a large number of small, independent, random effects are superimposed, then, regardless of their individual distribution, the sum of these components is approximately normally distributed. For the same reason, the central limit theorem is used as an approximation theorem in identification theory.

Using an assumed white-noise process $\{v_k\}$ as input to linear systems of the type

$$\begin{aligned} x_{k+1} &= \Phi x_k + \Gamma v_k \\ y_k &= C x_k + v_k \end{aligned} \tag{3.57}$$

provides new stochastic processes whose correlation properties are determined by the matrices Φ, Γ, and C. The class of random processes thus achieved is all stationary, and may well be amenable to modeling when the external perturbations and external actions $v(t)$ of a disturbance model are sufficiently regular and stationary in their behavior. However, it is certainly a source of problems in application that most methods rely on assumptions of stationary and normally distributed behavior. For instance, a system that is nonstationary due to parameter variation is difficult to model by such methods, which clearly limits the scope of their application.

As the elementary properties of stochastic processes will be known to the reader, this aspect will not be elaborated further, though some background material is available in Appendices B and D.

3.11 EXERCISES

3.1. Standard implementations of control systems have digital-to-analog converters with zero-order hold—*i.e.*, u is constant between the sampling instants. Show that the transfer function $G(s)$ corresponds to the following pulse transfer

$$H(z) = \frac{Y(z)}{U(z)} = (1 - z^{-1})\mathcal{Z}\{\mathcal{L}^{-1}\{G(s) \cdot \frac{1}{s}\}\} \tag{3.58}$$

in the case of zero-order hold.

3.2. Show that the coherence spectrum fulfills $0 \leq \gamma^2(\omega) \leq 1$.

3.3 Develop an expression similar to Eq. 3.48, which is valid as a signal-to-noise ratio, also for the case where non-zero correlation obtains between the signal x and perturbation v.

3.4 Show that the spectral density of a scalar white-noise process $\{x_k\}$ with $x_k \in \mathcal{N}(0, \sigma^2)$ is constant over the whole frequency range.

3.5 Determine the power spectrum of the variables x_k and y_k of the process (3.57) with a white-noise input where $v_k \in \mathcal{N}(0, \sigma^2)$.

3.6 Assume that the covariance of the estimate $\widehat{\theta}$ of the parameter vector

$$\theta = \begin{pmatrix} \theta_1 & \theta_2 \end{pmatrix}^T = \begin{pmatrix} 1 & 2 \end{pmatrix}^T \tag{3.59}$$

is

$$\Sigma_\theta = \text{cov}\{\widehat{\theta}\} = c \begin{pmatrix} 1 & -\rho \\ -\rho & 1 \end{pmatrix} \tag{3.60}$$

where c and ρ are constants. What is the variance of $\widehat{\theta}_1 + \widehat{\theta}_2$? What choice of α minimizes the variance of linear combination $\cos\alpha \cdot \theta_1 + \sin\alpha \cdot \theta_2$? Evaluate the 2-norm and the Frobenius norm of the covariance matrix Σ_θ; see Appendix A. Give an interpretation of the *covariance ellipsoid*, *i.e.*, the level surface determined by the equation

$$\widetilde{\theta}^T \Sigma_\theta^{-1} \widetilde{\theta} = \text{constant} \tag{3.61}$$

3.7 Consider the problem formulation of Exercise 3.6 and assume $\widetilde{\theta} \in R^n$ to be normally distributed $\mathcal{N}(0, \Sigma_\theta)$. Show that

$$\widetilde{\theta}^T \Sigma_\theta^{-1} \widetilde{\theta} \in \chi^2(n) \tag{3.62}$$

where $\chi^2(n)$ denotes the χ^2-distribution with n degrees of freedom; see Appendix B. ∎

4

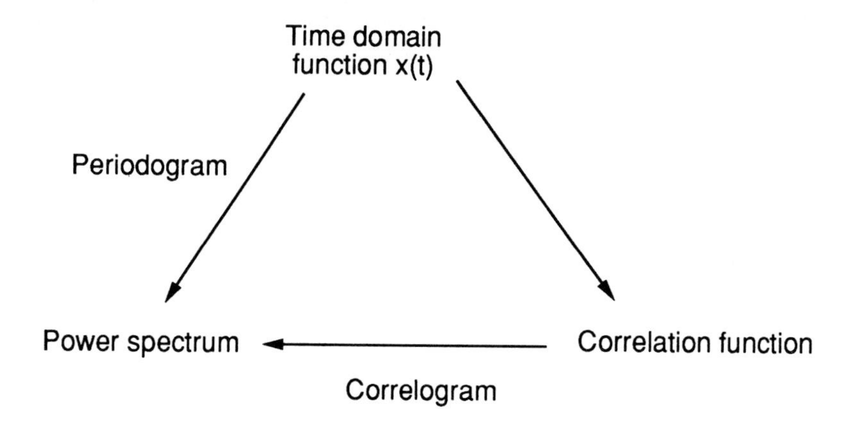

Spectrum Analysis

In previous chapters some notions and theories of signals and systems were introduced. We may now turn our attention to applications of these concepts in the analysis of experimental data by means of signal processing. This chapter covers spectral estimation techniques and some standard modifications of these techniques that are used to compensate for distortion due to discretization and finite measurement times. Covariance analysis and frequency response analysis are also treated.

Two important spectral estimation techniques based on Fourier transform operations have evolved: First, spectral estimation based on the direct approach *via* the discrete Fourier transform, usually referred to as a *periodogram*; and second, an important indirect approach, *the correlogram*, obtained by making a correlation estimate with a subsequent discrete Fourier transform based on the Wiener-Khintchine theorem. This methodology can be viewed as a generalization of frequency response analysis where multiple frequencies act simultaneously on the system input.

4.1 THE DISCRETE FOURIER TRANSFORM

Application of signal processing for spectrum analysis raises several important questions as to sampling strategy, quality of data, filtering, etc. The problems associated with finite sampling rates, finite measurement times, and aliasing naturally motivate the formulation of some sampling strategies for use under normal conditions. One method sometimes recommended for avoiding signal distortion is to sample the signal at a high rate with subsequent discrete-time filtering and data reduction, as analog filtering will not work with a linear phase shift which would be required for distortion-free filtering.

The discrete Fourier transform

As defined in Eq. (3.23), the discrete Fourier transform $X_N(i\omega)$ based on N measurements of a variable $x(t)$ is the sequence $\{X_k\}_{k=0}^{N-1}$ with the components

$$X_k = X_N(i\omega_k) = h\sum_{m=0}^{N-1} x(mh)e^{-i\omega_k mh} \tag{4.1}$$

defined at the discrete frequency points

$$\omega_k = \frac{2\pi}{Nh}k \quad \text{for} \quad k = 0, 1, \ldots, N-1 \tag{4.2}$$

The discrete Fourier transform thus evaluates the z-transform on the unit circle

$$X_k = hX_z(e^{i\omega_k h}) \tag{4.3}$$

Notice that the z-transform is generally an infinite series defined on the set of all complex-valued numbers for which it attains a finite value (*i.e.*, the region of convergence), whereas the discrete transform is only defined for a finite number of frequency points.

The *fast Fourier transform (FFT)* algorithm is an important implementation of the discrete Fourier transform, although often with the restriction that the

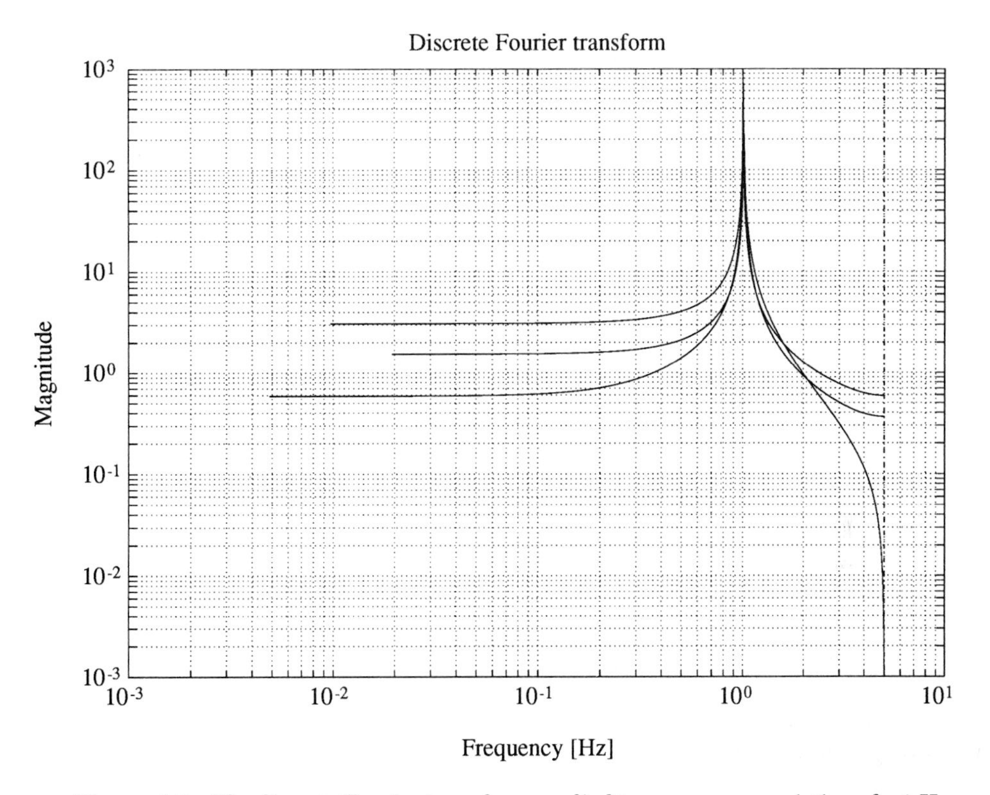

Figure 4.1 The discrete Fourier transform applied to sequences consisting of a 1-Hz sinusoidal signal sampled at 10 Hz with 512, 1024, and 2048 data points, respectively.

number of measurements should be a power of two, *i.e.*, $N = 2^\ell$ for some number $\ell = 2, 3, 4, \ldots$.

Example 4.1—FFT of a sinusoidal signal

A sinusoid of the frequency 1 Hz was sampled at the rate of 10 Hz. To each of three different sets of data for 512, 1024, and 2048 samples, respectively, the fast Fourier transform was applied (Fig. 4.1). Spectral leakage due to the finite measurement intervals is noticeable all over the spectrum. Moreover, the longer the duration of measurement the lower is the frequency at which the lower end frequency represented in the spectrum appears. ■

Remark: The definition (4.1) of the discrete Fourier transform does not immediately provide the spectrum in the desired frequency range determined by the Nyquist frequency ω_n, *i.e.*, in the interval $[-\omega_n, \omega_n]$. We accomplish this by shifting the periodic frequency scale on the unit circle so that $\omega_{-k} := \omega_{N-k}$ for $k = 1, 2, \ldots, N/2 - 1$; see Fig. 4.2.

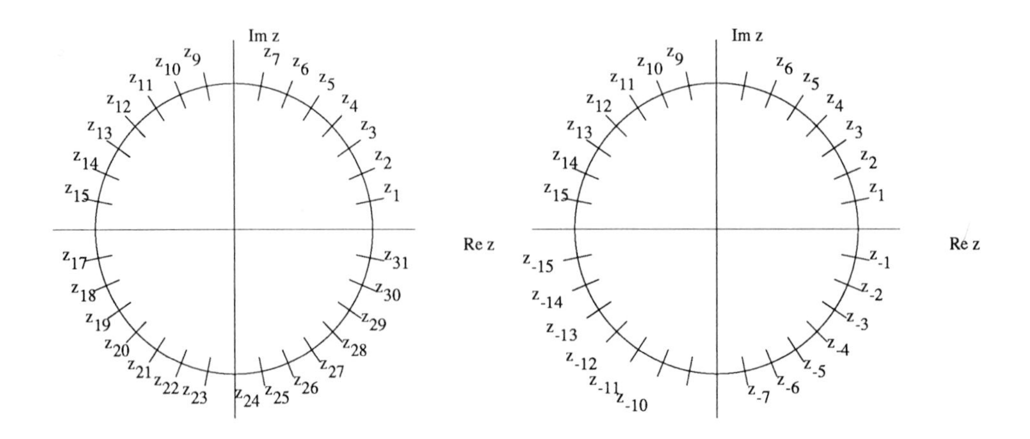

Figure 4.2 Circular permutation of frequency components $z_k = \exp(i\omega_k h)$ for $N = 32$ as applied in the discrete Fourier transform.

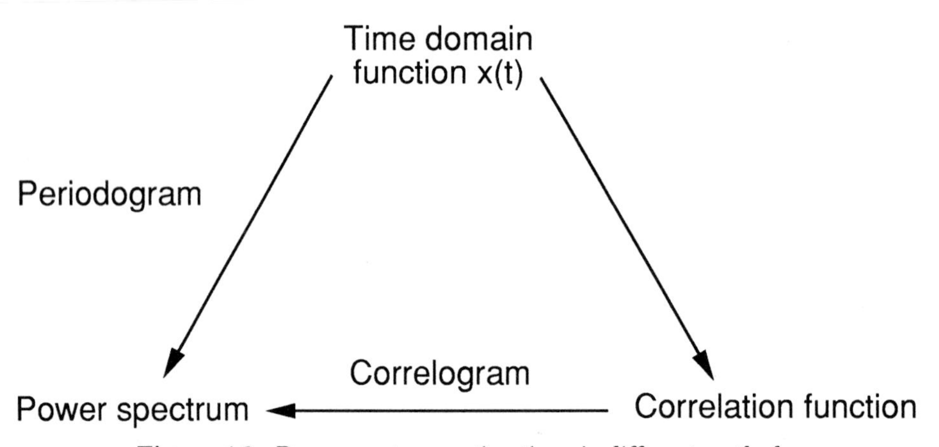

Figure 4.3 Power spectrum estimation via different methods.

Notice also that the frequency resolution of the spectrum is determined by the frequency points defined in Eq. (4.2). A longer sequence of data thus results in a finer resolution. ■

4.2 POWER SPECTRUM ESTIMATION

Two important spectral estimation techniques based on Fourier transform operations have evolved. The direct approach via the discrete Fourier transform is usually referred to as a *periodogram*, whereas an important indirect approach, the *correlogram*, is obtained from a correlation estimate with a subsequent discrete Fourier transform based on the Wiener-Khintchine theorem; see Fig. 4.3.

The periodogram

The periodogram or *sample spectrum* is defined as

$$\widehat{S}_{xx}(i\omega_k) = \frac{1}{Nh}|X_N(i\omega_k)|^2, \quad \text{for} \quad \omega_k = \frac{2\pi}{Nh}k \tag{4.4}$$

and is valuable for graphical inspection of the contents of a signal obtained from a system. In cases where an observed variable $x(t)$ consists of sinusoidal waves with superimposed white noise, the periodogram is effective in indicating the discrete frequencies of the sinusoids.

In effect, the periodogram $\widehat{S}_{xx}(i\omega)$ is a sampled version of a spectrum $S_{xx}(i\omega)$ convoluted with the transform of the rectangular window $\mathcal{F}\{\sqcap_T(t)\}$ that contains the data samples.

The correlogram

An important indirect approach, *the correlogram*, is obtained by means of a covariance estimate with a subsequent discrete Fourier transform based on the Wiener-Khintchine theorem. The estimated covariance functions are

$$\begin{aligned} \widehat{C}_{xx}(kh) &= \frac{1}{N-k}\sum_{\ell=k}^{N-1} x_\ell x_{\ell-k}^* \\ \widehat{C}_{xy}(kh) &= \frac{1}{N-k}\sum_{\ell=k}^{N-1} x_\ell y_{\ell-k}^*, \qquad k = 0, 1, \ldots, N-1 \end{aligned} \tag{4.5}$$

where the normalization factor $N - k$ in Eq. (4.5) is statistically motivated and results in unbiased estimates $\widehat{C}_{xx}$, $\widehat{C}_{xy}$ of the covariances C_{xx} and C_{xy}, respectively. The values of the covariance functions are often arranged in the matrix

$$\widehat{R}_{xx}(n) = \begin{pmatrix} \widehat{C}_{xx}(0) & \widehat{C}_{xx}(-1) & \cdots & \widehat{C}_{xx}(-n+1) \\ \widehat{C}_{xx}(1) & \ddots & \ddots & \vdots \\ \vdots & \ddots & \ddots & \widehat{C}_{xx}(-1) \\ \widehat{C}_{xx}(n-1) & \cdots & \widehat{C}_{xx}(1) & \widehat{C}_{xx}(0) \end{pmatrix} \tag{4.6}$$

which has a Toeplitz matrix structure, *i.e.*, the matrix is constant along its diagonals, a property that can be used for organizing efficient numerical algorithms; see Chapter 11. The subsequent power spectrum calculation is made according to the expressions

$$\begin{aligned} \widehat{S}_{xx}(i\omega_k) &= h\sum_{m=0}^{N-1}\widehat{C}_{xx}(mh)e^{-i\omega_k mh} \\ \widehat{S}_{xy}(i\omega_k) &= h\sum_{m=0}^{N-1}\widehat{C}_{xy}(mh)e^{-i\omega_k mh} \end{aligned} \quad \text{for } \omega_k = \frac{2\pi}{Nh}k \text{ and } k = 0,1,\ldots,N-1 \tag{4.7}$$

defined for discrete frequency points $\omega_k = (2\pi/Nh)\cdot k$ where $k = 0,1,\ldots,N-1$. The correlograms and periodograms for a pure sinusoid of 1 Hz and sampled at 10 Hz are compared in Fig. 4.4.

4.3 SPECTRAL LEAKAGE AND WINDOWING

All methods are based on a Fourier transform theory with infinite measurement sequences. For obvious reasons, the measurement is made on a finite-time interval only. A problem is that, as a measurement may not take infinite time, only N data points are assumed to be available, which causes problems of spectral leakage associated with the truncated measurements, which in turn entails a systematic distortion of calculated spectra; see Section 3.5.

The discrete Fourier transform based on N measurements as already formulated can be viewed as the truncated z-transform

$$Y_N(i\omega) = h\sum_{m=0}^{N-1} y(mh)e^{-i\omega hm} = h\sum_{-\infty}^{\infty} y(mh)\sqcap_N(m)e^{-i\omega hm} \tag{4.8}$$

where

$$\sqcap_N(k) = \begin{cases} 0, & k < 0 \\ 1, & 0 \le k < N \\ 0, & k \ge N \end{cases} \tag{4.9}$$

The Fourier transform of a product can be expressed as the convolution of Fourier transform of the two factors, respectively. Hence from Eq. (3.26) we have

$$Y_N(i\omega) = \mathcal{F}\{y(t)\cdot \text{Ш}_h(t)\cdot \sqcap_T(t)\} = Y(i\omega) * \mathcal{F}\{\text{Ш}_h(t)\sqcap_T(t)\} \tag{4.10}$$

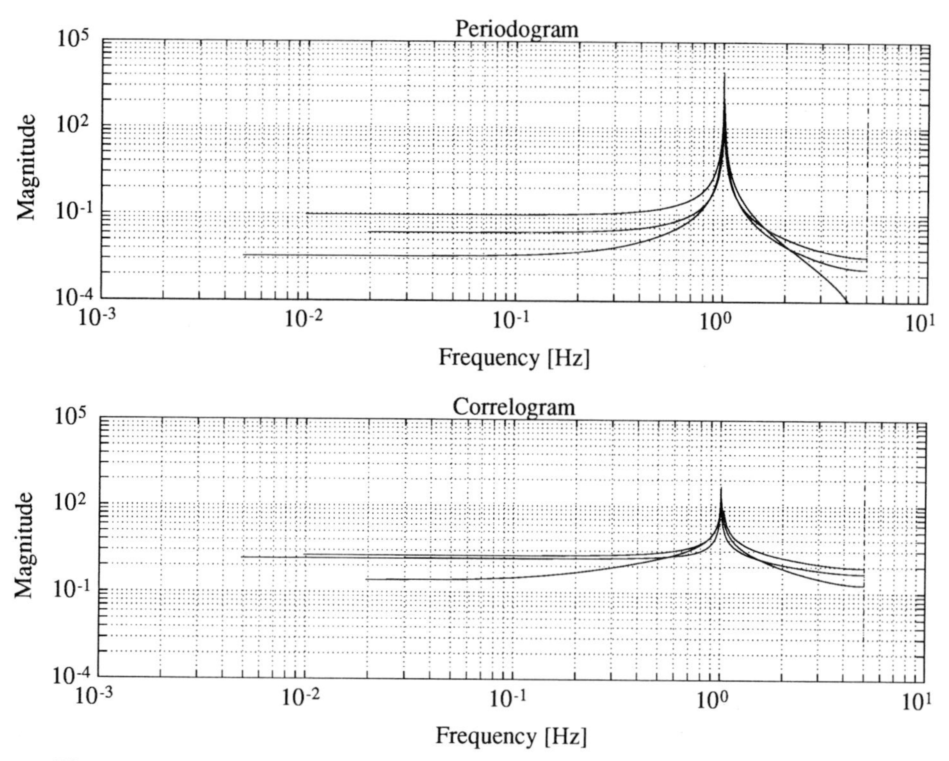

Figure 4.4 Periodograms (*upper*) and correlograms (*lower*) for a pure sinusoid of 1 Hz sampled at 10 Hz.

where the sampling-dependent final factor has the Fourier transform

$$\begin{aligned}\mathcal{F}\{\text{Ш}_h(t)\sqcap_T(t)\} &= \mathcal{Z}\{\sqcap_N(k)\}\,|_{z=exp(i\omega_k h)} \\ &= \frac{1-z^{-N}}{1-z}|_{z=exp(i\omega_k h)} = e^{-i\omega_k h(N-1)/2}\frac{\sin(\omega_k hN/2)}{\sin(\omega_k h/2)}\end{aligned} \tag{4.11}$$

The distortion arising due to convolution with the second factor is a systematic problem. The characteristic "sidelobes" of Eq. (4.11), *i.e.*, the Fourier transform of a rectangular time-domain window (Fig. 3.4), cause spectral leakage and a bias in the amplitude and the position of a harmonic estimate. In addition, spectral leakage has deleterious impact both on power spectrum estimation and on the detectability of sinusoidal components.

To reduce the effects of this bias, the window should exhibit low-amplitude sidelobes far from the main central lobe, and the transition to the low sidelobes should be very rapid. In fact, a given spectral component, say at $\omega = \omega_0$, will

Table 4.1 Application of spectrum estimation to experimental data

Discrete Fourier transform (N measurements)

$$X_N(i\omega_k) = h\sum_{m=0}^{N-1} x(kh)e^{-i\omega_k mh}, \quad \text{where} \quad \omega_k = \frac{2\pi}{Nh}k$$

Periodogram, sample spectrum

$$\widehat{S}_{xx}(i\omega) = \frac{1}{Nh}|X_N(i\omega)|^2$$

Correlogram

$$\widehat{S}_{yu}(i\omega_k) = h \sum_{m=-N/2}^{N/2-1} \widehat{C}_{yu}(mh)e^{-i\omega_k mh}$$

Correlogram with time window

$$\widehat{S}_{yu}(i\omega_k) = h \sum_{m=-N/2}^{N/2-1} \widehat{C}_{yu}(mh)w(mh)e^{-i\omega_k mh}$$

be observed at another frequency, say at $\omega = \omega_a$, according to the gain of a window obtained from (4.11), centered at ω_0 and measured at ω_a.

Example 4.2—Spectral leakage

Assume that

$$u(t) = u_1 \sin(t); \qquad N = 1024 \tag{4.12}$$

The spectral leakage in a spectrum obtained without windowing is clearly visible; see Fig. 4.5. In Fig. 4.5 the lower graphs show the reduced spectral leakage obtained by applying a Bartlett triangular window. Also notice the gross distortion of time-domain data that results from using the window functions. ■

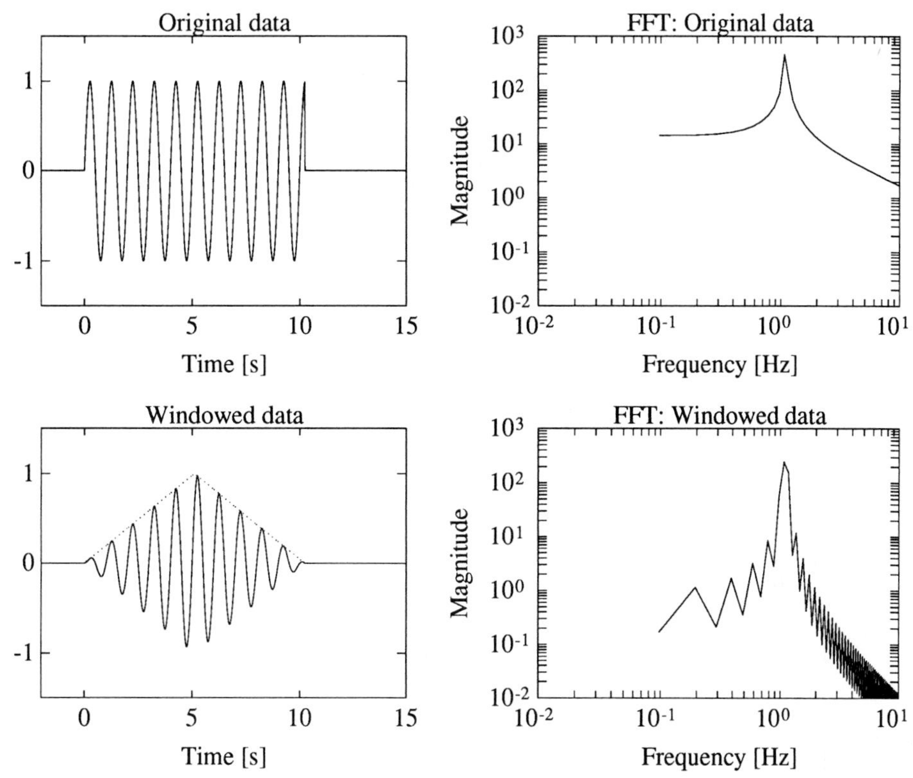

Figure 4.5 Discrete Fourier transform of a sinusoid (1 Hz) sampled with 100 Hz with $N = 1024$ samples. The spectral leakage in a spectrum obtained without windowing is clearly visible. The lower graphs show the reduced spectral leakage obtained by applying a Bartlett triangular window.

Window carpentry

A remedy to the spectral leakage problem is to introduce *window functions*, which are some weight functions applied to data to reduce the spectral leakage associated with the finite observation interval. Let the discrete Fourier transform include a window function $\{w_i\}_{i=0}^{N-1}$ in the form of a sequence that multiplies the data points.

$$Y_k = h \sum_{m=0}^{N-1} x_m w_m e^{-i\omega_k mh} \tag{4.13}$$

A major problem with using Fourier transforms based on finite measurements is that it forces a periodic extension to both the discretized data and the discretized transform values, whereas windowing presupposes that data outside the measurement interval are zero. The weighting function may be chosen as

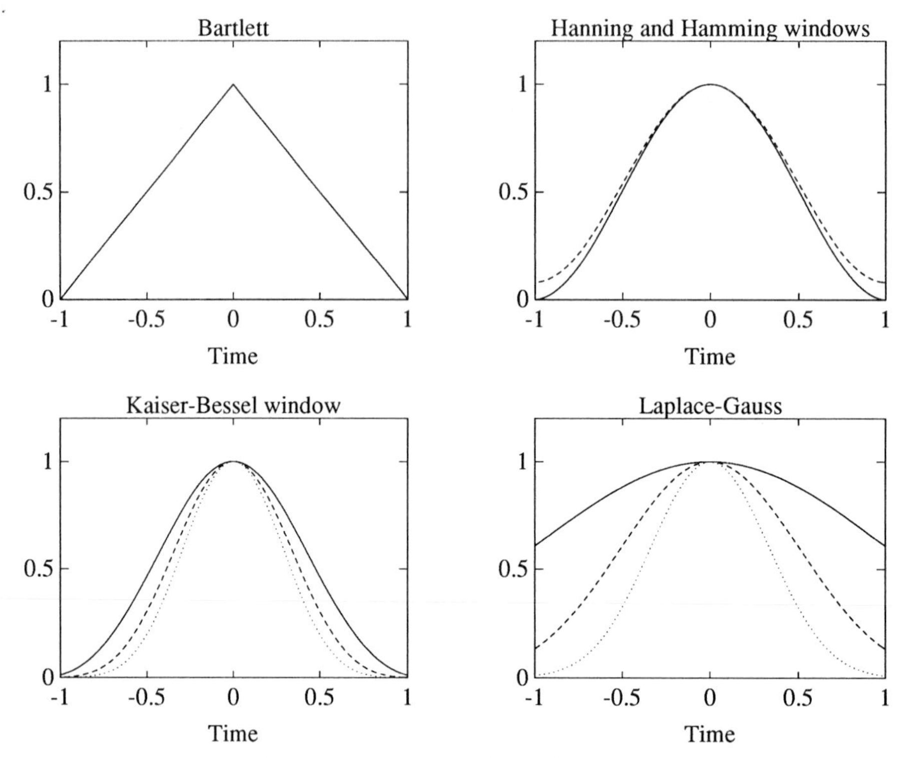

Figure 4.6 Some window functions commonly used to reduce spectral leakage. The time axis is expressed in normalized time, *i.e.*, τ/T_M takes on values over an interval $[-1, 1]$.

any function satisfying

$$w(\tau) = \begin{cases} 1, & \tau = 0 \\ 0, & \tau \text{ large} \end{cases} \tag{4.14}$$

Moreover, to minimize spectral leakage the window-function derivative should be small for large values of τ. The "time window" functions are often expressed in the frequency domain as $W(\omega)$. The rectangular window is the window that is always present in serial data recording, and that is defined by a finite number N of measurements; see Table 3.1.

$$w_{\sqcap}(\tau) = \begin{cases} 1, & |\tau| \leq T_M \\ 0, & |\tau| > T_M \end{cases} \tag{4.15}$$

Other commonly used windows are those of Bartlett, Hamming, Hanning, Kaiser-Bessel, Laplace-Gauss, and others (see Table 4.2 and Fig. 4.6).

Table 4.2 Some time windows commonly used in spectral analysis. Notice that all time windows are zero for time lag τ outside an interval $[-T_M, T_M]$.

	Time window $w(\tau)$	$\rightleftharpoons$	Spectral window $W(\omega)$
Rectangular	$w_{T_M}(\tau) = \begin{cases} 1, & \lvert\tau\rvert \le T_M \\ 0, & \lvert\tau\rvert > T_M \end{cases}$	$\rightleftharpoons$	$2T_M \dfrac{\sin \omega T_M}{\omega T_M} = W_\sqcap(\omega)$
Bartlett	$(1 - \dfrac{\lvert\tau\rvert}{T_M}) \cdot w_{T_M}(\tau)$	$\rightleftharpoons$	$T_M \dfrac{\sin(\omega T_M/2)^2}{(\omega T_M/2)^2}$
Hanning	$\dfrac{1}{2}\left(1 + \cos(\pi \dfrac{\tau}{T_M})\right) \cdot w_{T_M}(\tau)$	$\rightleftharpoons$	$\dfrac{1}{2} W_\sqcap(\omega) + \dfrac{1}{4} W_\sqcap(\omega + \dfrac{\pi}{T_M}) + \dfrac{1}{4} W_\sqcap(\omega - \dfrac{\pi}{T_M})$
Hamming	$\left(0.54 + 0.46 \cos(\pi \dfrac{\tau}{T_M})\right) \cdot w_{T_M}(\tau)$	$\rightleftharpoons$	$0.54 W_\sqcap(\omega) + 0.23 W_\sqcap(\omega + \dfrac{\pi}{T_M}) + 0.23 W_\sqcap(\omega - \dfrac{\pi}{T_M})$
Laplace-Gauss	$e^{\frac{-\zeta^2}{2} \cdot (\frac{\tau}{T_M})^2} w_{T_M}(\tau)$	$\rightleftharpoons$	$\dfrac{\sqrt{2}}{\zeta} T_M e^{-\frac{\omega^2}{8\pi^2 \zeta^2} T_M^2} * W_\sqcap(\omega)$
Kaiser-Bessel	$\dfrac{I_0(\beta(\sqrt{1 - (\tau/T_M)^2}))}{I_0(\beta)} w_{T_M}(\tau)$	$\rightleftharpoons$	(Numerical transform)

Example 4.3—Spectral leakage in frequency response analysis

Frequency response analysis (*cf.* Section 2.3) is also affected by spectral dis-

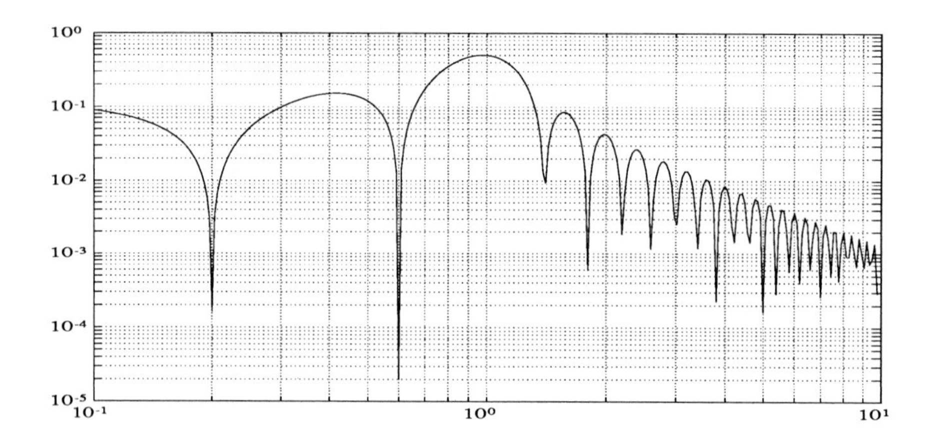

Figure 4.7 Transfer function of frequency response analysis with a fixed integration time T. The peak around the normalized test frequency $\omega = 1$ becomes more prominent for longer measurement times T. The frequency axis represents normalized frequencies.

tortion due to finite measurement time. In fact, the same analysis of spectral leakage and windowing applies to this method. Frequency response analysis may be viewed as a filter with certain rejection properties of periodic disturbances; see Fig. 4.7. In particular, the sensitivity to a periodic disturbance of the same frequency as the input test signal $u = \sin \omega_0 t$ is always important. The damping for frequencies different from the test frequency becomes better for longer measurement duration.

Assume that $y(t)$ is periodic for all t and let the integration time T be assumed to be fixed at $T = kh = k \cdot (2\pi/\omega_0)$. It is then possible to derive the following transfer functions from the system output y and the sine and cosine channels, respectively. Consider, for instance, the sine channel

$$s_T(\omega_0) = \int_0^T y(t) \sin(\omega_0 t) dt \tag{4.16}$$

and the convolution

$$y_s(t) = y(t) * g_s(t) = \int_0^t y(\tau) \sin(\omega_0 (t - \tau)) d\tau \tag{4.17}$$

Notice that $\sin \omega_0 t = \sin \omega_0 (T - t)$ for T such that $\omega_0 T = \pi, 3\pi, 5\pi, \ldots$. In fact, the functions $s_T(\omega_0)$ and $y_s(T)$ take on the same value for T such that $\omega_0 T = \pi, 3\pi, 5\pi, \ldots$. The integral $y_s(t)$ of Eq. (4.17) can be viewed as a

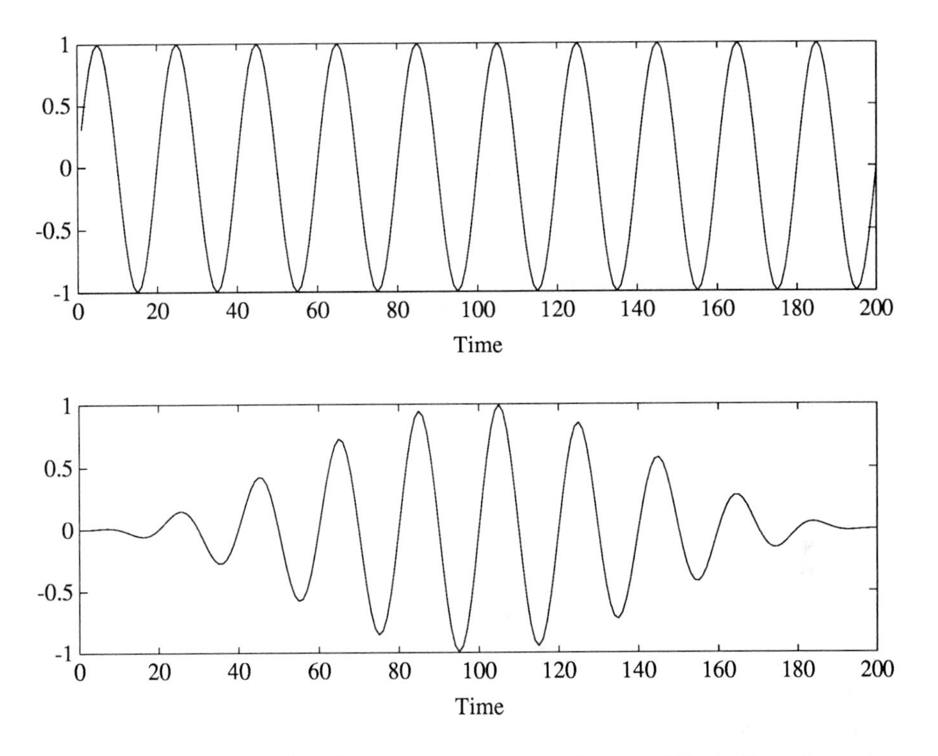

Figure 4.8 A test signal for frequency response analysis modified with a time window.

convolution between y and the function

$$g_s(t) = \sin(\omega_0 t) \cdot \sqcap_T(t) \tag{4.18}$$

The sine channel frequency properties may thus be understood in terms of a transfer function between input y and the frequency response

$$\mathcal{F}\{y_s(t)\} = G_s(i\omega) \cdot Y(i\omega) \tag{4.19}$$

where

$$G_s(i\omega) = \mathcal{F}\{g_s(t)\} = \mathcal{F}\{\sin(\omega_0 t) \cdot \sqcap_T(t)\} \tag{4.20}$$

This transfer function is obviously non-zero for $\omega \neq \omega_0$. Disturbances of any spectral composition entering into the frequency response analysis may therefore affect the result of frequency response analysis. However, the transfer function tends to depress frequencies different from the test frequency ω_0 for long experiment durations.

$$\frac{Y_s(i\omega)}{Y(i\omega)} = G_s(i\omega) = \mathcal{F}\{g_s(t)\} \tag{4.21}$$

Suppression of spectral leakage for frequencies $\omega \neq \omega_0$ of the transfer function in (4.21) may be enhanced by introducing a time window similar to that used in spectral analysis. For the sine channel we obtain

$$s_T(\omega_0) = \int_0^T y(t)w(t)\sin(\omega_0 t)dt, \qquad \omega_0 T = \pi, 3\pi, 5\pi, \ldots \tag{4.22}$$

An example of a window function applied to a test sequence is given in Fig. 4.8, where it is obvious that the modification of the test signal is considerable.

4.4 TRANSFER FUNCTION ESTIMATION

Estimation of transfer functions by methods of spectrum analysis is effective and can be viewed as a generalization of frequency response analysis in which the test signal is composed of multiple sinusoids at various frequencies with many frequencies acting simultaneously on the system input. As with spectrum estimation, there are two fundamental methods based on the Fourier transform that may be used to make spectral estimation of transfer functions. One method is based on the discrete Fourier transforms of input and output data. The transfer function estimate is then obtained according to the definition of transfer functions as a ratio between the discrete Fourier transforms of inputs and outputs.

$$\widehat{H}_1(e^{i\omega h}) = \frac{Y_N(i\omega)}{U_N(i\omega)} \tag{4.23}$$

The estimate $\widehat{H}_1(e^{i\omega h})$ is, of course, only defined for the discrete frequency points $\omega = \omega_k$ as obtained from the discrete Fourier transform.

A second method relies on the ratio between the input-output cross spectrum and the input autospectrum as obtained via the discrete Fourier transform of the cross-covariance and autocovariance functions.

$$\widehat{H}_2(e^{i\omega h}) = \frac{\widehat{S}_{yu}(i\omega)}{\widehat{S}_{uu}(i\omega)} \tag{4.24}$$

The noise characteristics of (4.23) and (4.24) are different, of course, as they depend on the noise spectrum and the input-disturbance cross-covariance, respectively. The expected contribution from the disturbance v in $\widehat{S}_{yu}(i\omega)$ is small in cases where the input and the disturbance are uncorrelated, whereas the disturbance contribution to $Y_N(i\omega)$ and thus $\widehat{H}_1(e^{i\omega h})$ might be considerable. Thus, the method (4.24) can be expected to yield better results than

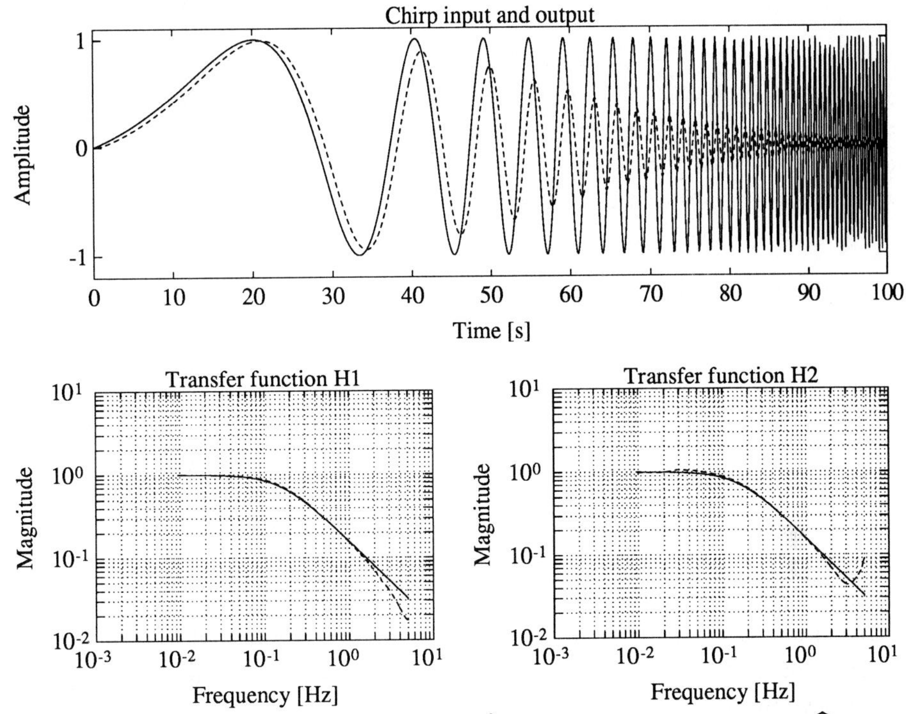

Figure 4.9 Transfer function estimates $\widehat{H}_1(i\omega) = Y_N(i\omega)/U_N(i\omega)$ and $\widehat{H}_2(i\omega) = \widehat{S}_{yu}(i\omega)/\widehat{S}_{uu}(i\omega)$ obtained from a system $G(s) = 1/(s+1)$. The input (*solid line*) is a swept-frequency sinusoid and is shown together with the output (*dashed line*) in the upper diagram.

(4.23) in cases where the input and the disturbance are uncorrelated. Generalizations of such averaging methods can be done by means of so-called *higher-order cumulants* or higher-order spectra.

Example 4.4—Transfer function estimates
A modified form of frequency response analysis is the sinusoidal input with a swept frequency (*cf.* Fig. 4.9). This input is known as a *chirp* signal and is effective in producing a rapid form of frequency response analysis. Obviously, both methods (4.23) and (4.24) work appropriately in frequency ranges with a non-zero input spectrum. On the other hand, problems can be expected to arise in frequency ranges where $U_N(i\omega)$ and $\widehat{S}_{uu}(i\omega)$ are of low amplitudes or where disturbances contribute to the spectral contents.

Evaluation of the coherence spectrum is valuable for the detection of disturbance levels prior to estimation of transfer functions. ■

Parametric models

Simple parametric models such as a rational transfer function may be obtained by manually fitting asymptotes to the Bode diagram. The procedure involves estimation of the low-frequency gain by fitting asymptotes to the gain plot. Each numerator factor $(s + q_1)$ contributes significant phase advance for $|s| = \omega > q_1$, whereas each denominator factor $(s + p_i)$ contributes significant phase lag from $|s| = \omega > p_i$.

Some important special features can be extracted from the Bode diagram such as static gain, resonances obtained from the gain diagram, or dead time as obtained from the phase plot.

Some statistical properties of transfer function estimates $\widehat{H}(e^{i\omega h})$

The quality of a transfer function estimate is dependent on many factors such as the statistical properties of covariance and spectrum estimates and the discrete Fourier transform, as well as the experiment that has generated the data used in identification. A technical problem is that the transfer function estimate is not one single number. Instead, the transfer function estimate is defined for a sequence of N frequency points ω_k for which the accuracy must be evaluated. Prior to estimation of transfer functions, it is valuable to evaluate the coherence spectrum for assessment of linear dependence and for detection of disturbance levels. It is therefore advisable to provide the coherence spectrum in support of an estimated transfer function.

Second, it is possible to derive statistical properties and criteria for evaluation of the quality of transfer function estimates based on spectrum analysis. Assuming the input to be periodic, and the measurement time N to be a multiple of the period, it can be shown that the following statistical properties hold:

- $\widehat{H}(e^{i\omega h})$ is only defined for a fixed number of points (which follows from Eq. [3.26]).
- $\widehat{H}(e^{i\omega h})$ is unbiased at these frequencies and its variance decreases as N grows.

If the input may be considered to be a stochastic process, the following important properties should be borne in mind:

- The estimate $\widehat{H}(e^{i\omega h})$ is asymptotically unbiased as the number of observations N increases.
- The variance of the transfer function estimate at a given frequency point does not decrease as N grows. Instead, the signal-to-noise ratio determines the accuracy at each frequency.

- The spectrum estimates at different frequencies are asymptotically uncorrelated.

Notice that the variance of $\widehat{H}(e^{i\omega h})$ does not decrease as the number of data points N grows, because there are as many independent estimates as there are data points (which means that there is no effective averaging as a result of chosing a longer data record). A remedy to this problem is to subdivide the measurement series into blocks and to average the results obtained from the spectral analysis of each block. The number of frequency points is then fixed, and an improved estimate may be obtained provided that $H(e^{i\omega h})$ has small variation within each block.

Inverse Laplace transformation

Impulse and step responses may be calculated from transfer function estimates via inverse Laplace transformation. However, the transfer function estimates always have a periodic extension to higher frequency ranges. Truncations of spectra and other operations on the spectral estimates may thus result in gross distortion of estimates of the impulse response obtained from inverse Fourier transforms. It is therefore useful to consider the application of spectral window functions prior to the calculation of the inverse Laplace transformation.

4.5 SMOOTHING OF SPECTRA

Statistically inconsistent results in spectrum analysis can occur, for instance, during estimation of a stochastic process generated by passing white noise through a linear system or filter, in which case the variance of the spectral estimates does not decrease. Therefore, there is a need for some sort of ensemble averaging or smoothing of the sample spectrum, and we distinguish among three different methods available for the purpose.

- Windowing offers a trade-off between spectral resolution and smoothness, the trade-off being dependent on the choice of window.
- Block segmentation of data includes splitting the data into segments, computing the periodogram for each segment, and averaging the periodograms for all the segments.
- *Zero padding* consists of making a DFT of a sequence of N data points extended with a $k \cdot N$ sequence of zeros. Transforming a data set with zeros only serves to interpolate additional spectrum values within the frequency

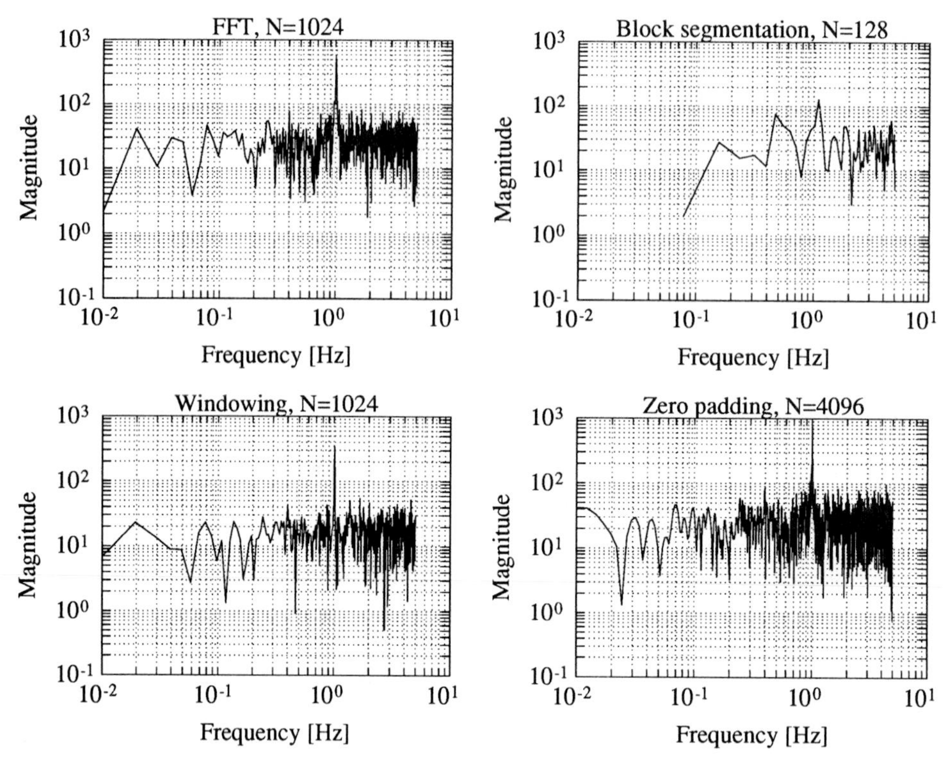

Figure 4.10 Smoothing of a spectrum of a signal consisting of a sinusoid $y(t) = \sin(t)$ confounded with noise with a signal-to-noise ratio $S/R = 1$. Smoothing can be obtained with block segmentation and windowing whereas higher spectral resolution can be obtained by windowing and zero padding.

interval $[-\omega_n, \omega_n]$. The additional values of the spectrum, computed by an **FFT** applied to the zero-padded data set, fill in the shape of the continuous-frequency periodogram. Zero-padding is useful for smoothing the appearance of the spectrum estimate and to resolve potential ambiguities, but it does not improve the fundamental frequency resolution—*i.e.*, the reciprocal of the measurement interval.

The three methods are demonstrated in Fig. 4.10 for a case with a sinusoid of frequency 1 Hz confounded with noise with a signal-to-noise ratio equal to one.

Another important aspect in this context is that the computation of $\widehat{S}_{yu}(i\omega)$ from $\widehat{C}_{yu}(\tau)$ has poor statistical and numerical properties. The deterioration in the quality of the estimate of $\widehat{C}_{yu}(\tau)$ that occurs as τ grows can be remedied by giving more weighting to the $\widehat{C}_{yu}(kh)$ for small τ values in the calculations than for large τ values.

$$\widehat{S}_{yu}(i\omega) = \frac{1}{2} \sum_{k=-N/2}^{N/2-1} \widehat{C}_{yu}(kh)w(kh)e^{-i\omega hk} \tag{4.25}$$

4.6 COVARIANCE ESTIMATES AND 'CORRELATION ANALYSIS'

Correlation analysis is a method based on statistical analysis for estimation of the weighting function $h(t)$. The input test signal is a white noise sequence, and the method applies to linear systems where the white-noise sequence is usually generated as a pseudorandom binary sequence (PRBS).

Consider the output y obtained as a convolution between the input u and the weight function $h(k)$ as

$$y(k) = \sum_{\ell=0}^{\infty} h(\ell)u((k-\ell)h) + e(kh) \tag{4.26}$$

The covariances may be obtained from the equation

$$C_{yu}(kh) = \sum_{k=0}^{\infty} h(k)C_{uu}((k-\ell)h) \tag{4.27}$$

The estimated counterparts based on N measurements are

$$\begin{cases} \widehat{C}_{yu}(kh) = \frac{1}{N-k} \sum_{\ell=k+1}^{N} y_\ell u_{\ell-k} \\ \widehat{C}_{uu}(kh) = \frac{1}{N-k} \sum_{\ell=k+1}^{N} u_\ell u_{\ell-k} \end{cases} \tag{4.28}$$

An estimate of $h(k)$ may thus be obtained from the equation

$$\widehat{C}_{yu}(kh) = \sum_{\ell=0}^{\infty} h(\ell)\widehat{C}_{uu}((k-\ell)h) \tag{4.29}$$

This infinite-dimensional equation is, in general, difficult to solve, but the special case of a white-noise input with

$$C_{uu}(kh) = \begin{cases} \sigma_u^2, & \text{if } k = 0 \\ 0, & \text{if } k \neq 0 \end{cases} \tag{4.30}$$

gives a simple estimate of the weighting function as

$$\widehat{h}(k) = \frac{1}{\sigma_u^2} \widehat{C}_{yu}(kh) \tag{4.31}$$

Thus a close connection exists between the cross-covariance function and the weight function $h(k)$. Unfortunately, the same poor numerical properties of the weight function also appear in the correlation analysis. The result $\widehat{C}_{yu}(kh)$ is in general a poor estimate of $C_{yu}(kh)$ for large k values, a problem analogous to that occurring in the impulse response test.

4.7 HISTORICAL REMARKS

The Fourier transform dates back about 200 years. Schuster (1898) developed the periodogram method for detecting hidden periodicities in sun-spot activity data. Yule (1927) and Walker (1931) pioneered work on autoregressive models. Wiener and Khintchine studied stochastic processes with methods of Fourier transform and established relations between autocorrelation functions and spectra. Blackman and Tukey provided practical implementation of Wiener's autocorrelation approach to power spectrum estimation in the 1950s. Cooley and Tukey presented the FFT (fast Fourier transform) in 1965.

4.8 BIBLIOGRAPHY

Early references are

- A. SCHUSTER, "On the Investigation of Hidden Periodicities with Application to a Supposed Twenty-Six-Day Period of Meteorological Phenomena." *Terr. Mag.*, Vol. 3, 1898, pp. 13–41.
- G.U. YULE, "On a Method of Investigating Periodicities in Disturbed Series with Special References to Wolfer's Sunspot Numbers." *Philos. Trans. R. Soc. London*, Ser. A, Vol. 226, 1927, pp. 267–298.
- G. WALKER, "On Periodicity in Series of Related Terms." *Proc. R. Soc.*, Ser. A, Vol. 313, 1931, pp. 518–532.

The standard treatment of spectral analysis is found in

- G.M. JENKINS AND D.G. WATTS, *Spectral Analysis and Its Applications*. San Francisco: Holden-Day, 1968.

– R. ISERMANN, *Theoretische Analyse der Dynamik Industrieller Prozesse, 1. Teil: Einführung*. Zürich: Bibliographisches Institut, 1971.

Modern and detailed descriptions of spectrum analysis are found in

– J.G. PROAKIS AND D.G. MANOLAKIS, *Introduction to Digital Signal Processing*. New York: Macmillan, 1989.

– S.M. KAY AND S.L. MARPLE, "Spectrum analysis: A modern perspective." *Proc. IEEE*, Vol. 69, 1981, pp. 1380–1419.

The systematic use of swept-frequency methods and the chirp-z transform was published in

– L.R. RABINER, R.W. SCHAFER, AND C.M. RADER, "The Chirp-z Transform Algorithm and Its Applications.", *Bell Syst. Tech. J.*, Vol. 48, 1969, pp. 1249–1292.

4.9 EXERCISES

4.1 Show the convolution property of the z-transform; *i.e.*, given that $X_1(z) = \mathcal{Z}\{x_1(k)\}$ and $X_2(z) = \mathcal{Z}\{x_2(k)\}$, show that

$$X(z) = \mathcal{Z}\{x_1(k) * x_2(k)\} = X_1(z)X_2(z) \tag{4.32}$$

4.2 A swept-frequency sinusoid of the type used as input in Example 4.4 can be represented by the sequence of complex exponentials

$$\{u_k\}_{k=0}^{N-1} = \{e^{i\omega_0 k^2/2}\}_{k=0}^{N-1}, \qquad \text{for some constant } \omega_0 \tag{4.33}$$

Show that the transfer function $H(z)$ of the input-output relationship

$$y_k = \sum_{j=0}^{N-1} h_j u_{k-j} \tag{4.34}$$

can be determined in a straightforward way by evaluating the z-transform

$$Y(z_n) = \sum_{k=0}^{N-1} y_k z_n^{-k}, \qquad \text{where} \quad z_n = e^{i\omega_0 n} \tag{4.35}$$

at a sequence of points $\{z_n\}_{n=0}^{N-1}$.

Remark. This algorithm, known as the *chirp-z transform* and developed by Rabiner *et al* (1969), thus evaluates the z-transform at another set of points than those used in the ordinary discrete Fourier transform. ■

5

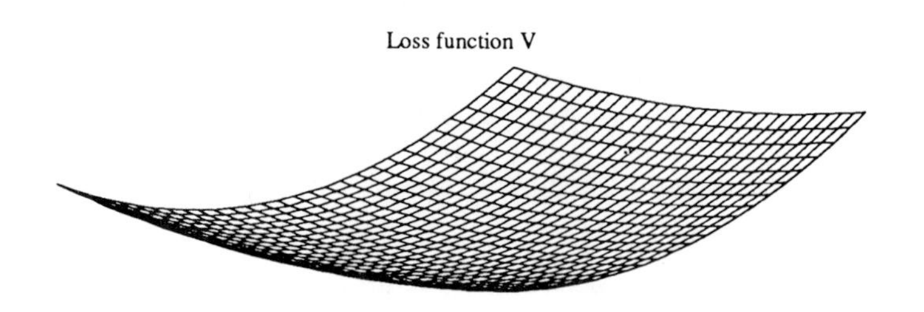

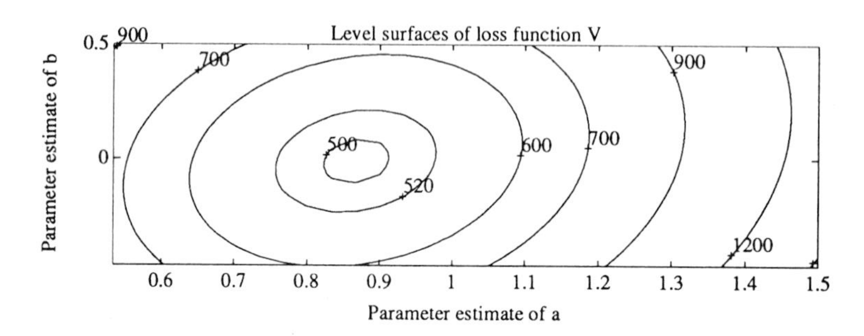

Linear Regression

5.1 INTRODUCTION

An important task in system identification and statistics is to find the relationships, if any, that exist in a set of variables when at least one is random or unknown, being subject to random (unknown) fluctuations and possible measurement errors. In regression, typically one of the variables, often called the *response* or dependent variable, is of particular interest and is denoted by y. The other variables $\phi_1, \phi_2, \ldots, \phi_p$, usually called explanatory, or independent variables, or *regressors*, are primarily used to predict or explain the behavior of y.

The appropriate mathematical relationship is sometimes known if the process is governed by accepted scientific laws or by a known physical process. It can happen that plots of data suggest some relationship to exist between y and the ϕ_i, a relationship we might attempt to express *via* some function

$$y = f(\phi_1, \ldots, \phi_p; \theta) + v \tag{5.1}$$

which is known except for some constants or coefficients θ called *parameters* and a possible disturbance v. For instance, the function f may be known except for the parameter vector $\theta = (\theta_1, \ldots, \theta_p)^T$, and a major aim of the investigation would then be to estimate the parameter vector θ of the process as precisely as possible. Finding a model is thus reduced to a question of estimating θ from experimental data. It is obviously important to use a parametric model flexible enough to represent or approximate a sufficiently wide range of behavior. An important special case is *linear regression analysis* based on the model

$$y(t) = \phi^T(t)\theta + e(t) \tag{5.2}$$

where additive errors e are used and where $\phi(t)$ is an n-dimensional vector. The regression vector $\phi = (\phi_1, \ldots, \phi_p)^T$ can, of course, include nonlinear functions of data such as squares, cross products, and transformations such as logarithmic, trigonometric, or exponential functions. The important requirement is that the expression (5.2) is *linear in parameters*.

Example 5.1—Linear regression models

The model

$$y = \theta_0 + \theta_1 u + \theta_2 u^2 = \begin{pmatrix} 1 & u & u^2 \end{pmatrix} \begin{pmatrix} \theta_0 \\ \theta_1 \\ \theta_2 \end{pmatrix} \tag{5.3}$$

is a linear regression model with respect to the parameters θ_i whereas the model

$$y = \theta_0 + \theta_1 e^{\theta_2 u} \tag{5.4}$$

is not a linear model. Linear regression models are often illustrated by means of straight lines in linear spaces; see Fig. 5.1. ■

It is often possible to convert nonlinear regression models to a form suitable for linear regression analysis as shown in the following example.

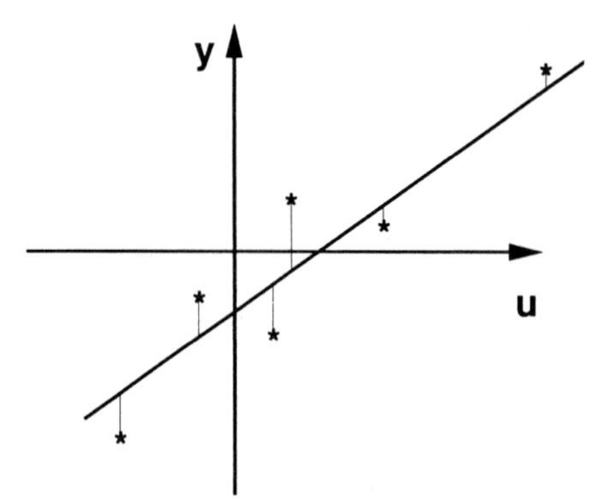

Figure 5.1 Least-squares estimation applied to fit a straight line to data points indicated by '*'. Notice that the least-squares fit minimizes the vertical deviation between the regression line and observed data.

Example 5.2—Nonlinear models

Consider a sample of gas kept at constant temperature with a volume V and a pressure p. Thermodynamical laws predict that

$$p \cdot V^{\gamma} = c \tag{5.5}$$

where γ and c are constants. Experimental verification of this law and determination of the parameters γ and c with linear regression methods directly applied to (5.5) are obviously not possible. The nonlinear model may, however, be transformed to a linear regression model by logarithmic transformation

$$\log p = -\gamma \log V + \log c = \begin{pmatrix} -\log V & 1 \end{pmatrix} \begin{pmatrix} \gamma \\ \log c \end{pmatrix} \tag{5.6}$$

The model (5.6) is now linear in the transformed parameters γ and $\log c$. Notice, however, that the statistical properties of the transformed model might be quite different from those of the original model. ■

As defined in Eq. (5.2) linear regression analysis is based on the model

$$y(t) = \phi^T(t)\theta + e(t) \tag{5.7}$$

with additive errors e. The observations y_k are assumed to be collected, with a constant sampling period h, together with the corresponding regression vectors $\{\phi_k\}$

$$\begin{aligned} y_k &= y(kh) \\ \phi_k &= \phi(kh) = \begin{pmatrix} \phi_{k1} & \phi_{k2} & \dots & \phi_{kp} \end{pmatrix}^T \end{aligned} \tag{5.8}$$

where the additive errors e are assumed to have the form

$$\mathcal{E}\{e_i\} = 0, \qquad \mathcal{E}\{e_i e_j\} = \sigma_e^2 \delta_{ij}, \qquad \forall i,j \tag{5.9}$$

Assuming that there are p parameters $\theta_1, \ldots, \theta_p$ to fit the model with $\theta \in R^p$ to N observations, the problem is to find an estimate $\widehat{\theta}$ of the parameter vector θ from the observed variables $\{y_k\}_{k=1}^N$ and $\{\phi_k\}_{k=1}^N$.

$$\begin{aligned} y_1 &= \phi_1^T \theta + e_1 \\ y_2 &= \phi_2^T \theta + e_2 \\ &\vdots \\ y_N &= \phi_N^T \theta + e_N \end{aligned} \tag{5.10}$$

In matrix notation we obtain the vector of observations

$$\mathcal{Y}_N = \begin{pmatrix} y_1 \\ y_2 \\ \vdots \\ y_N \end{pmatrix} \tag{5.11}$$

and the regressor matrix Φ_N and the error vector e where

$$\Phi_N = \begin{pmatrix} \phi_1^T \\ \phi_2^T \\ \vdots \\ \phi_N^T \end{pmatrix}, \quad \text{and} \quad e = \begin{pmatrix} e_1 \\ e_2 \\ \vdots \\ e_N \end{pmatrix} \tag{5.12}$$

Finally, the resulting estimation model for linear regression is

$$\mathcal{M}: \qquad \mathcal{Y}_N = \Phi_N \theta + e \tag{5.13}$$

The mismatch vector ε (also called *prediction errors*) between observations and the linear regression model for a certain parameter estimate $\bar{\theta}$ is

$$\varepsilon(\bar{\theta}) = \begin{pmatrix} \varepsilon_1 \\ \varepsilon_2 \\ \vdots \\ \varepsilon_N \end{pmatrix} = \mathcal{Y}_N - \Phi_N \bar{\theta} \tag{5.14}$$

The parameter vector $\bar{\theta} \in R^p$ often needs to be determined from a large set of observations. However, there is in general no satisfactory algebraic solution to the equation $\varepsilon = 0$, as this is an overdetermined linear equation system for $N > p$.

Several methods exist such as the deterministic criteria of least-squares estimation and L_1–estimation that involve minimization of criteria in the form of error functions (*loss functions*)

$$\min_{\bar{\theta}} \sum |y_i - \phi_i^T \bar{\theta}|^2, \quad \text{or} \quad \min_{\bar{\theta}} \sum |y_i - \phi_i^T \bar{\theta}|, \tag{5.15}$$

respectively. The aim is thus to find the model, of specified structure, which fits the observations best according to a deterministic measure of error between the model output and the observed output as considered over all observations.

Also, stochastic formulations such as maximum-likelihood and Bayesian estimation apply to obtain the "best" estimate $\hat{\theta}$. These methods are often somewhat more complicated as they entail a choice of the stochastic distribution of the observation errors involved; see Chapter 6.

5.2 LEAST-SQUARES ESTIMATION

Let $\bar{\theta}$ denote an arbitrary estimate of the parameter vector θ. The least-squares criterion aims to minimize the sum of the squared errors between the model output and the observations.

$$V(\bar{\theta}) = \frac{1}{2}\varepsilon^T \varepsilon = \frac{1}{2}\sum_{k=1}^{N} \varepsilon_k^2 = \frac{1}{2}(\mathcal{Y}_N - \Phi_N \bar{\theta})^T (\mathcal{Y}_N - \Phi_N \bar{\theta}) \tag{5.16}$$

with the minimum

$$\min_{\bar{\theta}} V(\bar{\theta}) = V(\hat{\theta}) \tag{5.17}$$

obtained for the optimal estimate

$$\hat{\theta} = (\Phi_N^T \Phi_N)^{-1} \Phi_N^T \mathcal{Y}_N \tag{5.18}$$

This can be seen by taking the gradient of the optimization criterion (5.16)

$$\frac{\partial V(\bar{\theta})}{\partial \bar{\theta}} = -\mathcal{Y}_N^T \Phi_N + \bar{\theta}^T (\Phi_N^T \Phi_N) \tag{5.19}$$

where the minimum $\partial V / \partial \bar{\theta} = 0$ provides the *normal equations*

$$-\mathcal{Y}_N^T \Phi_N + \bar{\theta}^T (\Phi_N^T \Phi_N) = 0 \tag{5.20}$$

The gradient takes on the value zero for $\bar{\theta} = \hat{\theta} = (\Phi_N^T \Phi_N)^{-1} \Phi_N^T \mathcal{Y}_N$, *i.e.*, when $\bar{\theta}$ is chosen as the least-squares estimate.

Notice that (5.19) and (5.20) are necessary conditions for obtaining a minimum. If the positive semidefinite matrix $\Phi_N^T \Phi_N$ is assumed to be invertible, then we can also show sufficiency by completing the squares of (5.16).

$$V(\bar{\theta}) = \frac{1}{2}(\mathcal{Y}_N - \Phi_N \bar{\theta})^T (\mathcal{Y}_N - \Phi_N \bar{\theta}) = \frac{1}{2} \mathcal{Y}_N^T (I - \Phi_N (\Phi_N^T \Phi_N)^{-1} \Phi_N^T) \mathcal{Y}_N + \\ + \frac{1}{2}(\bar{\theta} - (\Phi_N^T \Phi_N)^{-1} \Phi_N^T \mathcal{Y}_N)^T (\Phi_N^T \Phi_N)(\bar{\theta} - (\Phi_N^T \Phi_N)^{-1} \Phi_N^T \mathcal{Y}_N) \tag{5.21}$$

As the first term of Eq. (5.21) does not depend on $\bar{\theta}$, and as $\Phi_N^T \Phi_N$ is positive semidefinite and invertible by assumption, it may be concluded that $V(\bar{\theta})$ has a unique minimum at the least-squares solution

$$\hat{\theta} = (\Phi_N^T \Phi_N)^{-1} \Phi_N^T \mathcal{Y}_N \tag{5.22}$$

The first term of Eq. (5.21), then, represents the minimum value, *i.e.*, the residual sum.

Least-squares optimality has several attractive features for purposes of identification. First, large errors are heavily penalized. Second, the least-squares estimates can be obtained by straightforward matrix algebra. Third, the least-squares criterion is related to statistical variance, and the properties of the solution can be analyzed according to statistical criteria. Assuming that $\Phi_N^T \Phi_N$ is invertible, that the noise components are uncorrelated with the regressors, and that $\mathcal{E}\{e_i\} = 0$ and $\mathcal{E}\{e_i e_j\} = \sigma_e^2 \delta_{ij}$ for all i, j, it follows that the least-squares estimate has the following statistical properties:

i. $\hat{\theta}$ is an unbiased estimate of θ.

ii. The covariance matrix of $\hat{\theta}$ is

$$\sigma_e^2 (\Phi_N^T \Phi_N)^{-1} \tag{5.23}$$

iii. An unbiased estimate of σ_e^2 is

$$\hat{\sigma}_e^2 = \frac{2}{N - p} V(\hat{\theta}) \tag{5.24}$$

Proofs of the above statements *i–iii* are straightforward according to the following calculations all of which rely on the assumption of regressors uncorrelated with disturbances, *i.e.*, $\mathcal{E}\{\Phi_N^T e\} = 0$.

i. The least-squares estimate is unbiased when e_i are uncorrelated with the regressors so that

$$\widehat{\theta} = (\Phi_N^T\Phi_N)^{-1}\Phi_N^T(\Phi_N\theta + e) = \theta + (\Phi_N^T\Phi_N)^{-1}\Phi_N^T e \quad \Rightarrow \quad \mathcal{E}\{\widehat{\theta}\} = \theta \tag{5.25}$$

ii. When the e_i are uncorrelated with the regressors, it follows that

$$\begin{aligned}\mathcal{E}\{(\widehat{\theta}-\theta)(\widehat{\theta}-\theta)^T\} &= \mathcal{E}\{(\Phi_N^T\Phi_N)^{-1}\Phi_N^T e)e^T\Phi_N(\Phi_N^T\Phi_N)^{-1}\} = \\ &= (\Phi_N^T\Phi_N)^{-1}\Phi_N^T\mathcal{E}\{ee^T\}\Phi_N(\Phi_N^T\Phi_N)^{-1} = \\ &= \sigma_e^2(\Phi_N^T\Phi_N)^{-1}\end{aligned} \tag{5.26}$$

iii. The expected minimum of the least-squares loss function can be calculated using the relationships (5.13) and (5.20)

$$\begin{aligned}\mathcal{E}\{V(\widehat{\theta})\} &= \frac{1}{2}\mathcal{E}\{\mathcal{Y}_N^T\mathcal{Y}_N - \widehat{\mathcal{Y}}_N^T\widehat{\mathcal{Y}}_N\} = \frac{1}{2}\mathcal{E}\{e^T(I - \Phi_N(\Phi_N^T\Phi_N)^{-1}\Phi_N^T)e\} = \\ &= \frac{1}{2}\mathcal{E}\{\mathrm{tr}((I_{N\times N} - \Phi_N(\Phi_N^T\Phi_N)^{-1}\Phi_N^T)ee^T)\} = \\ &= \frac{1}{2}(\mathrm{tr}I_{N\times N} - \mathrm{tr}I_{p\times p})\sigma_e^2 = \frac{1}{2}(N-p)\sigma_e^2\end{aligned} \tag{5.27}$$

so that

$$\mathcal{E}\{\widehat{\sigma}_e^2\} = \mathcal{E}\{\frac{2}{N-p}V(\widehat{\theta})\} = \sigma_e^2 \tag{5.28}$$

which proves the statement. ■

Example 5.3—Least-squares estimation of coefficients
Consider the following data set of paired data

$$\mathcal{U} = \begin{pmatrix}1 & 2 & 3 & 4\end{pmatrix}^T, \quad \text{and} \quad \mathcal{Y} = \begin{pmatrix}6 & 17 & 34 & 57\end{pmatrix}^T \tag{5.29}$$

and assume that the following model should be fitted to data

$$\mathcal{M}: \quad y = \theta_0 + \theta_1 u + \theta_2 u^2 = \begin{pmatrix}1 & u & u^2\end{pmatrix}\begin{pmatrix}\theta_0 \\ \theta_1 \\ \theta_2\end{pmatrix} \tag{5.30}$$

Let the regressor matrix be

$$\Phi = \begin{pmatrix}1 & u_1 & u_1^2 \\ 1 & u_2 & u_2^2 \\ 1 & u_3 & u_3^2 \\ 1 & u_4 & u_4^2\end{pmatrix} = \begin{pmatrix}1 & 1 & 1 \\ 1 & 2 & 4 \\ 1 & 3 & 9 \\ 1 & 4 & 16\end{pmatrix} \tag{5.31}$$

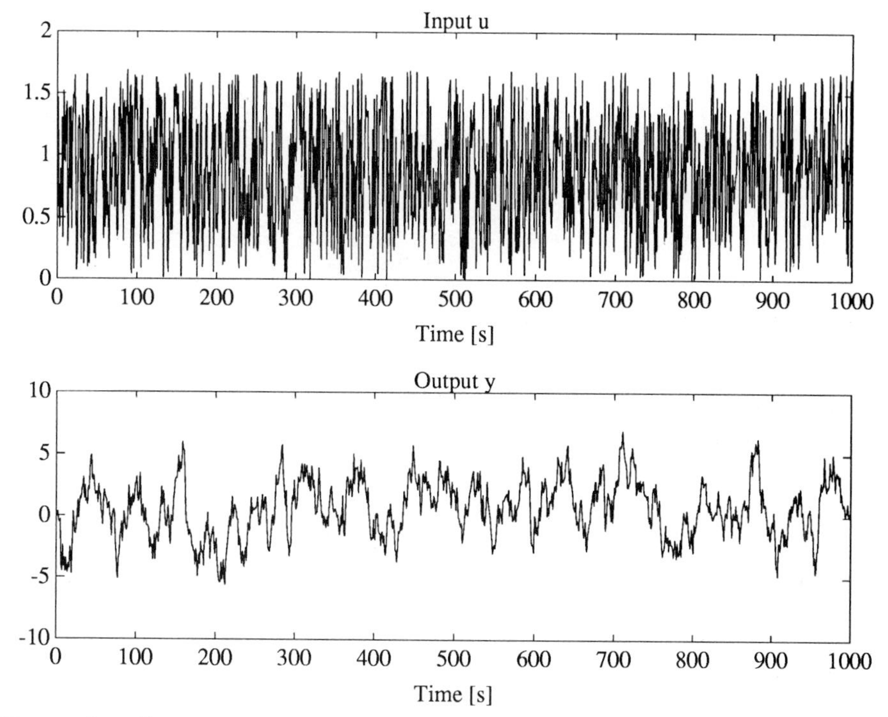

Figure 5.2 Input u and output y obtained from observation of a system $y_k = ay_{k-1} + bu_{k-1} + e_k$.

The least-squares solution $\widehat{\theta} = (\Phi^T\Phi)^{-1}\Phi^T\mathcal{Y}$ is then

$$\widehat{\theta} = \begin{pmatrix} \sum_{k=1}^{4} 1 & \sum_{k=1}^{4} u_k & \sum_{k=1}^{4} u_k^2 \\ \sum_{k=1}^{4} u_k & \sum_{k=1}^{4} u_k^2 & \sum_{k=1}^{4} u_k^3 \\ \sum_{k=1}^{4} u_k^2 & \sum_{k=1}^{4} u_k^3 & \sum_{k=1}^{4} u_k^4 \end{pmatrix}^{-1} \begin{pmatrix} \sum_{k=1}^{4} y_k \\ \sum_{k=1}^{4} u_k y_k \\ \sum_{k=1}^{4} u_k^2 y_k \end{pmatrix} = \begin{pmatrix} 1 \\ 2 \\ 3 \end{pmatrix} \tag{5.32}$$

which fits the polynomial to data without residual error. ■

Example 5.4—Least-squares identification of a first-order system
Assume 1000 samples of inputs u_k and outputs y_k to have been collected from a system described by the equation

$$\mathcal{S}: \qquad y_k = ay_{k-1} + bu_{k-1} + e_k \tag{5.33}$$

where the coefficients $a = 0.9$ and $b = 0.1$ are not known and need to be estimated. Artificial data have been generated with $a = 0.9$ and $b = 0.1$, and u and e as random variables with variances $\sigma_e^2 = \sigma_u^2 = 1$. The mean of the disturbance e is zero; see Fig. 5.2. It is natural to adopt the following linear

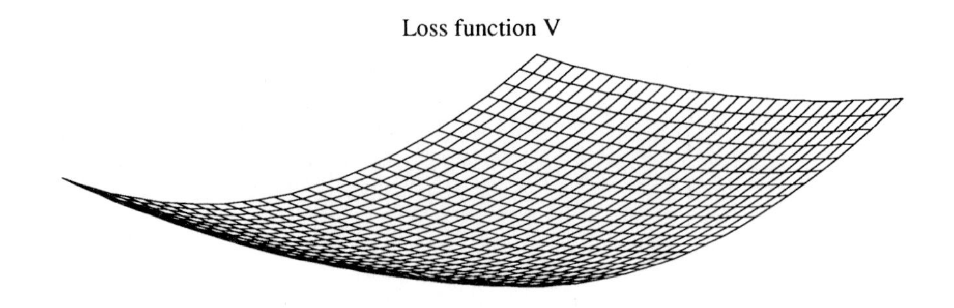

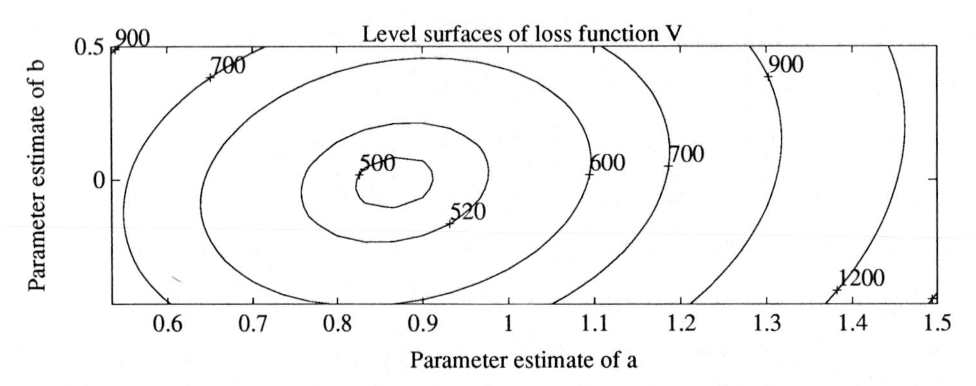

Figure 5.3 Level surfaces of the loss function V as obtained in linear regression from observations of a system $y_k = ay_{k-1} + bu_{k-1} + e_k$.

regression model

$$\mathcal{M}: \qquad y_k = \widehat{a}y_{k-1} + \widehat{b}u_{k-1} = \begin{pmatrix} y_{k-1} & u_{k-1} \end{pmatrix} \begin{pmatrix} \widehat{a} \\ \widehat{b} \end{pmatrix} = \phi_k^T \widehat{\theta} \tag{5.34}$$

The least-squares estimate is

$$\widehat{\theta} = \begin{pmatrix} \widehat{a} \\ \widehat{b} \end{pmatrix} = \begin{pmatrix} 0.8992 \\ 0.0899 \end{pmatrix} \tag{5.35}$$

The loss function and the variance estimate are

$$V(\widehat{\theta}) = 499.94, \quad \text{and} \quad \widehat{\sigma}_e^2 = 1.0019 \tag{5.36}$$

The estimated covariance matrix of $\widehat{\theta}$ is

$$\widehat{\sigma}_e^2(\Phi^T\Phi)^{-1} = \begin{pmatrix} 0.249 & -0.090 \\ -0.090 & 1.023 \end{pmatrix} \cdot 10^{-3} \tag{5.37}$$

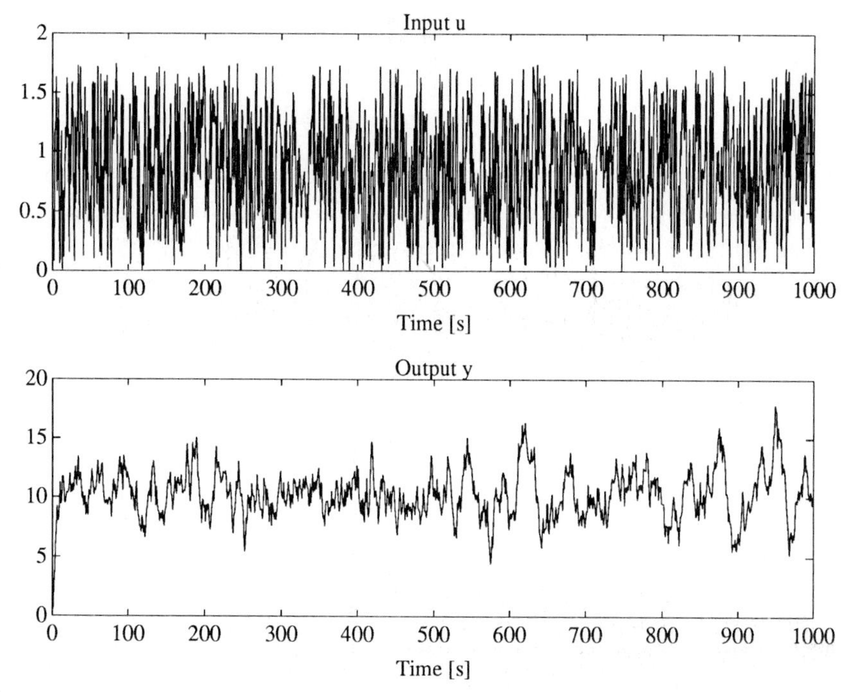

Figure 5.4 Input u and output y obtained from observation of a system $y_k = ay_{k-1} + bu_{k-1} + e_k$.

which provides good estimates considering the large disturbance level. The level surfaces of the loss function V as a function of the parameter estimates $\widehat{a}$ and $\widehat{b}$ are shown in Fig. 5.3.

■

Example 5.5—LS identification with non-zero disturbance mean
Assume 1000 samples of inputs u_k and outputs y_k have been collected from a system described by the equation

$$\mathcal{S}: \qquad y_k = ay_{k-1} + bu_{k-1} + e_k, \qquad \mathcal{E}\{e_k\} = 1 \tag{5.38}$$

where the coefficients $a = 0.9$ and $b = 0.1$ are not known and need to be estimated (Fig. 5.4). Artificial data have been generated with $a = 0.9$ and $b = 0.1$ with u and e as random variables with variances $\sigma_e^2 = \sigma_u^2 = 1$. The mean of the disturbance e is one, whereas it was zero in Example 5.4. By adopting the linear regression model

$$\mathcal{M}: \qquad y_k = \widehat{a}y_{k-1} + \widehat{b}u_{k-1} = \begin{pmatrix} y_{k-1} & u_{k-1} \end{pmatrix} \begin{pmatrix} \widehat{a} \\ \widehat{b} \end{pmatrix} = \phi_k^T \widehat{\theta} \tag{5.39}$$

and by fitting the least-squares estimate we have

$$\hat{\theta} = \begin{pmatrix} \hat{a} \\ \hat{b} \end{pmatrix} = \begin{pmatrix} 0.9829 \\ 0.1550 \end{pmatrix} \tag{5.40}$$

which provides poor estimates as compared to Example 5.4. The main difference between the conditions of Example 5.4 and this example is the non-zero mean of the disturbance in this example. Clearly, the disturbance properties in this example do not fulfill the condition of Eq. (5.9) and such disturbances are for this reason characterized as *colored noise*.

The loss function and the variance estimate are

$$V(\hat{\theta}) = 531.04, \quad \text{and} \quad \hat{\sigma}_e^2 = 1.0642 \tag{5.41}$$

The estimated covariance matrix of $\hat{\theta}$ is

$$\hat{\sigma}_e^2(\Phi^T\Phi)^{-1} = \begin{pmatrix} 0.0295 & -0.2734 \\ -0.2734 & 3.568 \end{pmatrix} \cdot 10^{-3} \tag{5.42}$$

which is close to being singular. Hence, *the least-squares solution is sensitive to colored noise.* ■

Example 5.6—Least-squares identification of a step response

Assume that 1000 samples of inputs u_k and outputs y_k obtained from a system described by the equation

$$\mathcal{S}: \qquad y_k = ay_{k-1} + bu_{k-1} + e_k \tag{5.43}$$

where the coefficients $a = 0.9$ and $b = 0.1$ are not known and need to be estimated (Fig. 5.5). (The data have been generated with $a = 0.9$ and $b = 0.1$ with u as unit step input under noise-free conditions, *i.e.*, $e_k = 0$ for all k.) The least-squares solution of the parameters a and b is obtained as

$$\hat{\theta} = \begin{pmatrix} \hat{a} \\ \hat{b} \end{pmatrix} = \begin{pmatrix} 0.9000 \\ 0.1000 \end{pmatrix} \tag{5.44}$$

which provides very good estimates. The loss function and the variance estimate are

$$V(\hat{\theta}) = 1.7844 \cdot 10^{-26} \quad \text{and} \quad \hat{\sigma}_e^2 = 3.5760 \cdot 10^{-29} \tag{5.45}$$

The covariance matrix of $\hat{\theta}$ is

$$\hat{\sigma}_e^2(\Phi^T\Phi)^{-1} = \begin{pmatrix} 0.520 & -0.514 \\ -0.514 & 0.512 \end{pmatrix} \cdot 10^{-28} \tag{5.46}$$

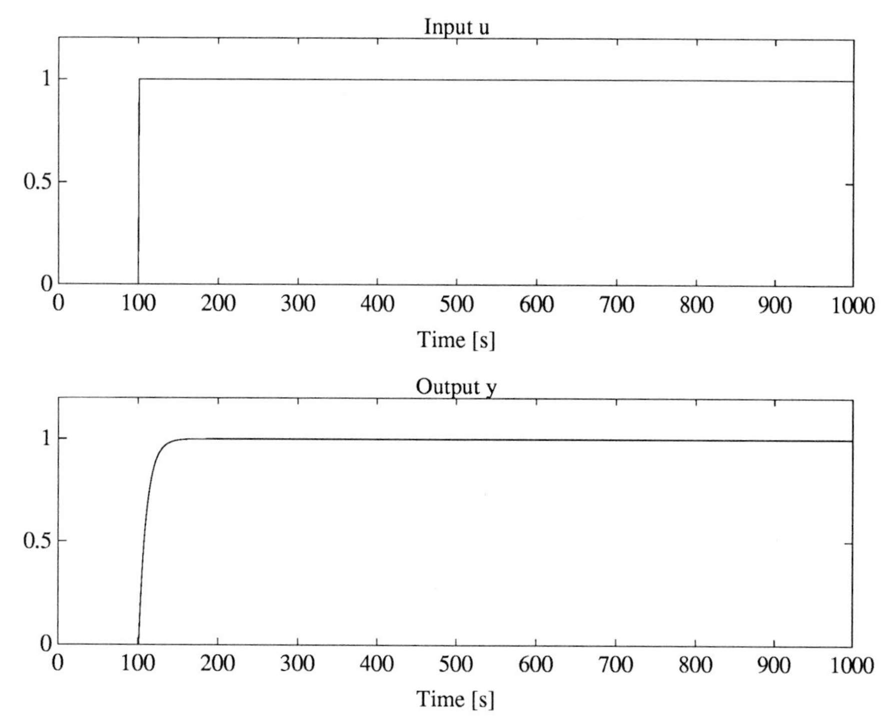

Figure 5.5 Input u and output y obtained from observation of a system $y_k = ay_{k-1} + bu_{k-1} + e_k$ during a noise-free step-response experiment.

where the eigenvalues

$$\begin{pmatrix} 1.0302 \\ 0.0015 \end{pmatrix} \cdot 10^{-28} \tag{5.47}$$

exhibit large differences in magnitude. The corresponding eigenvectors are the columns of the matrix

$$\begin{pmatrix} 0.7100 & 0.7042 \\ -0.7042 & 0.7100 \end{pmatrix} \tag{5.48}$$

The eigenvector associated with the small eigenvalue approximately corresponds to the linear combination $\widehat{a} + \widehat{b}$ for which the estimated variance is

$$\text{Var}\{\widehat{a}+\widehat{b}\} = \text{Var}\{\begin{pmatrix} 1 & 1 \end{pmatrix} \widehat{\theta}\} = \begin{pmatrix} 1 & 1 \end{pmatrix} \text{Var}\{\widehat{\theta}\} \begin{pmatrix} 1 \\ 1 \end{pmatrix} \approx 0.004 \cdot 10^{-28} \tag{5.49}$$

Hence, the sum $\widehat{a} + \widehat{b}$ is indicated to be very accurate, whereas the difference of the two estimates is much less accurate. This result is obviously related to the experimental condition of step response analysis. The general conclusion is, however, that parameter estimates are good when the noise level is low.

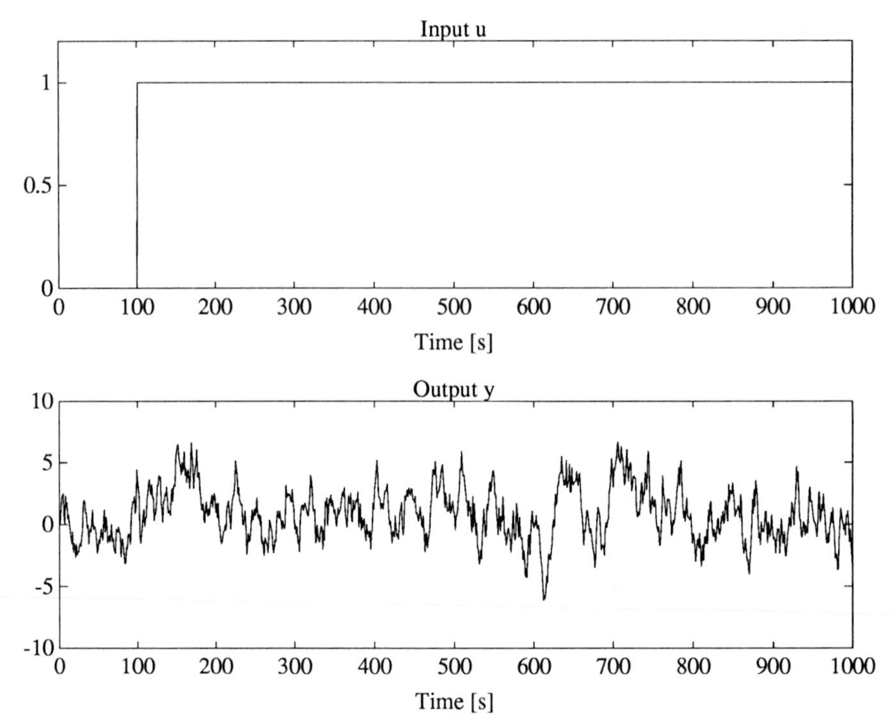

Figure 5.6 Input u and output y obtained from observation of a system $y_k = ay_{k-1} + bu_{k-1} + e_k$ during a step-response experiment with a high noise level.

The result of a similar experiment subject to a high noise level ($\sigma_e^2 = 1$) is shown in Fig. 5.6. The least-squares parameter estimates are

$$\widehat{\theta} = \begin{pmatrix} \widehat{a} \\ \widehat{b} \end{pmatrix} = \begin{pmatrix} 0.8836 \\ 0.1056 \end{pmatrix} \tag{5.50}$$

which provides good estimates despite the high noise level. The loss function and the variance estimate are

$$V(\widehat{\theta}) = 484.1 \quad \text{and} \quad \widehat{\sigma}_e^2 = 0.970 \tag{5.51}$$

The covariance matrix of $\widehat{\theta}$ is

$$\widehat{\sigma}_e^2(\Phi^T\Phi)^{-1} = \begin{pmatrix} 0.226 & -0.219 \\ -0.219 & 1.291 \end{pmatrix} \cdot 10^{-3} \tag{5.52}$$

with eigenvalues

$$\begin{pmatrix} 0.183 \\ 1.335 \end{pmatrix} \cdot 10^{-3} \tag{5.53}$$

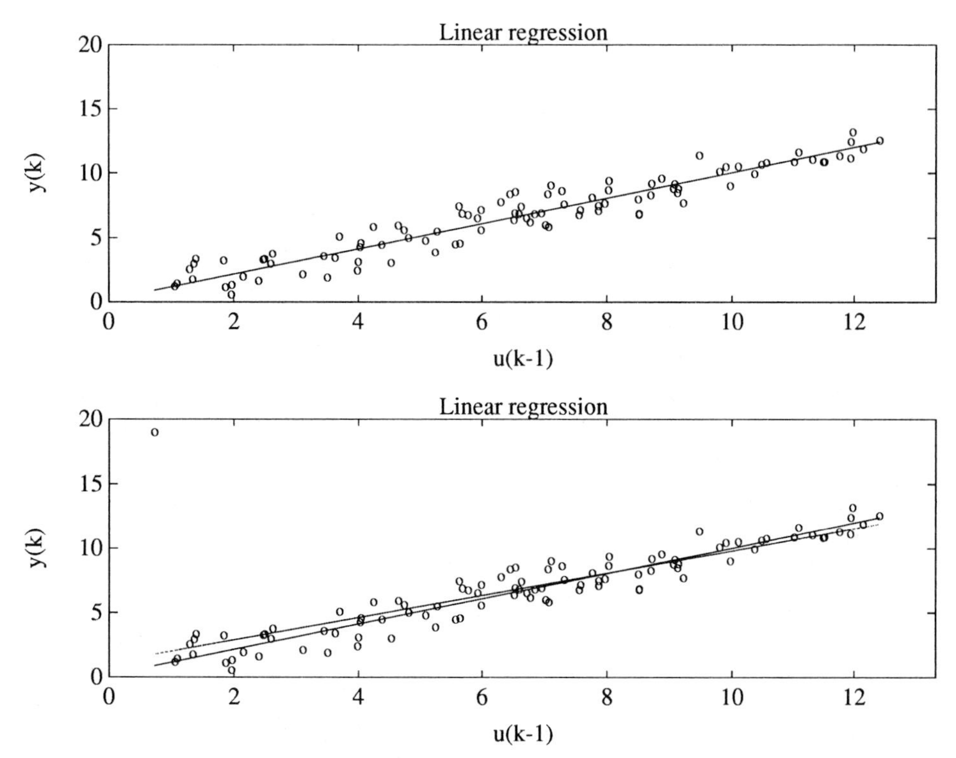

Figure 5.7 Linear regression and identification problems with offsets due to "outliers." Notice the outlier in the lower graph at $u \approx 0.8$ and $y \approx 19$.

The corresponding eigenvectors are the columns of the matrix

$$\begin{pmatrix} 0.981 & -0.194 \\ -0.194 & 0.981 \end{pmatrix} \tag{5.54}$$

where the first and second columns correspond to the linear combinations of small variance and large variance, respectively. This can be used to determine whether the linear combinations of estimated parameters are of small or large variance. ∎

A common flaw is that one residual is very much larger than any of the others. Such a residual is often called an *outlier* and generally causes problems in least-squares estimation.

Example 5.7—Sensitivity of the least-squares method to "outliers"
Assume 100 observations of inputs u_k and outputs y_k have been collected from the process

$$\mathcal{S}: \quad y_k = u_{k-1} + w_k; \qquad \mathcal{E}\{w_k\} = 0, \quad \mathcal{E}\{w_i w_j\} = \sigma^2 \delta_{ij} \tag{5.55}$$

with $\sigma^2 = 1$, and assume that the following model is fitted to the data

$$\mathcal{M}: \quad y_k = \begin{pmatrix} 1 & u_{k-1} \end{pmatrix} \begin{pmatrix} \theta_1 \\ \theta_2 \end{pmatrix} \tag{5.56}$$

The correct values $\theta = \begin{pmatrix} 0 & 1 \end{pmatrix}^T$ may be compared to the least-squares estimate

$$\widehat{\theta} = \begin{pmatrix} 0.1926 \\ 0.9832 \end{pmatrix} \tag{5.57}$$

with $\widehat{\sigma}^2 = 0.985$ and with the estimated covariance matrix

$$\widehat{\sigma}_e^2(\Phi^T\Phi)^{-1} = 0.985 \begin{pmatrix} 0.0526 & -0.0066 \\ -0.0066 & 0.0010 \end{pmatrix} = \begin{pmatrix} 0.0518 & -0.0065 \\ -0.0065 & 0.0010 \end{pmatrix} \tag{5.58}$$

Assume now that a data transmission error provides one abnormal data point at $u \approx 0.8$ and $y \approx 19$ (see Fig. 5.7). The least-squares estimate is now

$$\widehat{\theta} = \begin{pmatrix} 1.163 \\ 0.865 \end{pmatrix} \tag{5.59}$$

with $\widehat{\sigma}^2 = 4.277$. The estimated covariance matrix is

$$\widehat{\sigma}^2(\Phi^T\Phi)^{-1} = \begin{pmatrix} 0.2248 & -0.0281 \\ -0.0281 & 0.0043 \end{pmatrix} \tag{5.60}$$

Notice that the estimated variance has increased more than fourfold due to the single outlier. Clearly, it is a serious problem that the estimated covariance matrix (5.60) falsely indicates with some confidence that the parameter estimates are accurate. ■

Hence, the sensitivity of the least-squares estimate to disturbances other than white noise is a serious question, entailing careful examination of data in order to avoid and deal with possible outliers that would otherwise compromise the result.

5.3 OPTIMAL LINEAR UNBIASED ESTIMATORS

The least-squares estimator to determine parameters of a model $y_i = \phi_i^T\theta + e_i$ has valuable properties under the assumptions $\mathcal{E}\{e_i\} = 0$ and $\mathcal{E}\{e_ie_j\} =$

$\sigma_e^2\delta_{ij}$. Unfortunately, these assumptions are restrictive, and it is valuable to identify the class of all *linear estimates* of the form

$$\widehat{\theta} = T^T \mathcal{Y} \tag{5.61}$$

where $T \in R^{N \times p}$ is a matrix of suitable dimensions and where $\widehat{\theta}$ is a linear function of the data vector $\mathcal{Y}$. The corresponding parameter error is

$$\widetilde{\theta} = \widehat{\theta} - \theta = T^T \mathcal{Y} - \theta = (T^T \Phi - I_{p \times p})\theta + T^T e \tag{5.62}$$

The additional conditions $T^T \Phi = I$ and $\mathcal{E}\{T^T e\} = 0$ must be imposed to satisfy the extra condition of being an unbiased estimator

$$\mathcal{E}\{\widehat{\theta}\} = T^T \Phi \theta + \mathcal{E}\{T^T e\} = \theta \tag{5.63}$$

Determination of the best possible method involves minimization of the covariance

$$\mathrm{Cov}(\widehat{\theta}) = \mathcal{E}\{(\widehat{\theta} - \theta)(\widehat{\theta} - \theta)^T\} = \mathcal{E}\{(T^T \mathcal{Y} - \theta)(T^T \mathcal{Y} - \theta)^T\} = T^T R T \tag{5.64}$$

for $R = \mathcal{E}\{ee^T\}$. The Lagrangian associated with this constrained optimization problem is

$$L(T, \Lambda) = \widetilde{\theta}^T T^T R T \widetilde{\theta} + \mathrm{tr}\Lambda(T^T \Phi - I) \tag{5.65}$$

where the first term should be minimized for any $\widetilde{\theta}$. The second term consists of a matrix of Lagrange multipliers Λ multiplying the constraint $T^T \mathcal{Y} - I$ imposed to satisfy the requirement on unbiased estimates. The partial derivatives of the Lagrangian are

$$\begin{aligned} 0 &= \frac{\partial L}{\partial T} = 2RT\widetilde{\theta}\widetilde{\theta}^T + \Phi \cdot \Lambda \\ 0 &= \frac{\partial L}{\partial \Lambda} = T^T \Phi - I \end{aligned} \tag{5.66}$$

Multiplying the first equation above from the left by $\Phi^T R^{-1}$ and solving for Λ for any vector $\widetilde{\theta}$ gives

$$\Lambda = -2(\Phi^T R^{-1} \Phi)^{-1} \widetilde{\theta}\widetilde{\theta}^T \tag{5.67}$$

Substituting this in the first equation gives

$$2R(T - R^{-1}\Phi(\Phi^T R^{-1} \Phi)^{-1})\widetilde{\theta}\widetilde{\theta}^T = 0 \tag{5.68}$$

which determines T as

$$T = R^{-1}\Phi(\Phi^T R^{-1}\Phi)^{-1} \tag{5.69}$$

The optimal unbiased estimator

$$\hat{\theta} = T^T \mathcal{Y} = (\Phi^T R^{-1}\Phi)^{-1}\Phi^T R^{-1}\mathcal{Y} \tag{5.70}$$

is known as the *Markov estimate* with the covariance matrix

$$\text{Cov}\{\hat{\theta}\} = (\Phi^T R^{-1}\Phi)^{-1} \tag{5.71}$$

This estimator is also known as the *best linear unbiased estimate* or by the acronym *BLUE*. It is noteworthy that the same optimal estimate is obtained by optimizing the loss function

$$V(\hat{\theta}) = \frac{1}{2}(\mathcal{Y} - \Phi\hat{\theta})^T R^{-1}(\mathcal{Y} - \Phi\hat{\theta}) \tag{5.72}$$

or by completing the squares of (5.72)

$$\begin{aligned} V(\hat{\theta}) &= \frac{1}{2}(\mathcal{Y} - \Phi\hat{\theta})^T R^{-1}(\mathcal{Y} - \Phi\hat{\theta}) = \frac{1}{2}\mathcal{Y}^T(R^{-1} - R^{-1}\Phi(\Phi^T R^{-1}\Phi)^{-1}\Phi^T R^{-1})\mathcal{Y} \\ &+ \frac{1}{2}(\hat{\theta} - (\Phi^T R^{-1}\Phi)^{-1}\Phi^T R^{-1}\mathcal{Y})^T(\Phi^T R^{-1}\Phi)(\hat{\theta} - (\Phi^T R^{-1}\Phi)^{-1}\Phi^T R^{-1}\mathcal{Y}) \end{aligned} \tag{5.73}$$

5.4 LINEAR REGRESSION IN THE FREQUENCY DOMAIN

Frequency response fitting based on least-squares identification in the complex frequency domain is a natural idea. Let the polynomial ratio

$$\hat{G}(i\omega) = \frac{\hat{B}(i\omega)}{\hat{A}(i\omega)} = \frac{b_1(i\omega)^{n-1} + \cdots + b_n}{(i\omega)^n + a_1(i\omega)^{n-1} + \cdots + a_n} \tag{5.74}$$

denote a transfer function estimate to be fitted to the experimental data $G(i\omega_k)$ and known at the frequency points ω_k, $k = 1, 2, \ldots, N$. A natural goal of optimization is to minimize the error

$$\min_{a,b} \sum_k |G(i\omega_k) - \frac{\hat{B}(i\omega_k)}{\hat{A}(i\omega_k)}|^2 \tag{5.75}$$

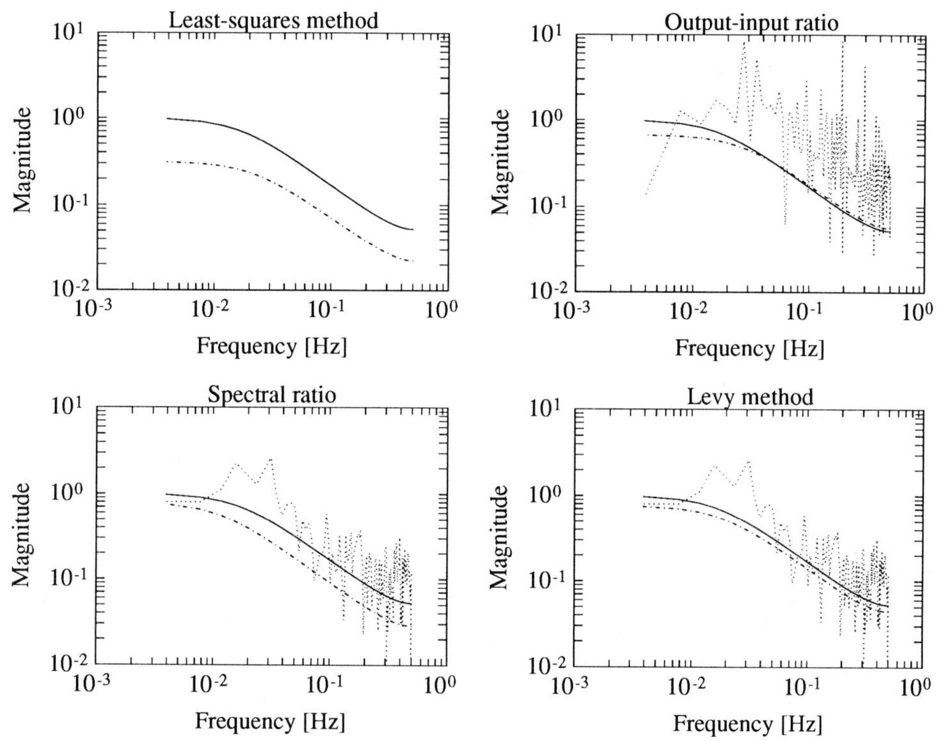

Figure 5.8 Transfer function estimation by least-squares method (*upper left*), input-output ratio (*upper right*), spectral ratio (*lower left*), and the Levy method (*lower right*). The nominal transfer function (*solid line*) is shown in all cases. Transfer function estimates obtained from spectral estimates are shown in dotted lines.

A problem of minimizing (5.75) is that the expressions involved are not linear in the parameters and thus this optimal estimation problem cannot be solved by ordinary least-squares identification. The Levy method (Fig. 5.8) partially resolves this difficulty by minimizing the error function

$$\min_{a,b} \sum_{k} |\widehat{A}(i\omega_k)G(i\omega_k) - \widehat{B}(i\omega_k)|^2 \tag{5.76}$$

This error function may be formulated as a standard least-squares problem by defining the vector

$$\mathcal{Y} = \begin{pmatrix} (i\omega_1)^n G(i\omega_1) \\ (i\omega_2)^n G(i\omega_2) \\ \vdots \\ (i\omega_N)^n G(i\omega_N) \end{pmatrix} \tag{5.77}$$

and the regressor matrix

$$\Phi = \begin{pmatrix} -(i\omega_1)^{n-1}G(i\omega_1) & \cdots & -G(i\omega_1) & (i\omega_1)^{n-1} & \cdots & i\omega_1 & 1 \\ -(i\omega_2)^{n-1}G(i\omega_2) & \cdots & -G(i\omega_2) & (i\omega_2)^{n-1} & \cdots & i\omega_2 & 1 \\ \vdots & & \vdots & \vdots & & \vdots & \\ -(i\omega_N)^{n-1}G(i\omega_N) & \cdots & -G(i\omega_N) & (i\omega_N)^{n-1} & \cdots & i\omega_N & 1 \end{pmatrix} \tag{5.78}$$

and parameter vector

$$\theta = \begin{pmatrix} a_1 & \cdots & a_n & b_1 & \cdots & b_n \end{pmatrix}^T \tag{5.79}$$

The least-squares solution minimizing (5.76) is then

$$\widehat{\theta} = (\Phi^*\Phi)^{-1}\Phi^*\mathcal{Y} \tag{5.80}$$

where Φ^* denotes the transpose and complex conjugate of Φ. This method is capable of fitting complicated frequency responses and has been widely used, although its use is associated with a number difficulties. One possible problem is that the method may fail to yield a good fit if data span several decades of frequency points — for instance, if the components of both $\mathcal{Y}$ and Φ differ widely in magnitude owing to their multiplication by $A(i\omega)$, as part of the the Levy approach, which entails heavy weighting for large values of ω_k. A remedy is to filter $\mathcal{Y}$ and Φ by some approximation to the filter $1/A(i\omega)$ and iterate this procedure.

*Least-squares properties of the discrete Fourier transform

Assume that the continuous variable $y(t)$ is sampled with $y_k = y(kh)$ and suppose that these data should be fitted to the series

$$y_k = y(kh) = \sum_{m=0}^{N-1} \widehat{\theta}_m \exp(i\omega_k mh) \tag{5.81}$$

with a set of coefficients $\{\theta_m\}_{m=0}^{N-1}$ and $\omega_k = k \cdot (2\pi/Nh)$ for $k = 0, 1, \ldots, N-1$. The approximating values $\widehat{y}_k$ can be expressed in matrix form as

$$\widehat{\mathcal{Y}} = \Phi\widehat{\theta} \tag{5.82}$$

where $\widehat{\mathcal{Y}} = \begin{pmatrix} \widehat{y}_0 & \widehat{y}_1 & \ldots & \widehat{y}_{N-1} \end{pmatrix}^T$ and $\widehat{\theta} = \begin{pmatrix} \widehat{\theta}_0 & \widehat{\theta}_1 & \ldots & \widehat{\theta}_{N-1} \end{pmatrix}^T$ and

$$\Phi = \begin{pmatrix} 1 & 1 & \cdots & 1 \\ e^{i\omega_0 h} & e^{i\omega_1 h} & \cdots & e^{i\omega_{N-1} h} \\ \vdots & \vdots & & \vdots \\ e^{i\omega_0 (N-1)h} & e^{i\omega_1 (N-1)h} & \cdots & e^{i\omega_{N-1}(N-1)h} \end{pmatrix} \tag{5.83}$$

A least-squares approach to estimate θ that applies here is the least-squares minimization problem

$$\min_{\widehat{\theta}} \sum_{k=0}^{N-1} |y_k - \widehat{y}_k|^2 = \min_{\widehat{\theta}} (\mathcal{Y} - \Phi\widehat{\theta})^*(\mathcal{Y} - \Phi\widehat{\theta}) \tag{5.84}$$

with the solution

$$\widehat{\theta} = (\Phi^*\Phi)^{-1}\Phi^*\mathcal{Y} \tag{5.85}$$

A simplification gives

$$(\Phi^*\Phi)^{-1} = \frac{1}{N}I \tag{5.86}$$

and

$$\widehat{\theta} = \frac{1}{N}\Phi^*\mathcal{Y} \tag{5.87}$$

so that

$$\widehat{\theta}_m = \frac{1}{N}\sum_{k=0}^{N-1} y_k \exp(-i\omega_m kh) \tag{5.88}$$

The Fourier transform may thus be viewed as a least-squares approximation to the fitting of data to a sum of complex exponentials. It is also worth noting that the power of the sinusoidal component at the frequency ω_m is

$$|\theta_m|^2 = |\frac{1}{N}\sum_{k=0}^{N-1} y_k e^{-i\omega_m kh}|^2 \tag{5.89}$$

This expression is proportional to the periodogram expression (4.4). Thus the periodogram may be viewed as a least-squares fit of a set of sinusoids to the data.

5.5 LEAST-SQUARES ESTIMATION WITH LINEAR CONSTRAINTS

It is sometimes desirable to find the $\widehat{\theta}$ that minimizes

$$\begin{gathered} V(\widehat{\theta}) = \frac{1}{2}(\mathcal{Y} - \Phi\widehat{\theta})^T R^{-1}(\mathcal{Y} - \Phi\widehat{\theta}) \\ \text{subject to } F\widehat{\theta} - G = 0 \end{gathered} \tag{5.90}$$

Problems of this form appear, for instance, when some measurement is considered to be of higher quality than ordinary data, or when there are physical reasons for some linear dependence between the parameters. As before it is assumed that $\mathcal{Y} = \Phi\theta + e$ with error covariance $\mathcal{E}\{ee^T\} = R$.

The Lagrangian for the constrained optimization problem (5.90) is

$$L(\hat{\theta}, \lambda) = \frac{1}{2}(\mathcal{Y} - \Phi\hat{\theta})^T R^{-1}(\mathcal{Y} - \Phi\hat{\theta}) + \lambda^T(F\hat{\theta} - G) \tag{5.91}$$

where λ is a vector of Lagrange multipliers. The gradient

$$\frac{\partial L(\hat{\theta}, \lambda)}{\partial \hat{\theta}} = \Phi^T R^{-1}\Phi\hat{\theta} - \Phi^T R^{-1}\mathcal{Y} + F^T\lambda \tag{5.92}$$

has an extremum at

$$\hat{\theta} = (\Phi^T R^{-1}\Phi)^{-1}(\Phi^T R^{-1}\mathcal{Y} - F^T\lambda) \tag{5.93}$$

By multiplying (5.92) by $F^T(\Phi^T R^{-1}\Phi)^{-1}$ and by substituting the constraint $F\hat{\theta} - G$, it is possible to solve for the Lagrange multiplier

$$\lambda = (F(\Phi^T R^{-1}\Phi)^{-1}F^T)^{-1}(F(\Phi^T R^{-1}\Phi)^{-1}\Phi^T R^{-1}\mathcal{Y} - G) \tag{5.94}$$

The estimate (5.93) contains one term that is identical to the unbiased optimal estimate (5.70) and a second correction term that is proportional to λ.

Example 5.8—Least-squares estimation with specified static gain
Consider the least-squares identification problem in Example 5.4 and assume that it is desirable to find the least-squares estimate with a specified static gain

$$\frac{\hat{b}}{1 - \hat{a}} = 1 \quad \Rightarrow \quad \begin{pmatrix} 1 & 1 \end{pmatrix} \begin{pmatrix} a \\ b \end{pmatrix} = 1 \tag{5.95}$$

Using the data from Example 5.4, the estimate is then modified to

$$\hat{\theta} = \begin{pmatrix} \hat{a} \\ \hat{b} \end{pmatrix} = \begin{pmatrix} 0.9008 \\ 0.0992 \end{pmatrix} \tag{5.96}$$

The loss function is $V(\hat{\theta}) = 500.07$ and it is straightforward to verify that this estimate satisfies the linear constraint (5.95). ■

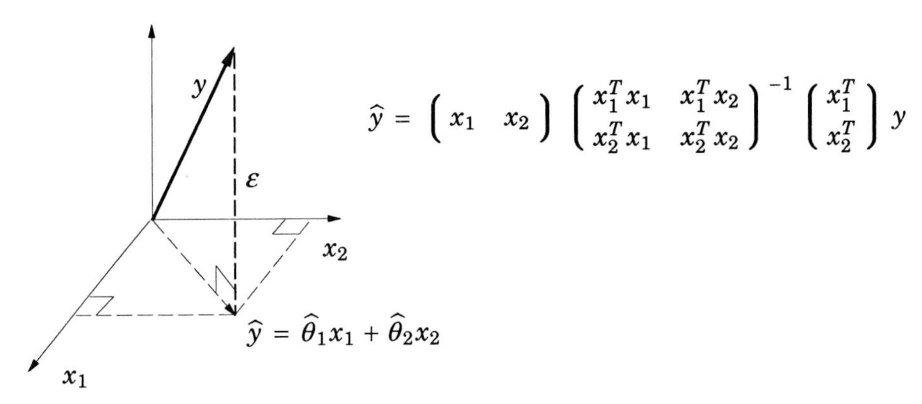

Figure 5.9 Orthogonal projection where $\hat{y}$ is the projection of y on the subspace spanned by x_1, x_2.

5.6 *A GEOMETRICAL INTERPRETATION

The mathematical solution to the least-squares problem offers some interesting geometrical interpretation. The least-squares estimate $\hat{\mathcal{Y}}_N$ can be regarded as the projection of $\mathcal{Y}_N$ on the subspace spanned by the columns of Φ_N with the projection matrix

$$P_N = \Phi_N(\Phi_N^T\Phi_N)^{-1}\Phi_N^T, \qquad \text{which satisfies} \begin{cases} P_N^T = P_N \\ P_N P_N = P_N \end{cases} \tag{5.97}$$

When P_N multiplies $\mathcal{Y}_N$, it provides the projection $\hat{\mathcal{Y}}_N = \Phi_N(\Phi_N^T\Phi_N)^{-1}\Phi_N^T\mathcal{Y}_N$ which is the linear combination of columns of Φ_N that is closest to $\mathcal{Y}_N$ (see Fig. 5.9). The corresponding minimum of the least-squares criterion is

$$V(\hat{\theta}_N) = \frac{1}{2}\varepsilon^T\varepsilon = \frac{1}{2}(\mathcal{Y}_N^T\mathcal{Y}_N - \hat{\mathcal{Y}}_N^T\hat{\mathcal{Y}}_N) \tag{5.98}$$

which indicates that a poorly fitted $\hat{\mathcal{Y}}_N$ tends to be small in magnitude. The relationship between the geometrical interpretation and the optimal estimate can be formulated as the following lemma.

Lemma 5.1—Principle of orthogonality
Let Y be a normed space, $y \in Y$, and let X be a subspace of Y. Then there is a $\hat{y} \in X$ minimizing

$$\min_{x\in X}(x - y)^T(x - y) = (\hat{y} - y)^T(\hat{y} - y) \tag{5.99}$$

if and only if

$$(\hat{y} - y)^T x = 0, \qquad \forall x \in X \tag{5.100}$$

Proof: Suppose that there exists a x such that $(\hat{y} - y)^T x = \alpha \neq 0$. Then for any scalar λ

$$(\hat{y} - y + \lambda x)^T (\hat{y} - y + \lambda x) = (\hat{y} - y)^T (\hat{y} - y) + \lambda^2 x^T x + 2\lambda\alpha \tag{5.101}$$

By choosing $\lambda = -\alpha / x^T x$ we find that $(\hat{y} - y + \lambda x)^T x = 0$ and

$$(\hat{y} - y + \lambda x)^T (\hat{y} - y + \lambda x) < (\hat{y} - y)^T (\hat{y} - y) \tag{5.102}$$

which shows the orthogonal property (5.100) to be a prerequisite of optimality. Now suppose that $(\hat{y} - y)^T x = 0$, for any $x \in X$. Then

$$(\hat{y} - y + \lambda x)^T (\hat{y} - y + \lambda x) = (\hat{y} - y)^T (\hat{y} - y) + \lambda^2 x^T x \geq (\hat{y} - y)^T (\hat{y} - y) \tag{5.103}$$

which proves the optimality. ■

An application of the principle of orthogonality to the least-squares parameter estimation problem gives the orthogonality condition $\Phi_N^T \varepsilon = 0$ (see Fig.5.9). Using the principle of orthogonality it is also possible to formulate a problem that is equivalent to the least-squares problem. Hence, find the optimal $\hat{\theta}_N$ that minimizes $\mathcal{Y}_N - \Phi_N \hat{\theta}_N = \varepsilon$ subject to the constraint $\Phi_N^T \varepsilon = 0$ — *i.e.*, prediction errors and regressors are required to be uncorrelated (or orthogonal). Combination of the two equations yields the normal equations $\Phi_N^T \Phi_N \hat{\theta}_N = \Phi_N^T \mathcal{Y}_N$. Moreover, these two equations can be formulated as the linear system of equations

$$\begin{pmatrix} I_{N\times N} & \Phi_N \\ \Phi_N^T & 0_{p\times p} \end{pmatrix} \begin{pmatrix} \varepsilon \\ \hat{\theta}_N \end{pmatrix} = \begin{pmatrix} \mathcal{Y}_N \\ 0_{p\times 1} \end{pmatrix} \tag{5.104}$$

The formulation (5.104) as a symmetric indefinite linear system is known as the *augmented system method* of solving least-squares estimation problems. The matrix in (5.104) is called the *augmented system matrix*, and there are special numerical methods developed to solve such systems of linear equations, see (Björck, 1991). Another approach is to use the pseudo-inverse based on the singular value decomposition, which provides the minimizer $(\varepsilon^T, \hat{\theta}_N^T)$ with the smallest 2-norm also in the case of an augmented system matrix with rank-deficit (see Appendix A).

5.7 *MULTIVARIABLE SYSTEM IDENTIFICATION

Consider a multi-input, multi-output system

$$\mathcal{S}: \quad A(z^{-1})Y(z) = B(z^{-1})U(z), \qquad \det A(z^{-1}) \neq 0 \tag{5.105}$$

with p inputs $u_k \in R^p$ and m outputs $y_k \in R^m$ and polynomial matrices

$$\begin{aligned} A(z^{-1}) &= I_{m\times m} + A_1 z^{-1} + \cdots + A_n z^{-n}, \qquad & A_1, \ldots, A_n \in R^{m\times m} \\ B(z^{-1}) &= B_1 z^{-1} + \cdots + B_n z^{-n} & B_1, \ldots, B_n \in R^{m\times p} \end{aligned} \tag{5.106}$$

A characteristic problem for multivariable linear systems is that, in general, there is no unique factorization $(A(z^{-1}), B(z^{-1}))$ that corresponds to a given transfer function $H(z^{-1}) = A^{-1}(z^{-1})B(z^{-1})$. For instance, the following two factorizations

$$\begin{aligned} \begin{pmatrix} 1 - z^{-1} + z^{-2} & -z^{-1} + z^{-2} \\ -z^{-1} & 1 - z^{-1} \end{pmatrix} Y(z) &= \begin{pmatrix} z^{-1} & z^{-1} \\ z^{-1} & -z^{-1} \end{pmatrix} U(z) \\ \begin{pmatrix} 1 - z^{-1} & 0 \\ -z^{-1} & 1 - z^{-1} \end{pmatrix} Y(z) &= \begin{pmatrix} z^{-1} + z^{-2} & z^{-1} - z^{-2} \\ z^{-1} & -z^{-1} \end{pmatrix} U(z) \end{aligned} \tag{5.107}$$

represent the same transfer function despite their different parametrizations. In fact, the second factorization is obtained from the first by multiplication of the polynomial matrices $A(z^{-1})$ and $B(z^{-1})$ from the left by the polynomial matrix

$$Q(z^{-1}) = \begin{pmatrix} 1 & z^{-1} \\ 0 & 1 \end{pmatrix}, \qquad \text{with} \quad \det Q(z^{-1}) = 1 \tag{5.108}$$

Hence, given a multivariable transfer function $H(z^{-1})$ it is suitable to define an equivalence class of factorizations $(Q(z^{-1})A(z^{-1}), Q(z^{-1})B(z^{-1}))$ of $H(z^{-1})$ for any stable, causal, and invertible polynomial matrix $Q(z^{-1})$. For any member of this equivalence class can be found the transfer function

$$(Q(z^{-1})A(z^{-1}))^{-1}Q(z^{-1})B(z^{-1}) = A^{-1}(z^{-1})B(z^{-1}) = H(z^{-1}) \tag{5.109}$$

Accordingly, any member of this equivalence class can be used to describe cross-coupling, delay, and other transfer function properties. An important conclusion is that assumptions on unique parametrizations are artificial assumptions that should be avoided unless there is an explicit *a priori* motivation. However, as for practical reasons it is desirable to use a finite number of well-defined parameters, it is often suitable to choose the parameter set with the smallest 2-norm. For the purpose of least-squares identification, then, it is suitable to organize model and data according to

$$
\begin{aligned}
y_k &= -A_1 y_{k-1} - \cdots - A_n y_{k-n} + B_1 u_{k-1} + \cdots + B_n u_{k-n}, & y_k &\in R^m \\
\phi_k &= \begin{pmatrix} -y_{k-1}^T & \ldots & -y_{k-n}^T & u_{k-1}^T & \ldots & u_{k-n}^T \end{pmatrix}^T, & \phi_k &\in R^{n(m+p)} \\
\theta &= \begin{pmatrix} A_1 & \ldots & A_n & B_1 & \ldots & B_n \end{pmatrix}^T, & \theta &\in R^{n(m+p)\times m}
\end{aligned}
\tag{5.110}
$$

which suggests the linear regression model

$$
\mathcal{M}: \quad \mathcal{Y}_N = \Phi_N \theta, \qquad \text{with } \mathcal{Y}_N = \begin{pmatrix} y_1^T \\ y_2^T \\ \vdots \\ y_N^T \end{pmatrix}, \quad \text{and } \Phi_N = \begin{pmatrix} \phi_1^T \\ \phi_2^T \\ \vdots \\ \phi_N^T \end{pmatrix} \tag{5.111}
$$

The normal equations of the associated least-squares estimation of θ will, as a result of the non-uniqueness of parameters, in general exhibit rank deficit. It is therefore natural to apply the least-squares solution

$$
\widehat{\theta}_N = (\Phi_N^T \Phi_N)^+ \Phi_N^T \mathcal{Y}_N \tag{5.112}
$$

where $(\Phi_N^T \Phi_N)^+$ denotes the matrix pseudo-inverse of $\Phi_N^T \Phi_N$; see Appendix A. The associated least-squares estimate then obtained has the smallest 2-norm of all possible minimizers of the least-squares criterion.

Example 5.9—Multivariable system identification

Consider the multivariable system

$$
\mathcal{S}: \quad y_k = \begin{pmatrix} 0.5 & 0.4 \\ 0.4 & 0.5 \end{pmatrix} y_{k-1} + \begin{pmatrix} 1 & 1 \\ 1 & -1 \end{pmatrix} u_{k-1}, \qquad u_k, y_k \in R^2 \tag{5.113}
$$

with data according to Fig. 5.10, which exhibit obvious cross-coupling properties. The estimate for model order $n = 1$ is

$$
\widehat{\theta}_N = \begin{pmatrix} \widehat{A}_1^T \\ \widehat{B}_1^T \end{pmatrix} = \begin{pmatrix} -0.5 & -0.4 \\ -0.4 & -0.5 \\ 1.0 & 1.0 \\ 1.0 & -1.0 \end{pmatrix}, \qquad \begin{cases} \|\widehat{\theta}_N\|_2 = 1.676 \\ \|\widehat{\theta}_N\|_F = 2.195 \end{cases} \tag{5.114}
$$

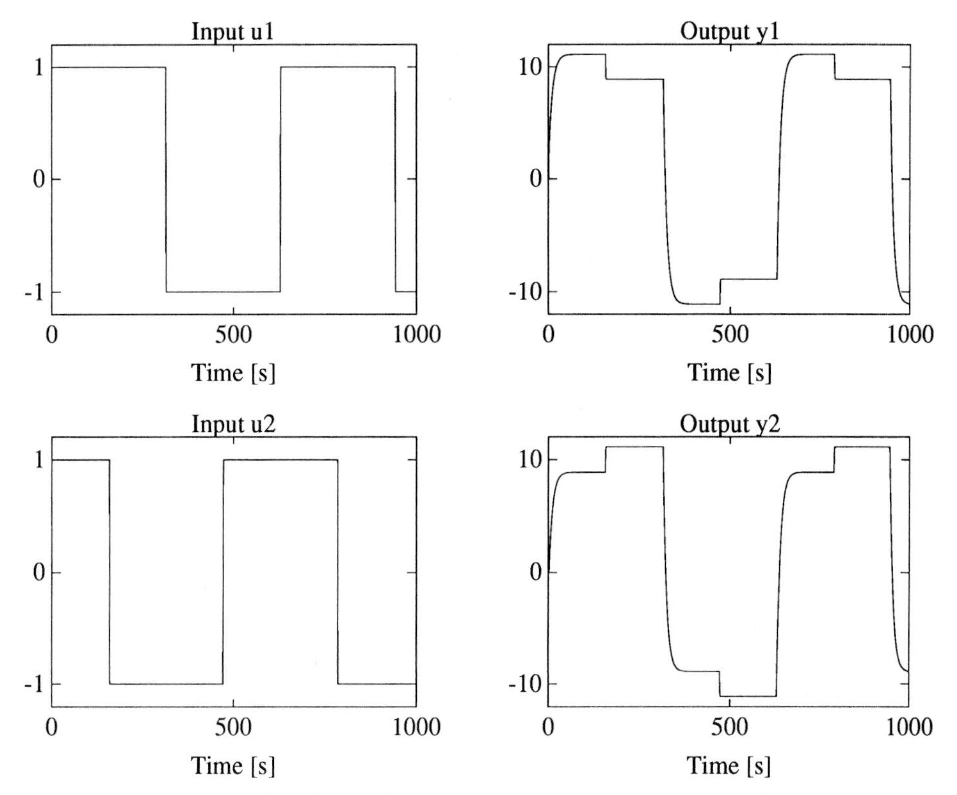

Figure 5.10 A multi-input, multi-output system with two inputs in the vector u_k and two outputs in the vector y_k from Example 5.9.

whereas an estimate for $n = 2$ is

$$\widehat{\theta}_N = \begin{pmatrix} \widehat{A}_1^T \\ \widehat{A}_2^T \\ \widehat{B}_1^T \\ \widehat{B}_2^T \end{pmatrix} = \begin{pmatrix} -0.5000 & -0.4000 \\ -0.2827 & -0.3534 \\ -0.0469 & -0.0587 \\ -0.0587 & -0.0733 \\ 1.0000 & 1.0000 \\ 1.0000 & -1.0000 \\ 0.1173 & 0.1466 \\ -0.1173 & -0.1466 \end{pmatrix}, \quad \begin{cases} \|\widehat{\theta}_N\|_2 = 1.641 \\ \|\widehat{\theta}_N\|_F = 2.168 \end{cases} \tag{5.115}$$

Notice the interesting property that the higher-order estimate for $n = 2$ gives a result with a lower 2-norm of the estimate than the estimate for $n = 1$. ■

5.8 CONCLUDING REMARKS

Two important prerequisites should be fulfilled for the application of the least-squares method without complications arising. These conditions are that $\Phi_N^T\Phi_N$ is invertible and that the noise is uncorrelated with the regressors Φ_N so that $\mathcal{E}\{\Phi_N^T e\} = 0$. It is often suitable to formulate this invertibility condition as a condition of the experimental procedure called *excitation*, and which can be measured in terms of the matrix rank, determinant, or singular values of $\Phi_N^T\Phi_N$. For instance, assuming the excitation to be characterized as $\text{rank}(\Phi_N^T\Phi_N) = m$ means that the variability of the input is such that at most m parameters can be uniquely determined as the result of a particular experiment. Thus, when fitting a model with p parameters where $p < m$, a low excitation results in an underdetermined set of normal equations. However, as the excitation properties of the experiment providing data for the least-squares estimation problem are straightforward to compute, such a check should become part of the experimental procedure. Failure to fulfill the condition of invertibility means that there is no unique solution to the least-squares problem. However, the underdetermined (or rank-deficient) least-squares problem may still be solved in a meaningful way, using the matrix pseudo-inverse (see Appendix A).

Unfortunately, the second assumption of uncorrelated regressors and disturbances with $\mathcal{E}\{\Phi_N^T e\} = 0$ is difficult to check either *a priori* or *a posteriori*, a circumstance which constitutes one of the major problems with least-squares identification (see Example 5.5). Violation of this correlation assumption may lead to inconsistent parameter estimates with a bias $(= (\Phi_N^T\Phi_N)^{-1}\Phi_N^T e)$.This is an important problem which is sometimes successfully resolved by making some reparametrization and which is further addressed in Chapter 6.

Linear estimates are attractive because they are simple to use in calculations owing to the availability of good software for linear algebra. Analytical expressions are thus possible for the unique optimal estimates as well as for covariance estimates. Methods of solving the normal equations include standard Cholesky-type linear equation solvers or the QR-factorization (see Appendix A) implemented with Householder orthogonalization, modified Gram-Schmidt methods, or Givens orthogonalization. Another important approach is to rely on the singular value decomposition and associated matrix pseudo-inverses which permit solution of the normal equations or the augmented system (5.104); see (Golub and Van Loan, 1989), for a systematic account of numerical issues. In addition, a number of special methods have been developed that exploit certain matrix structures of the normal equations, but these will be presented in Chapter 6 and recursive least-squares methods in Chapter 11.

Finally, suffice it to mention that it is important to check the matrix condition number of the normal equations as well as the least-squares residual sum, and that the occurrence of large numbers should alert one of impending numerical problems.

5.9 HISTORICAL REMARKS AND REFERENCES

The application of the least-squares method dates back to the beginning of the nineteenth century and the matematician C.F. Gauss, who developed a method of fitting an elliptical planetary orbit to few observations. He successfully used the method to predict the orbit of the asteroid Ceres, which was lost shortly after its discovery by J. Piazzi in 1801. The medical doctor and amateur astronomer W. Olbers succeeded in finding the lost planetoid by means of the orbit predictions provided by Gauss. Both Gauss's least-squares method and his results regarding error distribution, now known as the normal distribution, were published in

– C.F. GAUSS, *Theoria motus corporum coelestium in sectionibus conicis solem ambientium*. 1809.

A pioneering modern book is

– P. WHITTLE, *Prediction and Regulation by Linear Least Squares Methods*. New York: Van Nostrand, 1963.

Several computation methods that apply to solution of least-squares problems are to be found in

– G.H. GOLUB AND G.F. VAN LOAN, *Matrix Computations*, 2d ed. Baltimore: The Johns Hopkins University Press, 1989.
– Å. BJÖRCK, "Pivoting and stability in the augmented system method." Report Lith-Mat-R-1991-30, Dept. of Mathematics, Linköping University, Sweden.
– J.R. BUNCH AND L. KAUFMAN, "Some stable methods for calculating inertia and solving symmetric linear systems." *Mathematics of Computation*, Vol. 31, 1977, pp. 162–179.

Least-squares estimation in the frequency domain was introduced in

– E.C. LEVY, "Complex curve fitting." *IRE Trans. Autom. Control*, AC-4, 1959, pp.37–43.

5.10 EXERCISES

5.1 Show that the bias of the least-squares estimate is $(\Phi_N^T\Phi_N)^{-1}\Phi_N^T e$.

5.2 Calculate the least-squares parameter estimate using the matrix pseudo-inverse for subsets of data from Example 5.3 with rank-deficient $\Phi_N^T\Phi_N$, e.g., $\mathcal{U}_2 = (1,2)^T$ and $\mathcal{Y}_2 = (6,17)^T$. Verify by computation that $\|\widehat{\theta}_N\|_2 \leq \|\theta\|_2$ for the least-squares estimate as obtained for N=2.

5.3 Assume that the noise sequence $\{e_k\}_{k=1}^N$ consists of independent normally distributed components $e_k \in \mathcal{N}(0,\sigma^2)$. Show that the least-squares estimate $\widehat{\theta}_N$ of parameters of the model $y_k = \phi_k^T\theta + e_k$ is asymptotically normally distributed.

5.4 Show how the normal equations should be modified in order to solve the *weighted least-squares problem*

$$\min_{\bar{\theta}}(\mathcal{Y}_N - \Phi_N\bar{\theta})^T W(\mathcal{Y}_N - \Phi_N\bar{\theta}) \tag{5.116}$$

where W is a suitable weighting matrix.

5.5 Organize the weighted least-squares equations similar to Eq. (5.104).

5.6 Assume that the error components $\{e_k\}_{k=1}^N$ have zero mean and covariance matrix $\mathcal{E}\{e_i e_j^T\} = R\delta_{ij}$. What is the residual sum of the least-squares criterion for the optimal parameter estimate?

5.7 Show that it is possible to complete the squares of the Lagrangian (5.65) in a manner similar to (5.21) provided that $\Phi^T R^{-1}\Phi$ is invertible. What is the residual sum for the optimal constrained parameter estimate? How does this compare to the unconstrained estimate?

5.8 Show that the inverse of the augmented system matrix (5.104) is

$$\begin{pmatrix} I & \Phi_N \\ \Phi_N^T & 0 \end{pmatrix}^{-1} = \begin{pmatrix} I - \Phi_N(\Phi_N^T\Phi_N)^{-1}\Phi_N^T & \Phi_N(\Phi_N^T\Phi_N)^{-1} \\ (\Phi_N^T\Phi_N)^{-1}\Phi_N^T & -(\Phi_N^T\Phi_N)^{-1} \end{pmatrix} \tag{5.117}$$

when $\Phi_N^T\Phi_N$ is invertible.

5.9 Verify that the estimate (5.70) is optimal with respect to the criterion (5.72).

5.10 Formulate the least-squares criterion for multivariable data, *cf.* Section 5.7. Verify that minimization of this criterion yields the standard normal equations.

5.11 Assume that input-output data $\{u_k\}$ and $\{y_k\}$ are observed from the system

$$\mathcal{S}: \qquad y_k + ay_{k-1} = bu_{k-1} + w_k + cw_{k-1} \tag{5.118}$$

where $\{u_k\}$ and $\{w_k\}$ are independent white-noise sequences with variances $\sigma_w^2 = \sigma_u^2 = \sigma^2$. Assume that the model

$$\mathcal{M}: \qquad y_k + ay_{k-1} = bu_{k-1} \tag{5.119}$$

is fitted to data by means of least-squares identification. Show that the prediction error variance is minimized for

$$\widehat{\theta} = \begin{pmatrix} \widehat{a} & \widehat{b} \end{pmatrix}^T = \begin{pmatrix} a - c\sigma^2/\mathcal{E}\{y_k^2\} & b \end{pmatrix}^T \tag{5.120}$$

and that the prediction error variance is smaller than that for $\widehat{a} = a$ and $\widehat{b} = b$.

5.12 An impulse response test has given the result

Time t	0+	0.2	0.4	0.6	0.8
Impulse response $h(t)$	3.4	2.3	1.7	1.2	0.9

a. Determine a least-squares estimate of the parameters K and τ.

b. Determine an estimate of the parameters K and τ which minimizes the squared relative error.

c. Determine an estimate of the parameters K and τ by means of a least-squares fit of the transformed model

$$\log h(t) = \log K - \frac{t}{\tau} = \begin{pmatrix} 1 & -t \end{pmatrix} \begin{pmatrix} \log K \\ 1/\tau \end{pmatrix} \tag{5.121}$$

■

6

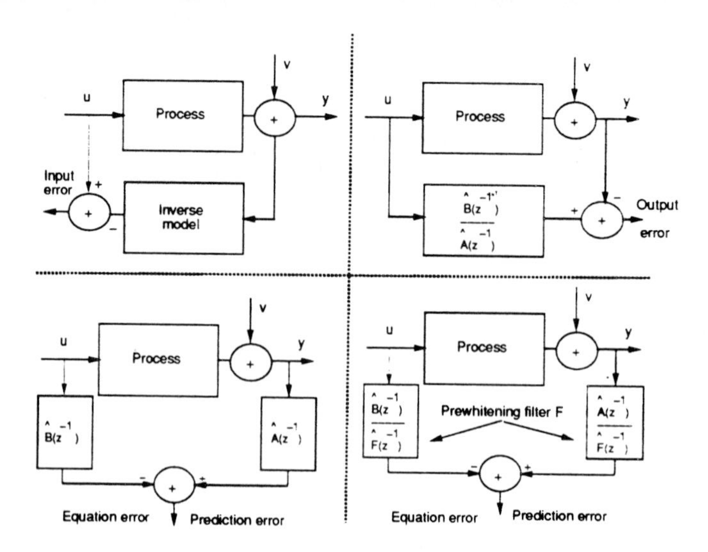

Identification of Time-Series Models

6.1 INTRODUCTION

The approaches to modeling presented thus far have been confined to black-box models of linear systems and linear regression models

$$y_k = \phi_k^T \theta + v_k \tag{6.1}$$

Linear regression identification consists in reformulating various estimation and prediction problems in the form (5.10), which for suitable definitions of the observations y_k, regressors ϕ_k, and disturbance v_k also applies directly to the model

$$A(z^{-1})y_k = B(z^{-1})u_k + v_k \tag{6.2}$$

where the discrete *time series* $\{y_k\}$ and $\{u_k\}$ provide data. The term *time series* here simply means a sequence of data ordered in time.

We now change the focus to models of the type (6.2), including their extension to more sophisticated disturbance models, which we call *time-series modeling* or *time-series analysis*. The corresponding identification problem consists in determining the model structure and parametric estimation of the polynomials involved. In many cases such parameters can be identified by using a linear regression approach. The least-squares solutions to the linear regression problems have excellent properties in cases where the disturbances at different times are uncorrelated. However, the presence of outliers or noise with non-zero mean or otherwise correlated disturbances may give rise to characteristic systematic errors and bias in the parameter estimates (see Example 5.7). The systematic errors in cases with more complicated spectral disturbance characteristics therefore constitute a significant problem. In particular, these systematic errors are unsatisfactory in the many cases where it is desirable to perform time-series modeling as well as careful spectrum analysis. Conversely, the presence of correlated disturbances of composition other than white noise renders bias reduction necessary. Efforts to solve such problems have given rise to several extensions of linear regression models. In particular, maximum-likelihood methods applied to estimation of autoregressive moving average models are of central importance in this context.

6.2 MODEL STRUCTURES

Time-series analysis is concerned with functions of time that exhibit random properties. In many situations it is of interest to consider a vector

$$x_k = \begin{pmatrix} x_{k1} & \dots & x_{km} \end{pmatrix}^T \tag{6.3}$$

of time series, in which case $\{x_k\}$ is said to be a *multivariate time series*. Sometimes the time series may be a function of several temporal and spatial variables, in which case it is said to be a *multidimensional time series*.

Model classification of time-series models distinguishes, for reasons of complexity, between single-input, single-output models and multi-input, multi-output models; linear and nonlinear models; and deterministic and stochastic models. As in many situations a deterministic model is inadequate to describe a system, it is natural in such cases to consider the system outputs as being a realization of a stochastic process that can only be described by statistical laws. Identification of time-series models offers several statistical approaches to model fitting in addition to the deterministic criteria used in least-squares identification. There are also several specialized topics such as detection of signals consisting of sinusoids confounded with noise (common in signal processing), and fitting of stochastic models with certain probability distribution functions. As the ultimate test of a model is, of course, its adequacy for a specific purpose, the model structure is usually a compromise between simplicity and the power to predict the observed behavior from given inputs to the system.

There are at least three important categories of times-series models:

- Difference equations and ARMAX models
- Transfer function models
- State-space models

In this context, we usually regard the true parameters as constant but unknown. In the case of time-varying parameters there is an additional problem in estimating instantaneous values of the parameters, a difficulty which calls for special methods. Also *aggregate parameters* or *lumped parameters* where two or more physical parameters aggregate and form a new parameter — *cf.*, Example 1.1—with transfer function time constant parameters RC, LC arising from electronic parameters R, C, L. Moreover, models arising from physical modeling may exhibit *distributed parameters* that originate from approximating partial differential equations by means of a discretized model or an ordinary differential equation.

Let us consider some important classes of linear discrete-time systems.

ARMAX models and difference equations

The ARMAX models (autoregressive moving average with exogenous input) constitute an important class of difference equations on the form

$$A(z^{-1})y_k = z^{-d}B(z^{-1})u_k + C(z^{-1})w_k \tag{6.4}$$

where d is a time delay and A, B, C are polynomials

$$\begin{aligned} A(z^{-1}) &= 1 + a_1 z^{-1} + \cdots + a_{n_A} z^{-n_A} \\ B(z^{-1}) &= b_0 + b_1 z^{-1} + \cdots + b_{n_B} z^{-n_B} \\ C(z^{-1}) &= 1 + c_1 z^{-1} + \cdots + c_{n_C} z^{-n_C} \end{aligned} \tag{6.5}$$

with the unknown parameters

$$\begin{pmatrix} a_1 & \dots & a_{n_A} & b_0 & b_1 & \dots & b_{n_B} & c_1 & \dots & c_{n_C} & \sigma_w^2 \end{pmatrix}^T \tag{6.6}$$

Notice that the only special case of the ARMAX model that admits a reformulation to the linear regression model is the controlled autoregressive model (ARX)

$$A(z^{-1})y_k = z^{-d}B(z^{-1})u_k + w_k \tag{6.7}$$

where w_k is white noise. Hence, there is no immediate reformulation of the general ARMAX model that results in a linear regression model $y = \phi\theta + v$ unless the disturbances $\{w_k\}$ are available to measurement.

The ARMAX models include several interesting special cases such as the autoregressive (AR) model

$$A(z^{-1})y_k = w_k \tag{6.8}$$

which is effective to model harmonics confounded with noise. The *moving average* (MA) model

$$y_k = C(z^{-1})w_k \tag{6.9}$$

is another type that is popular in signal processing as a basis for filter design (FIR) and the identification of truncated impulse responses. The ARMA model

$$A(z^{-1})y_k = C(z^{-1})w_k \tag{6.10}$$

is effective in modeling disturbance spectra with spectral peaks as well as zeros and is therefore used in model-based spectrum analysis.

The *prediction error methods* (*i.e.*, methods to predict y based on previous data and the identified model) are natural to use in the context of time-series models of prediction and filtering. It is sometimes argued that all model-fitting methods are prediction error methods, in the sense that they compare the behavior of a model and the model-based prediction with experimental data and try to adjust the model to obtain a better fit. In addition, the technical solutions to many estimation problems are based on similar numerical optimization methods. However, prediction error methods in a more restricted and adequate sense comprise a class of identification methods based on optimization criteria of the type

$$\min_{\bar{\theta}} \mathcal{E}\{\sum_{k=1}^{N-\tau}(\widehat{y}_{k+\tau|k}(\bar{\theta}) - y_{k+\tau})^2\} = \min_{\bar{\theta}} \mathcal{E}\{\sum_{k=1}^{N-\tau}\varepsilon^2_{k+\tau|k}(\bar{\theta})\} \tag{6.11}$$

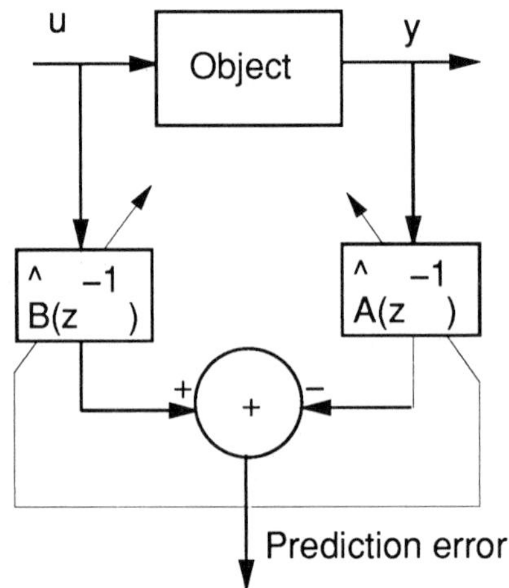

Figure 6.1 Prediction error method or "equation error" method based on the one-step-ahead prediction error $\varepsilon_k = \widehat{y}_{k|k-1} - y_k$.

These predictors minimize the variance of the prediction τ steps ahead of the output y where the prediction is based on present data; see Fig. 6.1.

Example 6.1—Prediction error for a first-order model
Consider the first-order model

$$y_k = -ay_{k-1} + bu_{k-1} + w_k + cw_{k-1}, \qquad \mathcal{E}\{w_k\} = 0, \quad \mathcal{E}\{w_k^2\} = \sigma^2, \tag{6.12}$$

If we consider any predictor $\bar{y}_{k|k-1}$ of y_k based upon data up to sample $k - 1$, then we can estimate its variance by

$$\mathcal{E}\{(y_k - \bar{y}_{k|k-1})^2\} = \mathcal{E}\{(-ay_{k-1} + bu_{k-1} + cw_{k-1} - \bar{y}_{k|k-1})^2\} + \mathcal{E}\{(w_k)^2\} \geq \sigma^2 \tag{6.13}$$

where the two terms depend on data up to time $(k - 1)$ and data at time k, respectively. As the first term is positive definite, we may conclude that Eq. (6.13) provides a lower bound on the prediction error variance, *i.e.*, the achievable accuracy of any predictor $y_{k|k-1}$.

As the noise $\{w_k\}$ is not measured we use the optimal predictor of $y_k(\theta)$ based on data $\{y_k\}$ and $\{u_k\}$, and the parameter vector θ is

$$\widehat{y}_{k|k-1}(\theta) = -c\widehat{y}_{k-1|k-2}(\theta) + (c - a)y_{k-1} + bu_{k-1} \tag{6.14}$$

The resulting prediction error for $|c| < 1$ is determined from the recursive equation

$$\varepsilon_k(\theta) + c\varepsilon_{k-1}(\theta) = y_k + ay_{k-1} - bu_{k-1} \tag{6.15}$$

Hence, the prediction error approach to identification is to determine θ so that the prediction error variance is minimized. ■

Transfer function models

A transfer function model that allows for deterministic and stochastic modeling is

$$y_k = H_u(z)u_k + H_v(z)v_k \qquad \mathcal{E}\{v_i v_j^*\} = \Sigma_v \delta_{ij}$$

where $\{v_k\}$ is discrete-time noise sequence, and $\{u_k\}$, $\{y_k\}$ are input and output data, respectively.

There are several algorithmic reasons to factorize transfer functions into numerator and denominator polynomials. In the context of identification, there are two popular transfer function models

$$\begin{aligned} y_k &= \frac{B(z^{-1})}{F(z^{-1})}u_k + v_k, && \text{Output error model} \\ y_k &= \frac{B(z^{-1})}{F(z^{-1})}u_k + \frac{C(z^{-1})}{D(z^{-1})}w_k, && \text{Box-Jenkins model} \end{aligned} \tag{6.16}$$

where the first model (*output error model*) contains no assumption on the spectrum of the disturbance sequence $\{v_k\}$, whereas the second model (sometimes called a *Box-Jenkins model*) contains a noise model with the white-noise sequence $\{w_k\}$ filtered through the transfer function C/D. An attractive property of the Box-Jenkins model is that it yields separate descriptions of the input-output relationship between u and y and the noise spectrum model described by C/D and $\{w_k\}$. This is obviously an advantage as compared to ARMAX models, which contain one description only with the same A-polynomial for both the poles of the transfer function and for spectral peaks of the noise model.

A description to treat transfer function models as well as difference equations is

$$A(z^{-1})y_k = \frac{B(z^{-1})}{F(z^{-1})}u_k + \frac{C(z^{-1})}{D(z^{-1})}w_k \tag{6.17}$$

where A, B, C, D, F are polynomials in z^{-1} of degrees n_A, n_B, n_C, n_D, n_F, respectively. The model set is described by the parameters

$$\begin{pmatrix} a_1 \dots a_{n_A} & b_1 \dots b_{n_B} & \dots & f_1 \dots f_{n_F} & \sigma_w^2 \end{pmatrix}^T \tag{6.18}$$

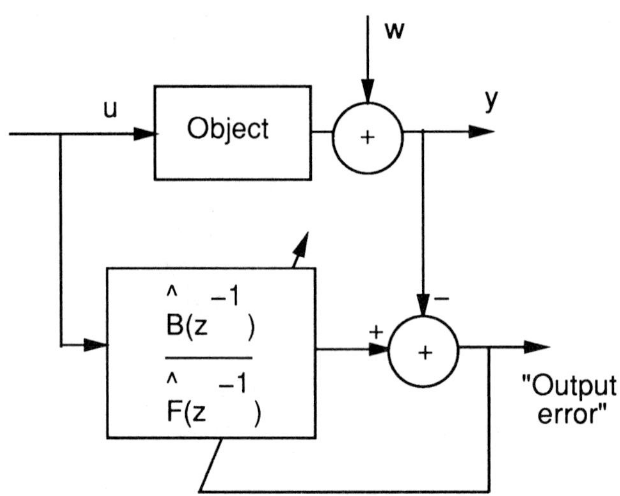

Figure 6.2 Output error identification where the "output error" provides a comparison between the system and model outputs.

The output error model is a special case of the general model (6.17) with $A = C = D = 1$, *i.e.*,

$$y_k = \frac{B(z^{-1})}{F(z^{-1})} u_k + w_k \tag{6.19}$$

With the output error method the identification effort is directed toward estimation of parameters of the $F-$, and $B-$polynomials

$$\min_{\widehat{B},\widehat{F}} \sum_{k=0}^{N-1} |y_k - \frac{\widehat{B}(z^{-1})}{\widehat{F}(z^{-1})} u_k|^2 \tag{6.20}$$

The criterion (6.20) is not exactly the same as a prediction error criterion (see Figs. 6.1 and 6.2). A major difference from prediction error methods is that the model response of the output error method, *i.e.*, $(\widehat{B}/\widehat{F})u$ of (6.20), is not a function of the value of y some steps back. Actually it is open to question whether the term "prediction error method" is justified for model-fitting methods of the type (6.20). Some of the differences between output error models and prediction error models appear in the following example.

Example 6.2—Difference between output error and prediction error The difference in model equations between equation error (prediction error) analysis and output error analysis can be demonstrated by the two predictors

$$\begin{aligned} \widehat{y}_k &= \widehat{a}\widehat{y}_{k-1} + \widehat{b}u_{k-1}, \quad \text{Output error method} \\ \widehat{y}_k &= \widehat{a}y_{k-1} + \widehat{b}u_{k-1}, \quad \text{Prediction error method} \end{aligned} \tag{6.21}$$

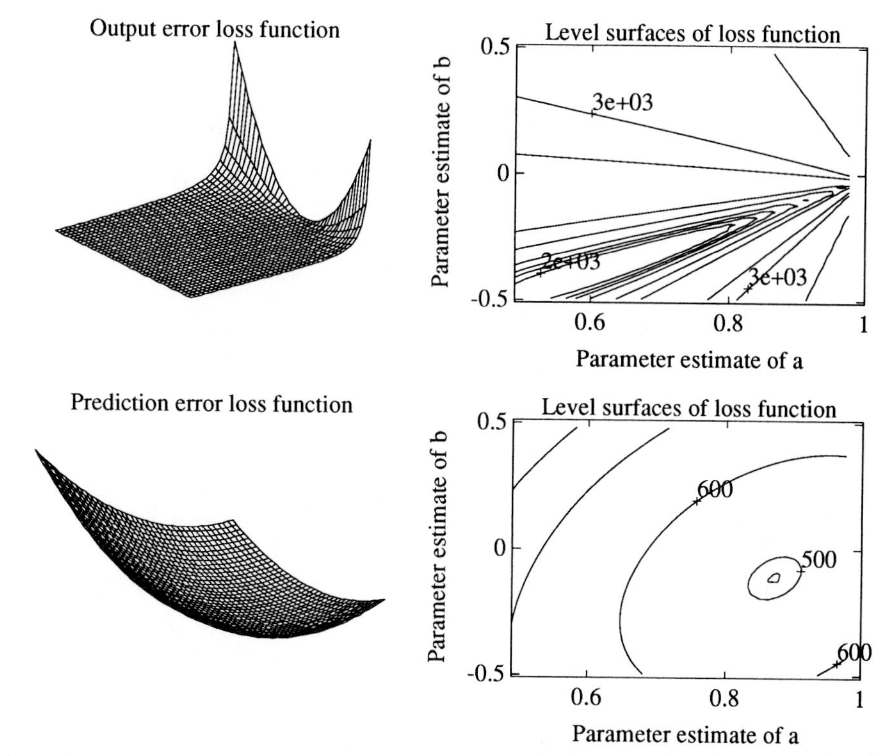

Figure 6.3 Loss functions with level surfaces of the output error method and the prediction error method as applied in Example 6.2. The true parameters of the first order system are $a = 0.9$ and $b = -0.1$. Notice the local minima.

The output error method thus relies more on the accuracy of future output modeling. The output error loss function (6.20) and the prediction error loss function (6.11) are both least-squares problems, although with fundamental differences inasmuch as the output error identification is a nonlinear estimation problem, whereas the equation error identification is a linear estimation problem. The two methods do have different properties with respect to parameter optimization where the output error method may exhibit local minima; *cf.* Fig. 6.3. ■

In the following example we demonstrate an implementation of an output error identification algorithm.

Example 6.3—An algorithm for output error identification
An iterative solution to the problem of output error identification starts by applying standard least-squares identification to find an initial estimate of F and B from the model

$$\mathcal{M}: \qquad F(z^{-1})y_k = B(z^{-1})u_k + v_k \tag{6.22}$$

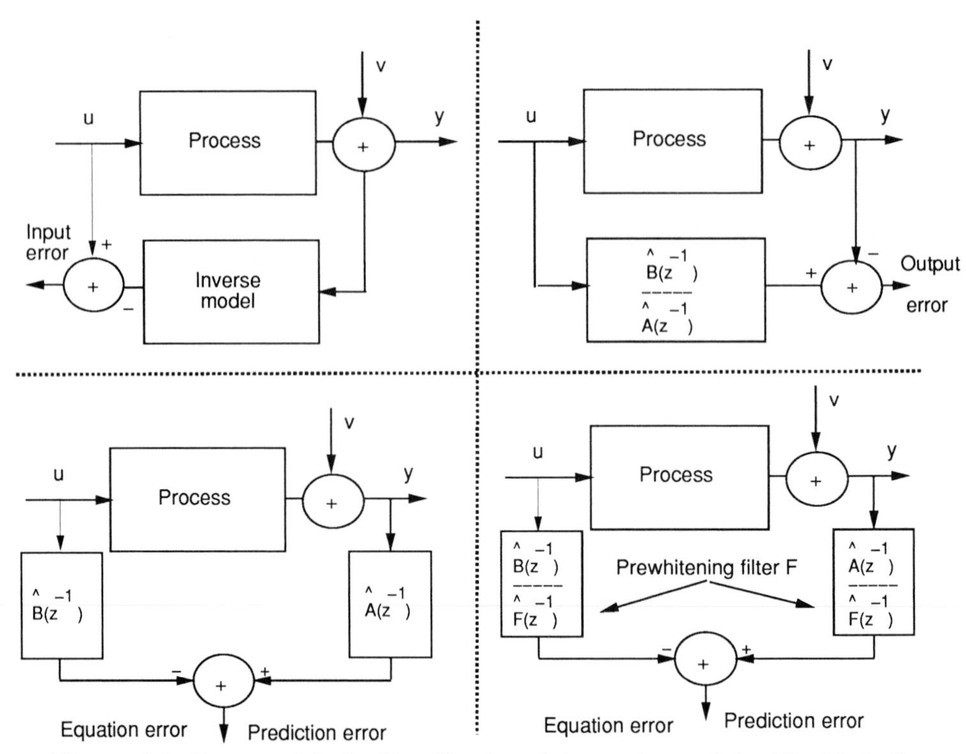

Figure 6.4 Error models for identification of time-series models. The block diagrams show input error (*upper left*), output error (*upper right*), prediction error or equation error (*lower left*), and the use of prewhitening filters (*lower right*).

A second step is to filter the data according to

$$\begin{cases} y_k^F = \dfrac{1}{\widehat{F}(z^{-1})} y_k \\ u_k^F = \dfrac{1}{\widehat{F}(z^{-1})} u_k \end{cases} \tag{6.23}$$

with a subsequent least-squares estimation of F and B from the model

$$\mathcal{M}' : \qquad F(z^{-1}) y_k^F = B(z^{-1}) u_k^F + v_k \tag{6.24}$$

The filtering and estimation steps are continued until the residual noise sequence $\{\widehat{y}_k - y_k\}$ is minimized in the least-squares sense, and until the parameter estimation has converged. ■

The procedure of regressor filtering is often called *prewhitening* with a filter $\widehat{F}(z^{-1})$ called *prewhitening filter* (see Fig. 6.4). The term is justified in cases where the filtering results in white-noise residuals. A standard use of prewhitening allows effective reduction of bias caused by correlated noise, and this principle of prewhitening is incorporated into several estimation algorithms. An example is found in Eq. (6.23) where the filters $\widehat{F}(z^{-1})$ act as prewhitening filters.

A modified form of output error identification, known as *input error identification*, models the system input $\{u_k\}$ (see Fig. 6.4). The method consists of fitting an *inverse model* by matching the modeled input to data with the system output $\{y_k\}$ as model input. The corresponding *input error* $\widetilde{u}_k = \widehat{u}_k - u_k$ thus yields the misfit between the modeled input and data. Input error identification is relevant for problems of input estimation, deconvolution, and other similar filtering and signal restoration problems. A problem in this context is how to handle the causality problems inherent in inverse modeling.

6.3 MAXIMUM-LIKELIHOOD IDENTIFICATION

In maximum-likelihood (ML) identification we select the estimate which renders the given observations $\mathcal{Y}$ most probable. This is accomplished by maximizing the likelihood function $p(\mathcal{Y}|\theta)$.

$$\max_{\bar{\theta}} p(\mathcal{Y}|\bar{\theta}) = p(\mathcal{Y}|\widehat{\theta}) \tag{6.25}$$

which yields the estimate $\bar{\theta} = \widehat{\theta}$ that maximizes $p(\mathcal{Y}|\bar{\theta})$. (A comparison between loss functions of maximum-likelihood method and the least-squares method is shown in Fig. 6.5.)

Example 6.4—Maximum-likelihood estimation and Markov estimates
Assume that $\mathcal{Y} = \Phi\theta + v$ with known mean $\mathcal{E}\{v\} = 0$ and known covariance $\mathcal{E}\{vv^T\} = \Sigma_v$. If v is assumed to be normally distributed and it has N elements, its probability density function is

$$p(v) = ((2\pi)^N \det \Sigma_v)^{-1/2} \exp(-\frac{1}{2}v^T\Sigma_v^{-1}v) \tag{6.26}$$

Under the assumption $\mathcal{Y} = \Phi\theta + v$ we have

$$p(v) = ((2\pi)^N \det \Sigma_v)^{-1/2} \exp(-\frac{1}{2}(\mathcal{Y} - \Phi\theta)^T\Sigma_v^{-1}(\mathcal{Y} - \Phi\theta)) \tag{6.27}$$

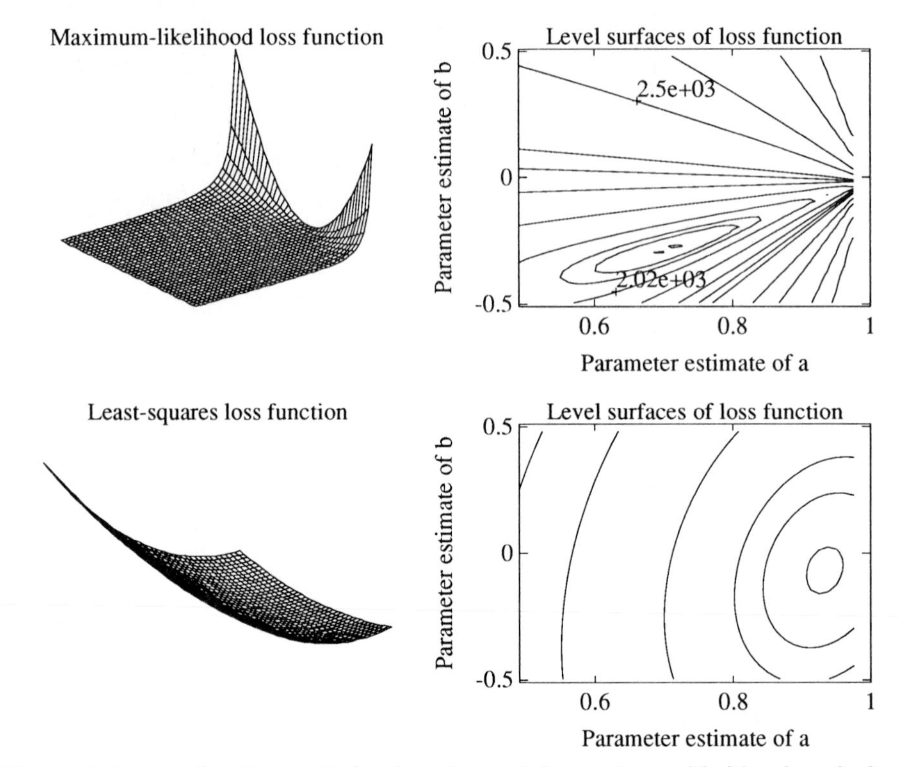

Figure 6.5 Loss functions with level surfaces of the maximum-likelihood method and the least-squares method.

which we call a *likelihood function*. We may choose to consider Eq. (6.27) as a function of a parameter vector $\bar{\theta}$ and prediction errors $\varepsilon(\bar{\theta}) = \mathcal{Y} - \Phi\bar{\theta}$. It is practical to take the logarithm of this function, and then we have

$$\log L(\bar{\theta}) = \log p(\varepsilon|\bar{\theta}) = -\frac{1}{2}\log((2\pi)^N \det \Sigma_v) - \frac{1}{2}(\mathcal{Y} - \Phi\bar{\theta})^T \Sigma_v^{-1}(\mathcal{Y} - \Phi\bar{\theta}) \quad (6.28)$$

Here we expect the log-likelihood function to have a maximum at $\bar{\theta} = \theta$ and $\varepsilon = v$. Hence, the maximum likelihood estimator for a model linear in parameters and with normally distributed white noise is identical to the Markov estimator; *cf.* Eq. (5.70). In the case of disturbance with the covariance matrix $\Sigma_v = \sigma^2 I$, it follows that the maximum likelihood estimator reduces to the least-squares estimator. ■

Remark: If it is required to optimize, say, the empirical prediction error covariance or the parameter error covariance by means of ordinary optimization methods, it is necessary to formulate a mapping of the covariance matrix to some scalar function with appropriate properties of optimality. Suitable optimization criteria are, for instance,

$$\begin{cases} J_1(\widehat{\theta}) = \ell_1(Q) = \log\det Q \\ J_2(\widehat{\theta}) = \ell_2(Q) = \mathrm{tr}(WQ) \end{cases} \quad \text{where} \quad Q(\widehat{\theta}) = \frac{1}{N}\varepsilon(\widehat{\theta})^T\varepsilon(\widehat{\theta}) \tag{6.29}$$

where $W = W^T > 0$ is some weighting matrix. Both functions ℓ_1, ℓ_2 have the property

$$\begin{aligned} \ell_1(Q) &\geq \ell_1(Q_0) \qquad \text{if } Q \geq Q_0 \in R^{n\times n} \\ \ell_2(Q) &\geq \ell_2(Q_0) \qquad \text{if } Q \geq Q_0 \in R^{n\times n} \end{aligned} \tag{6.30}$$

The proof is called for in Exercise 6.9. ■

There is actually a lower limit for the covariance of an unbiased estimator obtained by maximum-likelihood identification. This is known as the Cramér-Rao lower bound.

Theorem 6.1—The Cramér-Rao lower bound
Let $\mathcal{Y}$ be observations of a stochastic variable, the distribution of which depends on an unknown vector θ. Let $L(\mathcal{Y},\theta)$ denote the likelihood function, and let $\bar{\theta} = \bar{\theta}(\mathcal{Y})$ be an arbitrary unbiased estimate of θ. Then

$$\mathrm{Cov}(\bar{\theta}) \geq (\mathcal{E}\{(\frac{\partial \log L}{\partial\theta})(\frac{\partial \log L}{\partial\theta})^T\})^{-1} = -(\mathcal{E}\{\frac{\partial^2 \log L}{\partial\theta\partial\theta^T}\})^{-1} \tag{6.31}$$

(A proof is to be found in Appendix B.) ■

The matrix defining the Cramér-Rao lower bound is called the Fisher *information matrix*, and an estimate which achieves the lower bound is referred to as *efficient*.

Autoregressive moving average models with exogenous input (ARMAX) constitute a model set general enough to describe colored noise that cannot be described by the ARX models of the type (6.7). Similar to Eq. (6.4) we consider ARMAX models of the type

$$A(z^{-1})y_k = z^{-d}B(z^{-1})u_k + C(z^{-1})v_k \tag{6.32}$$

where the noise covariance matrix $\Sigma_v = \mathcal{E}\{vv^T\}$ is now assumed to be unknown. Formulation of a maximum-likelihood problem involves the calculation of a likelihood function $L(\bar{\theta})$. Considering the case of normally distributed noise, we have

$$L(\bar{\theta}) = \frac{1}{(2\pi)^{N/2}(\det\Sigma_v)^{1/2}}\exp(-\frac{1}{2}\varepsilon^T(\bar{\theta})\Sigma_v^{-1}\varepsilon(\bar{\theta})) \tag{6.33}$$

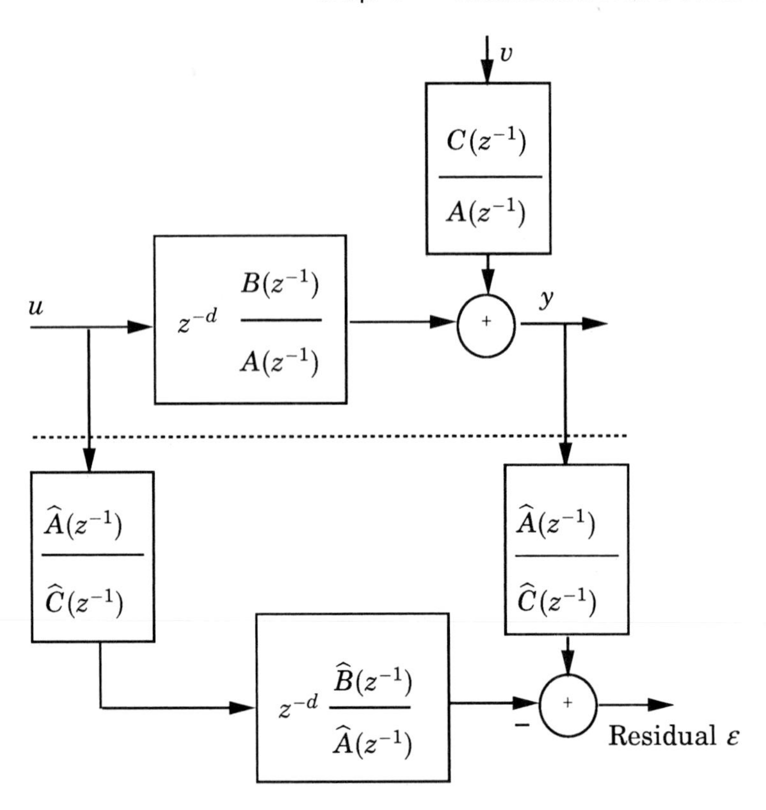

Figure 6.6 Block diagram showing the approximate maximum-likelihood estimation of an ARMAX model with a time delay of d samples. Minimization of the residuals and consequently the prediction error is accomplished by adjustment of $\widehat{A}$, $\widehat{B}$, and $\widehat{C}$.

or

$$\log L(\bar{\theta}) = -\frac{1}{2}\log((2\pi)^N \det \Sigma_v) - \frac{1}{2}\varepsilon^T(\bar{\theta})\Sigma_v^{-1}\varepsilon(\bar{\theta}) \tag{6.34}$$

with ε containing the components $\varepsilon_k = y_k - \phi_k^T\bar{\theta}$ and

$$\begin{aligned} y_k = & - a_1 y_{k-1} - \cdots - a_{n_A} y_{k-n_A} + \\ & + b_1 u_{k-d-1} + \cdots + b_m u_{k-d-n_B} + \\ & + v_k + c_1 v_{k-1} + \cdots + c_{n_C} v_{k-n_C} = \phi_k^T \theta + v_k \end{aligned} \tag{6.35}$$

with

$$\begin{aligned} \phi_k &= \left(-y_{k-1} \ldots - y_{k-n_A} \quad u_{k-d-1} \ldots u_{k-d-n_B} \quad v_{k-1} \ldots v_{k-n_C} \right)^T \\ \theta &= \left(a_1 \ldots a_{n_A} \quad b_1 \ldots b_{n_B} \quad c_1 \ldots c_{n_C} \right)^T \end{aligned} \tag{6.36}$$

where it is a problem that the regressor components $v_{k-1}, \ldots, v_{k-n_c}$ are not known. Another problem is that this optimization criterion is a function of θ and Σ_v and is thus not known. In the absence of the desired parameters, it is therefore only feasible to make approximate solutions by finding successively better estimates of the covariance matrix Σ_v and the parameters θ through some iterative procedure (*cf.* Fig. 6.6). In the important special case with normally distributed white noise with $\Sigma_v = \sigma_v^2 I$ and σ_v^2 unknown, we replace Eq. (6.34) by the empirical likelihood function

$$\begin{aligned} \log \widehat{L}(\bar{\theta}, \bar{\sigma}_v^2) &= -\frac{N}{2}\log(2\pi) - \frac{1}{2\bar{\sigma}_v^2}\sum_{k=1}^{N} \varepsilon_k^2(\bar{\theta}) - \frac{N}{2}\log\bar{\sigma}_v^2 \\ &= -\frac{N}{2}\log(2\pi) - \frac{1}{\bar{\sigma}_v^2} V_N(\bar{\theta}) - \frac{N}{2}\log\bar{\sigma}_v^2 \end{aligned} \tag{6.37}$$

where

$$V_N(\bar{\theta}) = \frac{1}{2}\sum_{k=1}^{N} \varepsilon_k^2(\bar{\theta}) = \frac{1}{2}\varepsilon^T(\bar{\theta})\varepsilon(\bar{\theta}) \tag{6.38}$$

The gradient and the second-order derivatives of $\log L(\bar{\theta})$ determine the extrema of $\log L$ according to the equations

$$\begin{aligned} 0 &= \frac{\partial}{\partial\bar{\theta}} \log \widehat{L}(\bar{\theta}, \bar{\sigma}_v^2) = -\frac{1}{\bar{\sigma}_v^2}\nabla V_N(\bar{\theta}) \\ 0 &= \frac{\partial}{\partial\bar{\sigma}_v^2} \log \widehat{L}(\bar{\theta}, \bar{\sigma}_v^2) = \frac{1}{\bar{\sigma}_v^4} V_N(\bar{\theta}) - \frac{N}{2\bar{\sigma}_v^2} \end{aligned} \tag{6.39}$$

with the solution

$$\begin{aligned} \widehat{\sigma}_v^2 &= \frac{2}{N} V_N(\widehat{\theta}) \\ \nabla V_N(\widehat{\theta}) &= 0 \end{aligned} \tag{6.40}$$

A numerical solution to the problem

$$\nabla V_N(\widehat{\theta}) = V_N'(\widehat{\theta}) = 0 \tag{6.41}$$

can be obtained as an iterative procedure via the Newton(-Raphson) method

$$\widehat{\theta}^{(i+1)} = \widehat{\theta}^{(i)} - \alpha_i (V''(\widehat{\theta}^{(i)}))^{-1} V'(\widehat{\theta}^{(i)}) \tag{6.42}$$

where α_i denotes the step length to choose and (i) the iteration order. The initial estimate $\widehat{\theta}^{(0)}$ is usually chosen as a least-squares estimate. The elements of this computation can be given the form

$$\begin{aligned} V_N(\bar{\theta}) &= \frac{1}{2}\sum_{k=1}^{N} \varepsilon_k^2(\bar{\theta}) \\ \psi_k(\bar{\theta}) &= -\nabla \varepsilon_k(\bar{\theta}) \\ \nabla V_N(\bar{\theta}) = V_N'(\bar{\theta}) &= -\sum_{k=1}^{N} \varepsilon_k(\bar{\theta})\psi_k^T(\bar{\theta}) \\ \nabla^2 V_N(\bar{\theta}) = V_N''(\bar{\theta}) &= \sum_{k=1}^{N} \psi_k(\bar{\theta})\psi_k^T(\bar{\theta}) + \sum_{k=1}^{N} \varepsilon_k(\bar{\theta})\nabla^2 \varepsilon_k^T(\bar{\theta}) \end{aligned} \tag{6.43}$$

The Newton method is a good numerical procedure with "quadratic" convergence properties; see Appendix C. The method must be modified, however, when V'' is not invertible, which may constitute a problem at some distance away from the solution. A similar problem arises when the model set is not uniquely parametrized or when a poor choice of optimization criterion has been made. It is therefore an advantage to start the iterative algorithm with good initial values obtained from some other numerical algorithm or from a least-squares estimate. For the same reason it is difficult to ensure global convergence by using the Newton method with arbitrary initial conditions.

Example 6.5—A comparison of LS- and ML-identification
Consider data generated by the system in Example 6.1

$$\mathcal{S}: \qquad y_k = ay_{k-1} + bu_{k-1} + w_k + cw_{k-1} \tag{6.44}$$

with $\mathcal{E}\{w_k\} = 0$ and $\mathcal{E}\{w_i w_j\} = \sigma^2\delta_{ij}$ with $N = 1000$ samples collected (see Fig. 6.7). Application of least-squares (LS) identification and maximum-likelihood (ML) identification yielded the following results:

	$\widehat{a}$	$\widehat{b}$	$\widehat{c}$	V	$\widehat{\sigma}^2$	$\|\|\widetilde{\theta}\|\|$
LS identification	0.9493	0.0398	–	723	1.446	0.0778
ML identification	0.8992	0.0857	0.7072	501	1.008	0.0143

where the value of $||\widetilde{\theta}||$ includes the errors of $\widehat{a}$ and $\widehat{b}$ only. These results suggest maximum-likelihood identification performs better than least-squares identification in the case of colored noise. ■

Example 6.6—Pseudolinear regression
Assume that data $\{y_k\}$, $\{u_k\}$ are generated from the system

$$\mathcal{S}: \qquad A(z^{-1})y_k = B(z^{-1})u_k + C(z^{-1})w_k \tag{6.45}$$

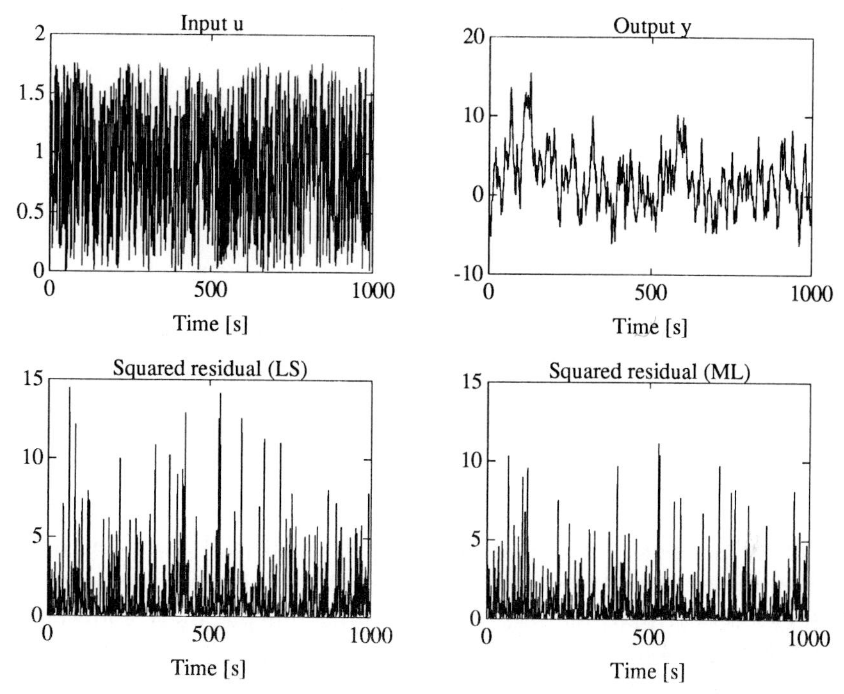

Figure 6.7 LS- and ML-identification of an object described by the ARMAX model $y_k + 0.9y_{k-1} = 0.1u_{k-1} + e_k + 0.7e_{k-1}$. The lower graphs (*left and right*) show the squared residuals ε_k^2 for the least-squares (LS) and maximum-likelihood estimation (ML) methods, respectively.

where $\{w_k\}$ is a white-noise sequence. An alternative to the iterative numerical optimization (6.43) is to estimate high-order polynomials A and B by means of least-squares identification which, in turn, allows the computation of $\{\varepsilon_k\}$. As the model order is high, it may be assumed that the computed residual sequence $\{\varepsilon_k\}$ yields a good approximation of the white-noise sequence $\{w_k\}$. A second step of least-squares identification can then be applied, using the extended regressors

$$\phi_k = \begin{pmatrix} y_{k-1} & \ldots & u_{k-1} & \ldots & \varepsilon_{k-1} & \ldots & \varepsilon_{k-n} \end{pmatrix}, \qquad k = 1, 2, \ldots, N \qquad (6.46)$$

by means of which the A-, B- and C-polynomials are estimated. If this method is applied to the data of Example 6.5 for a twentieth-order primary model and a second model according to Eq. (6.44), the competitive result is obtained:

	$\widehat{a}$	$\widehat{b}$	$\widehat{c}$	V	$\widehat{\sigma}^2$	$\|\|\widetilde{\theta}\|\|$
Two-step LS	0.9011	0.0914	0.6495	516	1.032	0.0086

The method is known as a *pseudolinear regression* or *two-step linear regression* approach. ∎

6.4 KALMAN FILTER

Now let us consider a general estimation and filtering problem. Consider the state-space model

$$\begin{aligned} x_{k+1} &= \Phi x_k + \Gamma u_k + v_k \\ y_k &= C x_k + D u_k + e_k \end{aligned} \tag{6.47}$$

with $\mathcal{E}\{v_k\} = 0$, $\mathcal{E}\{e_k\} = 0$, and

$$\begin{aligned} \mathcal{E}\{vv^T\} &= R_1 \\ \mathcal{E}\{ee^T\} &= R_2 \\ P(0) = \mathcal{E}\{x_0 x_0^T\} &= R_0 \end{aligned} \tag{6.48}$$

Assume the noisy input-output data $\{y_k\}$ and $\{u_k\}$ to be the only data available, and that the state x_k is not available for measurement. The problem of optimal estimation of x_k based on input-output data and knowledge of the model (6.47) can be solved by minimizing

$$J(\widehat{x}_k) = \mathcal{E}\{(\widehat{x}_{k+1|k} - x_{k+1})^2\}, \qquad \text{for} \quad k = 1, 2, 3, \ldots \tag{6.49}$$

subject to (6.47). The Kalman filter or Kalman-Bucy filter for prediction of x based on the present data at time k is

$$\begin{aligned} \widehat{x}_{k+1|k} &= \Phi \widehat{x}_{k|k-1} + \Gamma u_k + K_k (y_k - C\widehat{x}_{k|k-1}) \\ K_k &= \Phi P_k C^T (R_2 + C P_k C^T)^{-1} \\ P_{k+1} &= \Phi P_k \Phi^T + R_1 - \Phi P_k C^T (R_2 + C P_k C^T)^{-1} C P_k \Phi^T \end{aligned} \tag{6.50}$$

which is a *recursive* equation where the estimates are updated as soon as new input-output data are available. In particular, the Kalman filter minimizes Eq. (6.49) in cases where the noise components v_k and e_k are independent and normally distributed. The solution obtained is quite general and the method has a vast scope of application. A case of specific relevance for system identification is illustrated in the following example.

Example 6.7—Kalman filter for identification
The following formulation of the identification problem is useful for estimation of time-varying parameters:

$$\begin{aligned} \theta_{k+1} &= \theta_k + v_k, \qquad & \mathcal{E}\{v_k\} &= 0 \\ y_k &= \phi_k^T \theta_k + e_k, \qquad & \mathcal{E}\{e_k\} &= 0 \end{aligned} \tag{6.51}$$

with the standard covariance matrices $R_1 = \mathcal{E}\{v_k v_k^T\}$, and $R_2 = \mathcal{E}\{e_k e_k^T\}$. The Kalman filter for this problem is

$$\begin{aligned} \widehat{\theta}_{k+1} &= \widehat{\theta}_k + K_k(y_k - \phi_k^T \widehat{\theta}_k) \\ K_k &= P_k \phi_k (R_2 + \phi_k^T P_k \phi_k)^{-1} \\ P_{k+1} &= P_k + R_1 - P_k \phi_k (R_2 + \phi_k^T P_k \phi_k)^{-1} \phi_k^T P_k \end{aligned} \tag{6.52}$$

which is an excellent method for identification of time-varying systems. Notice that the variables R_1 and R_2 are important design parameters that should match the temporal variations of θ_k and the observation noise, respectively.

6.5 INSTRUMENTAL VARIABLE METHOD

Consider the linear regression model

$$\mathcal{Y} = \Phi\theta + v \tag{6.53}$$

with the disturbance vector $v \in R^N$, the regressor matrix $\Phi \in R^{N \times p}$, and the observations $\mathcal{Y} \in R^N$. Correlation between the regressors and the prediction error leads to bias of the parameter estimates $\widehat{\theta}$ obtained from least-squares solutions to the linear regression problem. Methods that replace the regressor Φ used in linear regression by some other variable Z are called *instrumental variable methods*, and the estimate takes the form

$$\widehat{\theta^z} = (Z^T \Phi)^{-1} Z^T \mathcal{Y} \tag{6.54}$$

There are two conditions to impose on the instrumental variables Z in order to make the estimator $\widehat{\theta^z}$ consistent. First, the instrumental variables Z should be uncorrelated with the disturbances so that $\mathcal{E}\{Z^T v\} = 0$. Second, the matrix $Z^T \Phi$ must be invertible. Provided these conditions are satisfied, the following covariance estimate can be justified (a proof is called for in Exercise 6.8 below):

$$\text{Cov}(\widehat{\theta^z}) = \mathcal{E}\{(\widehat{\theta^z} - \theta)(\widehat{\theta^z} - \theta)^T\} = (Z^T \Phi)^{-1} Z^T \Sigma_v Z (\Phi^T Z)^{-1} \tag{6.55}$$

Hence, in addition to the two previously imposed conditions it is necessary that $Z^T \Phi$ be "large" so that $\widehat{\theta^z}$ provides an efficient estimate. In other words, the instrumental variables should be chosen so that they are simultaneously uncorrelated with v and highly correlated with Φ. As we do not know v, it

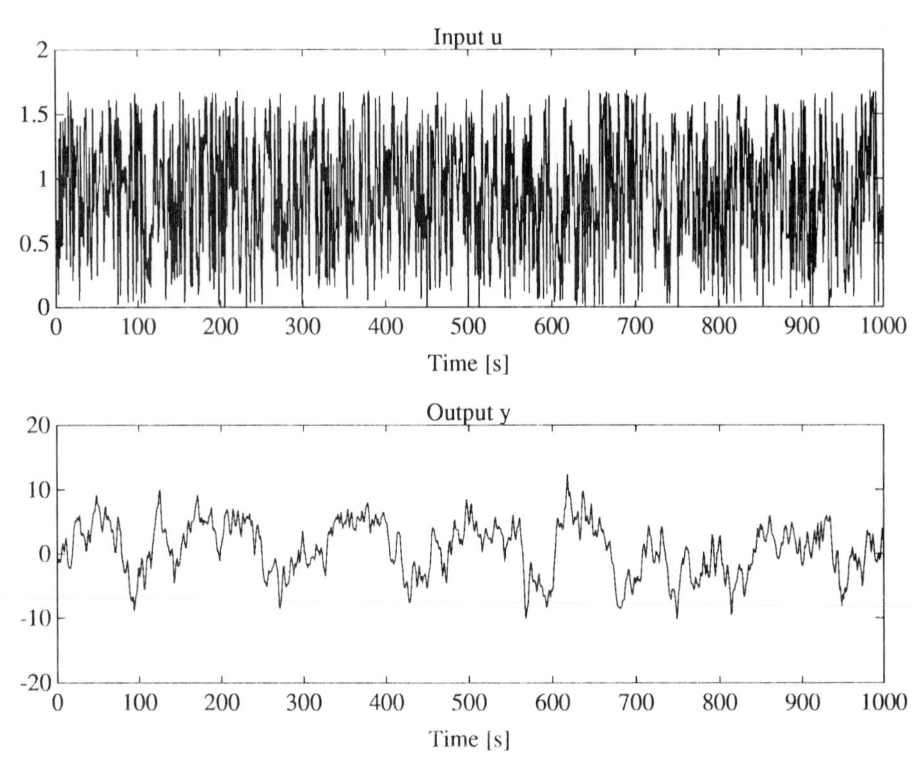

Figure 6.8 Input and output from a first-order system with correlated disturbance inputs.

is impossible to check how closely Z satisfies the requirement of uncorrelated behavior.

Example 6.8—Identification with instrumental variable method
Consider a case with data ($N = 1000$) (see Fig. 6.8) collected from the system

$$\mathcal{S}: \quad y_k = 0.9y_{k-1} + 0.1u_{k-1} + w_k + 0.7w_{k-1}; \qquad \mathcal{E}\{w_k\} = 0, \quad \mathcal{E}\{w_k^2\} = \sigma_w^2 \tag{6.56}$$

The parameter vector relevant for the transfer function $Y(z)/U(z)$ is

$$\theta = \begin{pmatrix} a \\ b \end{pmatrix} = \begin{pmatrix} 0.9 \\ 0.1 \end{pmatrix} \tag{6.57}$$

The biased least-squares estimate is

$$\begin{pmatrix} \widehat{a} \\ \widehat{b} \end{pmatrix} = (\Phi^T\Phi)^{-1}\Phi^T\mathcal{Y} = \begin{pmatrix} 0.957 \\ 0.047 \end{pmatrix} \tag{6.58}$$

and is based on the regressor matrix

$$\Phi = \begin{pmatrix} y_1 & u_1 \\ \vdots & \vdots \\ y_{N-1} & u_{N-1} \end{pmatrix}, \quad \text{and} \quad \mathcal{Y} = \begin{pmatrix} y_2 \\ \vdots \\ y_N \end{pmatrix} \tag{6.59}$$

Now introducing the variables we have

$$z_k = \hat{a} z_{k-1} + \hat{b} u_{k-1} \tag{6.60}$$

which is the predictor of y_k obtained from the least-squares estimate $\hat{a}, \hat{b}$. Note that owing to this special choice of instrumental variable the method is similar to the output error method. With these variables we can suggest the instrumental variable

$$Z = \begin{pmatrix} z_1 & u_1 \\ \vdots & \\ z_{N-1} & u_{N-1} \end{pmatrix} \tag{6.61}$$

The instrumental variable estimate gives

$$\hat{\theta}^z = (Z^T \Phi)^{-1} Z^T \mathcal{Y} = \begin{pmatrix} 0.918 \\ 0.075 \end{pmatrix} \tag{6.62}$$

which exhibits a reduced bias in both parameter estimates as compared to the least-squares estimate. ■

Example 6.9—Difficulties in choosing the instrumental variables
There is of course no guarantee that all choices of instrumental variables will provide good identification properties. Consider, for instance, the following instrumental variable applied to the data of the previous example.

$$Z = \begin{pmatrix} 0 & u_1 \\ u_1 & u_2 \\ \vdots & \vdots \\ u_{N-2} & u_{N-1} \end{pmatrix} \quad \Rightarrow \hat{\theta}^z = (Z^T \Phi)^{-1} Z^T \mathcal{Y} = \begin{pmatrix} 0.413 \\ 0.047 \end{pmatrix} \tag{6.63}$$

Clearly this is a poor estimate. It is, of course, also necessary to avoid degenerate cases such as

$$Z = \begin{pmatrix} u_1 & u_1 \\ \vdots & \vdots \\ u_{N-1} & u_{N-1} \end{pmatrix} \tag{6.64}$$

This is obviously not a feasible choice of instrumental variables as there is linear dependence between the two columns of Z so that $Z^T\Phi$ is not invertible.

■

A conclusion to be drawn from Example 6.9 is that it might be difficult to choose appropriate instrumental variables. As a result, it is standard procedure to use the instrumental variable method in an iterative manner that incorporates some kind of filtering. Moreover, certain established methods of other origin can be interpreted as instrumental variable methods — *e.g.*, the Yule-Walker equations for estimating the parameters of an autoregressive (AR) process.

Example 6.10—The Yule-Walker equations

Consider the AR process in the system

$$\mathcal{S}: \quad A(z^{-1})y_k = w_k; \qquad \mathcal{E}\{w_k\} = 0, \quad \mathcal{E}\{w_k^2\} = \sigma_w^2 \tag{6.65}$$

with $\deg A(z) = n_A$. The Yule-Walker equations elicit the relationships existing between the AR parameters and the autocovariance function.

$$\begin{aligned} C_{yy}(\tau) = \mathcal{E}\{y_k y_{k-\tau}^*\} &= \mathcal{E}\{(-\sum_{i=1}^{n_A} a_i y_{k-i} + w_k)y_{k-\tau}^T\} \\ &= -\sum_{i=1}^{n_A} a_i \mathcal{E}\{y_{k-i}y_{k-\tau}^T\} + \mathcal{E}\{w_k y_{k-\tau}^T\} \end{aligned} \tag{6.66}$$

so that

$$C_{yy}(\tau) = \begin{cases} -\sum_{i=1}^{n_A} a_i C_{yy}(\tau - i) + \sigma_w^2, & \tau = 0 \\ -\sum_{i=1}^{n_A} a_i C_{yy}(\tau - i), & \tau \neq 0 \end{cases} \tag{6.67}$$

Identification of the coefficients a_i can be approached by choosing numbers $M > n_A$ and $p > n_A$ and by introducing

$$\begin{aligned} \phi_k^T &= \begin{pmatrix} y_{k-1} & \cdots & y_{k-n_A} \end{pmatrix} \\ z_k^T &= \frac{1}{M}\begin{pmatrix} y_{k-2} & \cdots & y_{k-p-2} \end{pmatrix}; \qquad k = p, \ldots, M+p \end{aligned} \tag{6.68}$$

and the corresponding matrices

$$\Phi = \begin{pmatrix} \phi_p^T \\ \vdots \\ \phi_M^T \end{pmatrix}, \quad \text{and} \quad Z = \begin{pmatrix} z_p^T \\ \vdots \\ z_M^T \end{pmatrix} \tag{6.69}$$

so that

$$\begin{pmatrix} \widehat{C}_{yy}(i-1) & \widehat{C}_{yy}(i-2) & \cdots & \widehat{C}_{yy}(i-n_A) \\ \widehat{C}_{yy}(i) & \widehat{C}_{yy}(i-1) & \cdots & \widehat{C}_{yy}(i-n_A+1) \\ \vdots & & \ddots & \vdots \\ \widehat{C}_{yy}(i+p) & \widehat{C}_{yy}(i+p-1) & \cdots & \widehat{C}_{yy}(i+p-n_A+1) \end{pmatrix} \begin{pmatrix} a_1 \\ a_2 \\ \vdots \\ a_{n_A} \end{pmatrix} = \tag{6.70}$$

$$= (Z^T \Phi)\theta \approx - \begin{pmatrix} \widehat{C}_{yy}(i) \\ \widehat{C}_{yy}(i+1) \\ \vdots \\ \widehat{C}_{yy}(i+p) \end{pmatrix} = Z^T \mathcal{Y}$$

which provides an estimate of the AR parameters as

$$\widehat{\theta} = (Z^T \Phi)^{-1} Z^T \mathcal{Y} \tag{6.71}$$

■

6.6 SOME ASPECTS OF APPLICATION

An important problem is the sensitivity of the least-squares solutions to outliers and other abnormal data. A remedy is to inspect data and to exclude the outliers before further signal processing. A popular method in this context is application of a median filter applied to a sliding time window along the data series. The median filter thus elicits the median value from the set of data in the time interval under consideration. Unfortunately, however, the median filter may introduce artifacts in the form of time-delay variations in the input-output relationship, or even apparent noncausal behavior. The latter problem appears when a median filter is applied to rapidly varying data — *e.g.*, during the initial phase of a step response.

Prefiltering, smoothing, and prewhitening by data filtering are often effective in reducing bias provided that the filters are appropriately chosen with respect to the purpose of identification. A common problem is the compensation of known periodic variations as may occur due to seasonal effects in time-series data—*e.g.*, in economics, meteorology, hydrology, or ecology. Annual time-variations in monthly data can be compensated for by means of the following filtering

$$\begin{aligned} Y^f(z) &= (1 - z^{-12})Y(z) \\ U^f(z) &= (1 - z^{-12})U(z) \end{aligned} \tag{6.72}$$

and application of the modified model

$$\mathcal{M}': \qquad A(z^{-1})y_k^f = B(z^{-1})u_k^f + v_k \tag{6.73}$$

More generally, variation in sampled data with a periodic time-variant behavior and a period of d samples can sometimes be compensated by introducing the filter

$$F(z^{-1}) = 1 - z^{-d}, \qquad \text{d=period of trend} \tag{6.74}$$

Obviously this filter also removes constant offsets from data.

Bias reduction

As can be seen from Example 5.5 and Example 5.7 several methods are based on some kind of variance minimization that involves a trade-off between bias and variance minimization. Bias in the parameter estimates naturally constitutes a serious problem, and there are several *ad hoc* methods for bias reduction such as *trend elimination*, *differentiation of data*, and *bias estimation* implemented by estimation of an extra parameter.

Trend elimination of order n is made by subtracting a polynomial of order n from data. The polynomial of order n is most commonly found in such cases as a least-squares estimate adapted to data. Standard use of this method is usually limited to subtraction of the mean value of data (0th order trend elimination) or elimination of linear trends as obtained by linear regression.

Example 6.11—Trend elimination
Consider Example 5.5 where the noise offset $\mathcal{E}\{e_k\} = 1$ embodies a serious bias problem. Computation of the sample means

$$\begin{aligned} \bar{y} &= \frac{1}{N}\sum_{k=1}^{N} y_k \\ \bar{u} &= \frac{1}{N}\sum_{k=1}^{N} u_k \end{aligned} \tag{6.75}$$

and their subtraction from the sequences $\{y_k\}$ and $\{u_k\}$ results in the modified model

$$\mathcal{M}': \qquad A(z^{-1})(y_k - \bar{y}) = B(z^{-1})(u_k - \bar{u}) + v_k \tag{6.76}$$

Parameter estimation of a first-order model based on the same data as in Example 5.5 gives the result

$$\widehat{\theta} = \begin{pmatrix} \widehat{a} \\ \widehat{b} \end{pmatrix} = \begin{pmatrix} 0.9066 \\ 0.1143 \end{pmatrix} \quad \Rightarrow \quad \widetilde{\theta} = \widehat{\theta} - \theta = \begin{pmatrix} 0.0066 \\ 0.0143 \end{pmatrix} \tag{6.77}$$

which exhibits a clear improvement over Example 5.5 with respect to the bias magnitude. ■

Example 6.12—Differentiation of data
Consider the model of Example 5.5

$$\mathcal{M}: \qquad y_k = ay_{k-1} + bu_{k-1} + v_k; \qquad \mathcal{E}\{v_k\} = v_o \neq 0 \tag{6.78}$$

Straightforward application of discrete-time difference operator $\Delta = 1 - z^{-1}$ gives

$$\mathcal{M}': \qquad \Delta y_k = a\Delta y_{k-1} + b\Delta u_{k-1} + \Delta v_k, \quad \text{where} \quad \Delta = 1 - z^{-1} \tag{6.79}$$

The noise components now have the expected mean $\mathcal{E}\{\Delta v_k\} = \mathcal{E}\{v_k - v_{k-1}\} = 0$, which eliminates the offset in the noise. However, application of least-squares identification gives the disappointing result

$$\widehat{\theta} = \begin{pmatrix} \widehat{a} \\ \widehat{b} \end{pmatrix} = \begin{pmatrix} -0.0894 \\ 0.0852 \end{pmatrix} \tag{6.80}$$

which should be compared to the correct values $a = 0.9$ and $b = 0.1$. The differentiation solves one problem but, unfortunately, introduces new noise correlations and, in turn, causes new problems of bias in the parameter estimates. Maximum-likelihood identification of the model

$$\mathcal{M}'': \qquad \Delta y_k = a\Delta y_{k-1} + b\Delta u_{k-1} + w_k + cw_{k-1} \tag{6.81}$$

gives the result

$$\begin{pmatrix} \widehat{a} \\ \widehat{b} \\ \widehat{c} \end{pmatrix} = \begin{pmatrix} 0.8582 \\ 0.0522 \\ -0.9243 \end{pmatrix} \quad \Rightarrow \quad \widetilde{\theta} = \begin{pmatrix} -0.0418 \\ -0.0478 \\ -0.0757 \end{pmatrix} \tag{6.82}$$

which clearly provides improved accuracy as compared to the ordinary least-squares identification in Example 5.5 and, of course, better results than the estimate (6.80) obtained from straightforward differentiation and least-squares estimation. ■

Table 6.1 A comparison of various methods used in Example 6.13. Notice that there is no obvious relationship between $\widehat{\sigma}^2$ and $||\widetilde{\theta}||_2$.

	$\widehat{a}$	$\widehat{b}$	Offset	V	$\widehat{\sigma}^2$	$\|\|\widetilde{\theta}\|\|_2$
Ordinary LS	0.9829	0.1550	–	531	1.0642	0.0995
Differentiation+LS	-0.0894	0.0852	–	522	1.0467	0.9895
Differentiation+ML	0.8582	0.0522	–	515	1.0304	0.0635
Trend elimination	0.9066	0.1144	–	552	1.1073	0.0157
Offset estimation	0.9066	0.1143	0.9111	499	1.0015	0.0157

Example 6.13—Offset estimation via an extra parameter
Assume that the noise component $\{v_k\}$ can be decomposed as

$$v_k = v_o + w_k \tag{6.83}$$

where $\{w_k\}$ is a white-noise sequence and where $v_o = \mathcal{E}\{v_k\}$ represents the constant non-zero offset in the noise. Introduction of an extra parameter representing the offset yields the model

$$\mathcal{M}': \qquad y_k = -ay_{k-1} + bu_{k-1} + v_o + w_k = \begin{pmatrix} -y_{k-1} & u_{k-1} & 1 \end{pmatrix} \begin{pmatrix} a \\ b \\ v_o \end{pmatrix} + w_k \tag{6.84}$$

with the extended parameter vector $\theta' = \begin{pmatrix} a & b & v_o \end{pmatrix}^T$. The least-squares estimate of $\theta' = \begin{pmatrix} 0.9 & 0.1 & 1.0 \end{pmatrix}^T$ is

$$\widehat{\theta} = \begin{pmatrix} \widehat{a} \\ \widehat{b} \\ \widehat{v}_o \end{pmatrix} = \begin{pmatrix} 0.9066 \\ 0.1143 \\ 0.9111 \end{pmatrix} \quad \Rightarrow \quad \widetilde{\theta}' = \begin{pmatrix} 0.0066 \\ 0.0143 \\ -0.0889 \end{pmatrix} \tag{6.85}$$

where the basic idea is to "absorb" the bias due to the constant offset component v_o present in the noise sequence $\{v_k\}$.

As clearly demonstrated by the figures in Table 6.1, a comparison of the different methods suggests the offset estimation and trend elimination methods to be superior. It is obvious that differentiation as used in this example cannot be recommended in the context of least-squares identification, owing to the unavoidable noise correlation involved. Trend elimination and offset estimation both tend to give better results. ■

6.7 SOME REMARKS ON CONVERGENCE AND CONSISTENCY

Asymptotic convergence theory is concerned with the behavior of random variables and parameter estimates as the sample size tends towards infinity. For basic statistical definitions of convergence and consistency, the reader is referred to Appendix B which also contains important theorems such as the Cramér-Rao lower bound, the central limit theorem, and calculation rules for the probability limit. The application of these methods to convergence analysis can be illustrated, for instance, by means of the least-squares estimate

$$\widehat{\theta}_N = (\Phi_N^T \Phi_N)^{-1} \Phi_N^T \mathcal{Y} = \theta + (\Phi_N^T \Phi_N)^{-1} \Phi_N^T e \tag{6.86}$$

Assume that the disturbance sequence e fulfills the white-noise assumptions $\mathcal{E}\{ee^T\} = \sigma^2 I$. Then, if the regressors and disturbances are uncorrelated, using the probability limit we can show consistency in probability of least-squares identification according to the calculation

$$\text{plim } \widehat{\theta}_N = \theta + (\text{plim } \frac{1}{N} \Phi_N^T \Phi_N)^{-1} \cdot \text{plim } (\frac{1}{N} \Phi_N^T e) = \theta + 0 = \theta \tag{6.87}$$

with the covariance function

$$\begin{aligned} \text{Cov}\{\sqrt{N}(\widehat{\theta}_N - \theta)\} &= \text{plim } \{(\frac{1}{N} \Phi_N^T \Phi_N)^{-1} (\frac{N}{N^2} \Phi_N^T e e^T \Phi_N)(\frac{1}{N} \Phi_N^T \Phi_N)^{-1}\} = \\ &= \sigma^2 (\text{plim } \frac{1}{N} \Phi_N^T \Phi_N)^{-1} \end{aligned} \tag{6.88}$$

A problem related to consistency is the question of the limiting distributions of the parameter estimates. A valuable theorem in this context is the central limit theorem, see Appendix B, which can be applied to a consistent parameter estimate, if it can be regarded as the sum of independent stochastic variables. Direct application of the central limit theorem, to a consistent least-squares estimator, gives

$$\sqrt{N}(\widehat{\theta}_N - \theta) \overset{\text{dist.}}{\rightarrow} \mathcal{N}(0, \Sigma_\theta), \quad \text{where} \quad \Sigma_\theta = \sigma^2 \mathcal{E}\{\frac{1}{N} \Phi_N^T \Phi_N\}^{-1} \tag{6.89}$$

Hence, the limiting distribution for $\sqrt{N}(\widehat{\theta}_N - \theta)$ is normal. In addition, it turns out that the estimated parameters converge at a rate proportional to

$1/\sqrt{N}$, and that the corresponding covariance converges at a rate proportional to $1/N$.

Another valuable result for the analysis of limiting distributions is the following: Supposing $f(\theta) : R^n \rightarrow R^m$ to be a continuously differentiable function and that $\partial f/\partial \theta_i \neq 0$, for all $i = 1, 2, \ldots, n$, it can be shown that

$$\sqrt{N}(f(\widehat{\theta}_N) - f(\theta)) \overset{\text{dist.}}{\rightarrow} \mathcal{N}(0, (\frac{\partial f}{\partial \theta})^T P_\theta \frac{\partial f}{\partial \theta}) \tag{6.90}$$

If $\widehat{\theta}_N$ is found by means of the numerical method (6.43) when applied to an optimization criterion such as the log-likelihood function or some prediction error criterion based on the sample covariance, then we obtain the following Taylor series expansion at the minimum

$$0 = \nabla V_N(\widehat{\theta}_N) = \nabla V_N(\theta) + \nabla^2 V_N(\theta)(\widehat{\theta}_N - \theta) + \text{higher-order terms} \tag{6.91}$$

Consistency in probability for $\widehat{\theta}_N$ follows if plim $\{\nabla V_N(\widehat{\theta}_N)\} = 0$, provided that plim $\{\nabla^2 V_N(\widehat{\theta})\}$ exists and is invertible (*i.e.*, a unique parametrization is required). If the higher-order terms in the Taylor series expansion are neglected, we may also suggest

$$\text{plim } (\sqrt{N}(\widehat{\theta}_N - \theta)) \approx -(\text{plim}(\nabla^2 V_N(\theta)))^{-1}(\text{plim}(\sqrt{N}\nabla V_N(\widehat{\theta}))) \tag{6.92}$$

Provided that plim$\{\nabla^2 V_N(\widehat{\theta})\}$ exists and is invertible, *i.e.*, if V has a unique and global minimum at θ so that a unique parametrization is required, then it may be concluded

$$\sqrt{N}(\widehat{\theta}_N - \theta) \overset{\text{dist.}}{\rightarrow} \mathcal{N}(0, \Sigma_\theta) = \mathcal{N}(0, (\nabla^2 V_N(\theta))^{-1} P (\nabla^2 V_N(\theta))^{-1})$$

where

$$P = \lim_{N \to \infty} N\mathcal{E}\{(\nabla V_N(\widehat{\theta}))(\nabla V_N(\widehat{\theta}))^T\} \tag{6.93}$$

which can also be motivated by reference to Eq. (6.90). Moreover, by reference to Eq. (6.43), we may suggest the estimate

$$\begin{aligned} \nabla^2 V_N(\widehat{\theta}_N) &= \sum_{k=1}^{N} \psi_k(\widehat{\theta}_N)\psi_k^T(\widehat{\theta}_N) + \sum_{k=1}^{N} \varepsilon_k(\widehat{\theta}_N)\nabla^2 \varepsilon_k^T(\widehat{\theta}_N) \\ \widehat{P}_\theta(\widehat{\theta}_N) &= \frac{1}{N}\sum_{j=1}^{N}\sum_{k=1}^{N} \psi_j(\widehat{\theta}_N)\varepsilon_j(\widehat{\theta}_N)\varepsilon_k(\widehat{\theta}_N)\psi_k^T(\widehat{\theta}_N) \end{aligned} \tag{6.94}$$

With suitable approximations based on assumptions on uncorrelated residuals and by neglecting terms proportional to $\nabla^2 \varepsilon$, this can be simplified to

$$\begin{aligned} \widehat{\sigma}_N^2 &= \frac{1}{N} \sum_{k=1}^{N} \varepsilon_k^T(\widehat{\theta}_N)\varepsilon_k(\widehat{\theta}_N) \\ \widehat{\Sigma}_\theta &\approx \widehat{\sigma}_N^2 \Big(\frac{1}{N} \sum_{k=1}^{N} \psi_k(\widehat{\theta}_N)\psi_k^T(\widehat{\theta}_N)\Big)^{-1} \end{aligned} \tag{6.95}$$

as an estimate of the parameter covariance matrix. Similar results can be shown for almost sure consistency in the case of identification by means of ARMAX models with normally distributed disturbances with a unique parametrization. However, as all the results obtained with limiting distributions are valid only asymptotically, undue emphasis should not be placed on precise levels of significance for finite data series.

6.8 CONCLUDING REMARKS

It is obvious that differentiation as used in Example 6.12 cannot be recommended in the context of least-squares identification, owing to the unavoidable noise correlation involved. Trend elimination and offset estimation both tend to give better results. Notice that this is obvious from Example 6.13, despite the contrary evidence suggested by the least-squares estimation loss function. In this context it should be borne in mind that the parameter error $||\widetilde{\theta}||_2^2$ and the least-squares loss function are not always proportional in any simple manner. In fact, it is generally difficult to evaluate the quantitative behavior of convergence in mean square for standard linear autoregressive moving average models.

6.9 BIBLIOGRAPHY AND REFERENCES

The terminology of autoregressive and moving average stochastic processes was coined by H. Wold about 1938.

– H. WOLD, *A Study in the Analysis of Stationary Time Series*. Stockholm: Almqvist & Wiksell, 1938.

Early work on the maximum-likelihood method and its application to time-series analysis was done by the following:

- R.A. FISHER, "On the mathematical foundations of theoretical statistics." *Philos. Trans. R. Soc. London Ser. A*, Vol. 222, 1922, pp. 309–368.
- H.B. MANN AND A. WALD, "On the statistical treatment of linear stochastic difference equations." *Econometrica*, Vol. 11, 1943, pp. 173–220.
- K.J. ÅSTRÖM AND T. BOHLIN, "Numerical identification of linear dynamic systems from normal operating records." *IFAC Symposium Self-Adaptive Control Systems*, Teddington, England, 1965.

The instrumental variable method also has its origins in econometrics and was introduced in

- O. REIERSØL, "Confluence analysis by means of lag moments and other methods of confluence analysis." *Econometrica*, Vol. 9, 1941, pp. 1–23.

Time-series analysis and its applications are presented in these important works

- G.E.P. BOX AND G.M. JENKINS, *Time Series Analysis: Forecasting and Control*. San Francisco: Holden-Day, 1970.
- A.V. OPPENHEIM AND R.W. SCHAFER, *Discrete-Time Signal Processing*. Englewood Cliffs, NJ: Prentice-Hall, 1989.

Output error methods are treated in the paper

- L. DUGARD AND I.D. LANDAU,"Recursive output errors identification algorithms — Theory and evaluation." *Automatica*, Vol. 16,1980, pp. 443–462.

Convergence analysis including results on almost sure consistency for the special case of parameter identification using ARMAX models with a unique parametrization, normally distributed disturbances and appropriate excitation is to be found in the following works:

- P.E. CAINES AND J. RISSANEN, "Maximum-likelihood estimation of parameters in multivariate Gaussian stochastic processes." *IEEE Trans. Inf. Theory*, IT-20, 1974, pp. 102–104.
- E.J. HANNAN, *Multiple Time Series*. New York: John Wiley, 1970.
- E.J. HANNAN AND M. DEISTLER, *The Statistical Theory of Linear Systems*. New York: John Wiley, 1988.
- L. LJUNG, "Convergence analysis of parametric identification methods." *IEEE Trans. Automatic Control*, AC-23, 1978, pp. 770–783.

6.10 EXERCISES

6.1 Consider the system

$$\mathcal{S}: \qquad y_k + a y_{k-1} = b u_{k-1} + w_k + c w_{k-1} \tag{6.96}$$

where $\{u_k\}$ and $\{w_k\}$ are independent zero-mean white-noise processes with the variances $\mathcal{E}\{u_k^2\} = \sigma_u^2$ and $\mathcal{E}\{w_k^2\} = \sigma_w^2$, respectively. What are the asymptotic parameter estimates for a large number N of observations when fitting a model in the model set

$$\mathcal{M}: \qquad y_k + a y_{k-1} = b u_{k-1} \tag{6.97}$$

to data.

6.2. Show that the following numerical algorithms converge toward the minimum at $\widehat{x} = -Q_2^{-1} q_1$ for certain choices of step-length α_i when applied to a quadratic function

$$V(x) = \frac{1}{2} x^T Q_2 x + x^T q_1 + q_0 \tag{6.98}$$

where $x \in R^n$ and the matrix $Q_2 = Q_2^T > 0$ and $q_1 \in R^n$ and the scalar $q_0 \in R$.

i. The Newton-Raphson algorithm

$$x^{(i+1)} = x^{(i)} - \alpha_i (V''(x^{(i)}))^{-1} V'(x^{(i)}) \tag{6.99}$$

ii. The Levenberg-Marquardt algorithm

$$x^{(i+1)} = x^{(i)} - \alpha_i (\alpha_i I_{n \times n} + V''(x^{(i)}))^{-1} V'(x^{(i)}) \tag{6.100}$$

iii. The Gauss-Newton algorithm

$$x^{(i+1)} = x^{(i)} - \alpha_i R V'(x^{(i)}) \tag{6.101}$$

where R is a constant positive definite weighting matrix.

Determine conditions on R and α_i for the algorithm to converge.

Hint: Define the error at iteration i as $e^{(i)} = x^{(i)} - \widehat{x}$, and then consider the error norm

$$\|e^{(i)}\|_P^2 = (e^{(i)})^T P(e^{(i)}) \tag{6.102}$$

for some positive definite matrix P and take

$$\|e^{(i+1)}\|_P < \|e^{(i)}\|_P \tag{6.103}$$

as a criterion of convergence.

6.3 Consider a moving average (MA) process $y_k = b_1 u_{k-1} + \cdots + b_m u_{k-m} + v_k$, and show that the process parameters $b_1, \ldots, b_m$ can be consistently estimated using least-squares identification even in the presence of colored noise $\{v_k\}$.

6.4 Modify the maximum-likelihood identification method for statistically independent perturbations distributed according to the Rayleigh distribution with the asymmetric probability distribution function

$$f(x) = \frac{x}{\sigma^2} e^{-x^2/2\sigma^2}, \qquad x > 0 \tag{6.104}$$

6.5 Consider the system

$$\mathcal{S}: \qquad y_{k+1} = a(u_k + w_k), \qquad w_k \in \mathcal{N}(0, \sigma^2), \quad \mathcal{E}\{w_i w_j\} = \sigma^2 \delta_{ij} \tag{6.105}$$

where $\{w_k\}$ is a zero-mean white noise sequence and where $\{u_k\}$, $\{y_k\}$ are measured variables. How can a be estimated optimally?

6.6 Formulate the instrumentable variable equations on the augmented system form (5.104). Show that

$$\begin{pmatrix} I & \Phi_N \\ Z^T & 0 \end{pmatrix} \begin{pmatrix} \varepsilon \\ \widehat{\theta} \end{pmatrix} = \begin{pmatrix} \mathcal{Y}_\mathcal{N} \\ 0 \end{pmatrix} \tag{6.106}$$

6.7 Show that the instrumental variable estimate can be written

$$\begin{pmatrix} \varepsilon \\ \widehat{\theta} \end{pmatrix} = \begin{pmatrix} I - \Phi_N (Z^T \Phi_N)^{-1} Z^T & \Phi_N (Z^T \Phi_N)^{-1} \\ (Z^T \Phi_N)^{-1} Z^T & -(Z^T \Phi_N)^{-1} \end{pmatrix} \begin{pmatrix} \mathcal{Y}_\mathcal{N} \\ 0 \end{pmatrix} \tag{6.107}$$

6.8 Consider the instrumental variable method and show by means of Eq. (6.107) that

$$\mathrm{Cov}\{\begin{pmatrix} I & \Phi_N \\ Z^T & 0 \end{pmatrix}\begin{pmatrix} \varepsilon \\ \tilde{\theta} \end{pmatrix}\} = \mathrm{Cov}\{\begin{pmatrix} v \\ 0_{p\times 1} \end{pmatrix}\} = \begin{pmatrix} \Sigma_v & 0_{N\times p} \\ 0_{p\times N} & 0_{p\times p} \end{pmatrix} \tag{6.108}$$

Use this result to justify the covariance estimate

$$\begin{aligned}\mathrm{Cov}\{\begin{pmatrix} \varepsilon \\ \tilde{\theta} \end{pmatrix}\} &\approx \begin{pmatrix} I & \Phi_N \\ Z^T & 0 \end{pmatrix}^{-1}\begin{pmatrix} \Sigma_v & 0_{N\times p} \\ 0_{p\times N} & 0_{p\times p} \end{pmatrix}\begin{pmatrix} I & \Phi_N \\ Z^T & 0 \end{pmatrix}^{-T} = \\ &= \begin{pmatrix} P\Sigma_v P^T & P\Sigma_v R^T \\ R\Sigma_v P^T & R\Sigma_v R^T \end{pmatrix}\end{aligned} \tag{6.109}$$

for $P = (I - \Phi_N(Z^T\Phi_N)^{-1}Z^T)$ and $R = (Z^T\Phi_N)^{-1}Z^T$.

6.9 Consider the functions

$$\begin{aligned} f_1(A) &= \log\det A \\ f_2(A) &= \mathrm{tr}(WA) \end{aligned} \tag{6.110}$$

where W is a symmetric positive definite weighting matrix. Show that f_1 and f_2 are suitable as scalar optimization criteria when applied to sample covariance matrices. Show that these functions are such that $f_1(A) \geq f_1(A_0)$ and $f_2(A) \geq f_2(A_0)$ for square symmetric matrices A and A_0 satisfying the inequality $A \geq A_0$.

6.10 Consider the system

$$y_k = -ay_{k-1} + bu_{k-1} + v_k \tag{6.111}$$

where $\{v_k\}$ is a sequence of independent identically distributed stochastic variables, each with the probability density function $f(x) = \mu e^{-\mu x}$. Design a maximum-likelihood method that permits estimation of a and b. ■

7

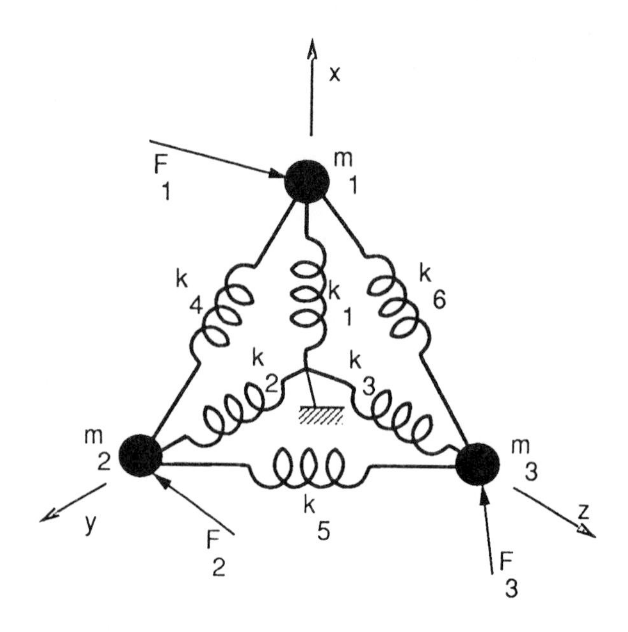

Modeling

7.1 INTRODUCTION

Although in a wider sense of the term, *modeling* is universally applicable not only in technology and such sciences as physics, chemistry, and biology, but also in fields such as economics and social sciences, this chapter will be confined to annotated examples drawn from the natural sciences. Moreover, several of the important modeling approaches that we have already will not be further commented on here. A suitable starting point is to consider some standard modeling procedures of importance in the context of identification.

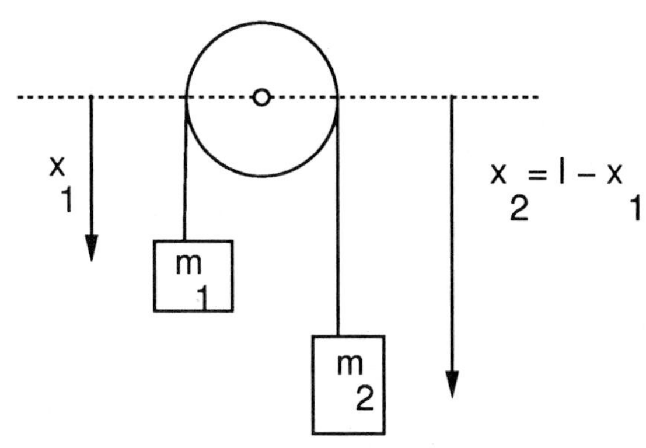

Figure 7.1 An Atwood machine with a frictionless pulley.

7.2 MECHANICAL SYSTEMS

The essential mechanics involved in the motion of a particle is contained in Newton's second law of motion, which defines force and mass. For a single particle this law is expressed thus:

$$F = \frac{dp}{dt} = \frac{d}{dt}(m\dot{q}) = m\ddot{q} \tag{7.1}$$

where p is the linear momentum of the particle, q its position, $\dot{q}$ its velocity, $\ddot{q}$ its acceleration, m its mass, and F the total force acting on the particle. (We use the standard notation used in mechanics where $\dot{q}$ and $\ddot{q}$ designate the time derivatives dq/dt and d^2q/dt^2, respectively.) An essential aspect is the *conservation of momentum and energy* (kinetic energy + potential energy) when no external forces are acting on the system. In generalizing these ideas to systems of particles and mechanical structures, it is necessary to distinguish between external forces acting on the particles from outside the system and internal forces exerted, say, on some particle i by all other particles in the system. Moreover, the presence of *constraints* such as fixed distances between some particles limits the motion.

Example 7.1—Newtonian mechanics

In Newtonian mechanics modeling starts with a force equation. Consider, for instance, an Atwood machine with two weights m_1 and m_2 and a frictionless pulley (Fig. 7.1). Let the vertical positions of the weights be denoted x_1 and

x_2 and let the tension of the rope be designated λ.

$$\begin{aligned} m_1\ddot{x}_1 &= m_1 g - \lambda \\ m_2\ddot{x}_2 &= m_2 g - \lambda \end{aligned} \tag{7.2}$$

The rope connecting the two weights constitutes a motion constraint such that

$$x_1 + x_2 - \ell = 0 \tag{7.3}$$

This constraint places a restriction on the velocities and accelerations of m_1 and m_2 such that $\dot{x}_1 = -\dot{x}_2$ and $\ddot{x}_1 = -\ddot{x}_2$. Elimination of $\ddot{x}_1$ and $\ddot{x}_2$ from Eq. (7.2) gives the tension of the rope

$$\lambda = \frac{2m_1 m_2}{m_1 + m_2} g \tag{7.4}$$

To surmount the problem that forces of constraint are unknown *a priori*, it is desirable to formulate the mechanics so that the forces of constraint are zero (d'Alembert's principle). This leads, *via* variational calculus based on modeling of energy, to Lagrangian mechanics. The following Euler-Lagrange equations may be obtained by considering the kinetic energy T, the potential energy U, and the *Lagrangian* $L = T - U$.

$$\frac{d}{dt}\left(\frac{\partial L}{\partial \dot{q}}\right) - \frac{\partial L}{\partial q} = \tau \tag{7.5}$$

where τ denotes external forces acting on the system and q the generalized coordinates. ■

Example 7.2—Lagrangian mechanics

Consider the kinetic energy T and the potential energy of the masses of the Atwood machine (Fig. 7.1)

$$\begin{aligned} U &= -m_1 g x - m_2 g(l - x) \\ T &= \frac{1}{2}(m_1 + m_2)\dot{x}^2 \end{aligned} \tag{7.6}$$

where x denotes the position of the center of mass. The Euler-Lagrange equations based on $L = T - U$ give

$$\begin{aligned} \frac{\partial L}{\partial x} &= (m_1 - m_2)g \\ \frac{d}{dt}\frac{\partial L}{\partial \dot{x}} &= \frac{d}{dt}((m_1 + m_2)\dot{x}) = (m_1 + m_2)\ddot{x} \end{aligned} \tag{7.7}$$

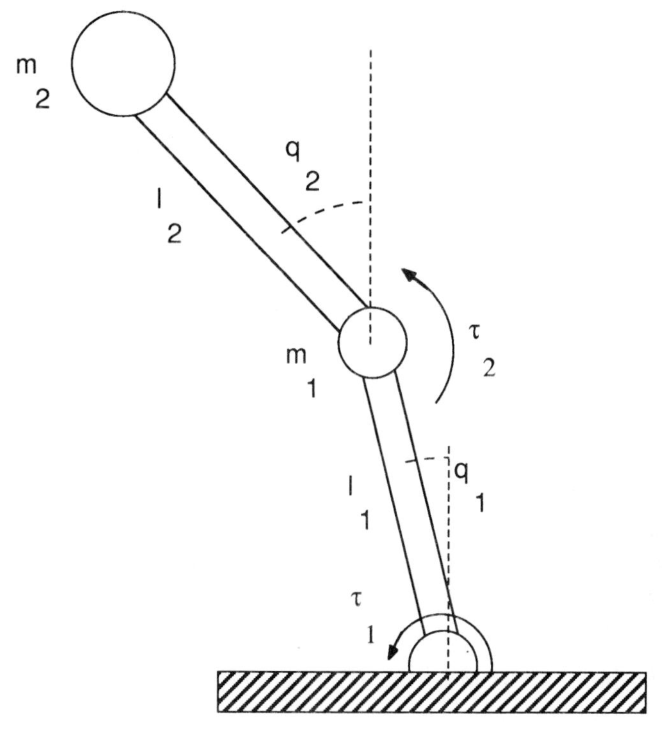

Figure 7.2 A robot with two arm segments of length ℓ_1 and ℓ_2, two point masses m_1 and m_2, and angular coordinates q_1 and q_2.

so that

$$\ddot{x} = \frac{m_1 - m_2}{m_1 + m_2} g \tag{7.8}$$

The motion constraint, *i.e.*, the tension of the rope, does not appear explicitly in these equations, which simplifies calculation. ■

Example 7.3—Lagrangian mechanics for modeling of a robot

Consider the robot in Fig. 7.2. The positions (x_1, y_1) and (x_2, y_2) of the segment end points expressed in the Cartesian coordinates x and y are

$$\begin{aligned} x_1 &= -l_1 \sin q_1 \\ y_1 &= l_1 \cos q_1 \\ x_2 &= -l_1 \sin q_1 - l_2 \sin q_2 \\ y_2 &= l_1 \cos q_1 + l_2 \cos q_2 \end{aligned} \tag{7.9}$$

where x is the horizontal coordinate and y the vertical coordinate with their origin at the first joint. The corresponding velocities v_1, v_2 are

$$
\begin{aligned}
v_1 &= \begin{pmatrix} \dot{x}_1 \\ \dot{y}_1 \end{pmatrix} = \begin{pmatrix} -l_1 \cos q_1 \cdot \dot{q}_1 \\ -l_1 \sin q_1 \cdot \dot{q}_1 \end{pmatrix} \\
v_2 &= \begin{pmatrix} \dot{x}_2 \\ \dot{y}_2 \end{pmatrix} = \begin{pmatrix} -l_1 \cos q_1 \cdot \dot{q}_1 - l_2 \cos q_2 \cdot \dot{q}_2 \\ -l_1 \sin q_1 \cdot \dot{q}_1 - l_2 \sin q_2 \cdot \dot{q}_2 \end{pmatrix}
\end{aligned} \tag{7.10}
$$

Introduce the following shorter notation $c_1 = \cos q_1$, $s_2 = \sin q_2$ etc., and $q = \begin{pmatrix} q_1 & q_2 \end{pmatrix}^T$. The potential energy is then

$$
U(q) = m_1 g l_1 c_1 + m_2 g (l_1 c_1 + l_2 c_2) \tag{7.11}
$$

The kinetic energy is

$$
T = \frac{1}{2} \begin{pmatrix} v_1^T & v_2^T \end{pmatrix} \begin{pmatrix} m_1 & 0 \\ 0 & m_2 \end{pmatrix} \begin{pmatrix} v_1 \\ v_2 \end{pmatrix} \tag{7.12}
$$

which can be expressed in the angular coordinates as

$$
\begin{aligned}
T(q, \dot{q}) &= \frac{1}{2} \dot{q}^T M(q) \dot{q} = \\
&\frac{1}{2} \begin{pmatrix} \dot{q}_1^T & \dot{q}_2^T \end{pmatrix} \begin{pmatrix} (m_1 + m_2) l_1^2 & m_2 l_1 l_2 (c_1 c_2 + s_1 s_2) \\ m_2 l_1 l_2 (c_1 c_2 + s_1 s_2) & m_2 l_2^2 \end{pmatrix} \begin{pmatrix} \dot{q}_1 \\ \dot{q}_2 \end{pmatrix}
\end{aligned} \tag{7.13}
$$

where the matrix $M(q)$ is the *inertia matrix.* Application of the Euler- Lagrange equations gives

$$
M(q)\ddot{q} + C(q, \dot{q})\dot{q} + G(q) = \tau \tag{7.14}
$$

where $M(q)$ is the inertia matrix, $C(q, \dot{q})\dot{q}$ the centripetal and Coriolis forces

$$
C(q, \dot{q})\dot{q} = m_2 l_1 l_2 (c_1 s_2 - s_1 c_2) \begin{pmatrix} -\dot{q}_2^2 + \dot{q}_1 \dot{q}_2 \\ \dot{q}_1^2 - \dot{q}_1 \dot{q}_2 \end{pmatrix} \tag{7.15}
$$

and $G(q)$ the gravitation forces involved

$$
G(q) = \frac{\partial U}{\partial q} = g \begin{pmatrix} -(m_1 + m_2) l_1 s_1 \\ -m_2 l_2 s_2 \end{pmatrix} \tag{7.16}
$$

■

Lagrangian mechanics can be modified to include constraint forces, *i.e.*, constraints that can be expressed in the form

$$c(q) = 0, \quad \Rightarrow \quad (\frac{\partial c}{\partial q})^T \dot{q} = 0 \qquad \text{(holonomic constraints)} \tag{7.17}$$

This type of constraint depends on the position q only; such constraints are called *holonomic constraints*. The force generated by the constraint is

$$F = \frac{\partial c}{\partial q} \tag{7.18}$$

at the constrained surface $c(q) = 0$ and the work done by these forces is $dW = F^T \dot{q} dt = 0$, *i.e.*, the constraint forces neither generate nor dissipate any energy. Holonomic constraints thus generate position-dependent workless contact forces.

7.3 THERMODYNAMIC MODELING

Thermodynamic modeling is widely used in chemistry, chemical engineering, and mechanical engineering to model heat transfer, fluid mechanics, and chemical reactions. According to thermodynamic theory, there exist some fundamental thermodynamic laws:

0. If two systems are in thermal equilibrium with a third system, they must be in thermal equilibrium with each other.

1. An equlibrium state of a system can be characterized by a quantity E called *internal energy*, which has the property that for an isolated system $E =$constant.

2. An equilibrium state of a system can be characterized by a quantity S called *entropy* which, in any process in which a thermally isolated system goes from one state to another, characteristically tends to increase, *i.e.*, $dS \geq 0$. If the system is not isolated and undergoes a process in which it absorbs heat dQ, then

$$dS = \frac{dQ}{T} \tag{7.19}$$

where T is the absolute temperature of the system.

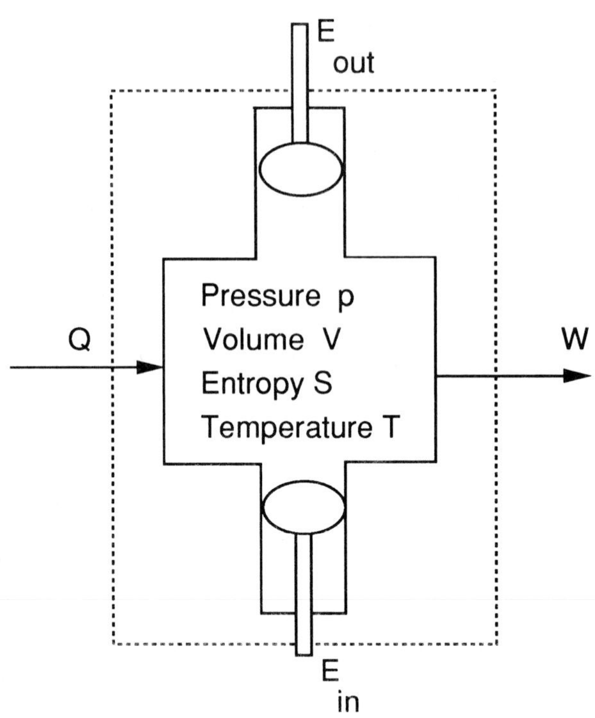

Figure 7.3 A thermodynamic process where Q is the heat absorbed by the system, W the work done by the system, and E_{in} and E_{out} the internal energy of the inflow and outflow, respectively.

3. The entropy S of the system has the property that $S \to S_0$ when $T \to 0_+$ where S_0 is a constant.

Let Q denote the heat absorbed by the system and W the work done by the system. The system interacts with the environment by absorbing (or emitting) heat and by performing mechanical work according to the relationship

$$dQ = TdS = dE + dW \tag{7.20}$$

Hence, in the relationship (7.20) the total energy is split into a part W due to mechanical interaction and a part Q due to thermal interaction. If for a given system the values of the state variables are independent of time, the system is said to be in thermodynamic equilibrium.

The number of state variables required to describe the process may be larger than the number required to describe the system at thermodynamic equilibrium, and there are several interdependent internal states (pressure p, volume V, entropy S, temperature T) whose evaluation is determined by the conversion of various forms of energy. Transformation from one subset of variables to some other subset can be accomplished by using Legendre transformations.

Internal energy E	$E(S,V)$,	$dE = TdS - pdV$
Enthalpy H	$H(S,p) = E + pV$,	$dH = TdS + Vdp$
Helmholtz free energy F	$F(T,V) = E - TS$,	$dF = -SdT - pdV$
Gibbs free energy G	$G(T,p) = E - TS + pV$,	$dG = -SdT + Vdp$

(7.21)

For instance, if there is no thermal interaction with the environment, the work that can be done by the system in Fig. 7.3 is limited by the enthalpy relationship

$$W = (E_{in} + p_{in}V_{in}) - (E_{out} + p_{out}V_{out}) \tag{7.22}$$

where enthalpy is the energy concept in which entropy and pressure are considered to be the independent variables.

7.4 COMPARTMENT MODELS

Material flow and storage in biological systems is often modeled with a compartmental system, which can be defined as a system made up of a finite number of macroscopic subsystems, called *compartments* or *pools*, each of which is homogenous and well mixed, that interact by exchanging materials. There may be inputs from the environment into one or more of the compartments, and there may be outputs (excretion) from one or more compartments into the environment. The compartments are depicted as boxes, and the exchange of materials from one compartment to another by an arrow indicating direction of flow. *Exchange rates* between compartments are often indicated adjacent to the arrows. As compartments need not be physically separated, compartment models are also useful for modeling chemical reactions between components A, B, and C such as

$$2H_2O \rightleftharpoons 2H_2 + O_2 \tag{7.23}$$

where the exchange rates represents chemical reaction rates. It is an excellent means of modeling the dynamics of such agents as radioactive tracers mixed into some normal substance. The tracers should be chosen such that the system is unable to distinguish between the normal material and the tracer and such that it does not affect the steady state and the exchange rates of the carrier substance. The methodology is an established approach in many natural

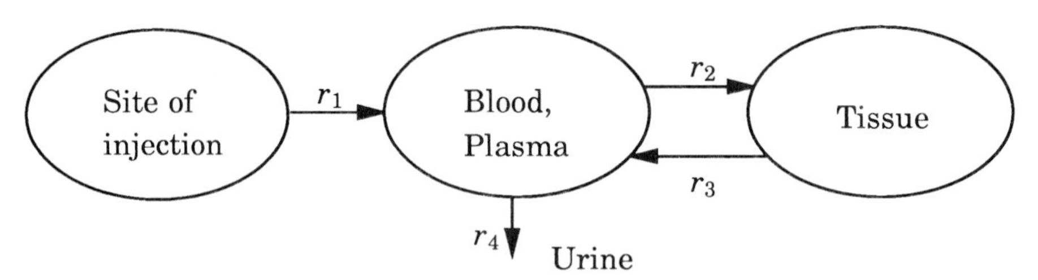

Figure 7.4 Distribution of a drug when the drug is injected at an intramuscular site.

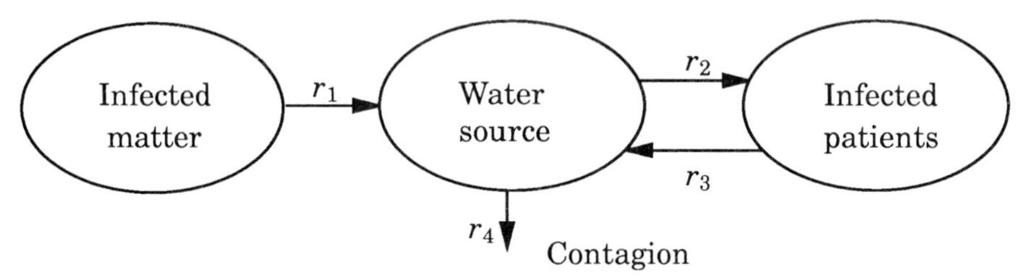

Figure 7.5 Transfer of a contagious disease in a contaminated environment.

sciences such as chemistry, biochemistry, physiology and medicine, pharmacology, and the study of ecosystems in ecology. It is also used in demography to describe population growth and migration and the spread of epidemics.

Example 7.4—State-space model and compartment model
Consider the example in Fig. 7.4 of a model of the distribution of a drug injected at an intramuscular site. A state-space model with states x_i; $i = 1, 2, 3$ and exchange rates or *transfer coefficients* r_i can be suggested as

$$\begin{aligned} \frac{d}{dt}\begin{pmatrix} x_1 \\ x_2 \\ x_3 \end{pmatrix} &= \begin{pmatrix} -r_1 & 0 & 0 \\ r_1 & -r_2 - r_4 & r_3 \\ 0 & r_2 & -r_3 \end{pmatrix}\begin{pmatrix} x_1 \\ x_2 \\ x_3 \end{pmatrix} \\ y &= \begin{pmatrix} 0 & r_4 & 0 \end{pmatrix}\begin{pmatrix} x_1 \\ x_2 \\ x_3 \end{pmatrix} \end{aligned} \tag{7.24}$$

The state-space model (7.24) would, of course, with another interpretation of the exchange rates r_i, also express the compartment model of Fig. 7.5 which has the same number of compartments and similar interactions as that of Fig. 7.4. ■

The dynamics of the system is typically studied as the transient responses from initial states (see Fig. 7.6). By adding a tracer to one compartment, and recording the responses from each compartment available to measurement,

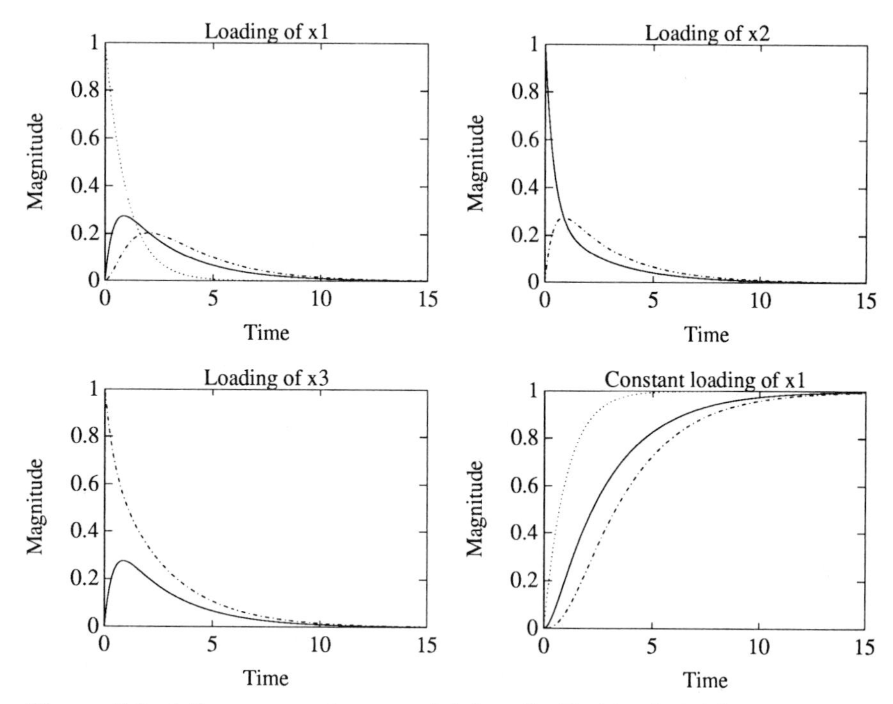

Figure 7.6 A three-compartment model described by Eq. (7.24) ($r_1 = \cdots = r_4 = 1$) with transient responses after loading of each one of the three compartments. The lower right diagram shows the response to a constant loading of compartment x_1.

it is possible to study the transfer between and among the compartments, and thus determine the exchange parameters of the system. The associated impulse response $g(t)$, then, can be interpreted as the elimination rate at time t of the tracer material that entered the system at time $t = 0$. Assume the total amount of tracer to be m, the distribution volume for the tracer in the compartment systems to be V, the constant outflow rate i_{out}, and the outflow concentration of the tracer to be $c(t)$. The case of a compartment system with one outflow and no consumption or production of material can be expressed thus:

$$\begin{aligned} &\int_0^\infty g(t)dt = 1, \qquad g(t) \geq 0, \quad \forall t \\ &\int_0^\infty tg(t)dt = V/i_{out}, \quad \text{Stewart-Hamilton equation} \end{aligned} \tag{7.25}$$

The inequality $g(t) \geq 0$ in (7.25) is motivated by the assumption that no tracer material enters the system *via* the outflow. The first equation is motivated by the fact that all the injected material will eventually leave the system *via* the outflow, whereas the second equation expresses the mean duration of the injected tracer's presence in the compartment system. The tracer elimination rate is then

$$i_{out}c(t) = mg(t), \quad \text{so that} \quad \begin{cases} i_{out} = \dfrac{m}{\int_0^\infty c(t)dt} \\ g(t) = i_{out}c(t)/m = \dfrac{c(t)}{\int_0^\infty c(t)dt} \end{cases} \tag{7.26}$$

where Eq. (7.25) has been used. Aside from the impulse response $g(t)$ it is thus possible to calculate the outflow rate i_{out} provided that the output concentration can be measured over a sufficently long time interval. Such a calculated flow rate is called *clearance* in biomedicine.

7.5 PRINCIPLES OF MODELING

Although modeling is in many respects a craft which is strongly dependent on the purpose of the model, there exist some general principles that can be used as guidelines for sound modeling. As modeling techniques would seem to have reached their highest level of maturity in mechanics and thermodynamics, the following guidelines derive from these branches of science.

Physical balance equation

The conservation of mass, energy, and momentum are fundamental principles of physical modeling and apply to systems modeled by mechanical, electrical, chemical, and thermodynamical laws. The balance equation of some subsystem takes the form

$$\text{accumulation flow} = \text{inflow} - \text{outflow} \tag{7.27}$$

which can be suitably modified to model production (*sources*) and consumption (*sinks*). In some instances there might be no justification for assuming the existence of a storage element (*cf.* Kirchhoff's law of electrical circuits), in which case the *static* equilibrium takes the form

$$0 = \text{inflow} - \text{outflow} \tag{7.28}$$

Flow variables thus add to zero at some connection point which does not contain any storage element. The presence of storage elements of mass, energy, momentum or some other variable leads to *state equations*. Storage elements of mass, energy, etc., are sometimes called containers, buffers, or compartments, and are ruled by equations of the form

$$\frac{d}{dt}(\text{state variable}) = \text{inflow} - \text{outflow} \tag{7.29}$$

When there are several subsystems exchanging mass, momentum, or energy among the subsystems, the balance and state equations give rise to a set of interacting equations of the type in Eq. (7.29).

Nature often acts so as to minimize energy, and many modeling problems describe a condition of equilibrium associated with a minimum of energy. The system strives towards a state of equilibrium, and a state of equilibrium is successful only if it is stable. Hence, a system would require input of energy to move away from equilibrium, and the systems often oscillate around the equilibrium when energy is conserved. Movement will grow if the system releases rather than consumes energy, such as occurs in unstable systems.

Potentials, gradients, and flows

The fluxes of a system are determined by certain *potentials* such as electric voltage, temperature, pressure, or concentration, and their *gradients*. Closely related to the notion of potential are energy concepts as used in physics. In contrast to flow variables, which add to zero at a connection point, the gradient variables are equal at connection points.

Nonequilibrium conditions are characterized by a flow in directions relative to the gradient. The flows involved disappear in equilibrium conditions, but may also reach a *steady-state* of time-invariant flows and gradients. In physics and chemistry, there are several *phenomenological laws* describing the relationship between flow and gradient variables such as:

- FICK'S law: The mass flow of diffusion is proportional to the mass concentration gradient in a physical medium.
- FOURIER's law: Heat flow is proportional to the temperature gradient in a medium.

$$J = -kA\frac{\partial T}{\partial x} \tag{7.30}$$

  where J is the heat-transfer rate, and $\partial T/\partial x$ is the temperature gradient with respect to a spatial coordinate x. The positive constant k is the thermal conductivity, and A the cross-section area of the thermal flow.
- OHM's law: Electric current is proportional to the electrical voltage.

Another example is Newton's law of the relationship between shearing force and velocity gradient. Similarly, the rate of chemical reaction is proportional to the concentration of each component of the reaction.

The coefficients between the gradients of the potentials and the corresponding flow variables during steady-state conditions are called *conductivity* (heat conductivity, electrical conductivity, admittance etc.), the reciprocal coefficient being *resistance* (*e.g.*, heat flow resistance, aerodynamical resistance, electrical resistance or *impedance*).

Entropy conditions

Most processes with energy storage also satisfy other conditions as not all possible behavior associated with a given state satisfies certain *entropy* conditions, which restrict the behavior of physical processes to be *irreversible.* Entropy describes the dispersal of energy, the natural tendency of spontaneous change toward states of higher entropy; *cf.* Eq. (7.19) and the second law of thermodynamics.

Entropy calculations are often used in statistical mechanics and thermodynamics, and have also been applied to communication theory in order to model information flow (Shannon) and coding.

*Onsager reciprocal relations

An important law describing the relationship between flows, gradients, and entropy in physics is the *Onsager reciprocal relations*. Onsager derived his principle from statistical mechanics considerations under the assumption of microscopic reversibility, that is, the symmetry of all mechanical equations of motion of individual particles with respect to time where a proper choice is made of fluxes J (heat flow, electrical current, chemical reaction rate, momentum, etc.) and generalized forces F (temperature gradient, electric potential gradient, chemical affinity, mechanical force, etc.) such that the entropy production per unit time may be written as

$$\frac{dS}{dt} = J^T F = F^T C F > 0 \tag{7.31}$$

for some matrix C. At thermodynamic equilibrium, all processes stop and we have simultaneously

$$J = 0, \quad \text{and} \quad F = 0 \tag{7.32}$$

If the fluxes J and forces F are related by linear phenomenological relationships

$$J = CF, \quad \text{or} \quad J_i = \sum_{j=1}^{n} C_{ij} F_j, \quad i = 1, \dots, n \tag{7.33}$$

then according to Onsager reciprocal relations the coefficient matrix C is symmetric.

Example 7.5—Thermodiffusion
Consider the case of one-dimensional thermodiffusion along a spatial coordinate x, and assume the temperature to be $T(x)$. For simple heat conduction we have for the heat flow and the temperature gradient

$$\begin{cases} J_1 = \frac{h}{T} \\ F_1 = -\frac{1}{T}\frac{\partial T}{\partial x} \end{cases} \tag{7.34}$$

Now include the diffusion and consider the modified gradient-flow relationship

$$\begin{cases} J_1 = C_{11}F_1 + C_{12}F_2 \\ J_2 = C_{21}F_1 + C_{22}F_2 \end{cases}, \quad \text{where} \quad \begin{cases} F_1 & \text{temperature gradient} \\ F_2 & \text{mass concentration gradient} \end{cases} \tag{7.35}$$

where C_{11} is the heat conductivity (Fourier's law), and C_{22} the diffusion coefficient (Fick's law). The coefficients C_{21}, C_{12} describe interference of the two irreversible processes of heat conduction and diffusion, respectively, C_{21} representing the appearance of a concentration gradient when a temperature gradient is imposed (Soret effect) and C_{12} the converse situation (Dufour effect). According to the Onsager relation we have in this case $C_{12} = C_{21}$. ■

7.6 PHYSICAL PARAMETRIZATIONS

Modeling methods based on physical balance equations are often effective in describing system behavior, although there are always subsystems or cases that require some identification or measurement in order to complete the modeling. In such cases it is advantageous to stick to the context of a physical model when performing parameter estimation. Identification of such physical parameters poses a number of special problems as demonstrated in the following example.

Example 7.6—A model of a DC-motor drive
Consider the motor drive of Fig. 7.7 with the input torque u as the control variable and the angular velocity $y = \dot{q}_1$ as the variable available to measurement. The motor drive with moment of inertia J_1 is coupled to a load *via* a torsion spring with spring constant k. The load is assumed to have a moment of inertia J_2 and a damping d_2. There is also a random load moment v acting on the system. The angular positions of the shafts are designated q_1 and q_2.

$$\begin{aligned} J_1\ddot{q}_1 &= k(q_2 - q_1) + u \\ J_2\ddot{q}_2 &= -k(q_2 - q_1) - d_2\dot{q}_2 + v \end{aligned} \tag{7.36}$$

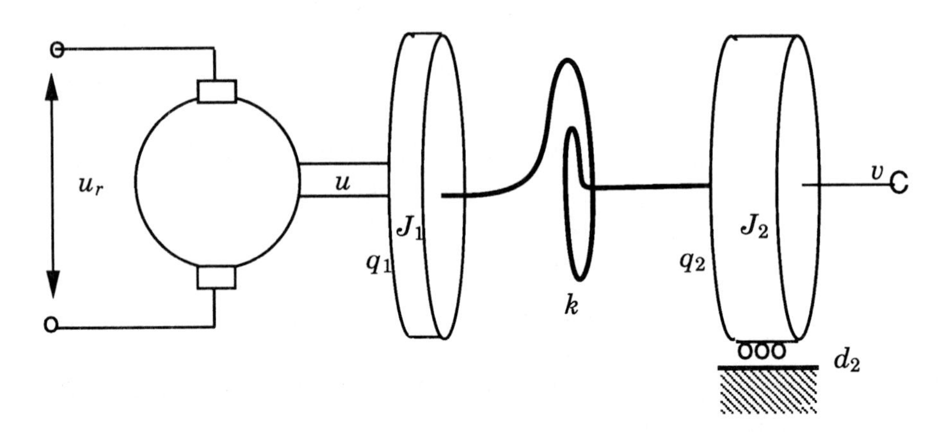

Figure 7.7 A DC motor and a load J_2 with a flexible coupling modeled by a torsion spring with a spring constant k.

For modeling purposes we introduce the state vector

$$\begin{pmatrix} x_1 \\ x_2 \\ x_3 \end{pmatrix} = \begin{pmatrix} \dot{q}_1 \\ \dot{q}_2 \\ q_2 - q_1 \end{pmatrix} \tag{7.37}$$

which evolves according to the system equations

$$\begin{aligned} \frac{dx(t)}{dt} &= \begin{pmatrix} 0 & 0 & k/J_1 \\ 0 & -d_2/J_2 & -k/J_2 \\ -1 & 1 & 0 \end{pmatrix} x(t) + \begin{pmatrix} 1/J_1 \\ 0 \\ 0 \end{pmatrix} u(t) - \begin{pmatrix} 0 \\ 1/J_1 \\ 0 \end{pmatrix} v(t) \\ y(t) &= \begin{pmatrix} 1 & 0 & 0 \end{pmatrix} x(t) \end{aligned} \tag{7.38}$$

with the transfer function

$$\mathcal{S}: \qquad G_u(s) = \frac{Y(s)}{U(s)} = \frac{J_2 s^2 + d_2 s + k}{J_1 J_2 s^3 + J_1 d_2 s^2 + k(J_1 + J_2)s + k d_2} \tag{7.39}$$

The physical parameters in the simulation model were $J_1 = 0.1$, $J_2 = 0.1$, $k = 10$, $d_2 = 1$, which leads to the discrete-time transfer function

$$\mathcal{S}: \qquad H_u(z) = \frac{B(z)}{A(z)} = \frac{0.848z^2 - 0.627z + 0.313}{z^3 - 0.986z^2 + 0.888z - 0.368} \tag{7.40}$$

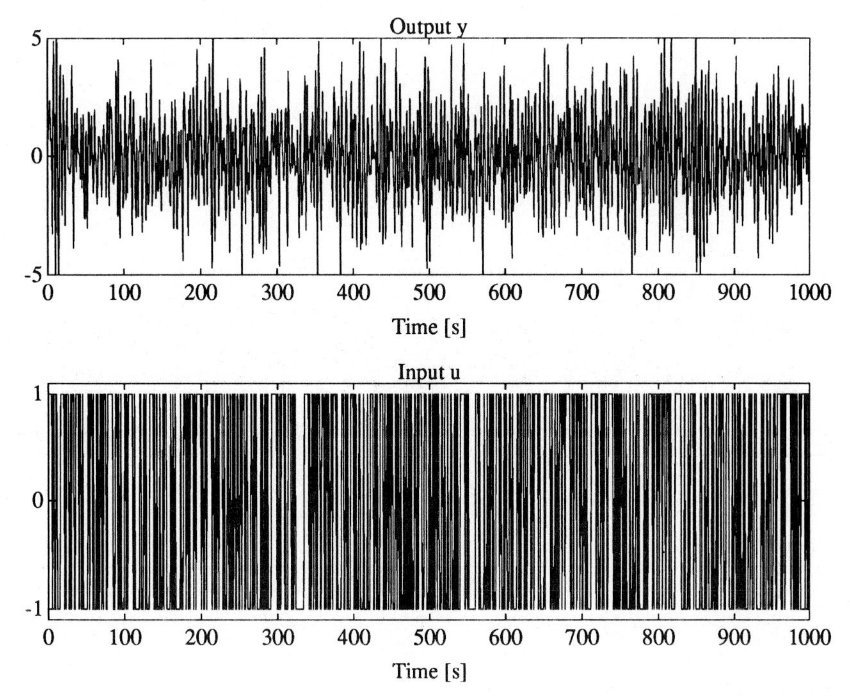

Figure 7.8 Data from a DC-motor drive with input u and output y as described in Example 7.6

in cases where the input u is constant over each sampling interval h. Assume the noise model to be expressed thus:

$$H_v(z) = \frac{C(z)}{A(z)} = \frac{z^3 - 1.8z^2 + 0.97z}{z^3 - 0.986z^2 + 0.888z - 0.368} \tag{7.41}$$

with a noise variance $\mathcal{E}\{v^2\} = \sigma^2 = 1$ and $\mathcal{E}\{u^2\} = 1$, so that the signal-to-noise ratio of inputs is equal to one. Artificial data obtained by simulation of the system in Eq. (7.40) and Eq. (7.41) when sampled with $h = 0.1$ are shown in Fig. 7.8 and the spectrum analysis of data is shown in Fig. 7.9.

A noteworthy feature is that the continuous-time representations (7.38) and (7.39) both have an advantage over the discrete-time model in that the parameters have a clear physical interpretation. We call such a representation *physical parametrization*. Nevertheless, the parameters are not present in the transfer function in a completely original form but as products, ratios, sums, or other aggregate parameters.

The identifiability of each individual physical parameter therefore depends upon two conditions:

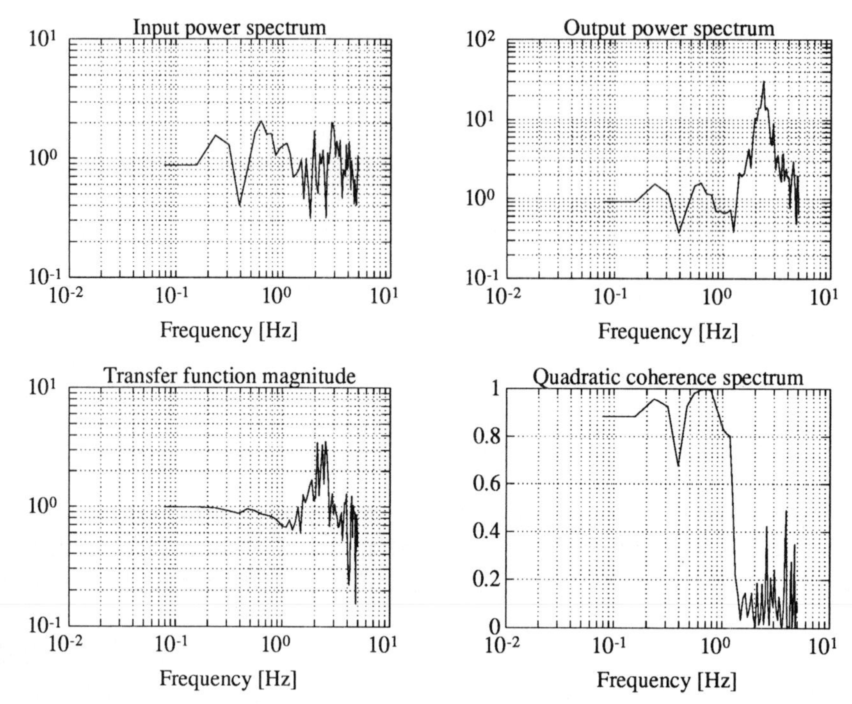

Figure 7.9 Coherence spectrum between input and output of the DC-motor drive of Example 7.6

i. The identifiability of each coefficient of the transfer function

ii. Determination of the physical parameters from the transfer function coefficients

Notice that determination of physical parameters *via* identification of the discrete-time coefficients requires one additional transformation, namely that from a discrete-time model to a continuous-time model. This is a nontrivial problem.

7.7 NETWORK MODELS

As mentioned earlier we distinguish between the potentials (force, voltage, pressure) and flows (displacement, momentum, current, flow rate). The manner in which individual processes interconnect one with another and act upon one another gives an organizational pattern to the overall process. Graphical descriptions are invaluable for network modeling, and it is natural to introduce *nodes* (or *vertices*), which define the connections of the components or

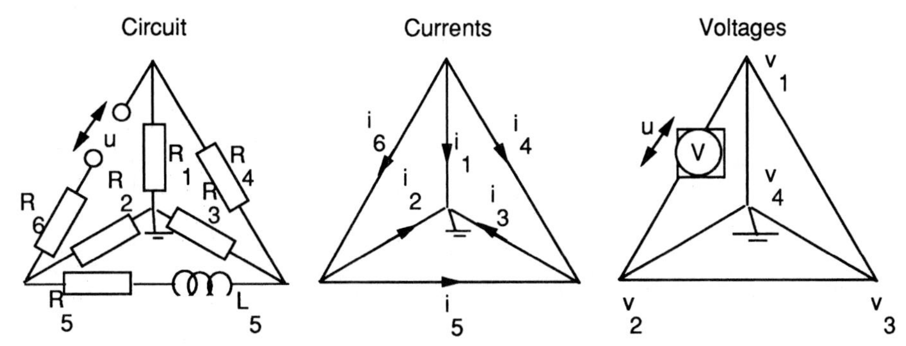

Figure 7.10 A circuit diagram and the corresponding currents and voltages.

subsystems, and *edges* (or *branches*), which always lie between two distinct nodes.

The nodes are associated with some gradient variable, whereas the edge variables are associated with flow variables. In network and graph models it is customary to use the synonyms of *node* variable for potential and *edge* variable for flow. An advantage is that such model descriptions allow analysis by methods of graph theory or linear systems. A determining factor is the difference between potentials that connect the edge variables to the node variables.

Example 7.7—An electrical network

Consider the circuit diagram in Fig. 7.10. The Kirchhoff current law provides the algebraic constraint

$$0 = A^T i = \begin{pmatrix} 1 & 0 & 0 & 1 & 0 & 1 \\ 0 & 1 & 0 & 0 & 1 & -1 \\ 0 & 0 & 1 & -1 & -1 & 0 \end{pmatrix} \begin{pmatrix} i_1 \\ i_2 \\ i_3 \\ i_4 \\ i_5 \\ i_6 \end{pmatrix} \tag{7.42}$$

The matrix A is called the *connectivity matrix* (or *incidence matrix*) because it is determined from the connections of the network. Kirchhoff's voltage law evaluated along the branches of the network provides the equations

$$Av = \begin{pmatrix} 1 & 0 & 0 \\ 0 & 1 & 0 \\ 0 & 0 & 1 \\ 1 & 0 & -1 \\ 0 & 1 & -1 \\ 1 & -1 & 0 \end{pmatrix} \begin{pmatrix} v_1 \\ v_2 \\ v_3 \end{pmatrix} = \begin{pmatrix} i_1 R_1 \\ i_2 R_2 \\ i_3 R_3 \\ i_4 R_4 \\ i_5 R_5 + L_5 \frac{di_5}{dt} \\ i_6 R_6 + u \end{pmatrix} \tag{7.43}$$

or

$$Av = \begin{pmatrix} R_1 & 0 & 0 & 0 & 0 & 0 \\ 0 & R_2 & 0 & 0 & 0 & 0 \\ 0 & 0 & R_3 & 0 & 0 & 0 \\ 0 & 0 & 0 & R_4 & 0 & 0 \\ 0 & 0 & 0 & 0 & R_5 & 0 \\ 0 & 0 & 0 & 0 & 0 & R_6 \end{pmatrix} \begin{pmatrix} i_1 \\ i_2 \\ i_3 \\ i_4 \\ i_5 \\ i_6 \end{pmatrix} + \begin{pmatrix} 0 \\ 0 \\ 0 \\ 0 \\ 0 \\ 1 \end{pmatrix} u + \begin{pmatrix} 0 \\ 0 \\ 0 \\ 0 \\ L_5 \frac{di_5}{dt} \\ 0 \end{pmatrix} \tag{7.44}$$

Let the first term be denoted R_i, the second term $G_u u$, and the third term $L(di/dt)$. These equations can then be organized as follows

$$\begin{pmatrix} L & 0 \\ 0 & 0 \end{pmatrix} \begin{pmatrix} di/dt \\ dv/dt \end{pmatrix} = \begin{pmatrix} -R & A \\ A^T & 0 \end{pmatrix} \begin{pmatrix} i \\ v \end{pmatrix} - \begin{pmatrix} G_u \\ 0 \end{pmatrix} u \tag{7.45}$$

which takes on the familiar form $E\dot{x} = Fx + Gu$ with x as the state variables and u as the external input variable. The network models are thus suitable both for graph theory analysis and for linear system analysis such as state-space models and transfer function models.

*Analysis of $E\dot{x} = Fx + Gu$

Application of the singular value decomposition $U\Sigma V$ to the matrix E gives

$$U\Sigma V\dot{x} = Fx + Gu \tag{7.46}$$

Introducing the transformed variable $z = Vx$ gives

$$\Sigma\dot{z} = U^{-1}FV^{-1}z + U^{-1}Gu = F'z + G'u \tag{7.47}$$

where

$$\Sigma = \begin{pmatrix} \Sigma_1 & 0 \\ 0 & 0 \end{pmatrix} \tag{7.48}$$

The matrix Σ_1 is a full rank diagonal matrix containing the nonzero singular values of E. The rank deficit of E is reflected in the diagonal components of Σ, which suggest the state-space decomposition $z = \begin{pmatrix} z_1^T & z_0^T \end{pmatrix}^T$ and

$$F' = U^{-1}FV^{-1} = \begin{pmatrix} F_{11} & F_{10} \\ F_{01} & F_{00} \end{pmatrix}, \quad \text{and} \quad G' = U^{-1}G = \begin{pmatrix} G_1 \\ G_0 \end{pmatrix} \tag{7.49}$$

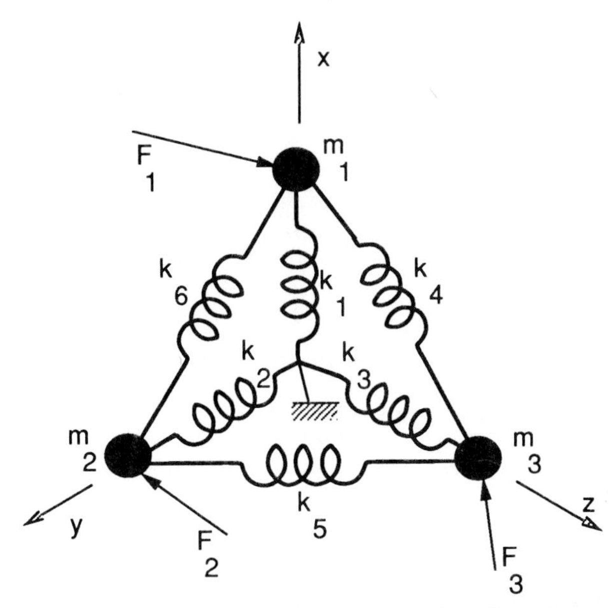

Figure 7.11 A mechanical network with springs on the edges (the spring constants are denoted k_i) and masses (m_j) at the nodes.

so that

$$\Sigma \dot{z} = \begin{pmatrix} \Sigma_1 & 0 \\ 0 & 0 \end{pmatrix} \begin{pmatrix} \dot{z}_1 \\ \dot{z}_0 \end{pmatrix} = \begin{pmatrix} F_{11} & F_{10} \\ F_{01} & F_{00} \end{pmatrix} \begin{pmatrix} z_1 \\ z_0 \end{pmatrix} + \begin{pmatrix} G_1 \\ G_0 \end{pmatrix} u \qquad (7.50)$$

Such a system is sometimes called a *differential-algebraic system* because of the decomposition into an algebraic equation for z_0 and a differential equation for z_1. There is thus a static relationship which determines z_0 as a function of z_1 and the input variable u so that

$$F_{01}z_1 + F_{00}z_0 + G_0 u = 0 \quad \Rightarrow \quad z_0 = -F_{00}^{-1}(F_{01}z_1 + G_0 u) \qquad (7.51)$$

Substitution of z_0 of Eq. (7.51) in Eq. (7.50) gives the unconstrained dynamical system

$$\dot{z}_1 = \Sigma_1^{-1}(F_{11} - F_{10}F_{00}^{-1}F_{01})z_1 + \Sigma_1^{-1}(G_1 - F_{10}F_{00}^{-1}G_0)u \qquad (7.52)$$

which can be further analyzed or simulated with standard methods. The original state-space variable may be recovered as

$$x = V^{-1} \begin{pmatrix} z_1 \\ z_0 \end{pmatrix} = V^{-1}\left(\begin{pmatrix} I \\ -F_{00}^{-1}F_{01} \end{pmatrix} z_1 + \begin{pmatrix} 0 \\ G_0 \end{pmatrix} u \right) \qquad (7.53)$$

■

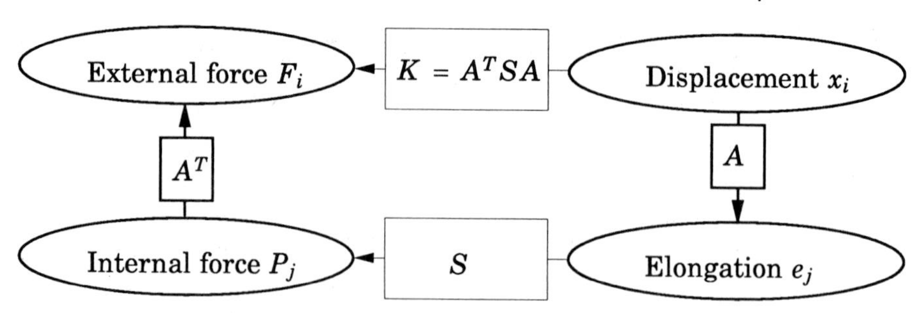

Figure 7.12 Relationships between external forces and displacements in mechanical structures with structural stiffness K.

Structure of the network models

The algebraic structure of the network models offers good opportunities for analysis as exemplified in the following:

Example 7.8—The connectivity matrix and the stiffness matrices
Consider the network in Fig. 7.11 with springs on the edges and masses at the nodes. Forces F_i acting at the network nodes will result in deformation of the structure and the elongation e_j in the jth spring in the three spatial components (x, y, z) is determined by the positions x_i of the masses m_i

$$e = \begin{pmatrix} e_{1x} \\ e_{1y} \\ e_{1z} \\ e_{2x} \\ \vdots \\ e_{6x} \\ e_{6y} \\ e_{6z} \end{pmatrix} = Ax = \begin{pmatrix} I_{3\times3} & 0 & 0 \\ 0 & I_{3\times3} & 0 \\ 0 & 0 & I_{3\times3} \\ I_{3\times3} & 0 & -I_{3\times3} \\ 0 & I_{3\times3} & -I_{3\times3} \\ I_{3\times3} & -I_{3\times3} & 0 \end{pmatrix} \begin{pmatrix} x_{1x} \\ x_{1y} \\ x_{1z} \\ x_{2x} \\ x_{2y} \\ x_{2z} \\ x_{3x} \\ x_{3y} \\ x_{3z} \end{pmatrix} \tag{7.54}$$

where the matrix A is called a connectivity matrix (or incident matrix) because it is determined from the edges of the network. The action of the external forces F_i results in internal forces P_j in the jth spring, and is proportional to the elongations e_j, *i.e.*,

$$P = Se = \begin{pmatrix} k_1 I_{3\times3} & 0 & \cdots & 0 \\ 0 & k_2 I_{3\times3} & \ddots & \vdots \\ \vdots & \ddots & \ddots & 0 \\ 0 & \cdots & 0 & k_6 I_{3\times3} \end{pmatrix} e \tag{7.55}$$

where the diagonal matrix S is composed of stiffness coefficients according to Hooke's law. The work done by the external forces is absorbed by the internal work of stretching the springs so that $x^T F = e^T P = x^T A^T P$ if there is energy conservation in the network. The external forces F in the springs thus balance the internal forces P for

$$F = A^T P = \begin{pmatrix} I_{3\times3} & 0 & 0 & I_{3\times3} & 0 & I_{3\times3} \\ 0 & I_{3\times3} & 0 & 0 & I_{3\times3} & -I_{3\times3} \\ 0 & 0 & I_{3\times3} & -I_{3\times3} & -I_{3\times3} & 0 \end{pmatrix} P \tag{7.56}$$

The potential energy is determined by the expression

$$V = \frac{1}{2} x^T A^T S A x - x^T F \tag{7.57}$$

where the first term is the strain energy in the set of springs, whereas the second term is the potential energy of the external forces F. The kinetic energy of the network is

$$T = \frac{1}{2}\dot{x}^T M \dot{x} = \frac{1}{2}\dot{x}^T \begin{pmatrix} m_1 I_{3\times3} & 0 & \cdots & 0 \\ 0 & m_2 I_{3\times3} & \ddots & \vdots \\ \vdots & \ddots & \ddots & 0 \\ 0 & \cdots & 0 & m_6 I_{3\times3} \end{pmatrix} \dot{x} \tag{7.58}$$

The total energy is then $H = T + V$ and dynamic equilibrium in case of energy conservation is obtained for $0 = dH/dt = (\partial V/\partial x)^T \dot{x} + (\partial T/\partial \dot{x})^T \ddot{x}$ where the gradients are

$$\begin{aligned} \frac{\partial V}{\partial x} &= A^T S A x - F \\ \frac{\partial T}{\partial \dot{x}} &= M \dot{x} \end{aligned} \tag{7.59}$$

The Euler-Lagrange equations then yield the motion equations

$$M\ddot{x} = -\frac{\partial V}{\partial x} = -A^T S A x + F \tag{7.60}$$

The static equilibrium occurs where $\partial V/\partial x = 0$, which is obtained for $F = Kx$ where $K = A^T S A$ is the *structural stiffness matrix*; see Fig. 7.12. The corresponding transfer function relationship between $F(s)$ and $X(s)$ is

$$X(s) = (Ms^2 + K)^{-1} F(s) \tag{7.61}$$

Identification and analysis of the natural frequencies and other properties of the transfer function of Eq. (7.61) are known as *modal analysis* in the domain of mechanical engineering. The eigenvectors and eigenvalues of the generalized eigenvalue problem

$$\det(M\lambda + K) = 0 \tag{7.62}$$

determine the resonance modes $\pm\sqrt{\lambda}$ of the network and provide important information about how to detect and avoid vibration and resonance. Modal analysis also makes use of spectrum analysis; see Chapter 4.

Constraints

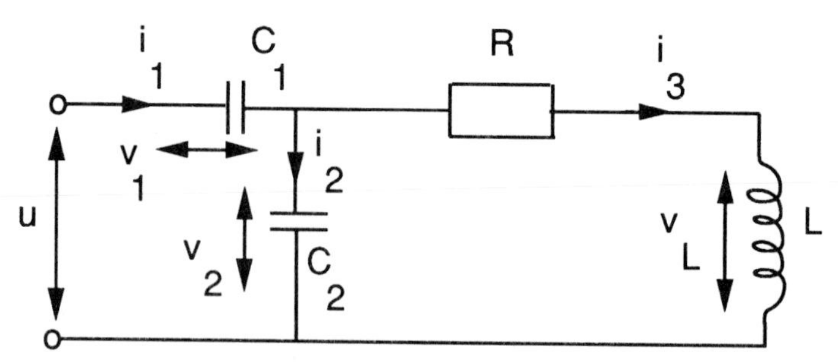

Figure 7.13 A constrained network.

Consider the electrical network in Fig. 7.13. The voltages of capacitors C_1 and C_2 and the current i_3 constitute a dynamical system with certain static constraints.

Kirchhoff's current law	$i_1 = i_2 + i_3$
Kirchhoff's voltage law	$v_2 = Ri_3 + v_L$
Kirchhoff's voltage law	$u = v_1 + v_2$

The balance equations determine a state-space equation with certain static constraints

$$\begin{pmatrix} C_1 & -C_2 & 0 \\ 0 & 0 & L \\ 0 & 0 & 0 \end{pmatrix} \frac{d}{dt} \begin{pmatrix} v_1 \\ v_2 \\ i_3 \end{pmatrix} = \begin{pmatrix} 0 & 0 & 1 \\ 0 & 1 & -R \\ -1 & -1 & 0 \end{pmatrix} \begin{pmatrix} v_1 \\ v_2 \\ i_3 \end{pmatrix} + \begin{pmatrix} 0 \\ 0 \\ u \end{pmatrix} \tag{7.63}$$

This system equation is of the form

$$E\dot{x} = Fx + Gu \tag{7.64}$$

with the matrix E as a singular matrix whereas F is a full-rank matrix. The system (7.63) is an example of a differential-algebraic system with system dynamics determined by a set of differential equations and constrained by algebraic equations. The transfer function from u to i_3 is

$$
\begin{aligned}
G(s) = \frac{I_3(s)}{U(s)} &= \begin{pmatrix} 0 & 0 & 1 \end{pmatrix} (sE - F)^{-1} \begin{pmatrix} 0 \\ 0 \\ 1 \end{pmatrix} = \\
&= \frac{sC_1}{s^2 L(C_1 + C_2) + sR(C_1 + C_2) + 1}
\end{aligned}
\tag{7.65}
$$

which is a second-order linear system, instead of a third-order system, which might be expected in view of the multiplicity of components. In such systems, the question therefore arises as to what variables should be chosen as the independent variables and what variables as the dependent variables. ■

Passivity

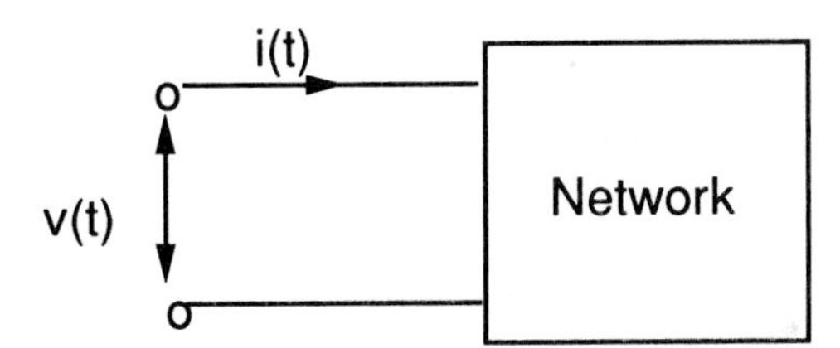

Figure 7.14 A network with the input $v(t)$ and the output $i(t)$.

Consider the network in Fig. 7.14. with input $v(t)$ and output $i(t)$. A *passive network* is characterized by the following relationship between the input $v(t)$ and the output $i(t)$.

$$
E = \int_{-\infty}^{t} i(s)v(s)ds \geq 0, \qquad \forall t \tag{7.66}
$$

The calculated variable (7.66) represents an interaction energy between input and output (see Section 3.7), and this quantity has the physical interpretation of energy in cases where the input is a gradient variable (voltage) and the output a flow variable (current). *Passivity* is a property of resistors because

$$
E_R = \int_{-\infty}^{t} i(s)v(s)ds = \int_{-\infty}^{t} Ri^2(s)ds \geq 0 \tag{7.67}
$$

Passivity is also a property of inductors and capacitors as

$$E_L = \int_{-\infty}^{t} i(s)v(s)ds = \int_{-\infty}^{t} i(s)L\frac{di(s)}{ds}ds = \frac{1}{2}Li^2(t) \geq 0$$

$$E_C = \int_{-\infty}^{t} i(s)v(s)ds = \int_{-\infty}^{t} C\frac{dv(s)}{ds}v(s)ds = \frac{1}{2}Cv^2(t) \geq 0$$

The passivity conditions are one important way of characterizing a system or subsystem as devoid of energy sources while allowing storage elements. Passivity has proved to be an important factor in stability analysis.

7.8 HISTORICAL AND BIBLIOGRAPHICAL REMARKS

Speculation as to the nature of mathematical models has a long history. For instance, the ancient Pythagorean natural philosophy believed that the "real" was the mathematical harmony present in nature. According to this philosophy, mathematical relationships that fit natural phenomena constitute valid explanations of why things are as they are. The alternative view is that mathematical models are computational devices that should be distinguished from theories about physical structure.

Another time-honored issue is model complexity. The fourteenth-century English philosopher William of Occam used simplicity as a criterion of concept formation and modeling. In his view it is desirable to eliminate superfluous concepts so that the simpler of two theories that account for a particular phenomenon is to be preferred. This methodological principle favoring low model complexity is often referred to as "Occam's razor."

Modeling and identification as a methodology dates back to Galileo (1564–1642), who is also important as the founder of dynamics. Modeling of phenomena and comparison with experimental data also stimulated the development of statistics, as a means of accounting for inexact measurement and uncertainty in data. Much successful modeling has thus been done using deterministic modeling of an object and statistical modeling of environmental perturbations. However, the practice of extensive statistical modeling in system identification has occasionally been challenged on the grounds that it might be obstructive to the analysis of trade-offs between approximation and model complexity. It should be borne in mind that a substantial part of statistical terminology can be viewed as a set of modeling assumptions with valuable abstractions such as the notions of uncertainty, probability distribution functions, discrete-time white noise, uncorrelated independent variables with spectra, and autocorrelation functions. White noise in continuous-time modeling is more difficult to treat, and is thus less popular because it presumes an infinite power that is clearly unrealistic and in contradiction to the precepts of standard physical modeling. Another exciting area of research in control theory and statistics is that of *discrete event models*, which are used to model sequence control and manufacturing systems. Another branch of statistical modeling is *game theory*—*e.g.*, differential games that are used to model hostile environments.

The theory of stochastic processes based on the ideas of white noise and Brownian motion is to be found, for instance, in

- R.S. LIPTSER AND A.N. SHIRYAEV, *Statistics of Random Processes. Part I: General Theory. Part II: Applications.*. New York: Springer-Verlag, 1977.

Modeling with the methods of classical mechanics dates back to Newton in the seventeenth and early eighteenth centuries, and is probably the oldest modeling methodology in use. The energy methods of Lagrange, Hamilton, and others date back to the eighteenth century. A good theoretical source is the work:

- R. ABRAHAM AND J.E. MARSDEN, *Foundations of Mechanics* 2d ed., Reading, MA: Addison-Wesley, 1978.

Several modeling techniques including networks, variational methods, and bond graphs are presented in

- P.E. WELLSTEAD, *Introduction to Physical System Modelling*. London: Academic Press, 1979.
- F.E. CELLIER, *Continuous System Modeling*. Berlin and New York: Springer-Verlag, 1990.

Thermodynamic theory and modeling are described in

- F. REIF, *Fundamentals of Statistical and Thermal Physics*. New York: McGraw-Hill, 1965.
- R. HAASE, *Thermodynamics of Irreversible Processes*. Reading, MA: Addison-Wesley, 1969; New York: Dover, 1990.

Variational methods, partial differential equations, approximation theory, and numerical methods are presented in a more academic style in the following work:

– R. DAUTRAY AND J.-L. LIONS, *Analyse Mathématique et Calcul Numerique*. Paris: Masson, 1984.

All practical modeling involves some kind of simulation which provides experience in experimental dynamics. Simulation of such physical processes as airflow over an aircraft wing poses challenges to modeling with many difficulties even in the case with good computation power. A present focus of interest is the chaotic behavior of dynamical systems, which has been discovered in simulation of certain nonlinear differential equations. Special topics in modeling such as chaos and catastrophe theory are to be found in the following books:

– J. GUCKENHEIMER AND P. HOLMES, *Nonlinear Oscillations, Dynamical Systems, and Bifurcations of Vector Fields*. New York: Springer-Verlag, 1983.

– R. THOM, *Stabilité Structurelle et Morphogenèse*. Paris: InterEdition, 1977.

Compartmental models began to be used in the 1940s with the use of tracer experiments in natural sciences, and such models are currently in routine use in pharmacokinetics and ecological sciences.

– C. COBELLI AND K. THOMASETH, "Optimal input design for identification of compartmental models—Theory and application to a model of glucose kinetics". *Math. Biosciences*, Vol. 77, 1985, pp. 267–286.

– J.A. JACQUEZ, *Compartmental Analysis in Biology and Medicine*, New York: Elsevier, 1972.

The technology of modal testing as used in mechanical engineering is covered in

– D.J. EWINS, *Modal Testing: Theory and Practice*, New York: John Wiley, 1984.

7.9 EXERCISES

7.1 Determine the elimination coefficient r in the compartmental model

$$\begin{aligned} \dot{x}(t) &= -rx(t) \\ y(t) &= cx(t) \end{aligned} \tag{7.68}$$

7.2 The serum concentration of an antibiotic drug exhibits the following course after oral intake:

Dose	Time (in hours)								
	0.5	1	1.5	2	3	4	6	8	12
200mg	0.45	0.65	0.55	0.45	0.30	0.25	0.20	0.15	0.10
400mg	0.85	1.25	1.25	1.00	0.70	0.65	0.45	0.30	0.20
600mg	1.45	1.75	1.85	1.45	1.05	0.85	0.55	0.40	0.25

The serum concentration after a 30-minute infusion, as measured from the end of infusion is as follows:

Dose	Time (in minutes and hours)									
	5'	10'	15'	30'	45'	1	2	4	8	12
25mg	0.75	0.60	0.55	0.40	0.35	0.30	0.20	0.15	0.05	0.05
50mg	1.55	1.20	1.05	0.85	0.65	0.60	0.45	0.30	0.15	0.05

The maximal serum concentration after an oral dose is thus with good approximation proportional to the dose given (see Fig. 7.15). Formulate a compartmental model that reproduces the above data.

7.3 Consider the following relationship (sometimes called a *logistic curve*)

$$y(u) = x(u) + \varepsilon = \frac{1}{1 + \exp(-(\alpha + \beta u))} + \varepsilon \tag{7.69}$$

between the measured variables u and y. The disturbance ε with the statistic mean $\mathcal{E}\{\varepsilon\} = 0$ is not available to measure. Show that the transformation

$$\log \frac{x}{1 - x} \tag{7.70}$$

is helpful in order to formulate the estimation of α and β from experimental data y, u as a linear regression problem.

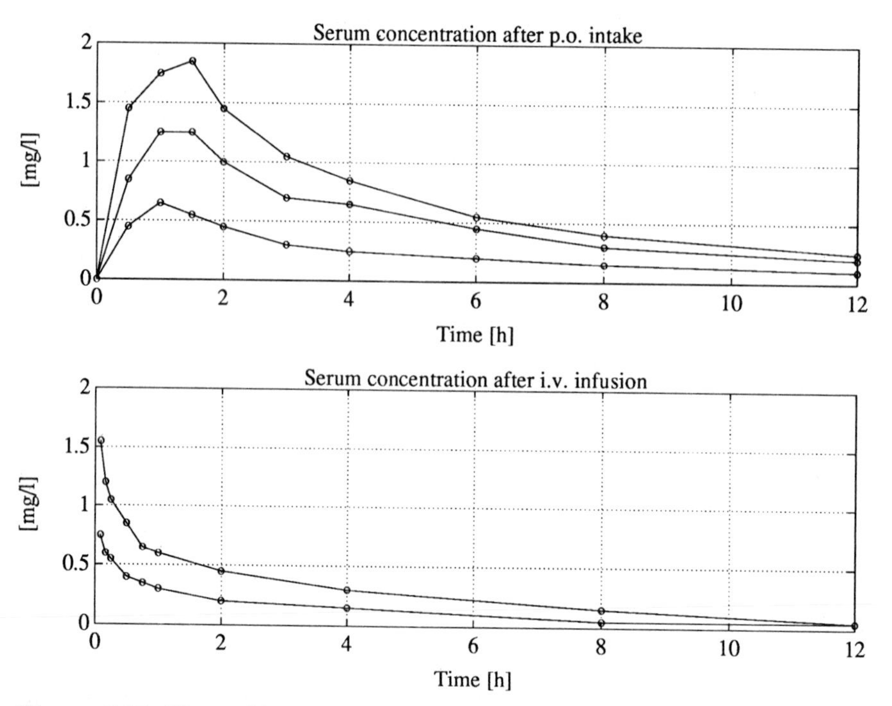

Figure 7.15 Upper diagram shows the serum concentration after oral intake of the drug (200mg, 400mg, and 600mg). Lower graphs show the serum concentration after an i.v. infusion (25mg and 50mg). All graphs versus time.

7.4 A model of a water tank with a cross sectional area A and an outlet area a is given by

$$A\frac{dh}{dt} = q_{in} - a\sqrt{2gh}$$

where q_{in} is the inflow, and h the water level.

a. Assume that there is no inflow, i.e., $q_{in} = 0$. Show that $h(t)$ is then given by

$$h(t) = h_0(1 - \frac{t}{T})^2, \quad 0 \le t \le T$$

where h_0 is the water level at time $t = 0$, and $T = \frac{A}{a}\sqrt{\frac{2h_0}{g}}$.

b. Devise a model of the form $y(t) = \varphi(t)\theta$ for the estimation of $\theta = T$.

c. Assume that $h_0 = 10$. Make a least-squares estimation of T based on the following data:

t	1	2	3	4
$h(t)$	8.9	7.4	6.3	5.5

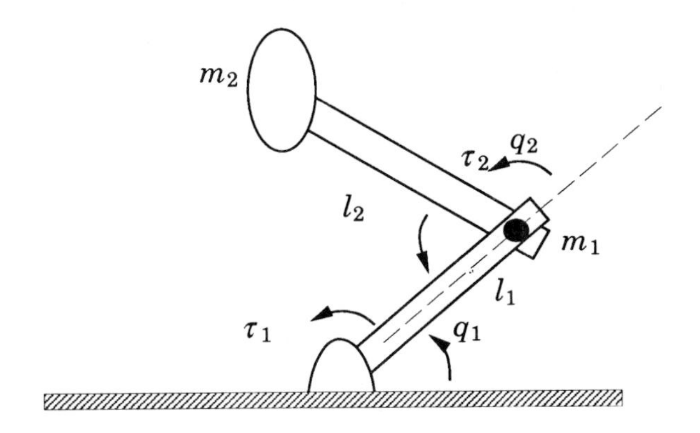

Figure 7.16 The two-link robot in Example 7.5.

7.5

The rigid body dynamics for a two link robot (see Fig. 7.16) are given by

$$\begin{aligned}\tau_1 &= m_2 l_2^2(\ddot{q}_1 + \ddot{q}_2) + m_2 l_1 l_2 c_2(2\ddot{q}_1 + \ddot{q}_2) + (m_1 + m_2) l_1^2 \ddot{q}_1 - m_2 l_1 l_2 s_2 \dot{q}_2^2 \\ &\quad - 2 m_2 l_1 l_2 s_2 \dot{q}_1 \dot{q}_2 + m_2 l_2 g c_{12} + (m_1 + m_2) l_1 g c_1 \\ \tau_2 &= m_2 l_1 l_2 c_2 \ddot{q}_1 + m_2 l_1 l_2 s_2 \dot{q}_1^2 + m_2 l_2 g c_{12} + m_2 l_2^2(\ddot{q}_1 + \ddot{q}_2)\end{aligned} \tag{7.71}$$

where m_i and l_i are the masses and the lengths of the links, and the notation $c_i = \cos(q_i)$, $s_i = \sin(q_i)$, $c_{12} = cos(q_1 + q_2)$, etc. is used.

a. Assume that $\ddot{q}_i$, $\dot{q}_i$, q_i, and τ_i are measurable, and that l_1 and l_2 are known. Devise a model for identification of the parameters m_1 and m_2.

b. Devise an identification model for the case where only $\dot{q}_i$, q_i, and τ_i are measurable.

c. Is it possible to identify the parameters l_1, l_2, m_1, and m_2? Assume that the measurable variables are as in *a*.

7.6 The dissolved oxygen dynamics in an open aerator of an activated sludge system are given by

$$\dot{y}(t) = \alpha(t)u(t)\left(c(t) - y(t)\right) - R(t) \tag{7.72}$$

where y is the dissolved oxygen concentration, R the respiration rate, αu the oxygen transfer rate, and c is a time-varying coefficient. A typical goal is to estimate α and R on-line during control of y *via* the input u. To simplify the problem, consider only discrete time estimation of R.

a. Suppose that the sampling period h is chosen such that α, u, c, and R can be regarded as constant between the sampling instants. Devise a model of the form

$$\dot{y}(kh) = \alpha(kh)u(kh)\left(c(kh) - y(kh)\right) - R(kh) \tag{7.73}$$

where $\dot{y}(kh)$ can be computed exactly from $y(kh)$, $y(kh+h)$, $\alpha(kh)$, and $u(kh)$.

b. Compare the formula for computation of $\dot{y}(kh)$ with a forward Euler approximation, *i.e.*, where $\dot{y}(kh) \approx (y(kh+h) - y(kh))/h$. Discuss the choice of sampling period for the two cases.

7.7

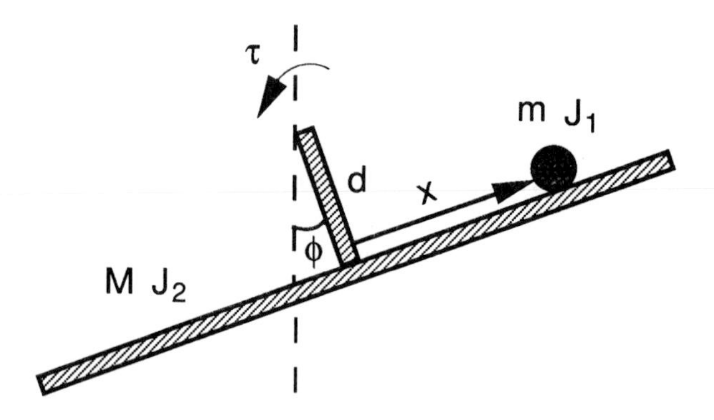

Figure 7.17 The ball and beam process in Exercise 7.7.

For the ball and beam process shown in Fig. 7.17,

a. Devise a dynamical model for the system. The mass of the ball is m and the moment of inertia $J_1 = \alpha m r^2$ where r is the radius of the ball. The mass of the beam is M and the moment of inertia J_2. The applied torque is denoted τ.

b. Formulate a model for the identification of m. Assume that x, $\dot{x}$, φ, and τ are measurable, and that M, J_2, and d are known.

7.8 In contrast to other network models it appears that compartment models are described only in terms of the transfer of material between the compartments. In the terminology of network analysis such transfer corresponds to a set of flow variables $J = \dot{x} = Ax$ with a matrix A containing the transfer coefficients. Show that one can associate to the compartment model a potential $V(x) = x^T P x$ with $P = P^T > 0$ and a set of gradient variables $F = -\partial V/\partial x$ and determine conditions for which $F^T J \geq 0$. ■

8

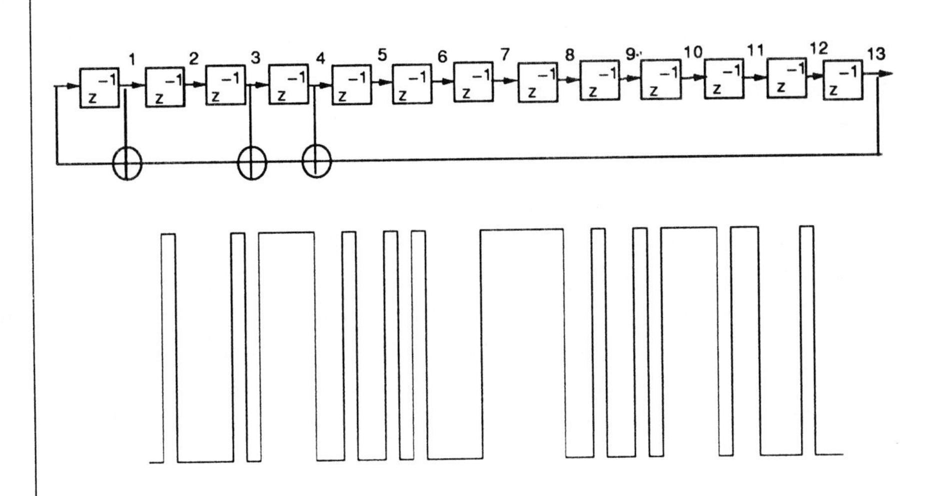

The Experimental Procedure

8.1 INTRODUCTION

This chapter discusses principles and methods to guide the experimental procedure. Problems of the experimental condition with respect to the choice of input and problems concerning identification of systems in closed-loop operation are treated. The end of this chapter gives some practical advice as to the planning of experiments.

8.2 THE EXPERIMENTAL CONDITION

There are several conditions that have to be imposed to obtain good experimental results and data that fulfill the prerequisites of the identification methods used. The choice of input is obviously of great importance for the outcome of the identification. Another important aspect is the presence of regulators in closed-loop operation and the complexity of such regulators. A major part of identification theory considers time-invariant methods for open-loop systems. There are much fewer methods for analysis of closed-loop systems and very few methods apply to identification of adaptive systems. A third aspect is the presence of subsystems in discrete-time operation, a fact that may result in interference with discrete-time measurement and data collection. All method prerequisites and circumstances of the identification procedures that have to be fulfilled at the time of the experiment are called the *experimental condition.* An objective of this chapter is to give some hints and guidelines as to the appropriate design of experiments.

8.3 IDENTIFICATION AND CLOSED-LOOP CONTROL

There are several interactions between identification and control of importance for the result of identification. For example, a variable time delay between measurement and control will result in variable phase delay which, in turn, may result in a variable model order of the identified model. A necessary requirement is therefore that there is a well-defined synchronization between discrete-time measurement and control.

Example 8.1—A system operating with feedback control
Consider a system according to Fig. 8.1 with the following signal relations

$$\begin{cases} x_k = H_P(z)u_k \\ y_k = x_k + v_k \end{cases} \tag{8.1}$$

with input u, disturbance v, and output y. Assume that the regulator set point is zero. The control variable is thus determined from

$$u_k = -H_R(z)y_k \tag{8.2}$$

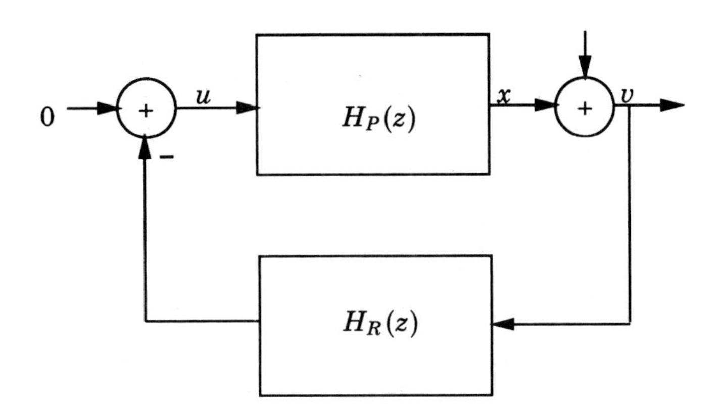

Figure 8.1 A system with feedback control.

and the resulting closed-loop system is

$$(1 + H_P(z)H_R(z))y_k = v_k \tag{8.3}$$

A simple but uninteresting relation between u and y can be obtained from the regulator equation (8.2)

$$y_k = -\frac{1}{H_R(z)}u_k \tag{8.4}$$

A conclusion is that there are potential problems of non-uniqueness for closed-loop identification with spectral estimation methods and other methods that do not respect causality. ■

Example 8.2—Time-series analysis of a closed-loop system
Consider a system described by the equation

$$y_{k+1} = -ay_k + bu_k + v_{k+1} \tag{8.5}$$

Assume that the regulator is a simple proportional controller with

$$u_k = -Ky_k \qquad \Rightarrow \qquad u_k + Ky_k = 0 \tag{8.6}$$

It is, of course, possible to add to (8.5) the zero-valued term $\lambda(u_k + Ky_k) = 0$ where λ is some arbitrary Lagrange multiplier so that

$$y_{k+1} = -ay_k + bu_k + v_{k+1} + \lambda(u_k + Ky_k) \tag{8.7}$$

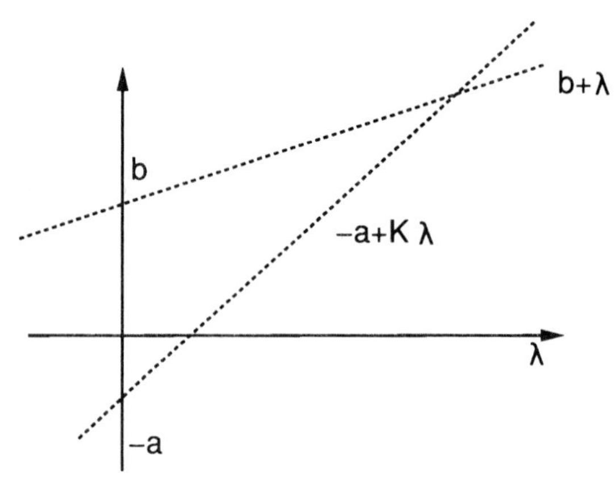

Figure 8.2 Subspaces of ARX-parameters with identified parameters on the ordinate as functions of the original parameters a and b and some arbitrary multiplier λ.

By collecting terms we obtain

$$y_{k+1} = (-a + K\lambda)y_k + (b + \lambda)u_k + v_{k+1} \tag{8.8}$$

The system equation (8.8) is thus no longer characterized by some unique system parameters a, b. Instead, any set of coefficients $-a + K\lambda$ and $b + \lambda$, where λ is arbitrary, is an adequate set of parameters (see Fig. 8.2).

An important observation is that it is not sufficient to know the regulator parameters and that the parametric model does not admit any unique solution even if the control law is explicitly known. ■

Example 8.3—Identification in a closed-loop system
Assume that the system dynamics is described by the equation

$$y_k + ay_{k-1} = bu_{k-1} + e_k \tag{8.9}$$

with the "white-noise" properties

$$E\{e_k\} = 0; \qquad E\{e_i e_j^T\} = \delta_{i,j} \tag{8.10}$$

and the regulator

$$u_k = -Ky_k \tag{8.11}$$

The transfer function between input u and output y may be computed as

$$H_P(e^{i\omega h}) = \frac{S_{yy}(i\omega)}{S_{uy}(i\omega)} \tag{8.12}$$

where the autospectrum S_{yy} and the cross spectrum S_{uy} are

$$\begin{aligned} S_{yy}(i\omega) &= \left|\frac{1}{1+(a+bK)e^{i\omega}}\right|^2 \\ S_{uy}(i\omega) &= \frac{1}{1+(a+bK)e^{-i\omega}} \cdot \frac{K}{1+(a+bK)e^{i\omega}} \end{aligned} \tag{8.13}$$

The estimated transfer function is

$$\widehat{H}_P(e^{i\omega h}) = \frac{\widehat{S}_{yy}(i\omega)}{\widehat{S}_{uy}(i\omega)} = \frac{1}{K} \tag{8.14}$$

The same conclusion as in Examples 8.1 and 8.2 applies to this example with problems of a calculated transfer function that does not estimate the control object. Instead, the estimated transfer function reflects the properties of the regulator and gives an estimate of the inverse of the regulator transfer function. ■

The examples given in this section indeed point to several difficult problems in the context of closed-loop control. However, all examples given rely on the fact that the external input u_c (see Fig. 8.3) was equal to zero, and it can be expected that better identifiability would result with a non-zero reference input.

Example 8.4—Identification of a system using a reference signal
Consider a system described by the equation

$$y_{k+1} = -ay_k + bu_k + v_{k+1} \tag{8.15}$$

Assume that the regulator is a simple proportional controller with

$$u_k = -K(y_k + u_c) \tag{8.16}$$

It is in this case no problem to find a sequence of u_c such that the matrix

$$\Phi = \begin{pmatrix} y_0 & u_0 \\ \vdots & \vdots \\ y_{N-1} & u_{N-1} \end{pmatrix} = \begin{pmatrix} y_0 \\ \vdots \\ y_{N-1} \end{pmatrix} \begin{pmatrix} 1 & -K \end{pmatrix} + \begin{pmatrix} 0 & Ku_{c0} \\ \vdots & \vdots \\ 0 & Ku_{cN-1} \end{pmatrix} \tag{8.17}$$

is a full-rank matrix (*i.e.*, rank $(\Phi^T\Phi)=2$) and such that there is a unique solution $\widehat{\theta}$ to the normal equations $(\Phi^T\Phi)\widehat{\theta} - \Phi^T\mathcal{Y} = 0$. ■

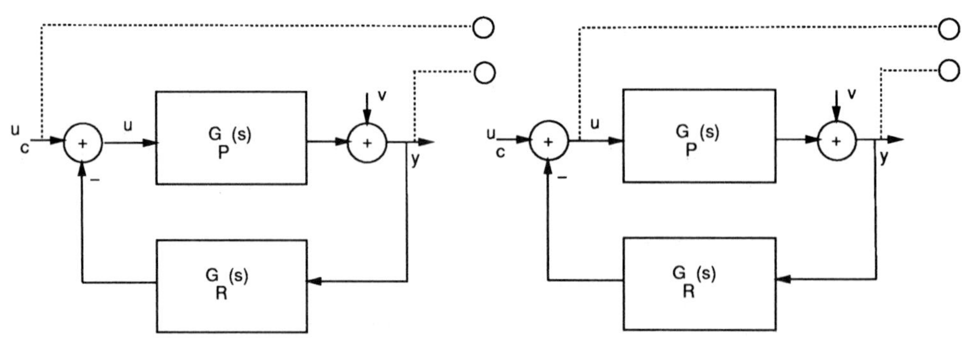

Figure 8.3 Indirect (*left*) and direct identification (*right*).

8.4 DIRECT OR INDIRECT IDENTIFICATION?

Identification of a transfer function between the control input u and the observation y of a system in closed-loop operation is called *direct identification* because the identification provides an estimate of the subsystem between control input and observations of output (see Fig. 8.3).

Identification may also proceed from external (reference) input u_c to the observations y (*indirect identification*) in cases where both the regulator and the closed-loop system are linear systems. If we assume that the feedback transfer function $G_R(s)$ is linear, time-invariant, and known, then we can suggest the procedure

1. Identify the closed-loop system from u_c to y.
2. Compute the open-loop transfer function.

$$\widehat{G}_P(i\omega) = \frac{\widehat{G}(i\omega)}{1 - G_R(i\omega)\widehat{G}(i\omega)} \tag{8.18}$$

Notice that this procedure presupposes a time-invariant regulator G_R which presents some problems as to the application of discrete-time regulators. A condition for indirect identification is therefore that there is well-defined synchronization between measurement and control. ■

Example 8.5—Spectral analysis in closed-loop systems
It is feasible to use indirect identification also in the context of spectral analysis. Consider the following example with a closed-loop operation according to Fig. 8.4. Assume that the noise w and the external input r are uncorrelated and that the control law is

$$U(s) = G_{FF}(s)R(s) - G_R(s)Y(s) \tag{8.19}$$

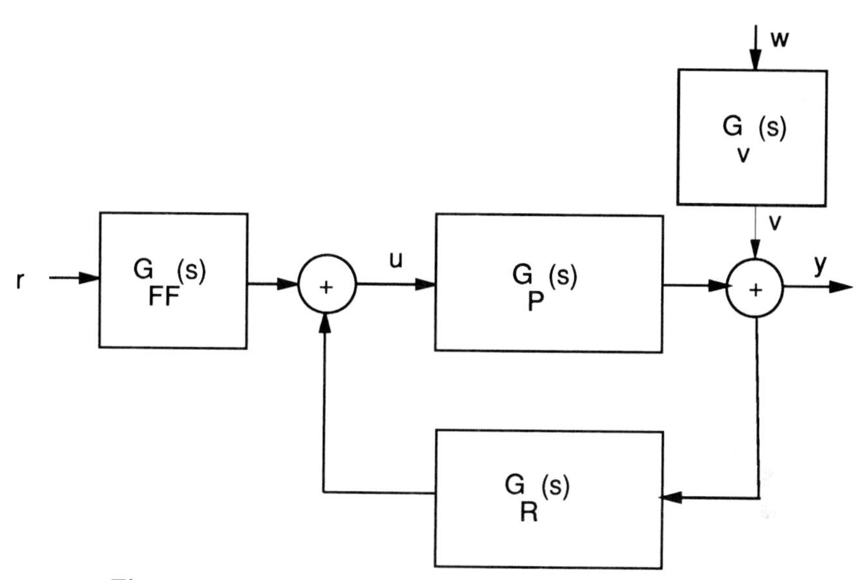

Figure 8.4 Indirect spectral analysis with reference signal.

where $R(s) = \mathcal{L}\{r(t)\}$ and where $U = \mathcal{L}\{u\}$ and $Y = \mathcal{L}\{y\}$ denote the control variable and the system output, respectively. The cross spectrum between the reference variable r and the observed variable y is

$$S_{yr}(i\omega) = G_P(i\omega)S_{ur}(i\omega) + G_v(i\omega)S_{wr}(i\omega) \tag{8.20}$$

If r and w are uncorrelated it holds that

$$S_{yr}(i\omega) = G_P(i\omega)S_{ur}(i\omega) \tag{8.21}$$

The input-output transfer function may thus be estimated as

$$\widehat{G}_P(i\omega) = \frac{\widehat{S}_{yr}(i\omega)}{\widehat{S}_{ur}(i\omega)} \tag{8.22}$$

■

8.5 CHOICE OF INPUT

The result of identification is contingent upon a careful choice of input to the system under investigation. For example, the final result of identification can be evaluated as the sum of residuals obtained by the least-squares method or the maximum-likelihood method. The sum of squared residuals is

$$\sum_{k=1}^{N} \varepsilon_k^2(\widehat{\theta}) = \sum_{k=1}^{N} (y_k - \phi_k^T \widehat{\theta})^2 = \frac{1}{2\pi} \sum_{k=1}^{N} |Y_k - \Phi_k^T(i\omega)\widehat{\theta}|^2 \tag{8.23}$$

where the last equality follows from the Parseval theorem. If we assume that the input-output relation is $Y(s) = G(s)U(s) + V(s)$ and $\widehat{Y}(s) = \widehat{G}(s)U(s)$ and if the noise $V(s)$ is of zero magnitude, then

$$\sum_{k=1}^{N} \varepsilon_k^2(\widehat{\theta}) = \frac{1}{2\pi} \sum_{k=1}^{N} |Y_k - \Phi_k^T(i\omega)\widehat{\theta}|^2 = \frac{1}{2\pi} \sum_{k=1}^{N} |(G(i\omega_k) - \widehat{G}(i\omega_k))U_k|^2 \tag{8.24}$$

for frequency points $\{\omega_k\}$ obtained from the discrete Fourier transform. It is thus obvious that the accuracy of the transfer function estimate is dependent upon the spectrum of the input. Frequency domain methods for design of the input thus consist of choice of a suitable input spectrum. For instance, a suitable input spectrum to be used in system identification for the purpose of control system design is such that there is much input energy (*i.e.*, $|U_k|^2$ is large) for frequencies ω_k around the bandwidth frequency of the investigated system. (This statement requires, of course, further theoretical motivation, which follows in the sensitivity analysis below.) Such an approach is by necessity an iterative method as the system bandwidth is not known *a priori*.

The idea of weighting in the frequency domain may also be applied to filtering of regressors and observations by means of data filters with high transmission at the desired frequency range. Such methods may often be reformulated *via* the Parseval relation as *weighted least-squares methods*.

Apart from frequency domain methods there is obviously a general interest to provide a measure on the input spectrum to determine if the input is of sufficient complexity. One such approach is the criterion of persistent excitation:

Definition 8.1—Persistency of excitation (Åström)
A signal u fulfils the condition of *persistent excitation* of order n if the following limits exist

$$\begin{aligned} \bar{u} &= \lim_{N\to\infty} \frac{1}{N} \sum_{k=1}^{N} u_k \\ \widehat{C}_{uu}(\tau) &= \lim_{N\to\infty} \frac{1}{N} \sum_{k=1}^{N} u_k u_{k-\tau}^T \end{aligned} \tag{8.25}$$

and if the correlation matrix

$$R_{uu}(n) = \begin{pmatrix} \widehat{C}_{uu}(0) & \widehat{C}_{uu}(1) & \cdots & \widehat{C}_{uu}(n-1) \\ \widehat{C}_{uu}(-1) & \widehat{C}_{uu}(0) & \cdots & \widehat{C}_{uu}(n-2) \\ \vdots & \vdots & \ddots & \vdots \\ \widehat{C}_{uu}(1-n) & \widehat{C}_{uu}(2-n) & \cdots & \widehat{C}_{uu}(0) \end{pmatrix} \tag{8.26}$$

is positive definite. ■

Persistent excitation is sufficient to obtain consistent estimates for the least-squares method and maximum-likelihood identification. The condition (8.25) implies a condition on the autocorrelation $C_{uu}(\tau)$ and thus on the autospectrum $S_{uu}(i\omega)$ for at least n different frequencies $\omega_1, \ldots, \omega_n$. In fact, as it is difficult to formulate criteria for experimental verification of persistency of excitation, this definition can be regarded as a stipulative definition of some sufficient conditions for consistency.

Methods for experimental verification of the number of identifiable parameters from a set of data organized as the regressor matrix

$$\Phi_N = \begin{pmatrix} \Phi_y & \Phi_u \end{pmatrix} \in R^{N\times(n_Y+n_U)} \tag{8.27}$$

include singular value decomposition (SVD)

$$\Phi_N^T \Phi_N = U \Sigma V^*, \qquad \Sigma = \text{diag}\begin{pmatrix} \sigma_{11} & \ldots & \sigma_{pp} & 0 & \ldots & 0 \end{pmatrix} \tag{8.28}$$

where the number of non-zero singular values determines the number of parameters for which the normal equations have a unique solution. The SVD may of course also be applied to the matrix $\Phi_u^T \Phi_u$ which, thus, provides a characterization of the input similar to persistency of excitation.

'Optimal' input signals

A relevant question is if there is any 'optimal' input signal—in some meaningful sense—to choose for identification purposes. One approach to the formulation of suitable conditions of optimality is to start by considering the Cramér-Rao lower bound on the parameter covariance

$$\mathcal{E}\{(\widehat{\theta} - \theta)(\widehat{\theta} - \theta)^T\} \geq M^{-1} \tag{8.29}$$

where the Fisher information matrix M is defined as

$$M = \mathcal{E}\{\frac{\partial \log p(\mathcal{Y}|\theta)}{\partial \theta}(\frac{\partial \log p(\mathcal{Y}|\theta)}{\partial \theta})^T\} \tag{8.30}$$

Let us consider the case $\mathcal{Y} = \Phi\theta + v$ with $\mathcal{E}\{v\} = 0$. Assuming that v is normally distributed noise with covariance matrix $\Sigma = \mathcal{E}\{vv^T\} = \sigma^2 I$ this gives

$$\begin{aligned} p(\mathcal{Y}|\theta,\Sigma) &= (2\pi)^{-N/2}(\det\Sigma)^{-1/2}\exp(-\frac{1}{2}(\mathcal{Y}-\Phi\theta)^T\Sigma^{-1}(\mathcal{Y}-\Phi\theta)) = \\ = p(\mathcal{Y}|\theta,\sigma) &= (2\pi)^{-N/2}(\det\sigma^2 I)^{-1/2}\exp(-\frac{1}{2\sigma}(\mathcal{Y}-\Phi\theta)^T(\mathcal{Y}-\Phi\theta)) \end{aligned} \tag{8.31}$$

and

$$\begin{pmatrix} \frac{\partial \log p(\mathcal{Y}|\theta,\sigma^2)}{\partial\theta} \\ \frac{\partial \log p(\mathcal{Y}|\theta,\sigma^2)}{\partial\sigma^2} \end{pmatrix} = \begin{pmatrix} \frac{1}{\sigma^2}(\mathcal{Y}-\Phi\theta)^T\Phi \\ -\frac{N}{2\sigma^2}+\frac{1}{2\sigma^4}(\mathcal{Y}-\Phi\theta)(\mathcal{Y}-\Phi\theta)^T \end{pmatrix} \tag{8.32}$$

Calculation of the Fisher information matrix for normally distributed noise $\Sigma = \sigma^2 I$ gives

$$M = \begin{pmatrix} \frac{1}{\sigma^2}\Phi^T\Phi & 0 \\ 0 & \frac{(N-1)^2}{2\sigma^4} \end{pmatrix} \tag{8.33}$$

An attractive approach is to choose the experiment input $\mathcal{U}$ so that the Cramér-Rao lower bound M^{-1} is minimized in some sense under the input constraint

$$\frac{1}{N}\sum_{k=1}^{N} u_k^2 = 1 \tag{8.34}$$

In order to formulate this as a well-posed optimization problem it is natural to minimize some scalar function or norm of M^{-1} under the constraint (8.34) and to choose the experiment input $\mathcal{U}$ that minimizes this scalar optimization criterion. Consider, for instance, optimization criteria of the type

$$\begin{aligned} J_1(\mathcal{U}) &= \operatorname{tr}(WM^{-1}) \\ J_2(\mathcal{U}) &= -\log\det M \end{aligned} \tag{8.35}$$

where W is some weighting matrix. As M is a function of both the inputs and outputs, it is obviously necessary to know the system—or to actually make the desired experiment—before it can be calculated. This unfortunate problem formulation presupposes data that are not available and the intended optimization of J_1 or J_2 of Eq. (8.35) has to be replaced by iterative or approximate methods. For instance, pseudorandom binary sequences (PRBS) are often close to optimizing the criteria of Eq. (8.35) under the constraint Eq. (8.34).

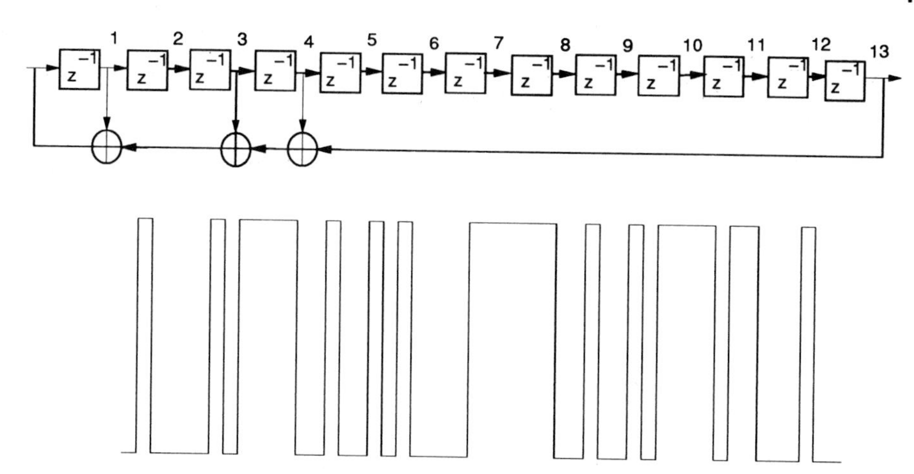

Figure 8.5 A circuit to generate pseudorandom binary sequences with low autocorrelation.

Example 8.6—How to generate pseudorandom binary sequences
One motivation is the need for special test signals with a given amplitude and with low autocorrelation and with a period that can be chosen arbitrarily long.

Such test signals can be accomplished by using feedback shift registers according to Fig. 8.5. These circuits operate in discrete time and the registers take on values 0 or 1. The addition operation $\oplus$ in the circuit is modulo-two-addition according to which

$$x \oplus y = \begin{cases} 0, & \text{if } x = 0, \quad y = 0 \\ 1, & \text{if } x = 1, \quad y = 0 \\ 1, & \text{if } x = 0, \quad y = 1 \\ 0, & \text{if } x = 1, \quad y = 1 \end{cases} \tag{8.36}$$

The interconnections and the feedback in the circuit are conveniently described by polynomials

$$P(z^{-1}) = 1 \oplus z^{-1} \oplus z^{-3} \oplus z^{-4} \oplus z^{-13} \tag{8.37}$$

where the polynomial degree is equal to the number of states in the shift register (*cf.* Fig. 8.5). Polynomials that give long sequences of low autocorrelation are generated by irreducible polynomials, *i.e.*, polynomials which cannot be divided by any other polynomial of degree greater than 0.

The interconnections of the circuit may also be described by an octal number or a binary number (*cf.* Fig. 8.5).

Power of 2	14	13	12	11	10	9	8	7	6	5	4	3	2	1	0
Coefficients															
Binary	0	1	0	0	0	0	0	0	0	0	1	1	0	1	1
Octal		2			0			0			3			3	

The following numbers describe in octal numbers the interconnections for some suitable (*i.e.*, irreducible) polynomials that generate PRBS.

n	Octal code
10	2011
11	4055
12	10123
13	20033
14	42103
15	100003
16	210013
17	400011
18	1000201
19	2000047
20	4000011

As the circuit used for generation of the PRBS operates without external inputs and because there is a finite number of states it is obvious that the PRBS generator will provide a periodic output. The period of the output depends upon n, *i.e.*, the number of shift register stages of the circuit. The period of the resulting autocorrelation function is

$$\text{Period} = (2^n - 1)h \tag{8.38}$$

which indicates how n should be chosen in order to provide a long enough sequence of low autocorrelation. ■

Example 8.7—Fitting of a weighting function

As an example of design of an optimizing input we consider the case of a finite-time impulse response described by a transfer function

$$H(z) = b_1 z^{-1} + \cdots + b_n z^{-n} \tag{8.39}$$

with the unknown coefficients

$$\theta = \begin{pmatrix} b_1 & \dots & b_n \end{pmatrix}^T \tag{8.40}$$

Let

$$U_i = \begin{pmatrix} u_{n-i} & \dots & u_{N-i} \end{pmatrix}^T, \qquad i = 0, 1, \dots, n \tag{8.41}$$

The linear regression problem $\mathcal{Y} = \Phi\widehat{\theta} + V$ may then be formulated with

$$\Phi = \begin{pmatrix} u_n & \dots & u_0 \\ u_{n+1} & \dots & u_1 \\ \vdots & & \vdots \\ u_N & \dots & u_{N-n} \end{pmatrix} = \begin{pmatrix} U_0 & \dots & U_n \end{pmatrix} \tag{8.42}$$

and

$$\Phi^T\Phi = \begin{pmatrix} U_0^T U_0 & \dots & U_0^T U_n \\ \vdots & & \vdots \\ U_n^T U_0 & \dots & U_n^T U_n \end{pmatrix} \tag{8.43}$$

Let the optimization problem be formulated as

$$\begin{aligned} \text{minimize} \quad & J_2(\mathcal{U}) = -\log\det M = -\log\det\Phi^T\Phi + \text{const} \\ \text{subject to} \quad & \frac{1}{N}\sum_{i=0}^{n} U_i^T U_i = 1 \end{aligned} \tag{8.44}$$

This problem has a solution—see proof below—which specifies that all eigenvalues of the matrix $\Phi^T\Phi$ should be equal in order to allow for an optimal solution to Eq. (8.44). A pseudorandom binary sequence with $U_i^T U_j \approx (N-n)\sigma_u^2\delta_{ij}$ for some magnitude σ_u thus provides a good approximate solution to the problem of finding an optimal input.

Proof: The magnitude constraint $\sum_{i=0}^{n} U_i^T U_i/N = 1$ can be reformulated to the equivalent condition

$$\mathrm{tr}(\Phi^T\Phi) = N \tag{8.45}$$

where *tr* denotes the matrix trace (see Appendix A). Now denote the eigenvalues of $\Phi^T\Phi$ by $\lambda_1, \dots, \lambda_n$ where all $\lambda_i > 0$. Then follows (except for a neglected constant term in J_2) that

$$\begin{aligned} J_2 &= -\log\det\Phi^T\Phi = -\log\prod_{i=1}^{n}\lambda_i = -\sum_{i=1}^{n}\log\lambda_i \\ N &= \mathrm{tr}(\Phi^T\Phi) = \sum_{i=1}^{n}\lambda_i \end{aligned} \tag{8.46}$$

The optimization criterion J_2 may thus be replaced by

$$J(\lambda_1, \dots, \lambda_m) = -\sum_{i=1}^{n}\log\lambda_i + \mu\Big(\sum_{i=1}^{n}\lambda_i - N\Big) \tag{8.47}$$

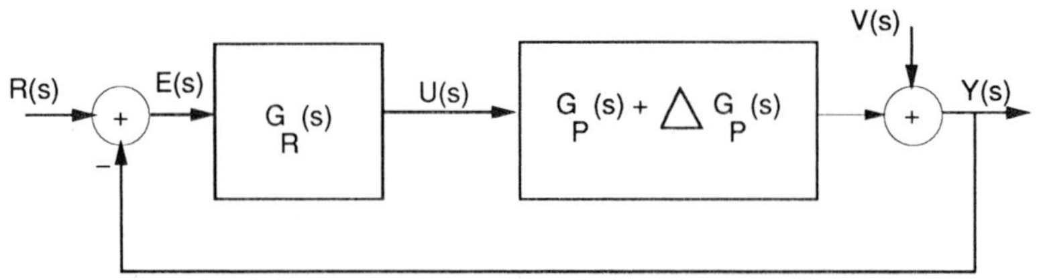

Figure 8.6 Transfer function uncertainty $\Delta G_P(s)$ of a control object in closed-loop control with the regulator $G_R(s)$.

where the constraint of Eq. (8.46) is adjoint by means of the Lagrange multiplier μ. Conditions for an extremum can be evaluated by means of the gradient

$$0 = \frac{\partial J}{\partial \lambda_i} = -\frac{1}{\lambda_i} + \mu, \quad \Rightarrow \quad \lambda_i = \frac{1}{\mu}, \quad i = 1, 2, \ldots, n \tag{8.48}$$

By substituting $\lambda_i = 1/\mu$ into J and evaluating the extremum with respect to μ, one finds the optimal $\mu = n/N$ and, thus, the optimal $\lambda_i = N/n$, $i = 1, 2, \ldots, n$. Finally, evaluation of $\partial^2 J/\partial\lambda_i\partial\lambda_j$ shows that the extremum is a minimum. ■

8.6 PARAMETER UNCERTAINTY AND CONTROL

The significance of parameter uncertainty in the case of regulator design is reflected in the concepts *gain margin* and *phase margin* which in terms of transfer function gain and phase describe the possible parameter variation or parameter uncertainty of a control object for which the closed-loop system maintains stability. Assume that the transfer function uncertainty may be described by

$$G_P(s) + \Delta G_P(s) = (I + L(s))G_P(s) \tag{8.49}$$

The transfer function uncertainty $L(s)$ expressed in Eq. (8.49) and in Fig. 8.6 is sometimes called an *unstructured multiplicative uncertainty* which attracts much interest in the domain of H^∞-robust control. Assume that the parameter uncertainty $L(s)$ can be characterized in the frequency domain by some bound $m(\omega)$ so that

$$\bar{\sigma}[L(i\omega)] \leq m(\omega) \tag{8.50}$$

where $\bar{\sigma}[L(i\omega)]$ denotes the maximum singular value for all ω of the multiplicative uncertainty $L(i\omega)$. Furthermore, let $S(s)$ denote the *sensitivity function*

$$S(s) = \frac{E(s)}{V(s)} = (I + G_P(s)G_R(s))^{-1} \tag{8.51}$$

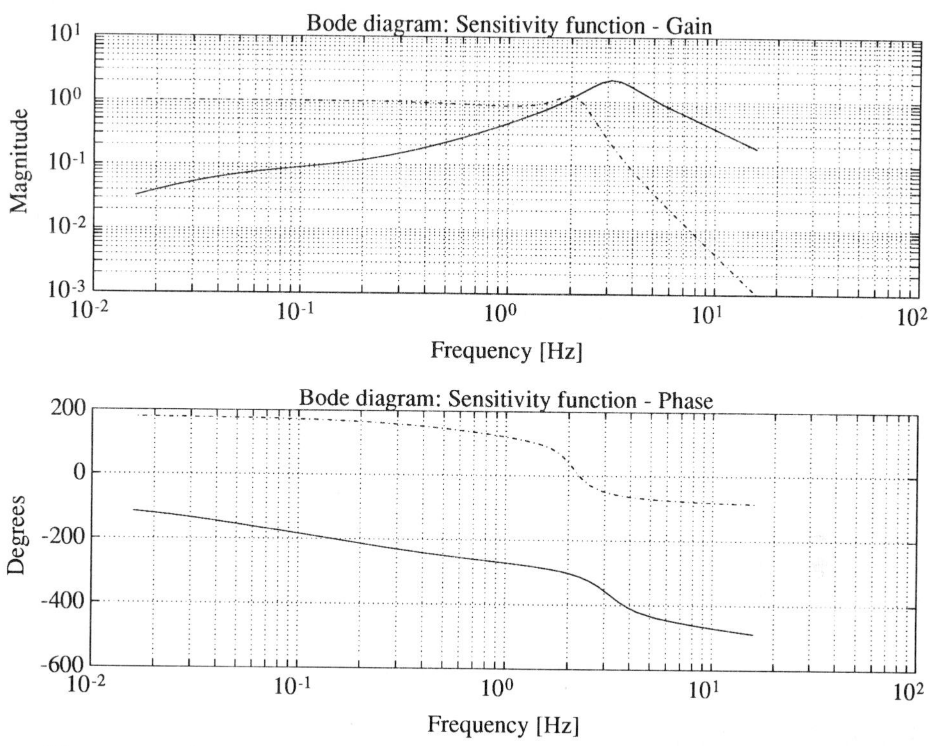

Figure 8.7 Sensitivity function $S(s) = E(s)/V(s)$ for the open-loop system (*dashed line*) and for the same system controlled by a PID-regulator (*solid line*) for the system (7.38). Notice that the sensitivity of the PID-controlled system is low in the low-frequency range and is highest around the bandwidth frequency where the closed-loop system is even more sensitive than the open-loop system.

which describes the transfer function between disturbance and output error. Let the *complementary sensitivity function* be defined as

$$T(s) = G_P(s)G_R(s)(I + G_P(s)G_R(s))^{-1} \tag{8.52}$$

Small errors in the presence of the reference signal $r(t)$ and measurement noise $v(t)$ are obtained if $S(s)$ is made "small" as the error $E(s)$ depends upon $S(s)V(s)$ and $S(s)R(s)$. In addition, stability of the closed-loop system is maintained in the presence of all possible uncertainties characterized by Eq. (8.50) if and only if the complementary sensitivity function satisfies

$$\bar{\sigma}[T(i\omega)] \leq \frac{1}{m(\omega)} \tag{8.53}$$

for all ω. An objective of control design is to minimize the sensitivity $S(s)$ to parameter uncertainties and parameter variations. This may be achieved by choosing the regulator transfer function $G_R(s)$ according to the following rules:

R1: Make $S(i\omega)$ small whenever $R(i\omega)$ or $V(i\omega)$ is large

R2: Keep $T(i\omega)$ small whenever $m(\omega)$ is large

The objective of identification is here to reduce the uncertainty bound $m(\omega)$ as much as possible. A frequency domain signal weighting is valuable in order to obtain a favorable accuracy distribution in the frequency domain.

Example 8.8—The sensitivity function of a motor drive
Consider the motor drive in Example 7.6 with the following system equations

$$\begin{aligned} \frac{dx(t)}{dt} &= \begin{pmatrix} 0 & 0 & k/J_1 \\ 0 & -d_2/J_2 & -k/J_2 \\ -1 & 1 & 0 \end{pmatrix} x(t) + \begin{pmatrix} 1/J_1 \\ 0 \\ 0 \end{pmatrix} u(t) - \begin{pmatrix} 0 \\ 1/J_1 \\ 0 \end{pmatrix} v(t) \\ y(t) &= \begin{pmatrix} 1 & 0 & 0 \end{pmatrix} x(t) \end{aligned} \tag{7.38}$$

The sensitivity function (*i.e.*, the transfer function from disturbance $V(s)$ to the error $E(s)$ according to Eq. (8.51)) of the open-loop system is shown in Fig. 8.7 (dotted line) and the sensitivity function of a closed-loop system controlled by a PID-regulator with state feedback and integral action

$$\begin{aligned} U(s) &= -\begin{pmatrix} 10 & 25 & -25 \end{pmatrix} X(s) + \frac{1}{s} X_1(s) \\ &\quad - \begin{pmatrix} 10 & 25 & -25 \end{pmatrix} X(s) - 10 Q_1(s) \end{aligned} \tag{8.54}$$

is shown in solid line. It is clear from the Bode diagrams that the PID-regulator causes the sensitivity to decrease for lower frequencies whereas it might even increase in the frequency range around the bandwidth of the system. For control systems analysis and design it is thus natural to develop dynamic models with good accuracy in intermediate frequency ranges. ■

A closer investigation shows that it is a good identification strategy to make $m(\omega)$ small in the frequency region where the sensitivity is large, *i.e.*, where $|G_P(i\omega)G_R(i\omega)| \approx 1$ and $\arg G_P(i\omega)G_R(i\omega) \approx -\pi$. One approach is to use this information to make a weighted least-squares solution that minimizes the Bode diagram uncertainty at the frequency ranges of high sensitivity. The sensitivity information can also be used for design of an input with much energy around the bandwidth frequency.

8.7 PLANNING AND OPERATION OF EXPERIMENTS

As some information is required to formulate good and precise experiments it is natural and often necessary to make iterated experiments. A standard procedure of identification is to start with some *first stage experiments* with simple experiments followed by continued experiments (*second stage experiments*) with careful motivations as to the choice of instruments, method prerequisites, choice of experimental inputs, and experiment duration. The sufficiency of the experiments made and the termination of the experimental procedure are usually determined by *model validation criteria* and/or by identification of major sources of error that may put limitations on the result of system identification. Details of these considerations are presented in the following sections.

First stage experiments

It is valuable to clarify and state the *purpose of the model* at the very first stage of identification. The requirements of model accuracy are different when the model is to be used, for instance, in control design, simulator design, fault detection, or simply for process knowledge. Also, the *prior knowledge* is valuable at early stages of identification as it gives clues to model complexity, error sources, time-variation of dynamics, and other important conditions.

The *practical considerations* involved are operating conditions of instruments, actuators, possible system limitations, and information about operation in closed-loop control.

Simple experiments such as logging of data during normal operation, step disturbances, step changes of available inputs, and impulse disturbances are natural initial experiments. Correlation analysis of signals from available inputs and outputs is often a good method to determine basic dependencies between the measured variables.

Evaluation of the simple experiments should focus on a basic qualitative understanding of the system under consideration. In particular, it is valuable to investigate the simple causal relations between available inputs and outputs. This can be done with coherence analysis to determine the presence of nonlinearities as well as disturbance magnitudes and spectral properties such as periodic disturbances. Also, to what extent a linear model may be assumed and in what frequency ranges such a model would be valid can be based on coherence analysis.

Dominating time constants in the output responses and low-frequency noise can be evaluated from the impulse and step response tests. Also, the problems of nonstationary, input-output relationships deriving from time-varying dynamics of the system under investigation or drift of instruments can often be detected at this stage of identification. In addition, system time delays and multivariable causal relations may often be determined from the simple impulse and step response experiments and from coherence analysis.

It should be borne in mind that operational records and other process documentation may reveal anomalities and undesirable events such as manual interventions or recalibration of measurement devices during experiments. The presence of such events may, if undetected, create great confusion in the interpretation of results.

Planning of the second stage experiment session

Before proceeding to the second phase of experiments it is valuable to make decisions concerning the sampling interval, signal filtering, and desirable signal levels with respect to nonlinearities, noise levels, and physical limitations and saturations. Also, operational problems such as manual interventions and calibration problems should be considered. It is important to interact with personnel operating the systems under investigation, to explain the purpose and needs of identification, and to discuss the experience obtained.

Continued experiments

The second stage of experiments is characterized by a systematic investigation of the subsystems involved in the identification and with design and execution of suitable experiments. A natural starting point is the instrument properties which involve accuracy, dynamics, noise, trends, calibration, and analog-to-digital converters. A second point of concern is actuator properties such as accuracy, dynamics, and limitations.

The first stage experiments may have suggested the use of certain identification methods. The necessary *method prerequisites* and other conditions necessary for application of available methods should be considered at this stage. The following tests are applicable to all parametric modeling.

Test of linearity can be approached by estimating models for different input amplitudes. Also, the symmetry of response for inputs of the same form but of opposite signs is valuable. Coherence tests may also be used to detect linearities in the frequency domain.

Test of time invariance can be approached *ad hoc* by evaluation of time-variant properties from recordings obtained at different occasions. The presence of trends in the data, however, is often a benign time variation that may be avoided with trend elimination.

Test of disturbance conditions and noise properties should focus on a characterization of the noise independent of the input. *Artifacts* such as abnormal data resulting from lost signals, outliers, and aliasing should be detected. Finally, *test of the experiment condition* involves an investigation to assure that input is correct and contains sufficient excitation with questions as to the presence of feedback and possible interference due to discrete-time control.

Choice of input

The *form* of the input is sometimes determined by the method, *e.g.*, with sinusoids as used in frequency response analysis, step functions and impulses in transient analysis, or PRBS in correlation analysis. The choice of input for model-based methods relies on criteria of sufficient excitation for determination of the spectral range of validity.

Amplitude of the input is chosen as a trade-off between signal-to-noise ratio and nonlinearities. A large amplitude should be avoided when fitting a linear model so that the test signal does not enter a nonlinear region of behavior. However, the amplitude should be large enough to ensure a good signal-to-noise ratio. A statistical motivation for a large amplitude is that the attainable parameter accuracy is

$$\text{Cov}\{\widehat{\theta}\} \quad \text{proportional to} \quad \frac{1}{\text{input power}} \tag{8.55}$$

A good rule of thumb for a minimum amplitude to choose is that the effect of the input on the output in a diagram need at least be perceptible to the eye.

Frequency domain characteristics of test signals are that the input should be sufficiently exciting with an autospectrum $S_u(i\omega)$ that is not "too small." In addition, the estimated input spectrum $\widehat{G}_u(i\omega)$ must not be "too small."

$$\widehat{G}_u(i\omega) = \frac{\widehat{S}_{yu}(i\omega)}{\widehat{S}_{uu}(i\omega)} \tag{8.56}$$

An idea that is seldom possible to realize is the choice of an "optimal input" which generally requires a good knowledge of the transfer funtion that should be identified.

Experiment duration

The sampling frequency can only be chosen appropriately if there exists some knowledge about the dynamics of the system. An important problem is to choose the sampling interval to avoid aliasing and interference with periodic discrete-time control. For a given sampling interval the "long" time constants will appear as integrators in the model or—simply—as trends in data, whereas the "short" time constants may remain undetected. The experiment duration also affects both the attainable parameter accuracy

$$\mathrm{Cov}\{\hat{\theta}\} \quad \text{proportional to} \quad \frac{1}{\text{experiment duration}} \tag{8.57}$$

and the spectral resolution. A reasonable rule of thumb is that the experiment duration need be chosen $\geq 5-10\times$ the longest time constant to be determined in the identification. Should the estimatated variance not decrease as the experiment duration increases, one might suspect that the parameter estimates are not statistically consistent. It should also be borne in mind that one experiment seldom results in knowledge of the transfer function over a wider frequency range than two or three decades. Finally, the experiment duration is also a trade-off between data economy and experiment economy.

As a general conclusion it might be stated that the purpose of the model affects the choice of experiment condition. Also, the experiment conditions should be chosen similar to those under which the model is intended to be used.

8.8 BIBLIOGRAPHY AND REFERENCES

Sensitivity functions and H^∞-robust control are described in

- G. STEIN AND M. ATHANS, "The LQG/LTR procedure for multivariable feedback control design." *IEEE Trans.Autom. Control, TAC-32*, 1987, pp. 105–114.
- J.M. MACIEJOWSKI, *Multivariable Feedback Design*. Reading, MA: Addisor Wesley, 1989.

An approach to identification theory directed toward support of robust control system design is found in

- G.C. GOODWIN, M. GEVERS, AND B. NINNESS, "Quantifying the error in estimated transfer functions with application to model order selection." *Trans. Automatic Control, Vol. 37*, 1992, pp. 913-928.

Properties of identification in closed-loop operation are described in

- P.E. WELLSTEAD AND J.M. EDMUNDS, "Least-squares identification of closed loop systems." *Int. J. Control, Vol. 21*, 1975, pp. 689–699.

– I. Gustafsson, L. Ljung, and T. Söderström, "Identification of processes in closed-loop — identification and accuracy aspects." *Automatica*, Vol. 13, 1977, pp. 59–75.

– B.D.O. Anderson and M.Gevers, "Representations of jointly stationary stochastic feedback processes", *Int. J. Control, Vol. 33*, pp. 777-809, 1981.

Several methods of identification of time-series are described in

– K.J. Åström "Maximum-likelihood and prediction error methods." *Automatica*, Vol. 16, 1980, pp. 551–57.,

Generation of pseudorandom binary sequences by polynomial methods are described in

– W.W. Peterson, *Error Correcting Codes*. New York: John Wiley, 1961.

8.9 EXERCISES

8.1 Excitation signals—Sinusoids

The transfer function of an unknown system is to be measured using a sinusoidal test signal. Both the creation of the test signal and the measurement are done using a computer. The sampled behavior of the computer will make the test signal differ from a perfect sinusoid. Determine how many samples N are needed per period in order to make the signal close to a sinusoid. Interpret *close* as

a. the energy in the strongest harmonic frequency component being less than 1% of the energy at the fundamental frequency.

b. having more than 99% of its energy at the fundamental frequency.

8.2 Excitation signals—White noise

One problem with using white (broad-band) noise as excitation signal is its amplitude. There is no guarantee that the signal level will be within fixed limits, and it is easy to get saturation of the process input signal. Suppose the input signal to a certain process is limited to the amplitude range -1.0 – 1.0. The input signal is chosen as a normally distributed white-noise sequence with zero mean and variance σ^2. What is the largest σ^2 one can tolerate if the risk for input signal saturation should be less than 0.01?

8.3 Excitation signals – pseudo random binary sequence (PRBS)

PRBS is an easily generated signal with almost "white" properties. A PRBS is generated using feedback around a shift register. The length of the shift register determines the period of the sequence. A length N register results in a sequence with period $2^N - 1$.

a. The sequences generated by shift registers of length 2 and 3, respectively, look as follows

$$\begin{aligned} &\text{length } 2: \quad , \underbrace{1, -1, 1,} \underbrace{1, -1, 1,} \\ &\text{length } 3: \quad , \underbrace{1, -1, -1, 1, 1, 1, -1,} \underbrace{1, -1, 1, 1, 1, 1, -1,} \end{aligned} \tag{8.58}$$

Determine the autocovariance functions $C_2(\tau)$ and $C_3(\tau)$ for the two sequences. What do you think $C_N(\tau)$ looks like?

Remark: Although the PRBS is a deterministic signal it is still possible to define the covariance function as

$$\begin{aligned} \mu &= \lim_{N\to\infty} \frac{1}{N} \sum_{t=1}^{N} u(t) \\ C_{uu}(\tau) &= \lim_{N\to\infty} \frac{1}{N} \sum_{t=1}^{N} (u(t+\tau) - m)(u(t) - m)^T \end{aligned} \tag{8.59}$$

where μ is the mean value. It is also possible to change the PRBS to a stochastic signal by introducing a stochastic phase (i.e. the signal form is known but the starting point in the sequence is unknown).

b. Determine the autospectra $S_2(i\omega)$ and $S_3(i\omega)$ of the two sequences in **a**. What do you think $S_N(i\omega)$ looks like?

8.4 If the PRBS in Exercise 8.3 is to be used as excitation signal it has to be fed out from the computer through the digital-to-analog converter. How does that change the spectrum of the signal?

8.5 Coherence function

The coherence function γ_{yu} between signals u and y is defined as

$$\gamma_{yu}(\omega) = \frac{|S_{yu}(i\omega)|}{\sqrt{S_{uu}(i\omega)S_{yy}(i\omega)}}. \tag{8.60}$$

Suppose we are going to identify the system G in Fig. 8.8. The input signal is u, the output signal is y, and n is a noise signal. Express S_{yu}

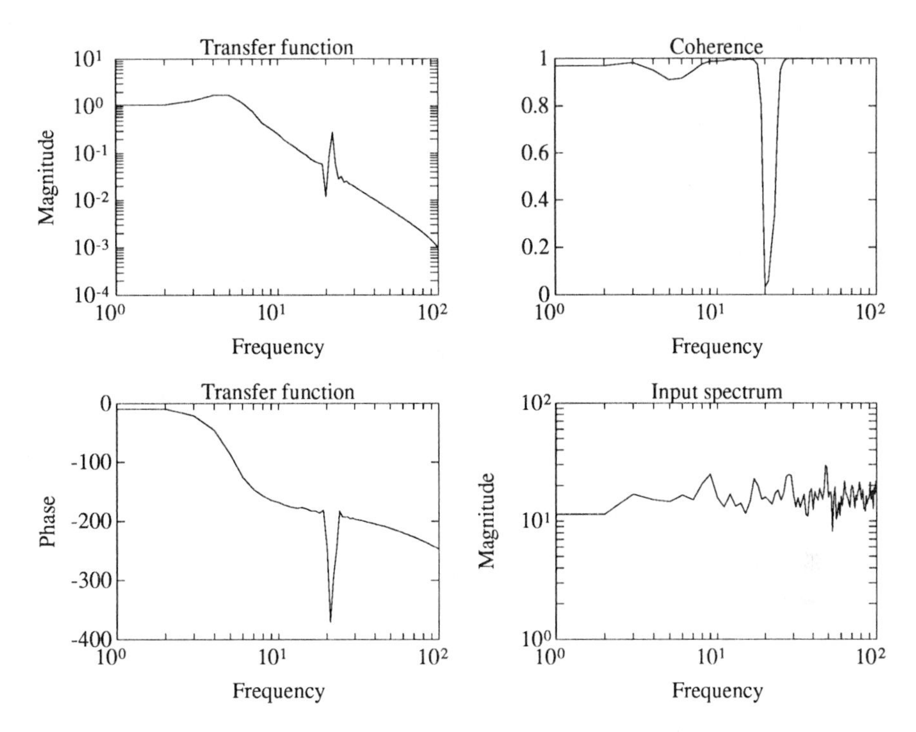

Figure 8.9 Transfer function and coherence function in Exercise 8.6.

in terms of G, S_{uu}, and S_{nn}, such that it is obvious that $\gamma_{yu}(\omega)$ can be used to judge if the excitation signal is sufficient for identification.

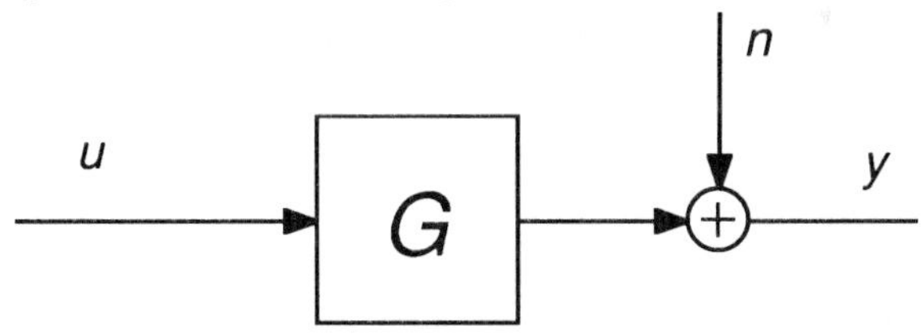

Figure 8.8 Identification experiment in Exercise 8.5.

8.6 Consider an identification experiment where a system was excited with a broad-band input signal. The transfer function was calculated using the spectra of the input u and output y. To be able to judge the quality of the estimate the coherence function $\gamma_{yu}(\omega)$ was calculated. Both the transfer function and the coherence function are shown in Fig. 8.9. Using this diagram it was concluded that the system is of low pass type, and that it has some sort of resonance at the frequency 20. Is this conclusion correct?

8.7 Consider the system

$$y(t) = b_1 u(t-1) + b_2 u(t-2) + b_3 u(t-3). \tag{8.61}$$

Determine if it is possible to identify all three parameters in the model by means of the input $u(t) = \sin(t)$.

8.8 Consider the system

$$\mathcal{S}: \qquad y_k + a y_{k-1} = b u_{k-1} + w_k + c w_{k-1} \tag{8.62}$$

where $\{w_k\}$ is a zero-mean white-noise process with the variance $\mathcal{E}\{w_k^2\} = \sigma^2$. Assume that the input $u_k = -K y_k$ so that identifiability of a and b is lost. What are the asymptotic parameter estimates for large N when fitting the model

$$\mathcal{M}: \qquad y_k = \alpha y_{k-1} \tag{8.63}$$

with data from $\mathcal{S}$.

8.9 Consider the system

$$\mathcal{S}: \qquad y_k + a y_{k-1} = b u_{k-1} + w_k \tag{8.64}$$

where $\{w_k\}$ is zero-mean white-noise with $\mathcal{E}\{w_i w_j\} = \sigma^2 \delta_{ij}$ and $u_k = 1$ for all $k \geq 0$. Show that the input is persistently exciting of order 1 but not of order 2 and determine under what circumstances the least-squares estimate of a, b might give consistent estimates.

8.10 Consider the system

$$\mathcal{S}: \qquad y_k + a y_{k-1} = w_k + c w_{k-1}, \qquad w_k \in \mathcal{N}(0, \sigma^2), \quad \mathcal{E}\{w_i w_j\} = \sigma^2 \delta_{ij} \tag{8.65}$$

where $a = -0.9$ and $c = 0.7$. How many samples would be required in order to achieve the variance 0.0001 of $\widehat{a}$ when fitting the ARMA model

$$\mathcal{M}: \qquad y_k + a y_{k-1} = w_k + c w_{k-1} \tag{8.66}$$

to data generated by the system (8.65).

8.11 Assume that the production of a continuous-flow fermentation process can be modeled by the equations

$$\begin{aligned} \dot{x} &= \mu x - i_{in} x \\ \dot{s} &= -R \mu x + i_{in}(s_{in} - s) \end{aligned} \tag{8.67}$$

where x is the produced biomass, s substrate concentration, s_{in} influent substrate concentration, i_{in} influent flow rate, R yield coefficient, μ specific growth rate. What identification problems can be anticipated and how should these be circumvented? *Hint:* This is model of growth. ■

9

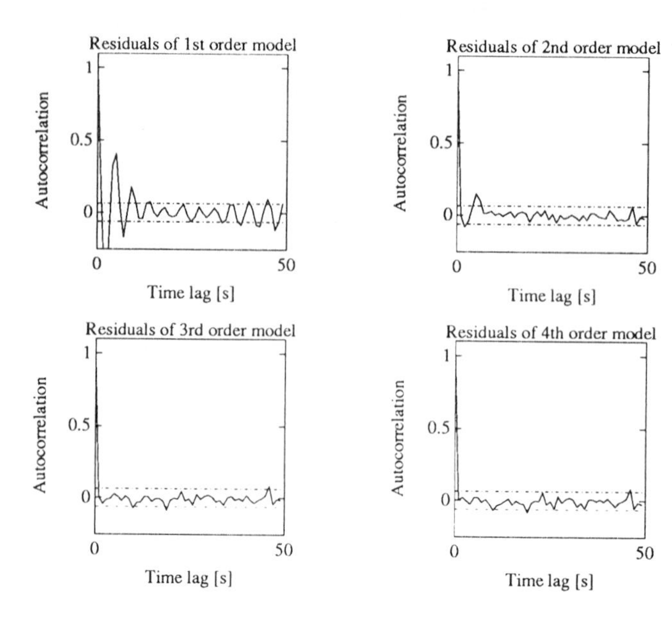

Model Validation

9.1 INTRODUCTION

The purpose of model validation is to verify that the identified model fulfills the modeling requirements according to subjective and objective criteria of good model approximation. A statistical approach not only involves hypothesis testing as to the complexity and model order of an estimated model, but also classification of models with equal model orders.

It is usually a major objective to obtain a model of least possible complexity within the limits of required model accuracy. In particular, it is necessary to distinguish between the lack of fit between model and data due to random processes, and that due to lack of model complexity. Mean square convergence is often a very attractive goal of validation because it unifies a statistical approach with a modeling and approximation approach. Hitherto, however, validation criteria have not quite accomplished this goal, and it is therefore not possible to advocate any such approach as orthodox procedure. Instead, many existing validation methods rely on various statistical assumptions and on statistical tests based upon such assumptions. For instance, model order tests are relevant for all identification of parametrized models and statistical decision criteria have been developed which take into consideration a number of aspects:

- Decrease of loss function as the number N of observations increases (F-test).
- Modified loss functions that penalize both model misfit and high model order. Examples of this type of criteria are the AIC, FPE, and MDL criteria presented in this chapter.
- Redundant parameters appearing in the estimated linear models which may suggest that the chosen model order is too high.
- Pole-zero cancellations and common factors appearing in transfer functions of linear systems.
- *A priori* knowledge and physical considerations which may suggest a certain model order.
- Model error and residual analysis.

The residuals represent misfit between data and model, and the presence of any information remaining in the residuals is a clue that the model might be insufficiently complex or otherwise inappropriate. Residual analysis comprises tests of such factors as:

- Independence of residuals
- Normal distribution of residuals
- Zero crossings (changes of sign) of the residual sequence
- Correlations between residuals and input

Subsequent to validation by means of residual analysis, it is necessary to evaluate the accuracy of the parameter estimates, for instance from the covariance matrix or by simulation. It is also relevant to check the following issues:

- Does the variance decrease as N increases? If the estimated variance fails to decrease as the number of measurements grows, it may indicate that the estimate is not statistically efficient;

- Is the parameter accuracy sufficient for the purpose of the model?
- Does stochastic simulation verify that the estimated model behaves qualitatively as expected?
- Does cross validation with data that have not been previously used verify that the estimated model is able to predict the behavior? This is often a very revealing test to pinpoint model anomalies.

Several validation methods and procedures are described in this chapter, in as systematic a presentation as possible. The first section covers some necessary background for meaningful validation, *i.e.*, that the method prerequisites are met. The role of coherence tests for examination of the signal conditions is described in Section 9.2. Structure determination by means of statistical tests is introduced in Section 9.3. Validation tests based on residuals are treated in some detail in Section 9.4 with the emphasis on tests of the assumed white-noise properties. Problems of model accuracy with some attention to physical parametrizations are discussed in Section 9.5. Finally, some methods for classification of the outcome of parameter estimation are presented.

9.2 METHOD PREREQUISITES

As model validation is predicated upon a correctly performed experimental procedure, it is necessary to confirm that the method prerequisites are fulfilled. Some of the relevant tests have been already mentioned in the context of experiment planning. Obviously, it is also a problem of validation to verify that these objectives have been met and that the method prerequisites allow for meaningful parameter estimation. An important check is to verify that the input has been correct and of sufficient excitation. Excitation properties can be investigated either by testing whether the persistent excitation criteria have been met, or by considering the autospectra of the input signals. An input spectrum with a non-zero level over a large spectral range generally ensures suitable experimental conditions offering good properties of identification. Prior knowledge about the presence of feedback is also important to document. As a whole, these tests provide a satisfactory check of the experimental conditions.

Elementary signal conditions and the impact of various artifacts are necessary to consider at this stage, and problems due to outliers, aliasing, or lost signals should be detected and circumvented. Another source of problems is interference from discrete-time control in some subsystem with non-harmonic distortion similar to aliasing.

Linearity can be tested visually by regarding signal behavior for different input amplitudes and by considering the symmetry in response to negative and positive input. The coherence spectrum is also valuable as a test of linearity, as explained in detail below.

Time-variant properties of the system under investigation are sometimes obvious from recordings obtained at different occasions. Simple time-variant properties such as trends can be elimininated by numerical differentiation of data or by subtracting some polynomial function of time fitted by linear regression. Another source of time-variant behavior is the effect of a non-zero initial condition with low damping. It should be borne in mind that an effect of low-frequency dynamics corresponding to a longer period than the measurement time may present itself as a trend. In such situations it is sometimes a task of identification to identify the low-frequency dynamics behind such trend behavior.

However, the use of trend elimination sometimes calls for certain precautions and bookkeeping in order to allow for determination of equilibrium points, static gain, and for data restoration.

Tests of disturbance conditions and noise properties can be made visually by examining records of data. It is usually valuable to determine whether noise is independent of input.

Coherence spectrum and test of linearity

The coherence spectrum was presented in Section 3.9 as a measure of dependence between two signals. An important use of the coherence spectrum is its application as a test of signal-to-noise ratio and linearity between one or more input variables and an output variable. Assume that there is some linear multi-input single-output relationship

$$Y(s) = G_0(s)U(s) + V(s) \tag{9.1}$$

that relates the output Y with the input vector U and some disturbance input V uncorrelated with U. Assume that the output y of a system has an autospectrum $S_{yy}(i\omega)$. The corresponding input u is assumed to have the autospectrum $S_{uu}(i\omega)$ and the cross spectrum between input and output is $S_{yu}(i\omega)$. The *(quadratic) coherence spectrum* between input and output is defined as

$$\Gamma_{yu}(i\omega) = \frac{S_{yu}(i\omega)S_{uu}^{-1}(i\omega)S_{uy}(i\omega)}{S_{yy}(i\omega)} \tag{9.2}$$

where the autospectrum of the output is

$$S_{yy}(i\omega) = G_0(i\omega)S_{uu}(i\omega)G_0^T(-i\omega) + S_{vv}(i\omega) \tag{9.3}$$

and the cross spectrum between input and output is

$$S_{yu}(i\omega) = G_0(i\omega)S_{uu}(i\omega) \tag{9.4}$$

The (quadratic) coherence spectrum is then

$$\Gamma_{yu}(i\omega) = \frac{S_{yu}(i\omega)S_{uu}^{-1}(i\omega)S_{uy}(i\omega)}{S_{yy}(i\omega)} = \frac{G_0(i\omega)S_{uu}(i\omega)G_0^T(-i\omega)}{G_0(i\omega)S_{uu}(i\omega)G_0^T(-i\omega) + S_{vv}(i\omega)} \tag{9.5}$$

The coherence function expresses the degree of linear correlation in the frequency domain between the input u and the output y. It should be stressed that there is no immediate counterpart in the time domain—*i.e.*, the inverse Laplace transform of Γ_{yu} has no interpretation. In a special case with no disturbances where $S_{vv} = 0$ it holds that

$$\Gamma_{yu}(i\omega) = \frac{S_{yu}(i\omega)S_{uu}^{-1}(i\omega)S_{uy}(i\omega)}{S_{yy}(i\omega)} = \frac{G_0(i\omega)S_{uu}(i\omega)G_0^T(-i\omega)}{G_0(i\omega)S_{uu}(i\omega)G_0^T(-i\omega)} = 1 \tag{9.6}$$

The coherence function may thus be viewed as a type of correlation function in the frequency domain where a coherence function not equal to 1 indicates the presence of one or more of the following:

- Disturbance affecting y
- Input not represented by u
- Nonlinearity, so that there is no linear relationship between u and y, *i.e.*, no transfer function between u and y
- Non-zero initial values with low damping.

A coherence test is valuable as a test of linearity and of the effect of disturbance at an early stage of identification. The empirical coherence function is then computed as

$$\widehat{\Gamma}_{yu}(i\omega) = \frac{\widehat{S}_{yu}(i\omega)\widehat{S}_{uu}^{-1}(i\omega)\widehat{S}_{uy}(i\omega)}{\widehat{S}_{yy}(i\omega)} \tag{9.7}$$

A value of coherence close to 1 usually gives promise of successful identification. It also indicates in what frequency ranges there is a good approximation

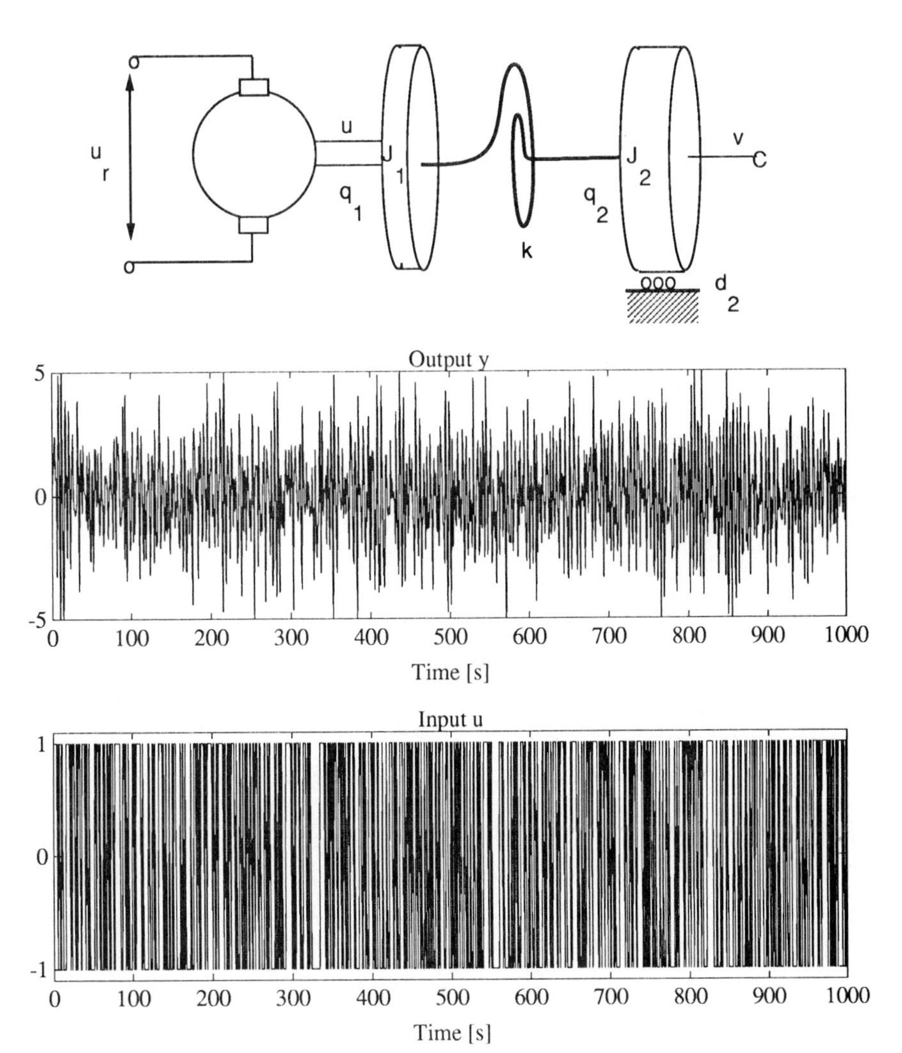

Figure 9.1 A DC motor and a load J_2 with a flexible coupling modeled by a torsion spring with spring coefficient k. Data are from a DC-motor drive with input u and output y as described in Examples 7.6 and 9.1.

with a linear model. Moreover, it indicates whether there is disturbance in special regions of frequency.

Example 9.1—A third-order model of a DC-motor drive
Consider the motor drive of Fig. 9.1 with the input torque u as the control variable and the angular velocity $y = \dot{q}_1$ as the variable available to measure-

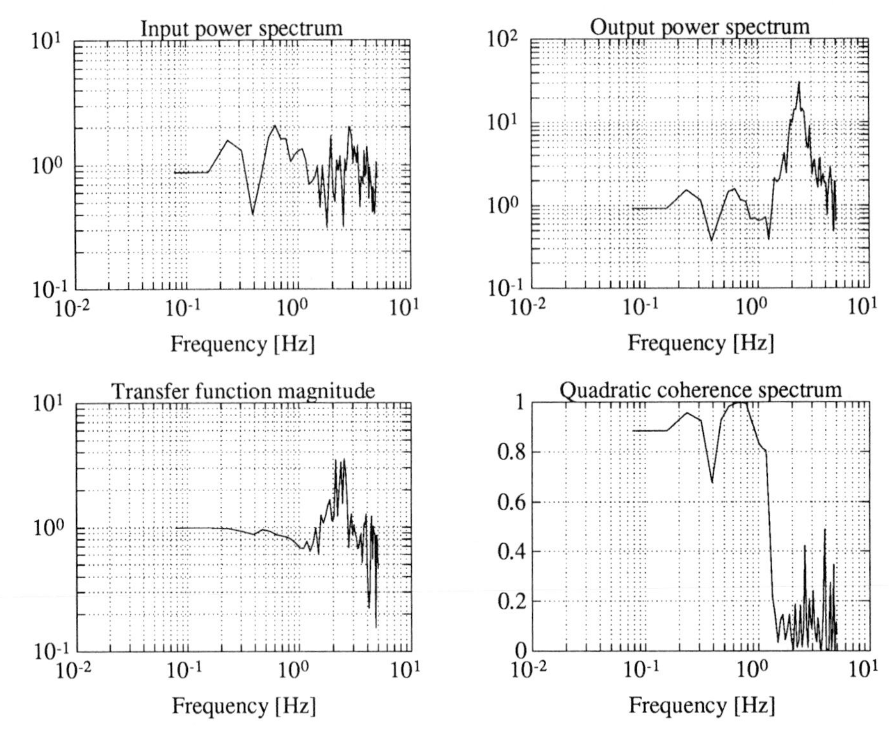

Figure 9.2 Input power spectrum (*upper left*), output power spectrum (*upper right*), transfer function (*lower left*), and coherence spectrum (*lower right*) based on input-output data from the DC-motor drive of Example 7.6. The coherence function may be used to determine weighting (or filtering) of data in the frequency domain, with a view to suppress any region where disturbance and nonlinearity are prominent.

ment. The motor drive with a moment of inertia J_1 is coupled to a load *via* a torsional flexibility with spring constant k. The load is assumed to have a moment of inertia J_2 and a damping d_2. There is also a load moment v acting randomly on the system. The angular positions of the shafts are designated q_1 and q_2, respectively. The motion equations are

$$\begin{aligned} J_1\ddot{q}_1 &= k(q_2 - q_1) + u \\ J_2\ddot{q}_2 &= -k(q_2 - q_1) - d_2\dot{q}_2 + v \end{aligned} \tag{9.8}$$

with a transfer function according to Eq. (7.39) and a state-space realization according to Eq. (7.38).

We use artificial data obtained by simulation of the system in Fig. 9.1 in order to allow for evaluation of the various methods by comparison with the correct values. Assume that the system is controlled and sampled with a sampling interval $h = 0.1$. The data collected from the input u and the output y thus

Table 9.1 Estimated coefficients with standard deviations and final prediction errors FPE (see for definition in Section 9.3) of various model orders for Example 9.1

Model order	A-coefficients		B-coefficients		C-coefficients		Estimated variance	Akaike FPE
$n = 1$	1.0		0		1.0		2.417	2.431
	−0.1897	±0.03	0.7613	±0.03187	−0.901	±0.01542		
$n = 2$	1.0		0		1.0		1.478	1.495
	−0.2801	±0.0375	0.7965	±0.03857	−0.812	±0.04952		
	0.5767	±0.02303	0.26	±0.05234	0.05104	±0.04705		
$n = 3$	1.0		0		1.0		1.026	1.046
	−1.064	±0.0381	0.8102	±0.03245	−1.884	±0.05084		
	0.9233	±0.03634	−0.5814	±0.06472	1.127	±0.0912		
	−0.4001	±0.02329	0.2372	±0.04354	−0.09273	±0.04885		
$n = 4$	1.0		0		1.0		1.004	1.029
	−0.02846	±0.05744	0.7876	±0.03189	−0.871	±0.06252		
	0.05379	±0.06262	0.2397	±0.05408	−0.6626	±0.1011		
	0.4431	±0.05577	−0.2215	±0.04507	0.8384	±0.07576		
	−0.2879	±0.03423	0.3217	±0.04106	0.03014	±0.04673		
$n = 5$	1.0		0		1.0		1.004	1.035
	0.8249	±0.07399	0.7995	±0.03203	−0.02023	±0.07774		
	−0.03971	±0.04821	0.9377	±0.05813	−1.454	±0.07542		
	0.4387	±0.03815	−0.07859	±0.06214	0.2981	±0.08395		
	0.1173	±0.04356	0.04999	±0.04758	0.8129	±0.07452		
	−0.2868	±0.03925	0.3013	±0.04329	−0.04175	±0.04696		
$n = 6$	1.0		0		1.0		1.004	1.041
	0.2846	±0.05793	0.7836	±0.03205	−0.5541	±0.062		
	1.039	±0.04985	0.4629	±0.04947	0.0615	±0.08111		
	0.4828	±0.08371	0.6614	±0.03544	−0.1888	±0.07864		
	−0.1133	±0.07164	0.5287	±0.04897	−0.4304	±0.0801		
	0.3537	±0.04634	−0.1281	±0.04217	0.8315	±0.06296		
	−0.2833	±0.03307	0.3193	±0.04173	0.05853	±0.04758		

correspond to data generated from a model with the discrete-time transfer function

$$\mathcal{S}: \qquad H_u(z) = \frac{B(z)}{A(z)} = \frac{0.848z^2 - 0.629z + 0.313}{z^3 - 0.986z^2 + 0.888z - 0.368} \tag{9.9}$$

obtained by means of a zero-order-hold assumption regarding the control input. The noise model is according to Fig. 9.1 with the transfer function

$$\frac{C(z)}{A(z)} = \frac{z^3 - 1.8z^2 + 0.97z}{z^3 - 0.986z^2 + 0.888z - 0.368} \tag{9.10}$$

where the coefficients of Eqs. (9.9) and (9.10) in turn corresponds to the physical parameters $J_1 = 0.1$, $J_2 = 0.1$, $k = 10$, $d_2 = 1$. The continuous-time noise variance $\mathcal{E}\{v^2\} = \sigma^2 = 1$ and $\mathcal{E}\{u^2\} = 1$ so that the signal-to-noise ratio for the input variables is equal to one. The diagrams of input and output data shown in Fig. 9.1 give no reason to suspect the presence of artifacts in the form of signal loss, outliers, or aliasing; neither is there any obvious time-dependence, nor are any trends perceptible in input-output data. The coherence spectrum in Fig. 9.2 between input u and output y with coherence close to 1 for frequency below 0.0 Hz verifies that there is good coherence between the two signals for the middle and lower frequency ranges. In all likelihood, therefore, the noise spectrum is of significant magnitude mainly in the higher frequency ranges, and the input apparently provides sufficient excitation over at least two decades of frequency.

Maximum-likelihood identification with $N = 1000$ samples of input and output provides parameter estimates according to Table 9.1 for model orders $n = 1-6$. The coefficients of the ARMAX model, their standard deviation, the estimated residual variance, and the final prediction error FPE (for definition see Section 9.3) are provided for all low-order models. In Table 9.1 it appears that the estimated residual variance does not decrease significantly for model order $n > 3$.

9.3 MODEL ORDER DETERMINATION

Unless *a priori* considerations in the field of application suggest that the model order is the correct one, an aim of identification may be to find at least an adequate if not the true model. As the correct model order is often not known *a priori*, it makes sense to postulate several different model orders. Based on these, one then computes some error criterion that indicates which model order to choose. One intuitive approach would be to construct ARMAX models of increasing order until the computed prediction error power reaches a minimum. However, as shown previously all least-squares estimation loss functions decrease monotonically with increasing model order. Typically, adding parameters to the model reduces the sum of squares of residuals whereas adding parameters that do not reduce it by much may be of little value. Thus, the prediction error or the loss functions alone might not be sufficient to indicate when to terminate the search for the correct model, and it is natural to adopt a statistical hypothesis testing approach to this problem (see Appendix B). Consequently, the experimenter adopts the hypothesis that the higher-order model has no ameliorating effect and tests whether data will disprove this. The decision should be supported by some *test statistic*, *i.e.*, a function of data. It is standard procedure to formalize this reasoning as a *null hypothesis* $\mathcal{H}_0$ and an alternative hypothesis $\mathcal{H}_A$. The essence of a statistical test is a decision rule that tells whether the collected data call for accepting or rejecting certain model hypotheses. We illustrate these ideas by presenting the following model order test.

F-test of system order n

Assume that the system $\mathcal{S}$ is of model order n_0 and is adequately described by p_0 parameters

$$\mathcal{S}: \quad Ay = Bu + Cw, \qquad w \in \mathcal{N}(0, \sigma^2) \tag{9.11}$$

Assume that there are two fitted models $\mathcal{M}_1$ and $\mathcal{M}_2$ of model orders n_1 and n_2 such that $n_0 \leq n_1 \leq n_2$ and assume that the number of estimated parameters are p_1 and p_2, respectively, with $p_0 \leq p_1 \leq p_2$. Let the minimum loss functions for parameter numbers p_1 and p_2 be denoted V^1 and V^2, respectively.

First suppose that we are to test the null hypothesis $\mathcal{H}_0$ that model $\mathcal{M}_1$ is correct, a model that is included as a special case of model $\mathcal{M}_2$. The alternative hypothesis would be that the true model is included in model $\mathcal{M}_2$ but is not as simple as $\mathcal{M}_1$. Let it be assumed in both models that the noise components are independent and normal with constant variance $\sigma_1^2 = \sigma_2^2 = \sigma^2$. Now we test the model order by adopting the null hypothesis

$$\mathcal{H}_0: \quad \mathcal{M}_1 \quad \text{and} \quad \sigma_1^2 = \sigma_2^2 \tag{9.12}$$

which means that if $\sigma_1^2 = \sigma_2^2$ then the simpler of the two models may be chosen. The alternative hypothesis is

$$\mathcal{H}_A: \quad \mathcal{M}_2 \supset \mathcal{M}_1 \quad \text{and} \quad \sigma_1^2 > \sigma_2^2 \tag{9.13}$$

which is predicated upon the possibility of reducing the variance by choosing $\mathcal{M}_2$ instead of $\mathcal{M}_1$. Thus, if the hypothesis tests lead us to reject $\mathcal{H}_0$, then we conclude that $\mathcal{M}_1$ has insufficient model complexity.

The loss function is also expected to decrease when the model order n increases beyond the appropriate system order. It is therefore relevant to ask, what is a significant decrease of the loss function V as the model order n increases? Now let N denote the number of observations. Under assumptions made in the context of prediction error methods that the loss function $V_N(\widehat{\theta}_N)$ represents a sum of squares of independent normally distributed random variables with a mean of zero, it follows from the Cochran theorem of statistics (see Appendix B), that

Table 9.2 Some percentiles for the χ^2–distribution. For degrees of freedom $k > 30$ it is possible to approximate the percentile as $\chi_p^2 = 0.5(z_p + \sqrt{2k-1})^2$ where z_p is the corresponding percentile of the standard normal distribution

Degrees of freedom	$\chi^2_{.005}$	$\chi^2_{.01}$	$\chi^2_{.025}$	$\chi^2_{.05}$	$\chi^2_{.95}$	$\chi^2_{.975}$	$\chi^2_{.99}$	$\chi^2_{.995}$
1	0.00	0.00	0.001	0.004	3.84	5.02	6.63	7.88
2	0.010	0.020	0.051	0.103	5.99	7.38	9.21	10.6
3	0.072	0.115	0.216	0.352	7.81	9.35	11.3	12.8
4	0.207	0.297	0.484	0.711	9.49	11.1	13.3	14.9
5	0.412	0.554	0.831	1.15	11.1	12.8	15.1	16.7
6	0.676	0.872	1.24	1.64	12.6	14.4	16.8	18.5
7	0.989	1.24	1.69	2.17	14.1	16.0	18.5	20.3
8	1.34	1.65	2.18	2.73	15.5	17.5	20.1	22.0
9	1.73	2.09	2.70	3.33	16.9	19.0	21.7	23.6
10	2.16	2.56	3.25	3.94	18.3	20.5	23.2	25.2
20	7.43	8.26	9.58	10.9	31.4	34.2	37.6	40.0
30	13.8	15.0	16.8	18.5	43.8	47.0	50.9	53.7
40	20.7	22.1	24.4	26.5	55.8	59.3	63.7	66.8
50	28.0	29.7	32.3	34.8	67.5	71.4	76.2	79.5

i. $2(V^1 - V^2)/\sigma^2$ is $\chi^2(p_2 - p_1)$–distributed under the model $\mathcal{M}_1$.

ii. $2V^2/\sigma^2$ is $\chi^2(N - p_2)$–distributed under the model $\mathcal{M}_2$.

iii. V^2 and $(V^1 - V^2)$ are independent under $\mathcal{M}_2$.

(9.14)

Assuming V^2 and $V^1 - V^2$ to be asymptotically statistically independent *(iii)* and using the χ^2–distributed properties (see Table 9.2) we can design a relevant test statistic. It is well known from statistical theory that a ratio of two χ^2–distributed variables is F-distributed with some degrees of freedom m_1 and m_2, respectively (see Table 9.3). A relevant test statistic for verification that the correct model order has been found is

$$\tau_F(p_2, p_1) = \frac{\widehat{\sigma}_1^2 - \widehat{\sigma}_2^2}{\widehat{\sigma}_2^2} = \frac{V^1 - V^2}{V^2} \cdot \frac{N - p_2}{p_2 - p_1} \in F(N - p_2, p_2 - p_1) \qquad (9.15)$$

which is an F-distributed variable with the degrees of freedom $N - p_2$ and $p_2 - p_1$ under the null hypothesis $\mathcal{H}_0$ expressed in Eq. (9.12). The criterion for rejection of the null hypothesis $\mathcal{H}_0$ is

$$\tau_F \geq F_\alpha(N - p_2, p_2 - p_1) \tag{9.16}$$

where α is the probability that the null hypothesis $\mathcal{H}_0$ is rejected when it is true (*i.e.*, the significance level of the test), and where F_α is the α-percentile of the F–distribution.

Example 9.2—F-test of model order

Consider the model of a DC drive in Example 9.1 and in Table 9.1. To test the hypothesis that the residual variances of two models are equal (*i.e.*, the null hypothesis $\mathcal{H}_0$), we compute the F-test statistic for $N = 1000$. As the ARMAX model of order n has $p = 3n$ estimated parameters and as we choose to test the model order n_1 against $n_2 = n_1 + 1$ with $p_1 = 3n_1$, $p_2 = 3(n_1 + 1)$, for $n_1 = 1, \ldots, 5$ we find by means of Table 9.1:

$$\begin{aligned}
n_1 &= 1, & \tau_F(6,3) &= \frac{2417 - 1478}{1478} \cdot \frac{1000 - 6}{6 - 3} \approx 210 \\
n_1 &= 2, & \tau_F(9,6) &= \frac{1478 - 1026}{1026} \cdot \frac{1000 - 9}{9 - 6} \approx 146 \\
n_1 &= 3, & \tau_F(12,9) &= \frac{1026 - 1004}{1004} \cdot \frac{1000 - 12}{12 - 9} \approx 7.21 \\
n_1 &= 4, & \tau_F(15,12) &= \frac{1004 - 1004}{1004} \cdot \frac{1000 - 15}{15 - 12} = 0 \\
n_1 &= 5, & \tau_F(18,15) &= \frac{1004 - 1004}{1004} \cdot \frac{1000 - 18}{18 - 15} = 0
\end{aligned} \tag{9.17}$$

This procedure generates the decision table

n_1	n_2	τ_F	$F_{0.05}(\infty, 3)$	accept	$F_{0.01}(\infty, 3)$	accept
1	2	210	8.53	$\mathcal{H}_A$	26.1	$\mathcal{H}_A$
2	3	146	8.53	$\mathcal{H}_A$	26.1	$\mathcal{H}_A$
3	4	7.21	8.53	$\mathcal{H}_0$	26.1	$\mathcal{H}_0$
4	5	0	8.53	$\mathcal{H}_0$	26.1	$\mathcal{H}_0$
5	6	0	8.53	$\mathcal{H}_0$	26.1	$\mathcal{H}_0$

Comparing the test statistics to the 95% percentile $F_{0.05}(991, 3) = 8.53$ we find that the null hypothesis $\mathcal{H}_0$ for model order $n_1 = 3$ is not rejected, *i.e.*, the third-order model is accepted.

The Akaike information criterion (AIC)

It is obvious from Table 9.1 that the estimated variance $\widehat{\sigma}^2$ and the loss function decrease as the model order increases. Assume that the loss function is

Table 9.3 Percentiles $F_{0.05}(m_1, m_2)$ of the F-distribution for degrees of freedom m_1 and m_2, respectively.

$m_2 \backslash m_1$	1	2	3	4	5	6	8	10	20	30	∞
1	161	200	216	225	230	234	239	242	248	250	254
2	18.5	19.0	19.2	19.2	19.3	19.3	19.4	19.4	19.4	19.5	19.5
3	10.1	9.55	9.28	9.12	9.01	8.94	8.85	8.79	8.66	8.62	8.53
4	7.71	6.94	6.59	6.39	6.26	6.16	6.04	5.96	5.80	5.75	5.63
5	6.61	5.79	5.41	5.19	5.05	4.95	4.82	4.74	4.56	4.50	4.36
6	5.99	5.14	4.76	4.53	4.39	4.28	4.15	4.06	3.87	3.81	3.67
7	5.59	4.74	4.35	4.12	3.97	3.87	3.73	3.64	3.44	3.38	3.23
8	5.32	4.46	4.07	3.84	3.69	3.58	3.44	3.35	3.15	3.08	2.93
9	5.12	4.26	3.86	3.63	3.48	3.37	3.23	3.14	2.94	2.86	2.71
10	4.96	4.10	3.71	3.48	3.33	3.22	3.07	2.98	2.77	2.70	2.54
20	4.35	3.49	3.10	2.87	2.71	2.60	2.45	2.35	2.12	2.04	1.84
30	4.17	3.32	2.92	2.69	2.53	2.42	2.27	2.16	1.93	1.84	1.62
∞	3.84	3.00	2.60	2.37	2.21	2.10	1.94	1.83	1.57	1.46	1.00

obtained from a least-squares model of order n with p estimated parameters $\widehat{\theta}_N$, and that the model is fitted with data from N samples

$$\widehat{\sigma}^2 = \frac{2}{N} V_N(\widehat{\theta}_N) = \frac{1}{N} \sum_{i=1}^{N} \varepsilon_i^2(\widehat{\theta}_N), \qquad \widehat{\theta}_N \in R^p \tag{9.18}$$

An interesting question is whether the loss function V can be replaced by some other relevant optimization index that also supports estimation of structural parameters such as model order or the number of model parameters. An optimization criterion that penalizes a high model order more effectively than the least-squares criterion function can be obtained by adding a term to the least-squares loss function such that the function grows as the model order n and the number p of model parameters increase.

One attempt to include both the estimated variance and the model complexity in one statistic is the *Akaike information criterion* (AIC), which decreases as the residual variance $\widehat{\sigma}^2$ decreases and which increases as the number of parameters p increases. As the expected residual variance decreases with increasing p for nonadequate model complexities, there should be a minimum around the correct number p. Let $\log L(\theta)$ denote the log-likelihood function

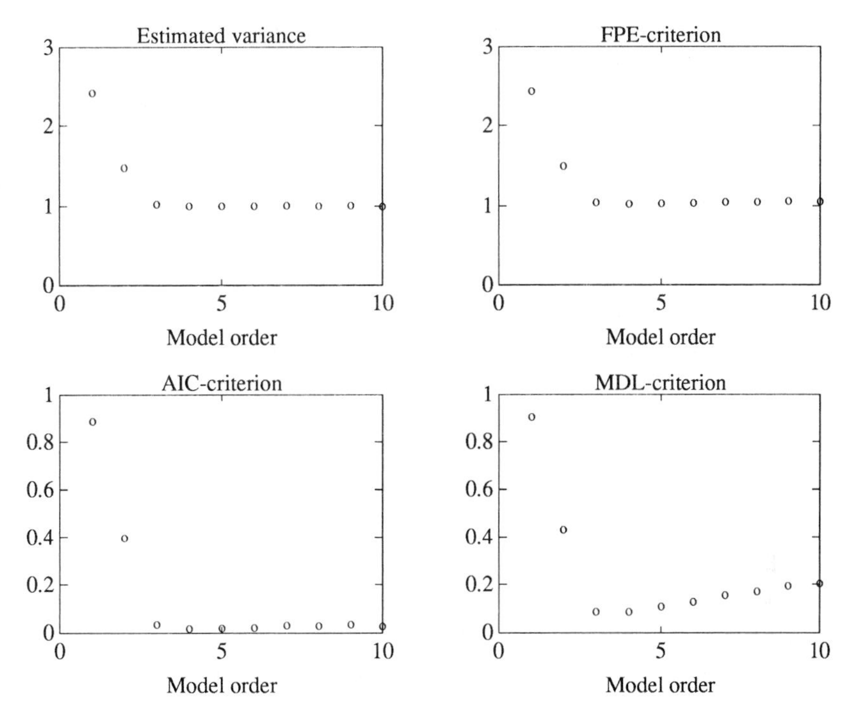

Figure 9.3 Some model validation criteria evaluated for the DC-motor example: the estimated variance (*upper left*); the Akaike model order tests FPE (*upper right*) and AIC (*lower left*); the MDL test (*lower right*). All tests support that a third-order model is the appropriate model order.

of θ based on N observations. By optimizing a measure of distance between the "true" likelihood distribution and the observed one in the form of the *Kullback-Leibler information*

$$J(\widehat{\theta}, \theta) = \mathcal{E}\{\log L(\theta) - \log L(\widehat{\theta}_N)\}, \qquad \widehat{\theta} \in R^p \tag{9.19}$$

it is possible to motivate the Akaike information criterion

$$\mathrm{AIC}(p) = \log \widehat{\sigma}^2(\widehat{\theta}_N) + \frac{2p}{N}, \qquad \widehat{\theta}_N \in R^p \tag{9.20}$$

which, if statistically consistent, would attain its minimum for the correct number of parameters. However, it can be shown the AIC is statistically inconsistent and gives an overestimated model order, that also motivates other criteria. An alternative is the minimum description length (MDL) statistic suggested by Rissanen.

$$\mathrm{MDL}(p) = \log \widehat{\sigma}^2 + \frac{p}{N} \log N + \frac{p}{N} \log ||\widehat{\theta}_N||_M \tag{9.21}$$

where M is the Fisher information matrix. The MDL is statistically consistent as $N \to \infty$ as a criterion for choosing the model order.

The result of the AIC and MDL criteria applied to model order determination Example 9.1 is shown in Fig. 9.3.

The Akaike final prediction error (FPE)

It was noticed that the average prediction error is expected to decrease as the number of estimated parameters increase. One reason for this behavior is that the prediction errors are computed for the data set that was used for parameter estimation. It is now relevant to ask what prediction performance can be expected when the estimated parameters are applied to another data set. Clearly, it might be suspected that a large overparametrized model might poorly predict the behavior of a new data set. In order to analyze this situation, we consider the expected prediction error based on the p parameter estimates $\widehat{\theta}$ based on N data fitted to some linear model $y_k = \phi_k^T \theta + w_k$ where

$$\begin{aligned} \mathcal{E}\{\varepsilon_k^2(\widehat{\theta})\} &= \mathcal{E}\{(y_k - \phi_k^T \widehat{\theta})^2\} = \mathcal{E}\{((y_k - \phi_k^T \theta) - (\phi_k^T \widehat{\theta} - \phi_k^T \theta))^2\} = \\ &= \mathcal{E}\{(y_k - \phi_k^T \theta)^2\} + \mathcal{E}\{(\phi_k^T \theta - \phi_k^T \widehat{\theta})^2\} = \\ &= \mathcal{E}\{w_k^2\} + \mathrm{tr}(\mathcal{E}\{\widetilde{\theta}\widetilde{\theta}^T \phi_k \phi_k^T\}) \approx \sigma^2 + \sigma^2 \frac{p}{N} \end{aligned} \tag{9.22}$$

where the term σ^2 derives from the noise variance properties, whereas the contribution $p\sigma^2/N$ derives from the parameter errors. Thus, the asymptotic prediction error decreases as the number of observations increases, whereas the prediction error variance increases as the number of estimated parameters increases. However, the expected loss function based on the null hypothesis $\mathcal{H}_0$ when estimating p parameters based on N observations is

$$\mathcal{E}\{V_N(\widehat{\theta}_N)\} = \frac{1}{2}\sigma^2(N-p) \tag{9.23}$$

which tends to decrease as the number of parameters increases and the *final prediction error criterion* (FPE) is estimated as

$$\mathrm{FPE}(p) = \widehat{\sigma}^2(1 + \frac{p}{N}) = \frac{N+p}{N-p}\frac{2}{N} V_N(\widehat{\theta}_N) \tag{9.24}$$

Hence, the quality of identification measured as expected prediction accuracy can be improved by introducing new parameters to be estimated as long as each new parameter can be accurately estimated. Thus we choose

$$\widehat{p} = \arg\min \mathrm{FPE}(p) \tag{9.25}$$

where the second factor increases as the number p of estimated parameters increases. The FPE criterion consists of choosing the model corresponding to the minimum FPE as the final prediction criterion and the corresponding number of parameters p and a corresponding correct model order. It is, however, sometimes observed that the FPE tends to underestimate the correct order of a system.

The decision criteria AIC, MDL, and FPE all include the model order or the number of parameters as a model parameter in the loss function to minimize. An attractive property is that the minimum of the test statistics indicates what model to choose so that these test statistics need not be compared against some statistical significance level; see Fig. 9.5.

In addition to the methods presented, there exist consistent criteria such as the Wald statistic for order estimation based on rank conditions. Another method for model order determination is based on singular value decomposition, but discussion of this must wait until the following chapter on model reduction.

9.4 RESIDUAL TESTS

Linear prediction error models of a system $\mathcal{S}$ are generally based on some transfer function relationship $Y(z) = H_u(z)U(z) + H_w(z)W(z)$. The residuals obtained as

$$\varepsilon(z) = \widehat{H}_w^{-1}(z)(Y(z) - \widehat{H}_u(z)U(z)) \tag{9.26}$$

represent a disturbance input or innovations that would explain the mismatch between the observed data and the behavior of the estimated model. A sequence of residuals which still exhibits some structure would then indicate that either the modeling or the identification is not complete. If the model is correct and if the method prerequisites are satisfied, then the residuals should be structureless; in particular, they should be uncorrelated to any other variable including inputs and outputs. This is the assumption upon which the tests known as *residual tests* are based.

A simple check is to plot the residual versus the fitted values; such a plot should not reveal any obvious pattern. Another valuable diagram is the histogram of the residual amplitudes, which reveals distributions that differ from normal distributions (see Fig. 9.4), and is a valuable complement to analysis of variance.

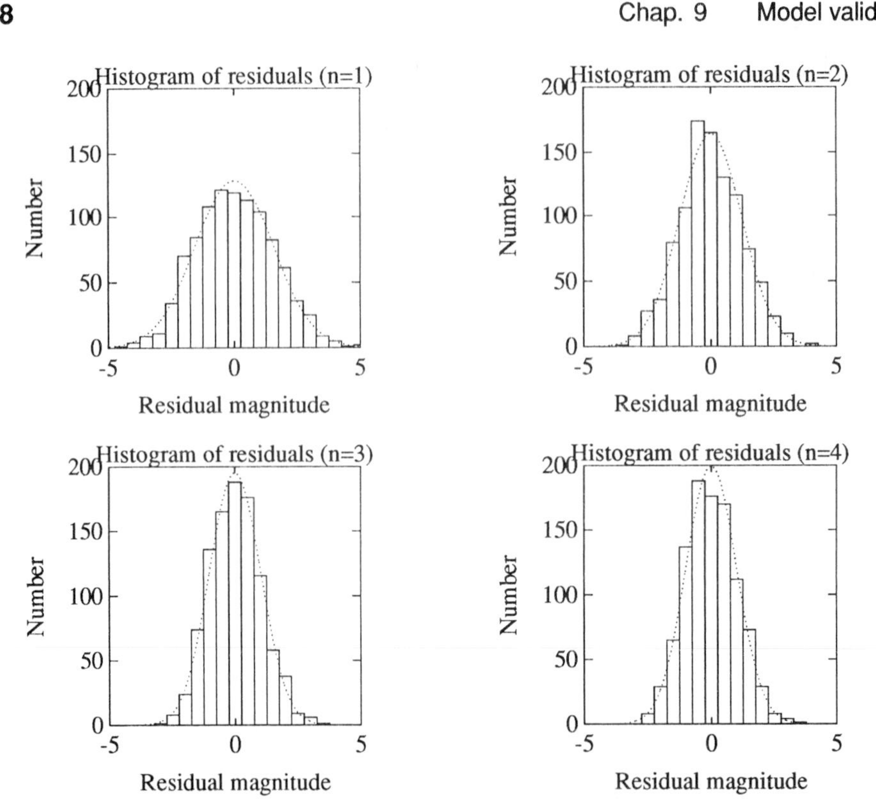

Figure 9.4 Histogram of residuals for DC-drive models of model orders $n = 1 - 4$. The normal distribution $\mathcal{N}(0, \widehat{\sigma}^2)$ (*dotted line*) is shown for comparison.

The following tests (as formulated here) apply to time-invariant systems and are based on statistical analysis of the residuals $\{\varepsilon_k(\widehat{\theta}_N)\}$. The *null hypothesis* $\mathcal{H}_0$ is

i. $\{\varepsilon_k\}$ constitute a white-noise process with mean 0

ii. $\{\varepsilon_k\}$ are normally distributed

iii. $\{\varepsilon_k\}$ are symmetrically distributed

iv. $\{\varepsilon_k\}$ are independent of previous inputs with $\mathcal{E}\{\varepsilon_i u_j\} = 0; \quad i > j$

v. $\{\varepsilon_k\}$ are independent of all inputs $\mathcal{E}\{\varepsilon_i u_j\} = 0; \forall i, j$ if there is no feedback

This null hypothesis $\mathcal{H}_0$ for the residuals can be used for several statistical tests of which we present the autocorrelation test, the cross-correlation test, test of normality, and test of the number of zero crossings (changes of sign).

Autocorrelation test

Consider the autocovariance function of residuals

$$\widehat{C}_{\varepsilon\varepsilon}(\tau) = \frac{1}{N-\tau} \sum_{k=\tau+1}^{N} \varepsilon_k \varepsilon_{k-\tau} \tag{9.27}$$

and define the vector of residual autocorrelations as

$$r_{\varepsilon\varepsilon} = \frac{1}{\widehat{C}_{\varepsilon\varepsilon}(0)} \begin{pmatrix} \widehat{C}_{\varepsilon\varepsilon}(1) & \dots & \widehat{C}_{\varepsilon\varepsilon}(m) \end{pmatrix}^T \tag{9.28}$$

for some number m. Under the null hypothesis $\mathcal{H}_0$ and according to the central limit theorem this statistic is asymptotically distributed as

$$\sqrt{N} r_{\varepsilon\varepsilon} \stackrel{\text{dist.}}{\rightarrow} \mathcal{N}(0, I_{m \times m}) \tag{9.29}$$

An autocorrelation test statistic can be formulated as the quantity

$$\tau_{\varepsilon\varepsilon} = N r_{\varepsilon\varepsilon}^T r_{\varepsilon\varepsilon} \stackrel{\text{dist.}}{\rightarrow} \chi^2(m) \tag{9.30}$$

which can be tested with standard analysis of variance. The null hypothesis $\mathcal{H}_0$ assumes that the residual mean is zero, which helps to avoid a reduction of the degrees of freedom of the χ^2 test. The decision criterion based on the null hypothesis is

$$\begin{aligned} \tau_{\varepsilon\varepsilon} &> \chi^2_\alpha(m), && \text{reject } \mathcal{H}_0 \\ \tau_{\varepsilon\varepsilon} &< \chi^2_\alpha(m), && \text{accept } \mathcal{H}_0 \end{aligned} \tag{9.31}$$

where α is the significance level. Rejection of $\mathcal{H}_0$ should imply that the model associated with $\mathcal{H}_0$ is refuted.

Example 9.3—Validation of a DC-drive model (cont'd.)
Computation of the autocorrelation test statistic for $m = 50$ gives the values

$$\begin{matrix} n & \tau_{\varepsilon\varepsilon} \end{matrix}$$
$$\begin{pmatrix} 1 & 802.2 \\ 2 & 70.9 \\ 3 & 40.9 \\ 4 & 45.5 \\ 5 & 44.4 \\ 6 & 46.8 \\ 7 & 42.3 \\ 8 & 41.8 \\ 9 & 36.6 \\ 10 & 44.2 \end{pmatrix} \tag{9.32}$$

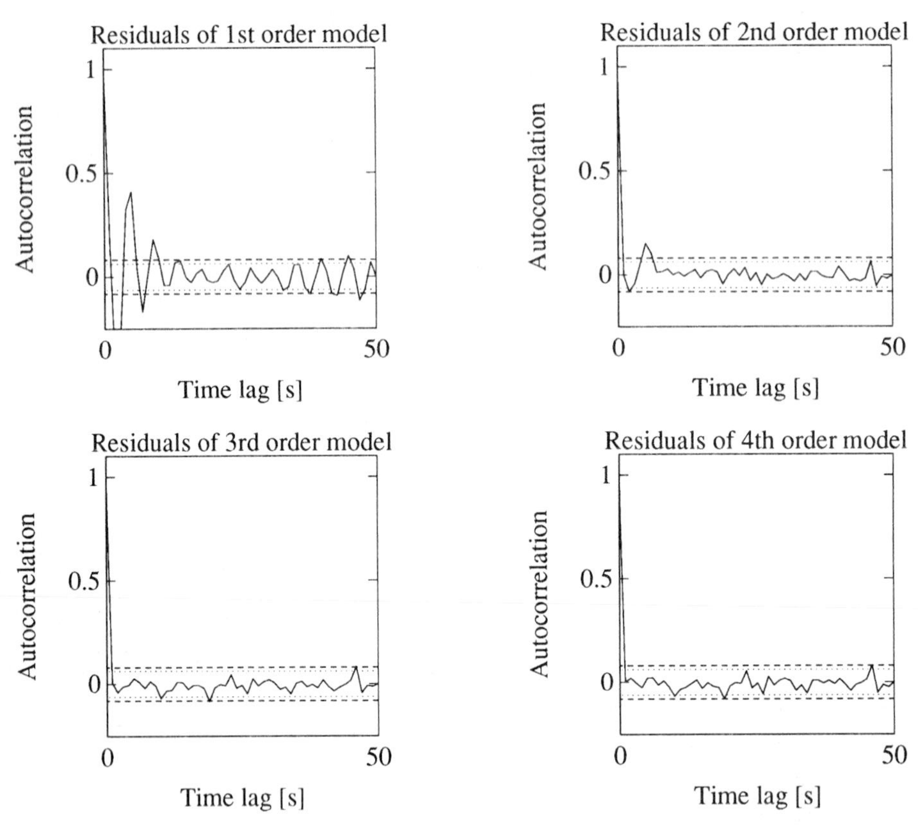

Figure 9.5 Residual autocorrelations for some model orders $n = 1 - 4$ with 95%-confidence interval limits (*dotted line*) and 99%-confidence interval limits (*dashed line*) for noncorrelated residuals.

The percentile $\chi^2_{0.95}(50) = 67.5$ and $N = 1000$, which means that the above test statistic $\tau_{\varepsilon\varepsilon} < 67.5$ suggests that the null hypothesis should be accepted for $n \geq 3$. ■

A standard way to present this decision problem in the case of interactive identification is to show a diagram of $r_{\varepsilon\varepsilon}$; see Fig. 9.5. The 95%-confidence interval for the asymptotic distribution of each component is $[-1.96/\sqrt{N},\ 1.96/\sqrt{N}]$, which is often drawn in the same diagram. A test of normality for each of the components of $r_{\varepsilon\varepsilon}(k)$, $1 \leq k \leq m$ can then be suggested as follows. If $r_{\varepsilon\varepsilon}(k)$ is within the indicated interval, then the null hypothesis $\mathcal{H}_0$ can be accepted for that value of k. Moreover, if the whole function $r_{\varepsilon\varepsilon}$ is within the indicated margins, then the null hypothesis $\mathcal{H}_0$ can be accepted (or more correctly, *i.e.*, it is not rejected).

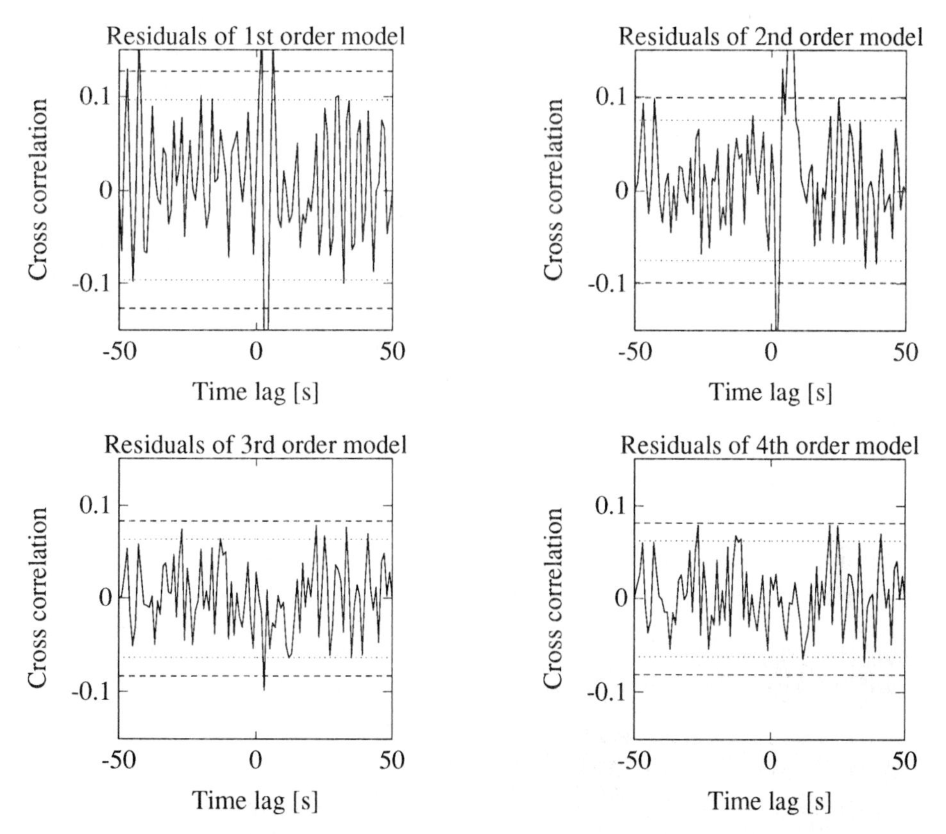

Figure 9.6 Residual cross correlations for some model orders $n = 1 - 4$ with 95%-confidence interval limits (*dotted line*) and 99%-confidence interval limits (*dashed line*) for noncorrelated residuals.

Cross-correlation test

A similar test for the independence of input u and residuals is based on the cross-covariance function

$$\widehat{C}_{u\varepsilon}(\tau) = \frac{1}{N-\tau} \sum_{k=\tau+1}^{N} \varepsilon_k u_{k-\tau} \tag{9.33}$$

Let m be the time interval (expressed as a number of samples) over which residual correlations should be investigated and define the vector

$$r_{u\varepsilon}(m) = \frac{1}{\sqrt{\widehat{C}_{\varepsilon\varepsilon}(0)}} \left(\widehat{C}_{u\varepsilon}(\tau+1) \quad \ldots \quad \widehat{C}_{u\varepsilon}(\tau+m) \right)^T \tag{9.34}$$

and the matrix

$$\widehat{R}_{uu}(m) = \frac{1}{N-m} \sum_{k=m+1}^{N} \begin{pmatrix} u_{k-1} \\ \vdots \\ u_{k-m} \end{pmatrix} \begin{pmatrix} u_{k-1} & \dots & u_{k-m} \end{pmatrix} \tag{9.35}$$

Under the null hypothesis $\mathcal{H}_0$ it is straightforward to verify that the asymptotic distribution is

$$\sqrt{N} r_{u\varepsilon} \overset{\text{dist.}}{\to} \mathcal{N}(0, \widehat{R}_{uu}) \tag{9.36}$$

Thus, we can formulate the cross-correlation test statistic

$$\tau_{u\varepsilon}(m) = N r_{u\varepsilon}^T \widehat{R}_{uu}^{-1} r_{u\varepsilon} \overset{\text{dist.}}{\to} \chi^2(m) \tag{9.37}$$

All these test quantities can be used for statistical hypothesis testing. The null hypothesis $\mathcal{H}_0$ is predicated upon a residual mean of zero, which helps to avoid a reduction of the degrees of freedom of the χ^2 test.

It is also valuable to inspect $r_{u\varepsilon}$ for negative m where non-zero values indicate that the residuals affect future inputs *via* some feedback mechanism (or that the system or model is noncausal in its behavior). The cross-correlation test may thus be used as an indicator of feedback.

Example 9.4—Validation of a DC-drive model (cont'd.)
Computation of the cross-correlation test statistic for $m = 50$ gives the result

n	$\tau_{u\varepsilon}$
1	127.7
2	164.4
3	65.1
4	63.1
5	60.6
6	56.9
7	59.9
8	56.4
9	57.5
10	59.1

(9.38)

The percentile $\chi^2_{0.95}(50) = 67.5$ suggests that the decision criterion for rejection of the alternative hypothesis based on the cross-correlation test statistic should be $\tau_{u\varepsilon} < 67.5$ — *i.e.*, that the null hypothesis should be rejected for $n \leq 2$ and that a third order may be adopted.

Results of these computations are often presented in a manner similar to that of the autocorrelation with 95%-confidence intervals $[-1.96/\sqrt{N},\ 1.96/\sqrt{N}]$ (see Fig. 9.6). If $r_{u\varepsilon}(k)$ is within the indicated interval then the null hypothesis can be accepted for that value of k. If the whole function $r_{u\varepsilon}$ is within the indicated margins then the null hypothesis can be accepted. ■

A common source of systematic failure of this test for all model orders is the presence of artifacts in data. A common defect is that one residual is very much larger than any of the others. Such a residual can be identified as an *outlier* and generally causes problems in the case of least-squares estimation. Problems due to outliers should be circumvented by eliminating such data before parameter estimation takes place.

Example 9.5—Validation of a DC-motor model (cont'd.)
Assume that data with input u and the output y are generated from a model with the transfer function

$$\mathcal{S}: \qquad H_u(z) = \frac{B(z)}{A(z)} = \frac{0.848z^2 - 0.627z + 0.313}{z^3 - 0.986z^2 + 0.888z - 0.370} \tag{9.39}$$

a noise model according to

$$H_w(Z) = \frac{C(z)}{A(z)} = \frac{z^3 - 1.8z^2 + 0.97z}{z^3 - 0.986z^2 + 0.888z - 0.370} \tag{9.40}$$

and a noise variance $\mathcal{E}\{w^2\} = \sigma^2 = 1$. Diagrams showing input and output data are found in Fig. 9.1. Maximum-likelihood identification with $N = 1000$ samples of input and output provides parameter estimates according to Table 9.1 for various model orders. The loss function (=estimated variance) is also available in Table 9.1, and is a continuously decreasing function of the model order n of the estimated model. The F-test gives a clear indication that there is no significant change of the loss function for a model order higher than $n = 3$. Thus far the validation procedure supports the choice of a linear model of model order $n = 3$. ■

Test of normality

One test of normality has been encountered already in the context of autocorrelation and cross correlation with a 95%-confidence interval indicated in the autocorrelation and cross correlation diagrams. Equation (9.29) can thus be used for design of tests of normality of the residuals. Also, the autocorrelation test statistic (9.30), which for large numbers m is well approximated

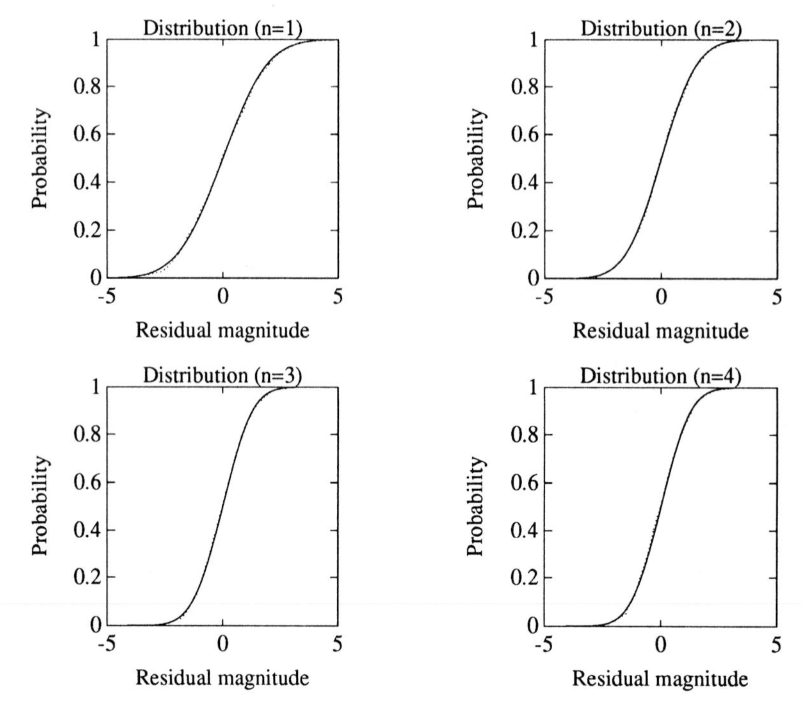

Figure 9.7 Empirical distribution of residuals for DC-drive models of model orders $n = 1 - 4$ (*solid line*). The corresponding normal distibution (*dotted line*) is indistinguishable from the empirical distributions except for $n = 1$.

by a normal distribution $\mathcal{N}(m, 2m)$, can be viewed as a test of normality, see Fig. 9.7.

A straightforward test of whether the distribution of residuals is normal is offered by the Kolmogorov-Smirnov test. The difference between the residual distribution function obtained and the assumed normal distribution function can be used as a statistic in determining whether or not to accept the null hypothesis as correct

$$\tau_{KS} = \sup_x |\widehat{F}_\varepsilon(x) - F_\varepsilon(x)|, \tag{9.41}$$

where the empirical distribution function is

$$\widehat{F}_\varepsilon(x) = \begin{cases} 0, & x < \varepsilon_{(1)} \\ k/N, & \varepsilon_{(k)} \le x \le \varepsilon_{(k+1)}, \quad k = 1, 2, \ldots, N-1 \end{cases} \tag{9.42}$$

and $\{\varepsilon_{(k)}\}$ is a permutation of the residuals $\{\varepsilon_k\}$ by sorting the components of the residual sequence in ascending order of magnitude. There are asymptotic

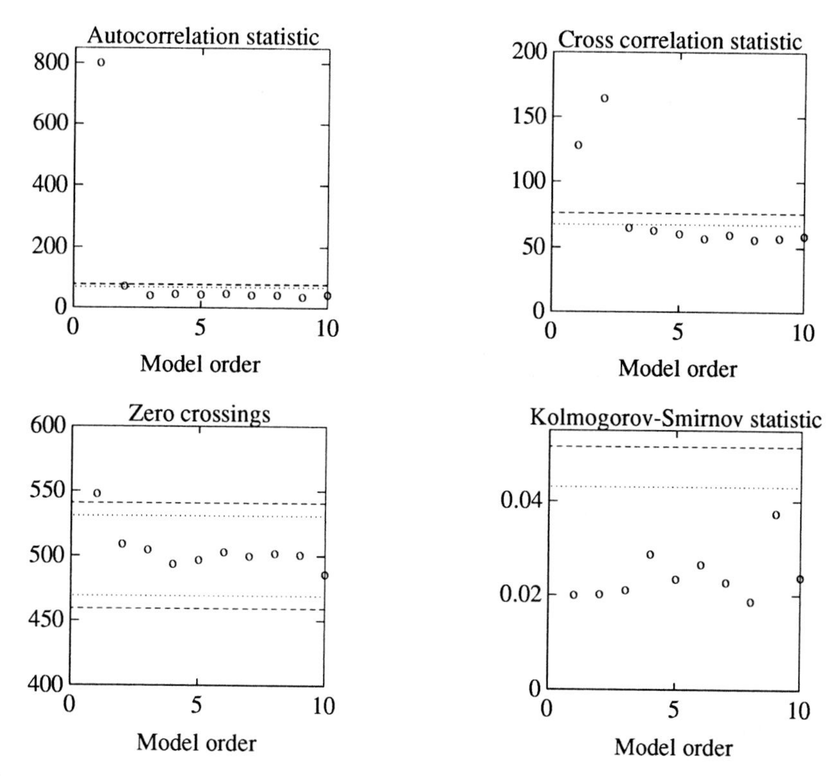

Figure 9.8 Test statistics and rejection thresholds for the autocorrelation, cross-correlation, zero crossings, and Kolmogorov-Smirnov tests. Confidence intervals at significance level $p < 0.05$ (*dotted line*) and $p < 0.01$ (*dashed line*) are indicated.

formulae for sample sizes $N > 100$

Significance level	Acceptance limit
$p = 0.05$	$\tau_{KS} \leq \frac{1.36}{\sqrt{N}}$
$p = 0.01$	$\tau_{KS} \leq \frac{1.63}{\sqrt{N}}$

(9.43)

Example 9.6—Validation of a DC-drive model (cont'd.)
The Kolmogorov-Smirnov test statistic was computed for various model orders, all of which fall within the margin of acceptance as shown in Fig. 9.8. Obviously, this statistic does not provide any means of distinguishing between models of different orders in this example. ■

Zero crossings

Consider the number of zero crossings of the residuals

$$\tau_x = \sum_{i=1}^{N-1} x_i \tag{9.44}$$

where a zero crossing x_k at time k is calculated as

$$x_k = \begin{cases} 1, & \text{if} \quad \varepsilon_k \varepsilon_{k+1} < 0 \\ 0, & \text{if} \quad \varepsilon_k \varepsilon_{k+1} > 0 \end{cases} \tag{9.45}$$

Assuming the x_k to be independent variables which take on the values 0 and 1 with equal probability under the null hypothesis $\mathcal{H}_0$, the test statistic τ_x for large N is then asymptotically distributed as

$$\tau_x \quad \in \quad \mathcal{N}(\frac{N}{2}, \frac{N}{4}) \tag{9.46}$$

A two-sided test for zero crossings at the significance level 0.05 is

$$-1.96 < \frac{\tau_x - N/2}{\sqrt{\frac{N}{4}}} < 1.96 \tag{9.47}$$

or

$$\frac{N}{2} - 1.96\sqrt{\frac{N}{4}} < \tau_x < \frac{N}{2} + 1.96\sqrt{\frac{N}{4}} \tag{9.48}$$

which leads to a test criterion stating that the number of zero crossings for $N = 1000$ with 95% probability should be in the interval $[459, 541]$.

Example 9.7—Validation of a DC-drive model (cont'd.)
The result of the zero crossings test applied to the model of the DC-motor drive is shown in Fig. 9.8. The estimated variance, AIC, FPE, and MDL as shown in Fig. 9.3 all point to the choice of a third- or fourth-order model. The following confidence intervals are given for the number of zero crossings

$$\text{Confidence intervals} \begin{pmatrix} p < 0.05^{*} \\ p < 0.01^{**} \\ p < 0.001^{***} \end{pmatrix} = \begin{pmatrix} 469 & 531 \\ 459 & 541 \\ 448 & 552 \end{pmatrix} \tag{9.49}$$

The number of zero crossings, the autocorrelation, and cross-correlation statistics shown in Fig. 9.8 all indicate that the third-order model is appropriate. The Kolmogorov-Smirnov test statistic does not distinguish between the different model orders. Also, the residual correlation tests in Fig. 9.5 and Fig 9.6 indicate that the first- and second-order models are insufficient to explain data but that the third-order model is the appropriate model order. ■

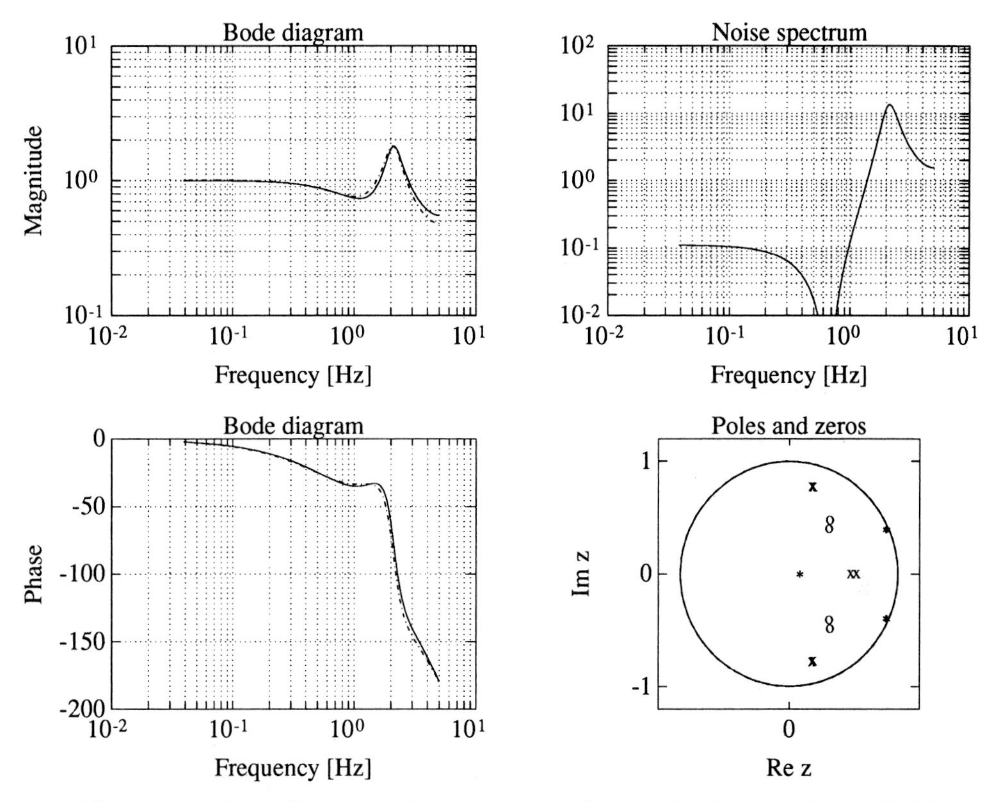

Figure 9.9 Bode diagram, noise spectrum, and zero-pole diagram of the DC-drive system (*solid line*) and the estimated model (*dashed line*). The zero-pole diagram contains the poles ('x'), and the zeros ('o','$*$') of the B- and C-polynomials, respectively.

9.5 MODEL AND PARAMETER ACCURACY

Several of the statistical methods already presented have been helpful in distinguishing the correct model order from other more or less correct models. However, it remains to be shown that the model chosen according to these methods is sufficiently accurate for the purpose of modeling. It is therefore indispensable to consider the model performance and behavior in comparison with real data. At least three different methods are relevant in this context.

First, a *stochastic simulation*, where both the deterministic input and the residuals of identification are used as inputs, is an effective means of checking whether the model reproduces the observed data. This test should give a close

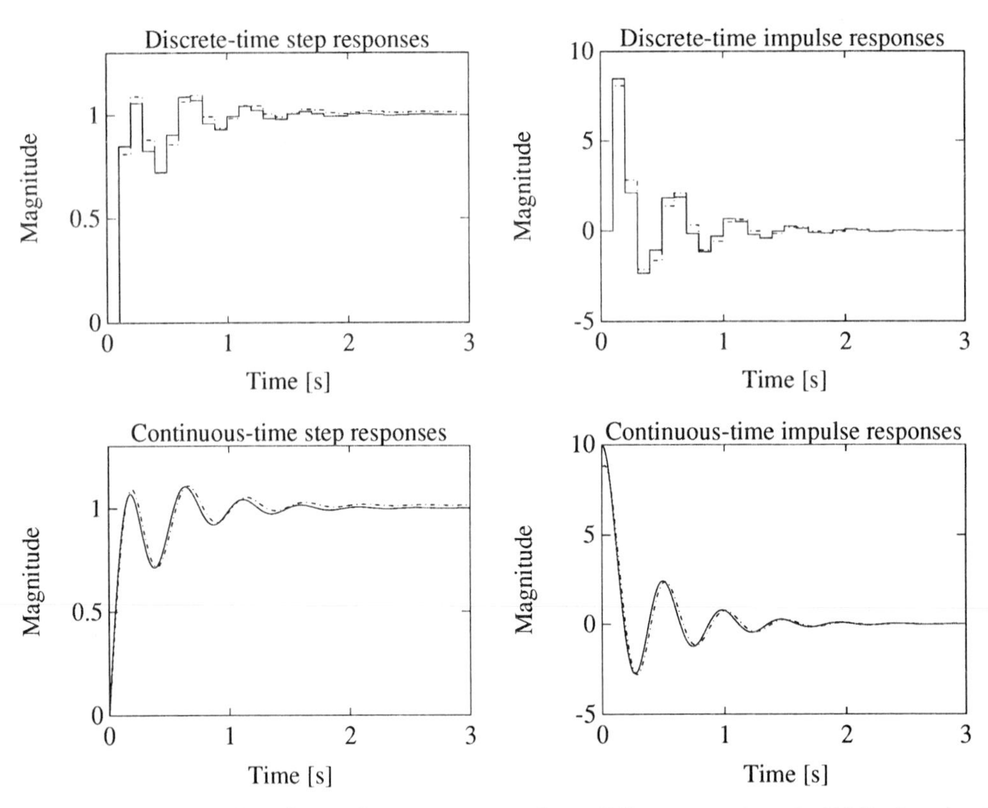

Figure 9.10 Impulse and step responses for a DC-motor system (*solid line*) and the identified model (*dashed line*) in both discrete time (*upper*) and continuous time (*lower*).

fit between observed and simulated data. A poor reproduction of data may indicate that the numerical procedures of parameter optimization have failed.

Second, a *deterministic simulation* can be used, where real data are compared with the model response to the recorded input signal used in the identification. This test should ascertain whether the model response is comparable to real data in magnitude and response delay. This test sometimes fails despite promising results having been obtained in previous statistical tests. Such a result makes sense in a case where the identification has determined an adequate stochastic model but where the input-output behavior has been compromised. A failing deterministic simulation often indicates that the input amplitude during experiments is inadequate; so the experiments that generate data may have to be redesigned. Another source of problems is model complexity, where a too simple model tends to give poor input-output behavior. Thus, recourse to nonlinear models, etc., often appears to be justified at this stage of identification.

Third, a *cross-validation simulation* can be made by applying input data not previously used in the identification. The comparison between the observed data and the model output is usually very revealing with regard to model anomalies not previously detected. A residual analysis applied to the misfit between model and data is also valuable in order to determine whether the model complexity is adequate. Failure to pass the cross-validation test may also indicate that the system is not time-invariant.

A problem that appears in physical modeling is whether the estimated model is compatible with *a priori* knowledge about the system behavior and its parameters. It is valuable in this context to evaluate information from the *zero-pole diagram* (see Fig. 9.9), and the model step and impulse responses (see Fig. 9.10).

Another problem is to ascertain whether the estimated parameters are statistically consistent. An empirical approach to this problem is to consider the parameter variance estimates as the number of observations used in the identification increases. A decrease of the estimated parameter variance as N increases supports the choice of model structure, whereas a constant or increasing parameter variance may indicate that the parameter estimates are statistically inconsistent. Theoretical approaches to justification for claims of consistency are based on normality assumptions and/or the central limit theorem both of which have a limited scope of application.

Continuous-time transfer functions and state-space realizations of the type (7.38) are examples of physical parametrizations. It is sometimes of interest to monitor individual physical parameters by means of identification. A problem in this context is that the physical parameters are present in the transfer functions as aggregate parameters which, of course, complicates interpretation of the estimated coefficients. For the same reason it is difficult to calculate precise estimates of the uncertainty of the physical parameter estimates obtained.

Example 9.8—Validation of a DC-drive model (cont'd.)

A deterministic simulation (Fig. 9.11) of the estimated model, using the recorded input as input to the model, shows reasonable agreement between the simulated output and the recorded output despite the unfavorable signal-to-noise ratio ($S/N = 1$). Also, the stochastic simulation in Fig. 9.11 with a noise input chosen as the computed residuals provides good agreement between original data and the behavior of the estimated model.

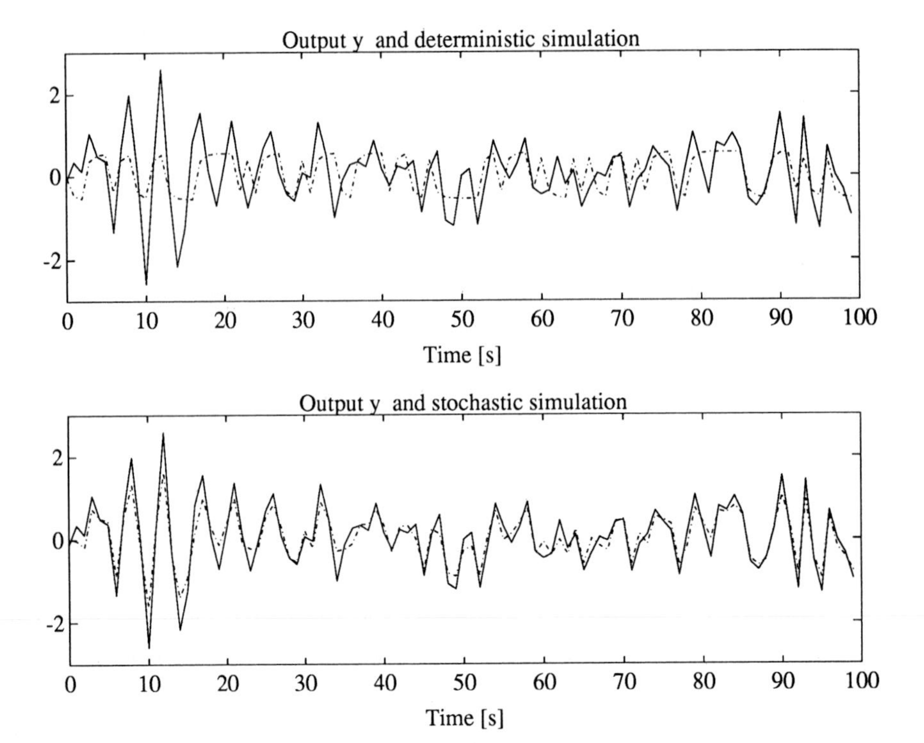

Figure 9.11 Deterministic and stochastic simulations (*dashed line*) of the recorded output by applying the recorded input sequence (and the residuals as noise input) to the estimated model. The recorded output is shown for comparison (*solid line*).

The discrete-time estimated model with sampling period $h = 0.1s$ is

$$\begin{aligned} Y(z) &= \frac{0.8102z^2 - 0.5814z + 0.2372}{z^3 - 1.0643z^2 + 0.9233z - 0.4001}U(z) \\ &+ \frac{z^3 - 1.8839z^2 + 1.1267z - 0.0927}{z^3 - 1.0643z^2 + 0.9233z - 0.4001}W(z) \end{aligned} \tag{9.50}$$

Transformation of this discrete-time model to a continuous-time model gives

$$\widehat{G}(s) = \frac{Y(s)}{U(s)} = \frac{8.57s^2 + 102s^2 + 836}{s^3 + 9.16s^2 + 192s + 824} \tag{9.51}$$

Matching this estimated transfer function with the analytically derived transfer function from a physical parametrization (7.38) would give

$$\begin{aligned} G(s) = \frac{Y(s)}{U(s)} &= \frac{J_2s^2 + d_2s + k}{J_1J_2s^3 + J_1d_2s^2 + k(J_1 + J_2)s + kd_2} = \\ &= \frac{10s^2 + 100s + 1000}{s^3 + 10s^2 + 200s + 1000} \end{aligned} \tag{9.52}$$

for

$$\begin{pmatrix} J_1 \\ J_2 \\ k \\ d_2 \end{pmatrix} = \begin{pmatrix} 0.1 \\ 0.1 \\ 10 \\ 1 \end{pmatrix} \tag{9.53}$$

The transfer function (9.52) is clearly a *lumped parameter* system that gives an overdetermined system of six nonlinear equations and four unknown variables J_1, J_2, k, and d to solve. A least-squares solution to this problem is

$$\begin{pmatrix} \widehat{J}_1 \\ \widehat{J}_2 \\ \widehat{k} \\ \widehat{d}_2 \end{pmatrix} = \begin{pmatrix} 0.111 \\ 0.100 \\ 9.39 \\ 0.984 \end{pmatrix} \tag{9.54}$$

which shows a close relationship to exist between original and estimated parameters. ■

An unsatisfactory point is clearly the complicated translation of the estimated parameters into physical parameters, and problems of this type are current issues in research. Some of these problems are addressed in the following chapters on continuous-time models and nonlinear identification.

9.6 CLASSIFICATION WITH THE FISHER LINEAR DISCRIMINANT

A standard problem of applied parameter estimation is that a parameter set θ may belong to any of a number of sets. Common examples are the classification of time-varying noise dynamics or changes in operating conditions which give rise to a number of different parameter sets. In such circumstances, it is therefore relevant to find criteria to verify whether θ belongs to an expected set or not.

Consider the problem of classification of a parameter estimate $\widehat{\theta}$ that might belong to either of two classes $\mathcal{A}$ and $\mathcal{B}$. Assume a parameter estimate $\widehat{\theta}$ to be an estimate of either the parameter vector $\theta_{\mathcal{A}} \in \mathcal{A}$ or $\theta_{\mathcal{B}} \in \mathcal{B}$. The two classes $\mathcal{A}$ and $\mathcal{B}$ may each contain elements with a certain parametric variability. Assume the mean m and the covariance R to be

$$\mathcal{E}\{\theta_i\} = m_i, \quad \text{and} \quad \mathcal{E}\{(\theta_i - m_i)(\theta_i - m_i)^T\} = R_i, \qquad i = \mathcal{A} \quad \text{or} \quad \mathcal{B} \tag{9.55}$$

One means of distinguishing between the two classes is the difference of the sample means. If we form a linear combination of the components $\widehat{\theta}$ of the form $\lambda^T\widehat{\theta}$, and we want the projections of $\widehat{\theta}$ from different classes falling on the line defined by λ to be well separated and the separation to reflect the sample means as well as the average variance

$$R = \frac{1}{2}(R_{\mathcal{A}} + R_{\mathcal{B}}) \tag{9.56}$$

then we can compare and classify the projected samples using a single number. To accomplish this we must determine a hyperplane λ that separates the two classes such that a given element can be classified either as belonging to $\mathcal{A}$ or $\mathcal{B}$. It is also of interest to use λ to determine a measure of separation μ

$$\mu = \lambda^T(m_{\mathcal{B}} - m_{\mathcal{A}}) \tag{9.57}$$

To determine λ, we minimize the function

$$J(\lambda) = \frac{(\lambda^T(m_{\mathcal{B}} - m_{\mathcal{A}}))^2}{\lambda^T R \lambda} \tag{9.58}$$

where the denominator serves to make J large with respect to the scattering of the projected samples. The *Fisher linear discriminant* is then

$$\lambda = R^{-1}(m_{\mathcal{B}} - m_{\mathcal{A}}) \tag{9.59}$$

and a parameter vector $\widehat{\theta}$ may now be classified by calculation of

$$\lambda^T\widehat{\theta} = (m_{\mathcal{B}} - m_{\mathcal{A}})^T R^{-1}\widehat{\theta} \tag{9.60}$$

Classification according to the Fisher linear discriminant is done by testing

$$\widehat{\theta} \in \begin{cases} \mathcal{B} & \text{if } \lambda^T\widehat{\theta} > \vartheta + \delta \\ \text{Uncertain} & \text{if } |\lambda^T\widehat{\theta} - \vartheta| \leq \delta \\ \mathcal{A} & \text{if } \lambda^T\widehat{\theta} < \vartheta - \delta \end{cases} \tag{9.61}$$

for some threshold ϑ and a region of uncertainty parametrized by δ. A possible choice of the threshold ϑ is

$$\vartheta = \frac{1}{\sqrt{(\lambda^T R_{\mathcal{A}}\lambda)} + \sqrt{(\lambda^T R_{\mathcal{B}}\lambda)}}\left(\sqrt{(\lambda^T R_{\mathcal{A}}\lambda)}\lambda^T m_{\mathcal{A}} + \sqrt{(\lambda^T R_{\mathcal{B}}\lambda)}\lambda^T m_{\mathcal{B}}\right) \tag{9.62}$$

The threshold value is then chosen in such a way that its distances to $\lambda^T m_{\mathcal{A}}$ and $\lambda^T m_{\mathcal{B}}$ are inversely proportional to the standard deviations as determined

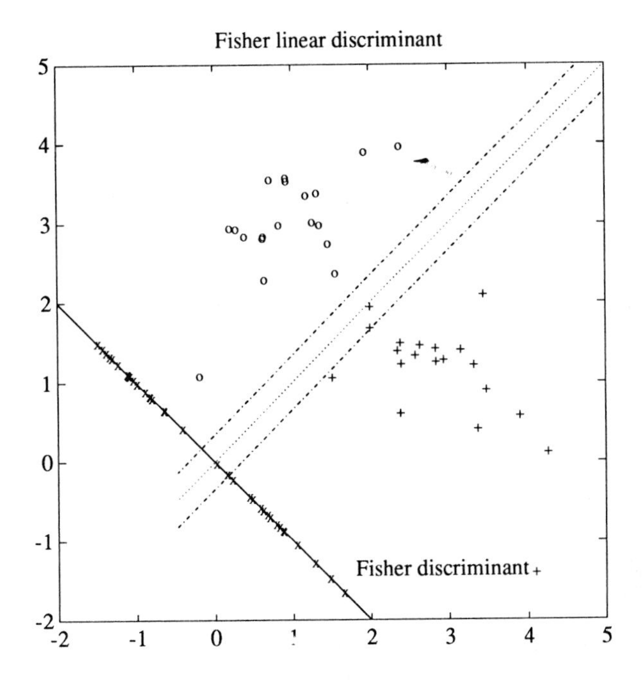

Figure 9.12 Fisher linear discriminant to separate two sets. The samples projected onto the line determined by the Fisher linear discriminant appear to be well separated.

by $R_{\mathcal{A}}$ and $R_{\mathcal{B}}$ of the two classes, respectively. The value of δ may be chosen so that classification is avoided for elements appearing in the interval of overlapping of the two distributions.

Example 9.9—Fisher linear discriminant

Consider the parameter classes $\mathcal{A}$ and $\mathcal{B}$ with mean values

$$m_{\mathcal{A}} = \begin{pmatrix} 1 \\ 3 \end{pmatrix}, \quad \text{and} \quad m_{\mathcal{B}} = \begin{pmatrix} 3 \\ 1 \end{pmatrix}$$

and the covariance matrices

$$R_{\mathcal{A}} = \begin{pmatrix} 1.5 & -0.5 \\ -0.5 & 1.5 \end{pmatrix}, \quad \text{and} \quad R_{\mathcal{B}} = \begin{pmatrix} 1.5 & 0.5 \\ 0.5 & 1.5 \end{pmatrix} \tag{9.63}$$

The Fisher linear discriminant to separate the two classses $\mathcal{A}$ and $\mathcal{B}$ is

$$\lambda = \begin{pmatrix} -1.33 \\ 1.33 \end{pmatrix} \tag{9.64}$$

and the threshold chosen is

$$\vartheta = 0.4575 \tag{9.65}$$

As shown in Fig. 9.12, the Fisher linear discriminant is effective in separating observations from the two classes. The samples and their projections falling on the Fisher linear discriminant are well separated except for two samples which fall within the uncertainty interval. ■

The method can be extended for classification into several classes.

9.7 *THE CONCEPT 'IDENTIFIABILITY'

It was shown in Examples 8.1 and 8.2 that there exist experimental conditions under which a system $\mathcal{S}$ cannot be uniquely identified. From different points of view this problem can be regarded as a problem of parametrization or as a problem originating from a poorly conducted experiment. Both aspects are sometimes discussed in terms of *identifiability*—a notion which has been a point of long and, as yet, unfinished discussion.

The question of whether a certain experiment is sufficiently informative for, say, estimation of p parameters can be approached by reference to the notion of excitation (see Chapter 5). The problem arising in Examples 8.1 and 8.2, for instance, can be explained by the fact that the regressor matrix does not have full rank. For the remaining part of this section, we assume that the experiment has been performed with an excitation that is sufficient relative to the desired model complexity. Thus we avoid further interpretation of poor identifiability due to lack of excitation.

Instead of postulating a set of linear systems, a model set $\mathcal{M}$ can be determined from physical modeling or—sometimes—from the gross behavior of data. For instance, the observation of an output delayed relative to the input, or the observation of an oscillative behavior, may be sufficient to suggest a certain model set. Such a practice, sometimes called *structural identification*, is predicated upon the idea that a model set containing as yet unspecified parameters can be postulated.

Identification methods are based on specific choices for a model set $\mathcal{M}$, its parametrization, and an identification method I determined by an optimization criterion. A particular model of the set $\mathcal{M}$, and described by a parameter vector θ may thus be designated $\mathcal{M}(\theta)$. The parametrization is said to be unique only if, for two parameter vectors θ_1 and θ_2, it holds that

$$\mathcal{M}(\theta_1) = \mathcal{M}(\theta_2) \quad \Rightarrow \quad \theta_1 = \theta_2 \tag{9.66}$$

An example of a unique parametrization is transfer function modeling of a single-input single-output system by means of a rational function in the indeterminate s with co-prime numerator and denominator polynomials so that

$$\mathcal{S} \in \mathcal{M} \quad \Rightarrow \quad \{\theta : \mathcal{S} = \mathcal{M}(\theta)\} = \{\theta_1\}, \quad \theta_1 \in R^p \tag{9.67}$$

where θ_1 is the unique parameter vector that parametrizes $\mathcal{S}$ as a model of the set $\mathcal{M}$.

Identifiablity is thus a property of a parametrization assuming that there is a unique *a priori* system representation which, of course, is independent of the experimental procedure. Identifiability in this form can be used as a notion to describe the ability to correctly estimate parameters in some process and is thus closely related to the idea of a unique model and an associated unique minimum of an identification criterion. Given a parametrized model set $\mathcal{M}$ with a unique parameter vector θ such that $\mathcal{S} = \mathcal{M}(\theta)$, it is necessary to formulate a suitable identification criterion and by extension an identification method that enables the estimation of θ. The system is said to be identifiable if it is possible to design such a procedure. Prediction error methods applied to single-input, single-output ARMAX models constitute one such category of procedures, in which the sample covariance matrix of the prediction error is minimized according to a suitable scalar optimization criterion. In this form, identifiability is closely related to to consistency properties of an estimation method. In fact, such identifiability can be regarded as a set of stipulative definitions of the conditions that imply consistency.

Assuming identifiability thus leads to powerful statements on consistency that are valuable for the analysis of limiting properties of parameter estimates. As assumptions on identifiability may also be restrictive, however, it is difficult to approach identification of state-space models and multi-input, multi-output systems for which, in general, there exists no unique parametrization.

Identifiability concepts also affect the validation procedure. Traditional approaches to solving the validation problem involve evaluation of the model misfit in the form of a residual sequence, assuming that the residual time series is a realization of a stochastic process. Statistical hypothesis testing of the stochastic nature of the residuals is then used, and the model is rejected when it has been refuted by data in a number of such statistical tests.

This circumstance raises yet another problem, as several models suggested for explanation of a data record will be unable to fit data in the statistical sense. In particular, it is necessary to distinguish between the lack of fit between model and data due to random processes and that due to lack of model complexity. It is, of course, also relevant to provide methods which give accurate results in the sense of some approximation criterion. This is a topic of Chapter 10. The combined problem of simultaneous model approximation and stochastic modeling remains to be solved. Among the methods already available, output error estimation seems to address this problem in the most straightforward manner.

9.8 CONCLUDING REMARKS

A difficult aspect of structure determination is to obtain a meaningful optimization criterion that provides a good compromise between simplicity and complexity, yielding a model that gives a better fit to data. The statistical tests presented are formulated as decision problems at a given significance level α, say 95%. Acceptability of a model is usually formulated as the decision problem of accepting or rejecting the null hypothesis $\mathcal{H}_0$. It is important to bear in mind that a serious problem is the risk of accepting $\mathcal{H}_0$ when it is not true, *i.e.*, when the alternative hypothesis should be accepted. As no statistical properties are tied to the alternative hypothesis, it is impossible to quantify the risk of accepting the null hypothesis, *i.e.*, the risk of accepting a wrong model. This is clearly an inherent weakness of the statistical tests presented. Acceptance of models based on statistical methods only must therefore be discouraged as there is no way of verifying the validity of the decision. For this reason it is important to pay some attention to the simulation performance and other methods that are of a different nature from those of statistical tests.

9.9 BIBLIOGRAPHY AND REFERENCES

Early work on statistical hypothesis testing is to be found in

– J. NEYMAN AND E.S. PEARSON, "On the problem of the most efficient tests of statistical hypotheses." *Trans. Roy. Soc. London, Series A*, Vol. 231, 1933, pp. 289–337.

Methods of statistical analysis are to be found in

– D.R. COX AND D.V. HINKLEY, *Theoretical Statistics*, London: Chapman and Hall, 1974.

– M.G. KENDALL AND A. STUART, *The Advanced Theory of Statistics, Vol. 1*, 3d ed., 1969 and *Vol. II, 3d ed.*, New York: Hafner Press, 1973.

Model validation in the context of ARMAX model identification is treated in

– T. SÖDERSTRÖM AND P. STOICA, *System Identification* London: Prentice-Hall, 1989.

Sources for the AIC and MDL model order tests are to be found in

– H. AKAIKE, "A new look at the statistical model identification." TAC-19, 1977, pp. 718–723.
– R. SHIBATA, "Asymptotically efficient selection of the order of a model for estimating parameters of a linear process." *Ann. Statistics*, Vol. 8, 1980, pp. 147–164.
– G. SCHWARTZ, "Estimating the dimension of a model." *Ann. Statistics*, Vol. 6, 1978, pp. 461–464.
– J. RISSANEN,"Modeling by shortest data description," *Automatica*, Vol. 14, 1978, pp. 465–471.
– D. BURSHTEIN AND E. WEINSTEIN, "On the application of the Wald statistic to order estimation of ARMA models." TAC-36, 1992, pp. 1091–1096.

The Fisher discriminant test was originally considered in

– R.A. FISHER, "The statistical utilization of multiple mesurements." *Ann. Eugen.*, Vol.8, 1938, pp. 376–386.

9.10 EXERCISES

9.1 Show that the covariance function of the least-squares estimate of the parameters of the linear model $\mathcal{Y}_N = \Phi_N\theta + e$ with $\mathcal{E}\{e\} = 0$ and $\mathcal{E}\{ee^T\} = \Sigma_e$ can be expressed as

$$\mathrm{Cov}\{\begin{pmatrix}\varepsilon \\ \tilde{\theta}_N\end{pmatrix}\} = \begin{pmatrix} I & \Phi_N \\ \Phi_N^T & 0\end{pmatrix}^{-1}\begin{pmatrix}\Sigma_e & 0 \\ 0 & 0\end{pmatrix}\begin{pmatrix} I & \Phi_N \\ \Phi_N^T & 0\end{pmatrix}^{-1} \tag{9.68}$$

Hint: Use the results previously derived for the augmented system matrix (5.104) and (5.117).

9.2 Determine the covariance between ε_N and $\tilde{\theta}_N$ for a least-squares estimate (*cf.* Exercise 9.1). ■

10

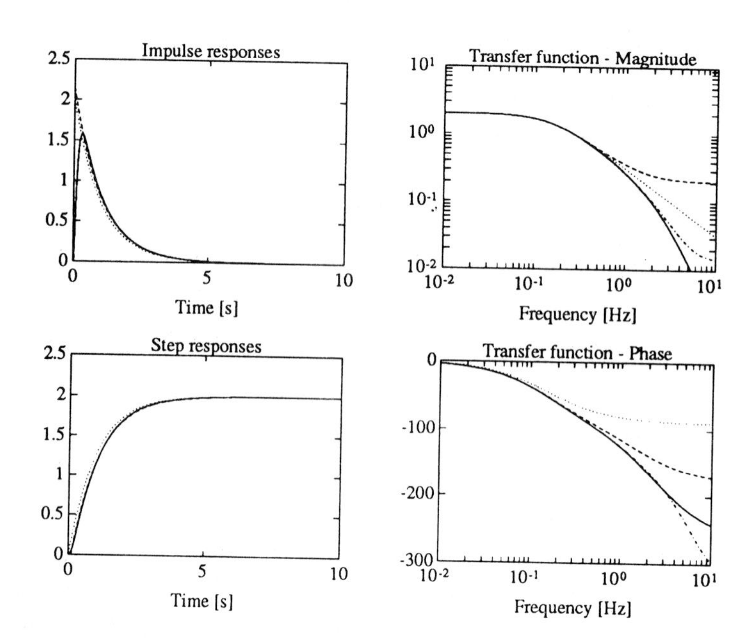

Model Approximation

10.1 INTRODUCTION

Model approximation and *model reduction* refer to methods for simplification, approximation, and order reduction of models of dynamic systems. Model approximation is of importance for extraction of dominant features of a model or to reduce a high-order time-series model to a lower-order structure motivated by physical considerations or for evaluation of the relative importance of subsystems in some large-scale system.

The term *model reduction* may cover several aspects such as (*1*) model order reduction in a linear system or (*2*) model approximation of a nonlinear differential equation by a linear system or (*3*) approximation of the nonlinear system by ignoring higher-order harmonics.

The first aspect is elaborated and we start this chapter by presenting some problems associated with heuristic model reduction methods. Then we proceed by stating some systematic methods of model reduction based on balanced realization. Other methods such as the Padé approximation, moment matching, and continued fraction approximations are presented with some attention to their shortcomings and drawbacks.

The second and third aspects above can be approached by various types of linearization techniques, *e.g.*, standard linearization and harmonic linearization. Model approximation of nonlinear systems is presented in the form of linearization and describing function analysis. Finally, the chapter includes a perspective on the use of model approximation methods in identification.

We start by considering some familiar model approximation methods that are basic in system analysis, *i.e.*, linearization and discretization.

Linearization

A fundamental method is *linearization* of nonlinear differential equation

$$\dot{x} = f(x, u) \tag{10.1}$$

which is replaced by the approximate equation

$$\dot{x} \approx f_0 + f_x(x_0, u_0)(x - x_0) + f_u(x_0, u_0)(u - u_0) \tag{10.2}$$

obtained from a truncated Taylor series expansion around a linearization point (x_0, u_0). The partial derivatives $f_x = \partial f/\partial x$ and $f_u = \partial f/\partial u$ are evaluated at the point (x_0, u_0).

Discretization

Discretization of system dynamics is a form of approximation where the approximated system and its approximate are equal or close at the time instants as defined by the discretization or measurement. A linear system with input $u(t)$ constant between the sampling instants, *i.e.*, step-response equivalenceor zero-order-hold input, can be discretized in time as

$$\begin{cases} \dot{x}(t) &= Fx(t) + Gu(t) \\ y(t) &= Cx(t) \end{cases} \quad \Rightarrow \quad \begin{cases} x_{k+1} &= \Phi x_k + \Gamma u_k \\ y_k &= Cx_k \end{cases} \tag{10.3}$$

where $x_k = x(t_k)$ at some sequence of time instants t_k. For equidistant time instants $t_k = kh$ there is a transformation

$$\Phi = e^{Fh}$$
$$\Gamma = \int_0^h e^{Fs} G ds \tag{10.4}$$

Model structures of the type found in Eq. (10.1) or Eq. (10.3) are often tacitly assumed in control systems analysis, although such models are indeed the result of modeling and identification.

Some heuristic model reduction methods

Popular methods of model order reduction are *polynomial truncation*, the *method of dominating poles*, and *pole-zero cancellations*, which are all applicable to linear systems only. Consider, for example, a linear system with the transfer function

$$H(z) = \frac{0.22z^{-1}}{1 - 0.7z^{-1} - 0.08z^{-2}} = \frac{0.22z^{-1}}{(1 - 0.8z^{-1})(1 + 0.1z^{-1})} \tag{10.5}$$

There is obviously a considerable difference in the two time constants. It is therefore reasonable to search for methods that reduce the model order without seriously affecting the input-output behavior. A natural but poor method of model reduction is simply to truncate the numerator and denominator polynomials (*polynomial truncation*) with a subsequent compensation of the static gain (*i.e.*, at $z = 1$)

$$H(z) \approx H_1(z) = \frac{0.3z^{-1}}{1 - 0.7z^{-1}} \tag{10.6}$$

It is easy to verify by impulse response and step response that H_1 is a poor approximation of $H(z)$; see Fig. 10.1. This approximation is poor also if the truncation is supported by compensation to maintain the same static gain. The reason is that the pole-zero location is very sensitive to the higher-order coefficients. (Similar problems appear in a realization of Eq. (10.5) by methods of limited parameter accuracy.)

It often makes more sense to keep the dominating poles and to eliminate fast modes while preserving the static gain (see Fig. 10.1). If this method is applied to Eq. (10.5) we have

$$H(z) \approx H_2(z) = \frac{0.2z^{-1}}{1 - 0.8z^{-1}} \tag{10.7}$$

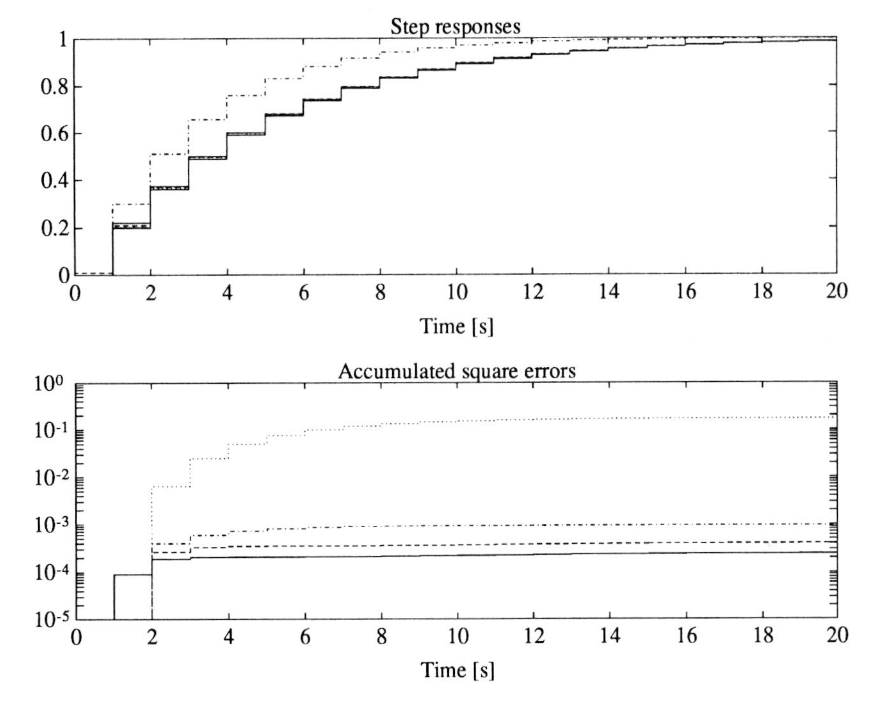

Figure 10.1 The upper graph shows the step responses of Eq. (10.6) and its reduced order models. The truncated pole polynomial model reduction exhibits the gross error. The lower graph shows the sum of squares of the model reduction error of the pole cancellation (*dotted line*), dominating pole method (*dotted and dashed line*), the moment matching (*dashed line*), and the balanced realization (*solid line*).

Pole-zero cancellation is often applied to transfer functions in the following manner

$$H_3(z) = \frac{z^{-1} - 0.78z^{-2}}{1 - 0.7z^{-1} - 0.08z^{-2}} \approx \frac{1.1z^{-1}}{1 + 0.1z^{-1}} \tag{10.8}$$

where a numerator factor $(z - 0.78)$ has been cancelled with a denominator factor $(z - 0.8)$ with a compensation to maintain the static gain.

10.2 BALANCED REALIZATION AND MODEL REDUCTION

As shown in Fig. 10.1 it is, however, easy to demonstrate serious shortcomings of the presented heuristic methods according to criteria of preserved static gain, step responses, impulse responses, or least-squares fitting. It is therefore desirable to derive methods for model approximation based on some sensitivity analysis of the input-output properties. An interesting approach in this context is the *balanced realizations* and model approximation methods based on this methodology.

Consider the state-space equation

$$\mathcal{S}: \quad \begin{cases} x_{k+1} &= \Phi x_k + \Gamma u_k, \qquad x_k \in R^n \\ y_k &= C x_k \end{cases} \tag{10.9}$$

A relevant question is to develop a quantitative measure on the observability and controllability of the states which can be reformulated by means of the following question: *What states can be reached with a given input energy assuming that* $x_0 = 0$*?* Consider for this reason the case of a finite input energy $J(u)$ where

$$J(u) = e_{uu} = \sum_{k=1}^{N} u_k^T u_k \leq 1 \tag{10.10}$$

Direct calculation of the state x_N from an input sequence $\{u_k\}_{k=0}^{N-1}$ *via* Eq. (10.9) gives

$$\begin{aligned} x_N &= \sum_{k=1}^{N} \Phi^{N-k} \Gamma u_{k-1} = \\ &= \begin{pmatrix} \Phi^{N-1}\Gamma & \Phi^{N-2}\Gamma & \ldots & \Gamma \end{pmatrix} \begin{pmatrix} u_0 \\ \vdots \\ u_{N-1} \end{pmatrix} = \psi_N U_N \end{aligned} \tag{10.11}$$

where ψ_N and U_N are defined from Eq. (10.11). As there are infinitely many control sequences U_N that result in the state x_N for $N > n$, it is suitable to choose the one with smallest 2-norm. The suitable control sequence $\{u_k\}_{k=0}^{N-1}$ that results in the desired state x_N and which is obtained by means of the pseudo-inverse of ψ_N is

$$U_N = \begin{pmatrix} u_0 \\ \vdots \\ u_{N-1} \end{pmatrix} = \psi_N^T (\psi_N \psi_N^T)^{-1} x_N \tag{10.12}$$

A control sequence chosen according to Eq. (10.12) yields a restriction of the input effort so that

$$1 \geq J(u^*) = \sum_{k=1}^{N-1} u_k^T u_k = U_N^T U_N = x_N^T P_N^{-1} x_N \tag{10.13}$$

where the *reachability Gramian* P_N is defined as

$$P_N = \psi_N \psi_N^T = \sum_{k=0}^{N-1} \Phi^k \Gamma \Gamma^T (\Phi^T)^k \geq 0 \tag{10.14}$$

As $P_N \geq 0$ is positive semidefinite, it can be concluded that Eq. (10.13) provides a quadratic form with a bound on the reachable states at time N and it is obvious that

$$\sum_{k=0}^{N} u_k^2 = U_N^T U_N < 1 \quad \Rightarrow \quad x_N^T P_N^{-1} x_N \leq 1, \quad \forall N \tag{10.15}$$

It is obvious from Eq. (10.14) that the reachability Gramian P_N satisfies the recursive equation

$$P_{N+1} = \Phi P_N \Phi^T + \Gamma \Gamma^T \tag{10.16}$$

The solution P_N for a stable matrix Φ approaches the solution to the Lyapunov equation

$$\Phi P \Phi^T - P + \Gamma \Gamma^T = 0 \tag{10.17}$$

and the asymptotic reachability Gramian $P = \lim_{N \to \infty} P_N$ satisfies the Lyapunov equation (10.17).

The observability Gramian

Again consider the state-space equation (10.9) and the following question: *What state energy is necessary for $u_k = 0$ in order to obtain a specified output energy?* A derivation analogous to that of the reachability Gramian gives

$$1 = \sum_{k=0}^{\infty} y_k^2 = x^T(0) Q x(0) \tag{10.18}$$

where the observability Gramian Q is defined as the infinite sum

$$Q = \sum_{k=0}^{\infty} (\Phi^T)^k C^T C \Phi^k \tag{10.19}$$

and satisfies the Lyapunov equation

$$\Phi^T Q \Phi - Q + C^T C = 0 \tag{10.20}$$

Balanced realization and model reduction

It is obvious that the reachability and observability Gramians P and Q define matrices that describe the sensitivities of the input-output map in different directions of state space (and independent of direct terms). Consider a state-space transformation $z_k = Tx_k$ and its state-space equations

$$S: \quad \begin{cases} z_{k+1} & = \Phi' z_k + \Gamma' u_k = T\Phi T^{-1} z_k + T\Gamma u_k \\ y_k & = C' z_k = CT^{-1} z_k \end{cases} \tag{10.21}$$

The Gramians for the system in Eq. (10.21) are

$$\begin{aligned} P_z &= TPT^T \\ Q_z &= T^{-T} Q T^{-1} \end{aligned} \tag{10.22}$$

Different state-space realizations may thus result in different Gramians, and it can be questioned whether there is any transformation T such that $P_z = Q_z$. In fact, this property is achieved by choosing a state-space representation z with equal and diagonal controllability and observability Gramians where

$$P_z = Q_z = \Sigma = \mathrm{diag}(\sigma_1, \sigma_2, \ldots), \quad \text{with} \quad \sigma_i = \sqrt{\lambda_i(PQ)} \tag{10.23}$$

where $\lambda_i(PQ)$ denotes the ith eigenvalue of the matrix PQ and Σ is a diagonal matrix with elements σ_i. One algorithm uses the Cholesky factors Q_1, U, Σ_1, of the matrices P, Q, Σ as intermediate results and determines the state-space transformation matrix T

$$\begin{aligned} Q &= Q_1^T Q_1 \\ Q_1 P Q_1^T &= U \Sigma^2 U^T \\ U^T U &= I \\ \Sigma &= \Sigma_1^T \Sigma_1 \\ T &= \Sigma_1^{-1} U^T Q_1 \end{aligned} \tag{10.24}$$

The resulting state-space realization is interesting because it has similar ("balanced") properties of reachability and observability, and the magnitude of the elements σ_i of the Gramian Σ expresses the relative importance of each state $(z_k)_i$ for the input-output behavior.

Consider the diagonal matrix Σ where large values σ_i represent essential states z_i whereas small σ_j represent states z_j that are less important for the input-output behavior. It is essentially straightforward to suggest elimination of rows j and columns j of a state-space realization with a small element σ_j of the Gramian. Let $\Phi' = T\Phi T^{-1}$ and $\Gamma' = T\Gamma$ denote the transformed system matrices of Eq. (10.21) and let the state vector $z_k = Tx_k$ be decomposed as

$$z_k = \begin{pmatrix} z_k^1 \\ z_k^0 \end{pmatrix} \tag{10.25}$$

where z_k^0 is the vector of components that are suggested to be eliminated. The state-space equation in Eq. (10.21) is then

$$\begin{aligned} z_{k+1} &= \begin{pmatrix} z_{k+1}^1 \\ z_{k+1}^0 \end{pmatrix} = \Phi' z_k + \Gamma' u_k = \begin{pmatrix} \Phi_{11} & \Phi_{10} \\ \Phi_{01} & \Phi_{00} \end{pmatrix} \begin{pmatrix} z_k^1 \\ z_k^0 \end{pmatrix} + \begin{pmatrix} \Gamma_1 \\ \Gamma_0 \end{pmatrix} u_k \\ y_k &= C' z_k = \begin{pmatrix} C_1 & C_0 \end{pmatrix} \begin{pmatrix} z_k^1 \\ z_k^0 \end{pmatrix} \end{aligned} \tag{10.26}$$

If we neglect the dynamics of z_k^0 by assuming that z_k^0 has no dynamics independent of z_k^1 and u_k and therefore eliminate $z_k^0 = (I - \Phi_{00})^{-1}(\Phi_{01} z_k^1 + \Gamma_0 u_k)$ from Eq. (10.26) we obtain the reduced order model dynamics

$$\begin{aligned} z_{k+1}^1 &= (\Phi_{11} + \Phi_{10}(I - \Phi_{00})^{-1}\Phi_{01}) z_k^1 + (\Gamma_1 + \Phi_{10}(I - \Phi_{00})^{-1}\Gamma_0) u_k \\ y_k &= (C_1 + C_0(I - \Phi_{00})^{-1}\Phi_{01}) z_k^1 + C_0(I - \Phi_{00})^{-1}\Gamma_0 u_k \end{aligned} \tag{10.27}$$

A model reduction guided by the magnitude of the singular values in Σ is a *balanced model reduction*. We illustrate the procedure with the following example.

Example 10.1—Model reduction from balanced realization
Consider the transfer function

$$H(z) = \frac{0.22z^{-1}}{1 - 0.7z^{-1} - 0.08z^{-2}} \tag{10.28}$$

with the controllable canonical realization

$$\begin{aligned} x(k+1) &= \begin{pmatrix} 0.7 & 0.08 \\ 1 & 0 \end{pmatrix} x(k) + \begin{pmatrix} 1 \\ 0 \end{pmatrix} u(k) \\ y(k) &= \begin{pmatrix} 0.22 & 0 \end{pmatrix} x(k) \end{aligned} \tag{10.29}$$

A balanced realization is

$$\begin{aligned} x(k+1) &= \begin{pmatrix} 0.7869 & 0.1079 \\ 0.1079 & -0.0869 \end{pmatrix} x(k) + \begin{pmatrix} 0.4579 \\ -0.1018 \end{pmatrix} u(k) \\ y(k) &= \begin{pmatrix} 0.4579 & -0.1018 \end{pmatrix} x(k) \end{aligned} \tag{10.30}$$

with

$$\Sigma = \begin{pmatrix} 0.5510 & 0 \\ 0 & 0.0169 \end{pmatrix} \tag{10.31}$$

and the state-space transformation matrix

$$T = \begin{pmatrix} 0.4579 & 0.0288 \\ -0.1018 & 0.1295 \end{pmatrix} \tag{10.32}$$

The elements of the diagonalized Gramian Σ are of different magnitudes, which indicates that a first-order model would be sufficient. Elimination of the second state vector component of Eq. (10.30) according to Eq. (10.27) results in balanced model reduction with the first-order model

$$\begin{aligned} x(k+1) &= 0.7976x(k) + 0.4478u(k) \\ y(k) &= 0.4478x(k) + 0.0095u(k) \end{aligned} \tag{10.33}$$

■

*Balanced model reduction for continuous-time systems

Assuming that the balanced state-space representation of a given continuous-time system is

$$\begin{aligned} \frac{d}{dt}\begin{pmatrix} x_1 \\ x_0 \end{pmatrix} &= \begin{pmatrix} A_{11} & A_{10} \\ A_{01} & A_{00} \end{pmatrix}\begin{pmatrix} x_1 \\ x_0 \end{pmatrix} + \begin{pmatrix} B_1 \\ B_0 \end{pmatrix} u \\ y &= \begin{pmatrix} C_1 & C_0 \end{pmatrix}\begin{pmatrix} x_1 \\ x_0 \end{pmatrix} + Du \end{aligned} \tag{10.34}$$

one can approach model reduction by approximating

$$\dot{x}_0 = 0, \quad \text{and} \quad x_0 = -A_{00}^{-1}A_{01}x_1 - A_{00}^{-1}B_0u \tag{10.35}$$

The reduced-order state-space model is then

$$\begin{aligned} \dot{x}_1 &= (A_{11} - A_{10}A_{00}^{-1}A_{01})x_1 + (B_1 - A_{10}A_{00}^{-1}B_0)u \\ y &= (C_1 - C_0A_{00}^{-1}A_{01})x_1 + (D - C_0A_{00}^{-1}B_0)u \end{aligned} \tag{10.36}$$

which contains a direct term from u to y also if the full-order direct term is zero. This reduction principle is natural in many applications as it eliminates the dynamics of x_0 while it preserves the essential low-frequency properties. This principle is at least natural in applications where it is important to have a good fit between the full-order model and the reduced-order model in the low-frequency range. Another choice is to put $x_0 \approx 0$, which results in the reduced-order model

$$\begin{aligned} \dot{x}_1 &= A_{11}x_1 + B_1 u \\ y &= C_1 x_1 + Du \end{aligned} \tag{10.37}$$

or more generally $\dot{x}_2 \approx \alpha x_2$ for some constant α which results in the model

$$\begin{aligned} \dot{x}_1 &= (A_{11} + A_{10}(\alpha I - A_{00})^{-1}A_{01})x_1 + (B_1 + A_{10}(\alpha I - A_{00})^{-1}B_0)u \\ y &= (C_1 + C_0(\alpha I - A_{00})^{-1}A_{21})x_1 + (D + C_0(\alpha I - A_{00})^{-1}B_0)u \end{aligned} \tag{10.38}$$

If one denotes the full-order transfer function $G(s)$ and the reduced-order transfer function as $G_{red}(s)$, it follows

$$\begin{aligned} \dot{x}_2 &= 0, && \Rightarrow && G_{red}(0) = G(0) \\ x_2 &= 0, && \Rightarrow && G_{red}(\infty) = G(\infty) \\ \dot{x}_2 &= \alpha x_2, && \Rightarrow && G_{red}(\alpha) = G(\alpha) \end{aligned} \tag{10.39}$$

from which it can be concluded that these model-reduction principles appear to have good fit at different points in the complex frequency domain.

Example 10.2—Rohrs' system
Consider the following third-order linear system

$$G(s) = \frac{2}{s+1} \cdot \frac{229}{s^2 + 30s + 229} \tag{10.40}$$

A balanced realization, *i.e.*, a state-space model of this transfer function, is

$$\begin{aligned} \dot{x} &= \begin{pmatrix} -0.6683 & -1.6355 & -0.6166 \\ 1.6355 & -8.2111 & -7.4027 \\ -0.6166 & 7.4027 & -22.1206 \end{pmatrix} x + \begin{pmatrix} 1.2136 \\ -1.3380 \\ 0.5634 \end{pmatrix} u \\ y &= \begin{pmatrix} 1.2136 & 1.3380 & 0.5634 \end{pmatrix} x \end{aligned} \tag{10.41}$$

with Gramian and its singular values

$$\Sigma = \begin{pmatrix} 1.1018 & 0 & 0 \\ 0 & 0.1090 & 0 \\ 0 & 0 & 0.0072 \end{pmatrix} \tag{10.42}$$

and the transformation matrix T between x and $z = Tx$.

$$T = \begin{pmatrix} 1.2136 & 38.6514 & 345.4191 \\ -1.3380 & -32.6780 & 26.8176 \\ 0.5634 & -5.6501 & 5.1795 \end{pmatrix} \tag{10.43}$$

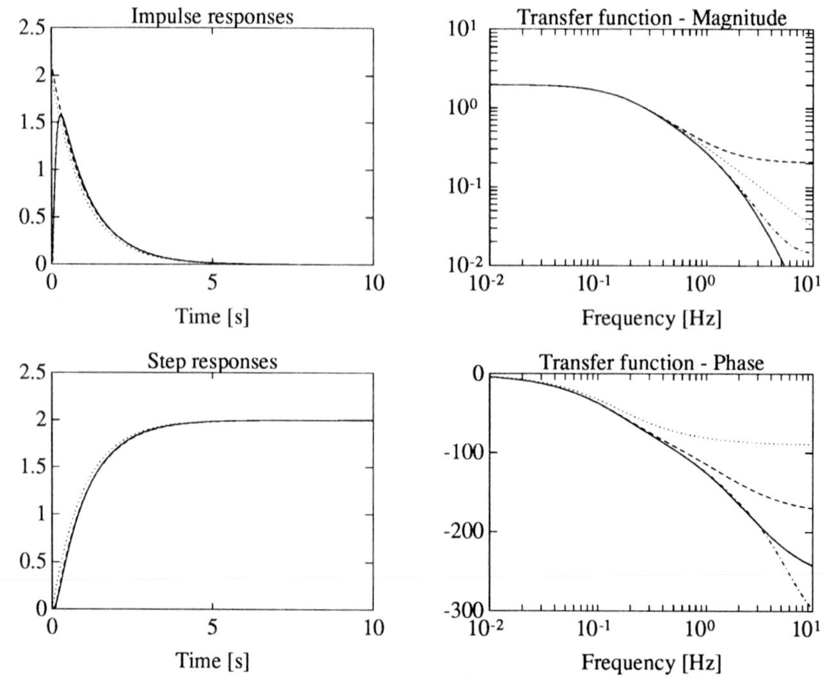

Figure 10.2 Step responses, impulse responses, and transfer function from Rohrs' system and from reduced-order models. The step responses of the balanced model reduction (*dashed lines*) are virtually indistinguishable from the original response (*solid line*). Responses of the heuristically reduced first-order model $2/(s+1)$ are shown by the dotted line.

The reduced second-order model is

$$
\begin{aligned}
\dot{x} &= \begin{pmatrix} -0.6512 & -1.8419 \\ 1.8419 & -10.6884 \end{pmatrix} + \begin{pmatrix} 1.1979 \\ -1.5266 \end{pmatrix} u \\
y &= \begin{pmatrix} 1.1979 & 1.5266 \end{pmatrix} x + 0.0144u
\end{aligned}
\tag{10.44}
$$

with the transfer function

$$
G_2(s) = \frac{-0.8955s + 20.5561}{s^2 + 11.3395s + 10.3523} + 0.0144 \tag{10.45}
$$

The reduced first-order model is

$$
\begin{aligned}
\dot{x} &= -0.9686x + 1.4610u \\
y &= 1.4610x - 0.2037u
\end{aligned}
\tag{10.46}
$$

Notice that the step responses and the impulse responses in Fig. 10.2 are well preserved through the model reduction as compared to heuristic model reduction to $G_1(s) = 2/(s+1)$. ■

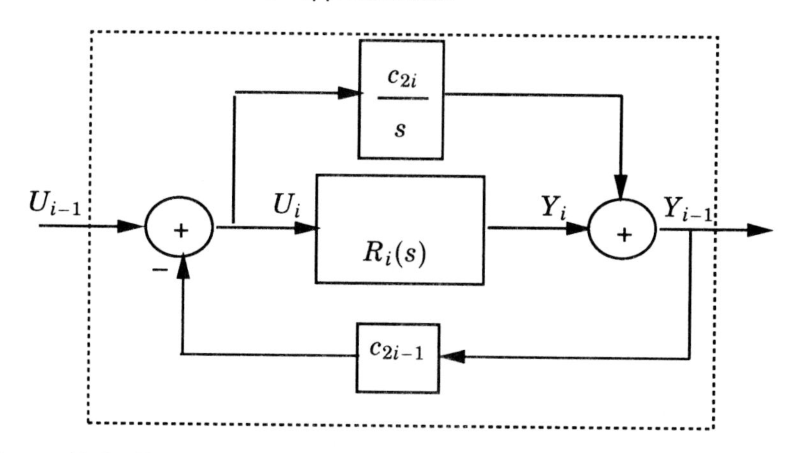

Figure 10.3 Expansion of a transfer function $R_{i-1}(s)$ into the coefficients c_{2i-1}, c_{2i} and a residual transfer function block $R_i(s)$ as practiced in continued fraction approximation.

10.3 CONTINUED FRACTION APPROXIMATION

Consider an asymptotically stable system with a transfer function $G(s)$ and develop the high-order transfer function

$$G(s) = \frac{B(s)}{A(s)} = \frac{1}{c_1 + \cfrac{1}{\cfrac{c_2}{s} + R_1(s)}} = \frac{1}{c_1 + \cfrac{1}{\cfrac{c_2}{s} + \cfrac{1}{c_3 + \cfrac{1}{\cfrac{c_4}{s} + R_2(s)}}}} \tag{10.47}$$

by expanding $G_0(s)$ into the coefficients $c_1, c_2, \ldots, c_{2m}$, and a residual transfer function block $R_m(s)$; see Fig. 10.3and Fig. 10.4. A model reduction method can be proposed by means of the approximation $R_m(s) = 0$, which truncates the sequence of coefficients after $2m$. This results in an mth reduced-order model with the approximating transfer function

$$G_m(s) = \frac{B_m(s)}{A_m(s)} \tag{10.48}$$

with coefficients obtained from calculations reminiscent of those for the Routh criterion of stability. Consider the numerator polynomial and denominator

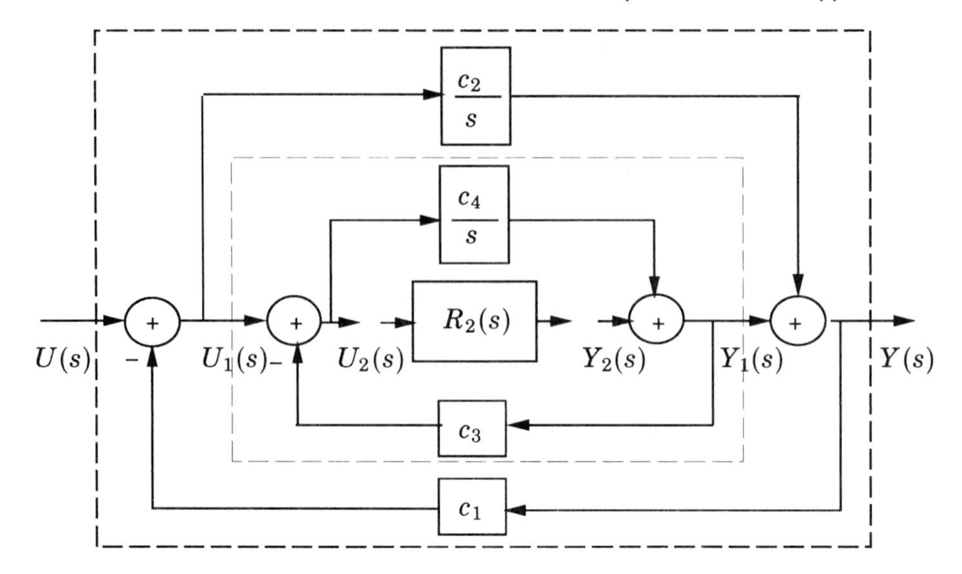

Figure 10.4 Interpretation of the continued fraction approximation as a recursive closed-loop system with subsystems G_i.

polynomials of the full-order transfer function $G(s)$ expressed as

$$\begin{aligned} A(s) &= a_0 + a_1 s + a_2 s^2 + \cdots = P^{(-2)}(s) \\ B(s) &= b_0 + b_1 s + b_2 s^2 + \cdots = P^{(-1)}(s) \end{aligned} \tag{10.49}$$

with polynomial coefficients entering the Routh array

$$\begin{array}{cccc} a_0 & a_1 & a_2 & \cdots \\ b_0 & b_1 & b_2 & \cdots \\ p_0^{(0)} & p_1^{(0)} & p_2^{(0)} & \cdots \\ p_0^{(1)} & p_1^{(1)} & p_2^{(1)} & \cdots \\ \vdots & & & \\ p_0^{(i)} & p_1^{(i)} & p_2^{(i)} & \cdots \\ \vdots & & & \end{array} \quad \Rightarrow \quad \begin{pmatrix} c_1 \\ c_2 \\ c_3 \\ \vdots \\ c_{i+2} \\ \vdots \end{pmatrix} = \begin{pmatrix} a_0/b_0 \\ b_0/p_0^{(0)} \\ p_0^{(0)}/p_0^{(1)} \\ \vdots \\ p_0^{(i-1)}/p_0^{(i)} \\ \vdots \end{pmatrix} \tag{10.50}$$

where the kth polynomial coefficient $p_k^{(i)}$ at iteration i is determined by the order recursive equation

$$p_k^{(i)} = p_{k+1}^{(i-2)} - p_0^{(i-2)} \frac{p_{k+1}^{(i-1)}}{p_0^{(i-1)}}, \qquad \begin{cases} k = 0, 1, 2, \ldots \\ i = 0, 1, 2, \ldots \end{cases} \tag{10.51}$$

The coefficients $\{c_i\}$ of the continued fraction expansion are found as successive ratios of the elements of the left-most column of the Routh array.

Example 10.3—Continued fraction approximation
Consider again Rohrs' transfer function

$$G(s) = \frac{458}{s^3 + 31s^2 + 259s + 229} = \frac{2}{1 + 1.131s + 0.1354s^2 + 0.0044s^3} \tag{10.52}$$

The Routh array of the coefficients of the denominator polynomial (*first row*) and the numerator polynomial (*second row*) is presented below where the coefficients c_i appear as the ratios of successive elements of the first column.

$$\begin{array}{llll} 1.0000 & 1.1310 & 0.1354 & 0.0044 \\ 2.0000 & 0.0 & 0.0 & 0.0 \\ 1.1310 & 0.1354 & 0.0044 & \\ -0.2394 & -0.0078 & & \\ 0.0986 & 0.0044 & & \\ 0.0029 & & & \\ \vdots & & & \end{array} \quad \Rightarrow \quad \begin{array}{l} c_1 = 0.5000 \\ c_2 = 1.7683 \\ c_3 = -4.7236 \\ c_4 = -2.4272 \\ c_5 = 34.029 \\ \vdots \end{array} \tag{10.53}$$

A first-order continued fraction approximation of $G(s)$ is then obtained as

$$G(s) \approx G_1(s) = \frac{1}{c_1 + \frac{s}{c_2}} = \frac{1.7683}{s + 0.8842} = \frac{2.0000}{1 + 1.1310s} \tag{10.54}$$

A second-order approximation $G_2(s)$ of $G(s)$ is obtained as

$$\begin{aligned} G_2(s) &= \frac{(c_2 + c_4)s + c_2c_3c_4}{s^2 + (c_1c_2 + c_1c_4 + c_3c_4)s + c_1c_2c_3c_4} = \\ &= \frac{-0.6588s + 20.2744}{s^2 + 11.1357s + 10.1372} = \frac{2.000 - 0.0650s}{1 + 1.0985s + 0.0986s^2} \end{aligned} \tag{10.55}$$

which reproduces the static gain of the full-order model (see Fig. 10.5). ■

Interpretation of continued fraction approximation

The continued fraction approximation is based on the expansion

$$\frac{B(s)}{A(s)} = \cfrac{1}{c_1 + \cfrac{1}{\cfrac{c_2}{s} + \cfrac{1}{c_3 + \cfrac{1}{\cfrac{c_4}{s} + \cdots}}}} \tag{10.56}$$

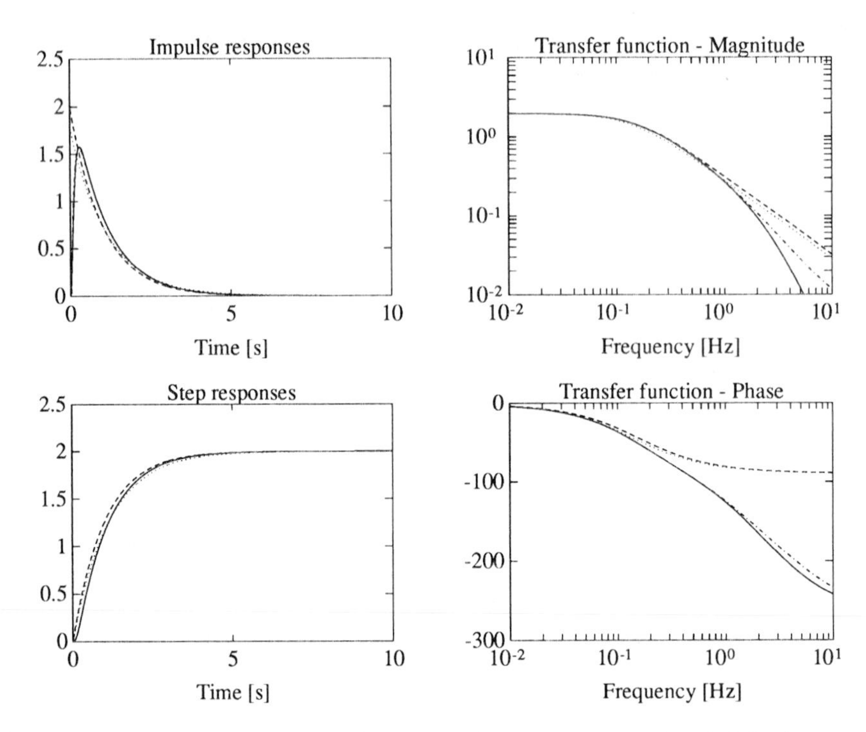

Figure 10.5 Step responses and transfer function magnitude of Rohrs' system (10.24) (*solid line*) and reduced-order models (*dashed lines*) by means of continued fraction approximation. The dotted line shows the result from the heuristic model reduction $G(s) = 2/(s + 1)$.

which may be interpreted graphically according to Fig. 10.3 and Fig. 10.4. The model reduction presupposes that the innermost transfer function block R_m may be eliminated, *i.e.*, $R_m(s) \approx 0$. Truncation of the continued fraction approximation after two and four coefficients provides the first- and second-order approximations, respectively. A condition for a good approximation is that the feedforward gain c_{2m}/s is much larger than the gain of the omitted transfer function block $R_m(s)$.

The continued fraction expansion may also be applied to discrete-time systems for which the order recursion and the forward shift properties interact in an interesting manner. Consider the input and output variables which exhibit the dependencies

$$\begin{aligned} Y_i(z) &= c_{2i}z^{-1}U_{i+1}(z) + Y_{i+1}(z) \\ U_{i+1}(z) &= -c_{2i-1}Y_i(z) + U_i(z) \end{aligned} \tag{10.57}$$

with a mixed backward/forward dependence in the order recursion; Fig. 10.6.

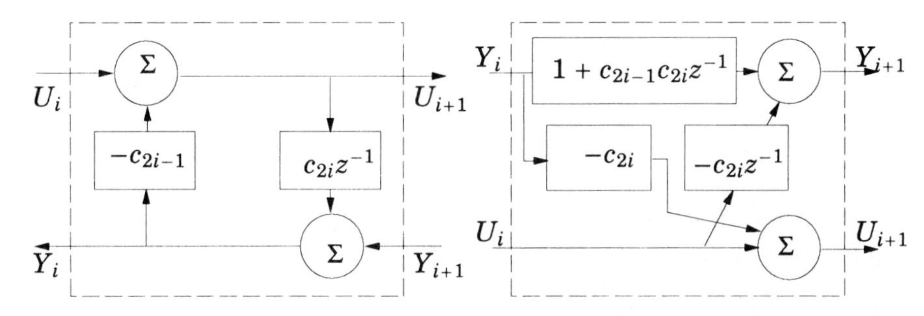

Figure 10.6 Interpretation of the continued fraction expansion as a recursive transfer function relationship in a forward (*right*) or a forward/backward manner (*left*).

A pure forward-order recursive equation is obtained as

$$\begin{pmatrix} Y_{i+1}(z) \\ U_{i+1}(z) \end{pmatrix} = \begin{pmatrix} 1 + c_{2i-1}c_{2i}z^{-1} & -c_{2i}z^{-1} \\ -c_{2i-1} & 1 \end{pmatrix} \begin{pmatrix} Y_i(z) \\ U_i(z) \end{pmatrix} \tag{10.58}$$

where the transfer matrix is unimodular, *i.e.*,

$$\det \begin{pmatrix} 1 + c_{2i-1}c_{2i}z^{-1} & -c_{2i}z^{-1} \\ -c_{2i-1} & 1 \end{pmatrix} = 1 \tag{10.59}$$

so that

$$\begin{pmatrix} Y_i(z) \\ U_i(z) \end{pmatrix} = \begin{pmatrix} 1 & c_{2i}z^{-1} \\ c_{2i-1} & 1 + c_{2i-1}c_{2i}z^{-1} \end{pmatrix} \begin{pmatrix} Y_{i+1}(z) \\ U_{i+1}(z) \end{pmatrix} \tag{10.60}$$

Some aspects of these nice algebraic properties are exploited in *lattice algorithms*; see Chapter 11.

10.4 MOMENT MATCHING

Consider a transfer function given as the (infinite) series

$$H(z^{-1}) = h_0 + h_1 z^{-1} + h_2 z^{-2} + \ldots \tag{10.61}$$

Matching of the reduced-order model B_m/A_m

$$H_m(z^{-1}) = \frac{B_m(z^{-1})}{A_m(z^{-1})} = \frac{b_0 + b_1 z^{-1} + \cdots + b_m z^{-m}}{1 + a_1 z^{-1} + \cdots + a_m z^{-m}} \tag{10.62}$$

can be made so that the moments

$$M_k = \sum_{n=0}^{\infty} n^k h_n \qquad k = 1, 2, 3, \ldots \tag{10.63}$$

match the original transfer function up to the $2m$th moment. Based on the observation that it is possible to formulate M_k as a weighted sum of the transfer function H and derivatives of H up to order k and evaluated at $z = 1$, it has been proposed to use the following "moment-matching" procedure

$$\begin{aligned} M_0 &= H|_{z=1} \\ M_1 &= \frac{dH}{dz^{-1}}|_{z=1} \\ M_2 &= \frac{d^2H}{dz^{-2}}|_{z=1} + \frac{dH}{dz^{-1}}|_{z=1} \\ &\vdots \end{aligned} \tag{10.64}$$

The resulting moment-matched and reduced-order rational transfer function can be viewed as an impulse response matching of H_m to H.

Example 10.4—Moment matching
Consider the discrete time transfer function

$$H(z^{-1}) = \frac{0.22z^{-1}}{1 - 0.7z^{-1} - 0.08z^{-2}} = 0 + 0.22z^{-1} + 0.1540z^{-2} + 0.1254z^{-3} + \cdots \tag{10.65}$$

that should match the reduced-order model

$$H_m(z^{-1}) = \frac{b_0 + b_1 z^{-1}}{1 + a_1 z^{-1}} \tag{10.66}$$

Moment matching by means of differentiation of H_m gives

$$\begin{aligned} H_m|_{z=1} &= \frac{b_0 + b_1}{1 + a_1} = 1 \\ \frac{dH_m}{dz^{-1}}|_{z=1} &= \frac{-a_1 b_0 + b_1}{(1 + a_1)^2} = 4.9091 \\ \frac{d^2H_m}{dz^{-2}}|_{z=1} &= 2\frac{(a_1^2 b_0 - a_1 b_1)}{(1 + a_1)^3} = 39.1074 \end{aligned} \tag{10.67}$$

Much work eventually gives the reduced-order transfer function

$$H_1(z^{-1}) = \frac{0.0149 + 0.1858z^{-1}}{1 - 0.7993z^{-1}} \tag{10.68}$$

10.5 THE PADÉ APPROXIMATION

The transfer function can be expanded to the polynomial series

$$G(s) = g_0 + g_1 s + g_2 s^2 + \dots \tag{10.69}$$

Assume that we truncate the polynomial series $G(s)$ after $2m$ terms and denote this truncated polynomial

$$G_m(s) = g_0 + g_1 s + \dots + g_{2m-1} s^{2m-1} \tag{10.70}$$

A way to match a rational function B_m/A_m to the truncated Taylor expansion G_m is to solve for the polynomial coefficients of the equation

$$B_m(s) = G_m(s)A_m(s) \tag{10.71}$$

The resulting reduced-order transfer function $B_m(s)/A_m(s)$ is known as the Padé approximation of the transfer function $G(s)$. We now give an example.

Example 10.5—Padé approximation applied to model reduction
Consider the transfer function

$$G_0(s) = \frac{Y(s)}{U(s)} = \frac{458}{s^3 + 31s^2 + 259s + 229} \tag{10.72}$$

with the Taylor series expansion

$$G_0(s) = 2.000 - 2.2620s + 2.2876s^2 - 2.2898s^3 + \dots \tag{10.73}$$

By matching the first few coefficients in the polynomial equation

$$B_m(s) = G_m(s)A_m(s) \tag{10.74}$$

for first- and second-order approximations of the type

$$\frac{B_1(s)}{A_1(s)} = \frac{b_0}{1 + a_1 s}, \quad \text{and} \quad \frac{B_2(s)}{A_2(s)} = \frac{b_0 + b_1 s}{1 + a_1 s + a_2 s^2} \tag{10.75}$$

it is straightforward to fit a first-order approximation

$$\begin{aligned} b_0 &= g_0 \\ 0 &= g_0 a_1 + g_1 = 0 \end{aligned} \quad \Rightarrow \quad \frac{B_1(s)}{A_1(s)} = \frac{2.0000}{1 + 1.1310s} \tag{10.76}$$

and a second-order approximation

$$\begin{aligned} b_0 &= g_0 \\ b_1 &= g_0 a_1 + g_1 \\ 0 &= g_0 a_2 + g_1 a_1 + g_2 \\ 0 &= g_1 a_2 + g_2 a_1 + g_3 \end{aligned} \quad \Rightarrow \quad \frac{B_2(s)}{A_2(s)} = \frac{2 - 0.0645s}{1 + 1.0987s + 0.0989s^2} \tag{10.77}$$

■

It is for two reasons necessary to state a warning against the Padé approximation. First, a serious problem is that an approximation of a stable transfer function may yield an unstable transfer function. A second problem is that the impulse response is poorly matched, which can be inferred from the following example.

Example 10.6—Unstable Padé approximations
Consider the transfer function

$$\frac{B_2(s)}{A_2(s)} = \frac{2s + 1}{s^2 + s + 1} = 1 + s + \cdots = g_0 + g_1 s + \cdots \tag{10.78}$$

A first-order transfer function approximation $B_1(s)/A_1(s) = b_0/(1 + a_1 s)$ can be obtained by application of Eq. (10.77)

$$b_0 = (g_0 + g_1 s)(1 + a_1 s) \quad \Rightarrow \quad \begin{cases} b_0 &= g_0 = 1 \\ a_1 &= -g_1/g_0 = -1 \end{cases} \tag{10.79}$$

which suggests the reduced-order model

$$\frac{B_1(s)}{A_1(s)} = \frac{1}{1 - s} \tag{10.80}$$

which is unstable and, of course, a very poor approximation of the original transfer function. ■

10.6 DESCRIBING FUNCTION ANALYSIS

The Laplace and Fourier transforms are powerful methods for application on linear systems, but few such methods are applicable to nonlinear systems. There are, however, attempts to extend the frequency domain analysis to

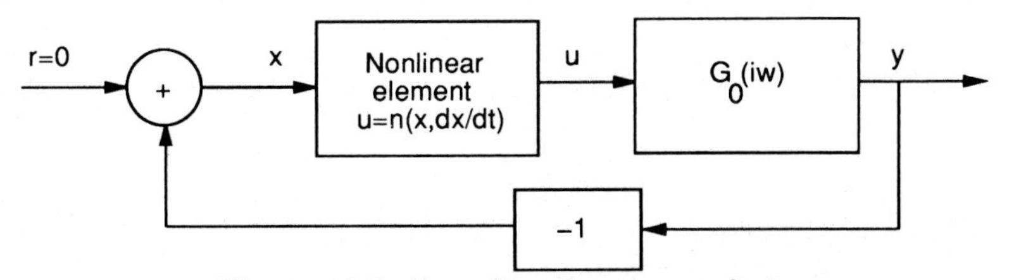

Figure 10.7 Describing function analysis

nonlinear systems, and one such method, *describing function analysis* (or *harmonic linearization*), is based on harmonic analysis. The method starts with an assumed periodic solution sufficiently close to a sinusoidal oscillation on the form

$$y(t) = C\sin(\omega t) \tag{10.81}$$

for a system with a nonlinear block, see Fig. 10.7. Static and dynamic blocks can be described by some function

$$u = n(x, \dot{x}) \tag{10.82}$$

It is also assumed that the input to the nonlinear element is close to a nonlinear oscillation although there are no precise quantitative criteria to establish the validity of this approximation.

If the forcing input x is a periodic function then the output too is a periodic function of time, and the output may thus be developed in a Fourier series expansion. For a periodic function $f(t) = f(t+T)$ (for all t and a period T) it holds that the Fourier series expansion is

$$f(x) = \frac{1}{2}a_0 + \sum_{k=1}^{\infty} a_k \cos(kx) + \sum_{k=1}^{\infty} b_k \sin(kx) \tag{10.83}$$

or

$$f(x) = \sum_{k=-\infty}^{\infty} c_k e^{i(kx+\varphi_k)} \tag{10.84}$$

with the coefficients

$$\begin{aligned} a_k &= \frac{2}{T}\int_{-T/2}^{T/2} f(t)\cos(kt)dt \\ b_k &= \frac{2}{T}\int_{-T/2}^{T/2} f(t)\sin(kt)dt \end{aligned} \tag{10.85}$$

or in polar coordinates

$$\begin{aligned} c_k &= \sqrt{a_k^2 + b_k^2} \\ \varphi_k &= \arctan\frac{a_k}{b_k} \end{aligned} \tag{10.86}$$

Real-valued functions thus take the form

$$\frac{1}{2}a_0 + \sum_{k=1}^{\infty} c_k \sin(kx + \varphi_k) \tag{10.87}$$

The analytic approach is to determine the conditions under which the expected oscillations occur. Consider a dynamic system with a periodic oscillation. This oscillation may be described with a Fourier series consisting of one fundamental frequency and higher harmonics. The amplification of the fundamental frequency (*i.e.*, ω) is the ratio of the first Fourier coefficient c_1 of the output of the nonlinearity and its input amplitude C

$$N(C) = \frac{c_1(C)}{C} e^{i(\varphi_1)} = \frac{b_1 + i a_1}{C} \tag{10.88}$$

where

$$\begin{aligned} a_1 &= \frac{1}{T}\int_{T/2}^{T/2} n(C\sin\omega t, C\omega\cos\omega t)\cos\omega t dt \\ b_1 &= \frac{1}{T}\int_{T/2}^{T/2} n(C\sin\omega t, C\omega\cos\omega t)\sin\omega t dt \end{aligned} \tag{10.89}$$

The describing function $N(C)$ is obtained as the amplitude dependent gain $|N(C)|$ and its phase shift $\varphi_1(\omega)$ for the nonlinear element. It is important to note that the describing function is related only to the nonlinearity $n(x, \dot{x})$ and not to the linear part of the system.

Harmonics caused by nonlinearities are ignored in describing function analysis and a balance for the fundamental Fourier component defines an equation for amplitude and frequency for a possible sustained oscillation.

Characteristic equation

Assume that the transfer function included in the control object is

$$G_0(s) = \frac{B(s)}{A(s)} \tag{10.90}$$

A differential equation related to the behavior of the closed-loop system in the absence of external inputs is

$$A(p)x(t) + B(p)n(x, \dot{x}) = 0, \qquad p \equiv \frac{d}{dt} \tag{10.91}$$

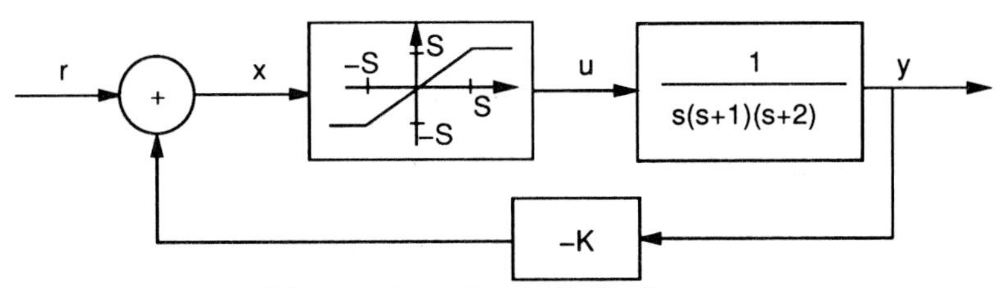

Figure 10.8 A rate-limited servo.

with a linear approximation on the form

$$(1 + G_0(p)N(C))x(t) = 0 \tag{10.92}$$

The characteristic equation has an approximate solution for ω and C satisfying the equation

$$G_0(i\omega) = -\frac{1}{N(C)} \tag{10.93}$$

This equation can be solved graphically by including the graph of $-1/N(C)$ in a Nyquist diagram. The crossing between the Nyquist curve $G_0(i\omega)$ and the describing function $-1/N(C)$ indicates the possible existence of a limit cycle with amplitude C and frequency ω_0 (see Fig. 10.9).

The methods of describing function analysis have been successful for analysis of *limit cycles* and sustained nonlinear oscillations. (The term "limit cycle" derives from phase plane analysis where the limit cycle describes an isolated path corresponding to the periodic solution.)

Example 10.7—A rate-limited servo

Consider the rate-limited servo in Fig. 10.8. The describing function for the saturating amplifier in Fig. 10.8 is

$$N(C) = \begin{cases} \frac{2}{\pi}(\arcsin\frac{S}{C} + \frac{S}{C}\sqrt{1 - \frac{S^2}{C^2}}), & C > S \\ 1, & C \leq S \end{cases} \tag{10.94}$$

where S is the saturation limit of the amplifier. The describing function is depicted in Fig. 10.9 on the form $-1/N(C)$, which reaches its maximum value -1 for small values of C, *i.e.*, in the linear range of $N(C)$.

Notice that the whole describing function has the phase $-\pi$ radians. The transfer function $G_0(i\omega)$ between u and y has a phase delay of $-\pi$ radians for $\omega = 1.414$ and the corresponding gain is $|G_0(i \cdot 1.414)| = 1/6$. From this information we expect the Nyquist contour and the describing function to cross

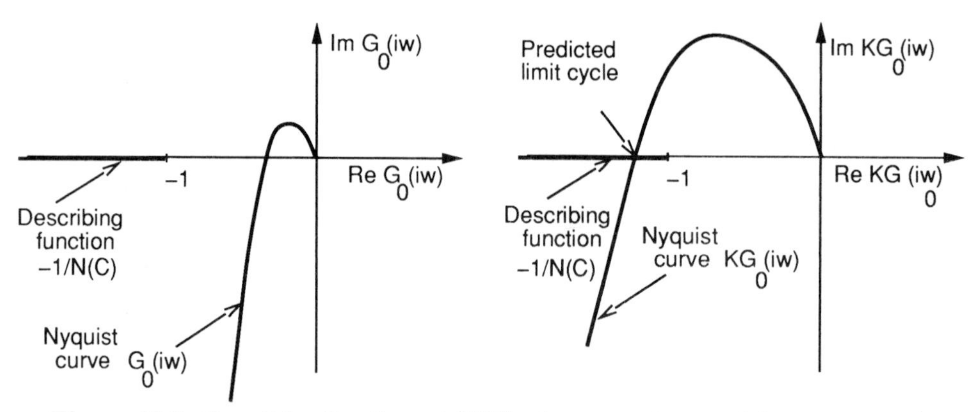

Figure 10.9 Describing function $-1/N(C)$ of a saturating amplifier nonlinearity in a Nyquist diagram of a DC motor with transfer function $G_0(s)$. The predicted limit cycle of amplitude C and frequency ω_0 is obtained from the crossing of the two curves.

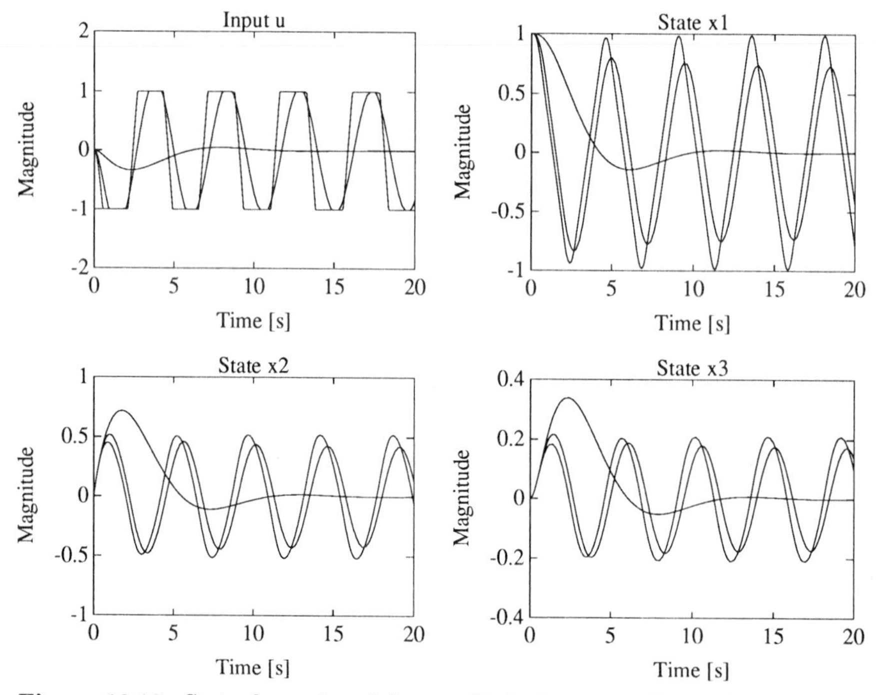

Figure 10.10 State dynamics of the rate-limited servo in Example 10.7 for $K = 1, 6, 12$.

for $K > 6$ for which we expect a periodic solution with period $2\pi/1.414 = 4.44$ s. Some simulations are shown in Fig. 10.10, which support the calculated values. ∎

Condition of application

The analysis is made under the assumption that the input to the nonlinearity is a sinusoid. The amplitude characteristic of the linear element has to be of low-pass nature so that it introduces a significant attenuation at the frequencies of higher harmonics, and the first condition below implies that the linear part indeed has low pass filter properties. The conditions may be stated as the following list:

i.

$$|G_0(ik\omega)| \ll |G_0(i\omega)|; \qquad k = 2, 3, \ldots$$

$$|G_0(ik\omega)| \to 0; \qquad k \to \infty$$

ii. $G_0(s)$ must not have any imaginary poles $s = \pm ik$.

iii. The function $n(x, \dot{x})$ should have finite partial derivatives with respect to x and $\dot{x}$ and must not be an explicit function of time.

A fourth condition applies to describing function analysis as presented here.

iv. The zero-order coefficient $c_0 = 0$ of $n(x, \dot{x})$.

The fourth condition excludes statically unbalanced systems and systems with rectifying properties. Another restriction is that the method is formulated without regarding the interference of external inputs or disturbances.

It is worth mentioning that the describing function analysis belongs to the class of methods which assume the existence of a solution and then proceed to show its characteristics. If the method prerequisites are not satisfied the describing function analysis may predict oscillations that do not exist and may fail to predict periodic solutions that indeed do exist.

It is clear from Example 10.7 that the describing function analysis determines the gain and phase of the transfer function at one point in the Nyquist diagram. This property can be exploited for a rudimentary form of frequency response analysis by using nonlinear elements in a feedback loop. The established limit cycle determines approximate values of phase and magnitude of the transfer function involved.

Example 10.8—Pulse-width modulation (PWM)

This type of modulation leads to asymmetric inputs and therefore needs a more complicated analysis with a nonzero constant term c_0 of the describing function.

Consider a pulse-width modulated (PWM) system that satisfies conditions *i*–*iii* cited previously. The nonlinearity is a switching element (see Fig. 10.11).

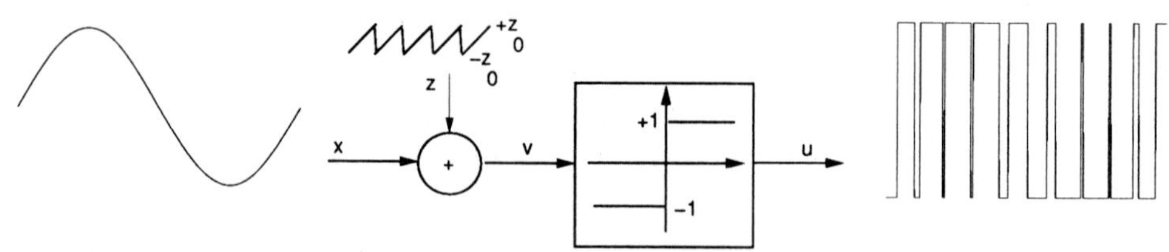

Figure 10.11 Describing function analysis applied to pulse width modulation.

Consider first a case with a nonzero modulation frequency $z(t) = z_1 \sin(\omega_z t)$ imposed on the slowly varying signal x. The pulse-width modulating signal z is often a sawtooth-shaped signal of high frequency. The pulse-width modulated signal is a square wave with a nonzero mean, which is high during a fraction of the modulating frequency.

A describing function to analyze the behavior of u is then

$$u(t) = n(v, \dot{v}) \approx \frac{1}{z_0} x(t) + n_1 z \tag{10.95}$$

where n_1 is the ordinary describing function terms. If the modulating frequency ω_z is very high compared to the transmission properties of $G_0(s)$, then the high-frequency components deriving from z are effectively absorbed by the low pass link G_0 so that

$$u(t) \approx \frac{1}{z_0} x(t) \tag{10.96}$$

The apparent linearization of the relay used for PWM is special for the sawtooth form of z and other wave forms give other function characteristics. The method of PWM analysis is relevant for modeling of thyristor-controlled devices and other actuator implementations and for control systems analysis and design. ■

10.7 BALANCED MODEL REDUCTION IN IDENTIFICATION

There are two obvious ways to use balanced model reduction in the context of modeling and identification. First, a linearized model may be further reduced in order. A second application is to reduce a linear model as obtained from identification.

This choice of model order is a possible alternative to statistical procedures based on loss function evaluation, AIC, FPE, etc., as practiced in identification

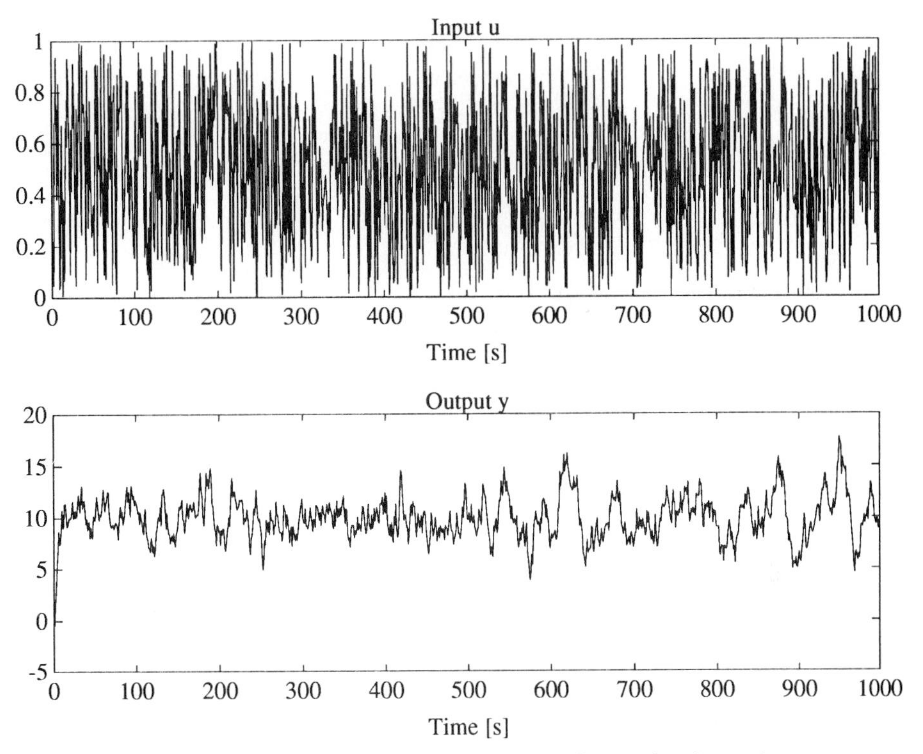

Figure 10.12 Graphs of input and output from a first-order dynamic system.

and which have problems with consistent model-order estimation. Identification assisted with model reduction comprises two steps with (*1*) estimation of a high-order model (provided that the excitation is sufficient to allow high-order model estimation) and (*2*) model reduction with elimination of less important states as indicated by the Gramian Σ obtained for the balanced realization.

Example 10.9—Model reduction in identification
Consider estimation of the first-order system

$$y_{k+1} = 0.9y_k + 0.1u_k + d \tag{10.97}$$

with data according to Fig. 10.12. The large constant $d = 1$ results in a clear bias of the first-order estimate

$$y_{k+1} = 0.9838y_k + 0.2638u_k \tag{10.98}$$

A better result (see Fig. 10.13) is obtained by estimating a tenth-order model

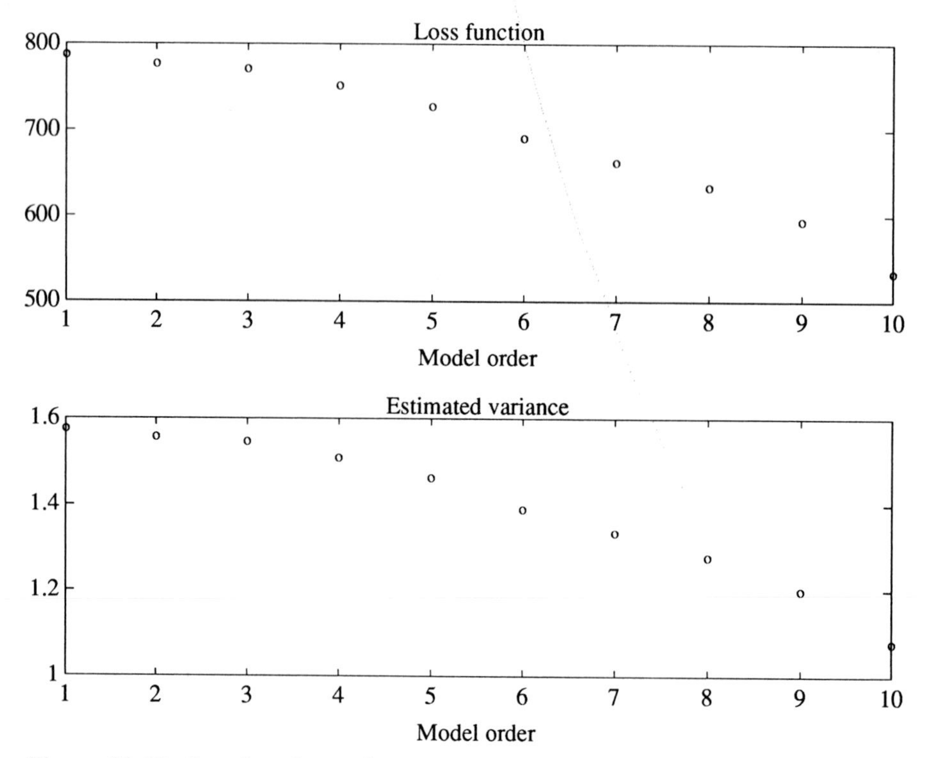

Figure 10.13 Loss function and estimated variance versus model order as obtained from identification of higher-order models.

and reducing the estimate into a first-order model

$$\begin{aligned} x_{k+1} &= 0.9466x_k + 0.2912u_k \\ y_k &= 0.2912x_k - 0.0192u_k \end{aligned} \quad \leftrightarrow \quad Y(z) = \frac{-0.0192z + 0.1029}{z - 0.9466} U(z) \qquad (10.99)$$

Step responses and impulse responses are shown in Fig. 10.14, and both have a clear bias that, however, is less prominent in magnitude. The step responses and impulse responses from the original system, the estimated first-order model Eq. (10.98), and the reduced-order first-order estimate Eq. (10.99) are shown in Fig. 10.14. The corresponding Bode diagrams are shown in Fig. 10.15.

Example 10.10—An impulse response test

Consider the data obtained from observation of the impulse response $\{g_i\}_{i=0}^{15}$ in Fig. 10.16.

$$g(t) = 1.2 \cdot 0.5^t - 0.2 \cdot 0.75^t \qquad (10.100)$$

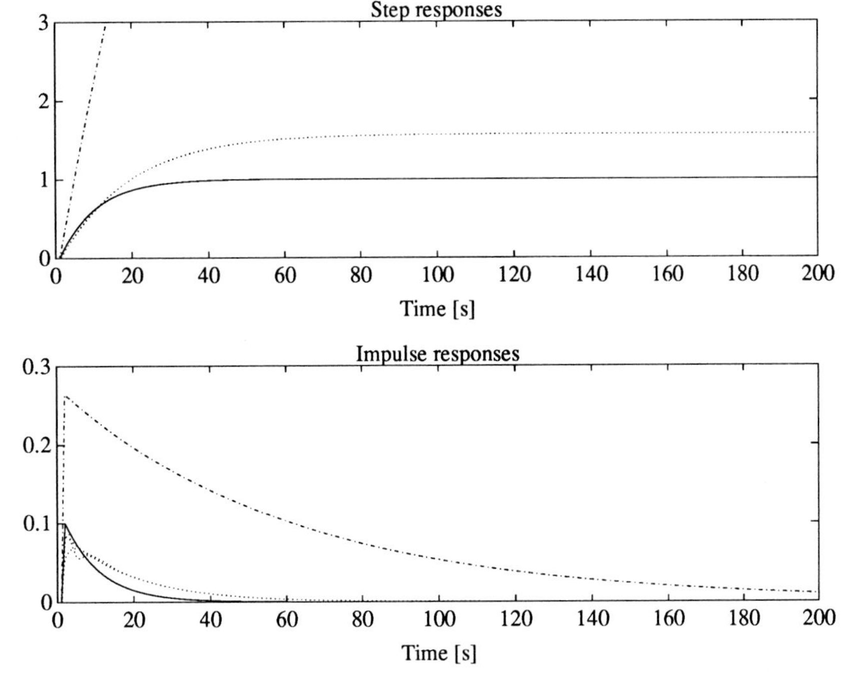

Figure 10.14 Step responses and impulse responses of the observed system (*solid line*), the first-order least-squares estimate (*dashed line*), and tenth- and reduced first-order models (*dotted line*). Notice that the least-squares estimate is poor due to the nonzero mean of the noise.

It is a problem that only few data points have been collected. Organization of the recorded impulse response g as the state-space model

$$x_{k+1} = \begin{pmatrix} 0 & 0 & \cdots & & 0 \\ 1 & 0 & \cdots & & 0 \\ 0 & 1 & \ddots & & \vdots \\ \vdots & \ddots & \ddots & & 0 \\ 0 & \cdots & 0 & 1 & 0 \end{pmatrix} x_k + \begin{pmatrix} 1 \\ 0 \\ 0 \\ \vdots \\ 0 \end{pmatrix} u$$

$$y(k) = \begin{pmatrix} g_0 & g_1 & \cdots & g_{n-1} & g_n \end{pmatrix} x_k$$

and application of balanced model reduction to this model permit determination of the appropriate model order $n = 2$ and the estimate of the impulse response

$$\widehat{g}(t) = 1.38 \cdot 0.523^t - 0.38 \cdot 0.703^t \tag{10.101}$$

It can be seen in Fig. 10.16 that the estimated impulse response and the recorded estimate are similar in magnitude and time course. However, the

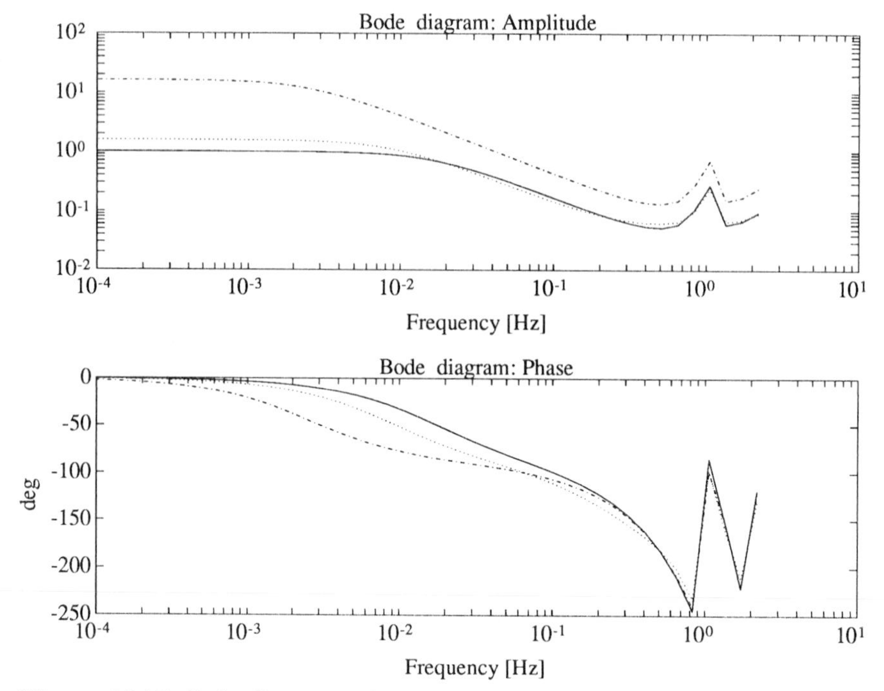

Figure 10.15 Bode diagrams of the observed system (*solid line*), the first-order least-squares estimate (*dashed line*), and tenth- and reduced first-order models (*dotted line*). Notice that the least-squares estimate is poor due to the nonzero mean of the noise.

parameter accuracy is not very good as can be expected from the few data and the incomplete impulse response. ■

10.8 BIBLIOGRAPHY AND REFERENCES

Model reduction can be studied in the following survey paper:

– M.J. Bosley and F.P. Lees, "A survey of simple transfer function derivations from high-order state variable models." *Automatica*, Vol. 8, 1972, pp. 765–775.

Balanced realization, model reduction, and its relation to numerical analysis were introduced in

– B.C. Moore, "Principal component analysis in linear systems: controllability, observability, and model reduction." *IEEE Trans. Autom. Control*, Vol. AC-26, 1981, pp. 17–32.

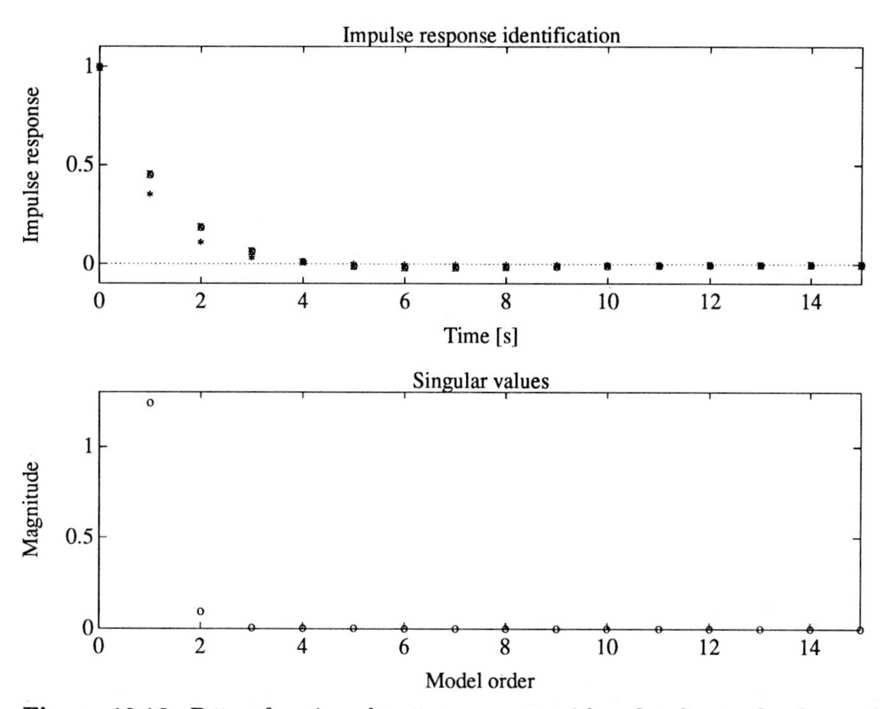

Figure 10.16 Data of an impulse response test with a fitted second-order model (*upper*). Notice that the data points ('o') and the impulse response ('x') of a second-order model fit closely whereas the first-order reduced model ('*') gives a less accurate fit. The lower diagram shows the singular values σ_i ($i = 1, 2, \ldots, 15$) obtained in the procedure to find the balanced realization.

- T. KAILATH, *Linear Systems.* Englewood Cliffs, NJ: Prentice-Hall, 1980, p. 619.
- A. LAUB, "On computing balancing transformations." *Proc. 1980 Joint Autom. Control Conf.*, San Francisco, August 1980.
- K. GLOVER, "All optimal Hankel-norm approximations of linear multivariable systems and their L^∞–error bounds." *Int.J.Control*, Vol. 39, 1984, pp. 1115–1193.
- L. PERNEBO AND L.M. SILVERMAN, "Model reduction via balanced state-space realizations." *IEEE Trans. Automatic Control*, Vol. 27, 1982, pp. 382–387.

Describing function analysis is presented in

- N.N. BOGOLIUBOV AND J.A. MITROPOLSKY, *Asymptotical Methods in the Theory of Nonlinear Oscillations*, Moscow: State Press for Physics and Mathematical Literature, 1963.
- N. MINORSKY, *Nonlinear Oscillations.* New York: Van Nostrand, 1962.

A theoretical basis of describing function analysis is found in the contributions of van der Pol and the "harmonic balance" of Krylov and Bogoliubov.

The continued fraction approximation method was introduced by Chen and Shieh. See

- C.F. CHEN AND L.S. SHIEH, "A novel approach to linear model simplification." *Int. J. Control*, Vol. 8, 1968. pp. 561–570.

The use of singular value decomposition techniques to determine the model order and related methods to solve for parameters is presented in

- J.A. CADZOW, "Spectral estimation: An overdetermined rational model approach." *Proc. IEEE*, Vol. 70, 1982, pp. 907–938.
- B. WAHLBERG, *On the Identification and Approximation of Linear Systems*. Phd-thesis 163, 1987, Linköping University, Sweden.

10.9 EXERCISES

10.1 Show that the Padé approximation

$$G_1(s) = \frac{b}{s+a} \tag{10.102}$$

of the transfer function

$$G_2(s) = \frac{s+\beta}{(s+1)(s+\alpha)} \tag{10.103}$$

is unstable for certain values of α and β.

10.2 The following model has been obtained from estimation of an ARX-model from input data u and output data y.

$$Y(z) = \frac{z-1}{z^2 - 1.79z + 0.792} U(z) \tag{10.104}$$

In consideration of model reduction the following balanced state-space model has been calculated

$$\begin{aligned} x_{k+1} &= \begin{pmatrix} 0.7910 & -0.0423 \\ 0.0423 & 0.9990 \end{pmatrix} x_k + \begin{pmatrix} 1.0001 \\ 0.0118 \end{pmatrix} u_k \\ y_k &= \begin{pmatrix} 1.0001 & -0.01198 \end{pmatrix} x_k \end{aligned} \tag{10.105}$$

The eigenvalues of the Gramians are

$$\begin{pmatrix} \sigma_1 \\ \sigma_2 \end{pmatrix} = \begin{pmatrix} 2.6837 \\ 2.4035 \end{pmatrix} \tag{10.106}$$

The transformation matrix is

$$T = \begin{pmatrix} 0.7363 & 22.3504 \\ -0.2638 & 22.3622 \end{pmatrix} \tag{10.107}$$

Is it advisable to reduce the model? If so, determine the reduced-order model. Otherwise, explain why the model reduction is not possible. ■

10.3 Show that there is a transformation $x = T\bar{x}$ so that the system

$$\begin{aligned} \bar{x}_{k+1} &= \bar{\Phi}\bar{x}_k + \bar{\Gamma}u_k \\ y_k &= \bar{C}\bar{x}_k \end{aligned} \tag{10.108}$$

has equal controllability and observability Gramians which are diagonal

$$\bar{P} = \bar{Q} = \mathrm{diag}(\sigma_1, \ldots, \sigma_n) \tag{10.109}$$

10.4 Make a geometric interpretation of balanced model reduction in Eq. (10.36) as a projection.

10.5 Consider a controllable state-space model

$$\dot{x} = Ax + Bu, \qquad x \in R^n \tag{10.110}$$

Let P be the solution of the Lyapunov equation

$$AP + PA^T = -BB^T \tag{10.111}$$

and let T be the Cholesky factor of P^{-1} so that $P^{-1} = T^T T$. Show that the components of $z = Tx$ for $u(t) = \delta(t)$ (*i.e.*, the impulse responses) are orthogonal in the sense that

$$\int_0^\infty z(t)z^T(t)dt = \int_0^\infty Te^{At}BB^Te^{A^Tt}T^Tdt = I_{n\times n} \tag{10.112}$$

How can this property be used in order to produce orthogonal regression variables for identification purposes?

■

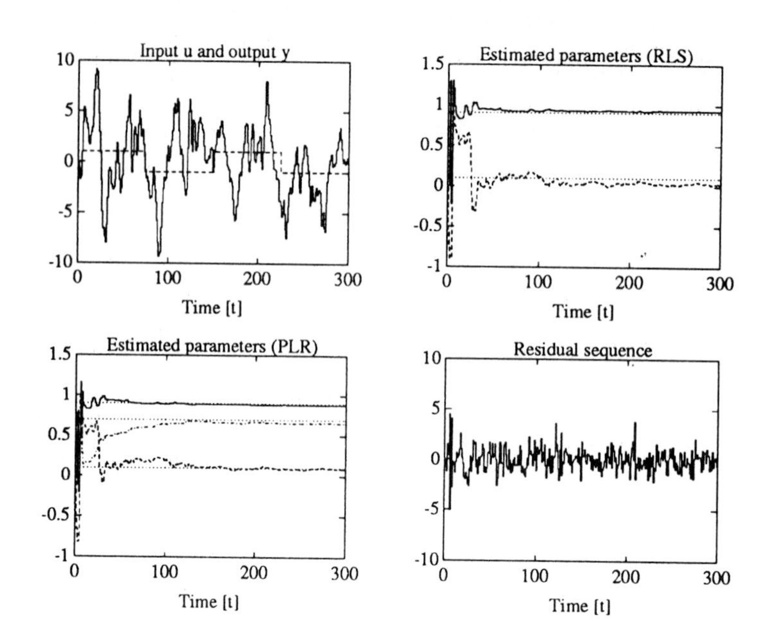

11 Real-time Identification

11.1 INTRODUCTION

Real-time application of identification algorithms is interesting for various purposes such as supervision, tracking of time-varying parameters for adaptive control, filtering, prediction, signal processing, detection, diagnosis, and artificial neural networks. However, most identification methods based on a set of measurements are not suitable for real-time application. It is therefore desirable to make a suitable reformulation of the algorithms in order to provide efficient procedures.

The recursive identification algorithms are estimators of the type

$$\begin{aligned}
\widehat{\theta}_k &= \widehat{\theta}_{k-1} + P_k \phi_k \varepsilon_k \\
\varepsilon_k &= y_k - \phi_k^T \widehat{\theta}_{k-1} \\
P_k &= P_{k-1} - \frac{P_{k-1}\phi_k \phi_k^T P_{k-1}}{1 + \phi_k^T P_{k-1}\phi_k}, \qquad P_0 \text{ given}
\end{aligned} \tag{11.1}$$

with the parameter estimate $\widehat{\theta}_k$, the regressor ϕ_k, the prediction error ε_k, and the matrix P_k, which are all evaluated at time $k = 1, 2, 3, \ldots$.

There are several attractive features of algorithms with an organization similar to Eq. (11.1). It is obviously suitable for real-time applications and only few data need to be stored. It is thus an organization which is attractive also as a computational organization of off-line algorithms. In particular, it provides a method for identification of systems with time-varying parameters.

There are also certain drawbacks such as the fact that the model structure is determined *a priori* and the fact that iterative solutions based on larger data sets may be difficult to organize. Thus, it is of some interest to consider the desirable modifications for real-time application of algorithms originally stated as off-line methods. The following example shows one such simple derivation.

Example 11.1—Recursive estimate of a constant
Consider the following noisy observation of a constant parameter

$$y_k = \theta + v_k, \qquad \mathcal{E}\{v_k\} = 0, \quad \mathcal{E}\{v_i v_j\} = \sigma^2 \delta_{ij} \tag{11.2}$$

which is on linear regression form $y_k = \phi_k \theta + v_k$ with $\phi_k = 1$ for all k. A least-squares estimate is found as the sample average

$$\widehat{\theta}_k = \frac{1}{k}\sum_{i=1}^{k} y_i \tag{11.3}$$

In order to avoid the summation at every instant of time it is natural to include previously made summations in some state which is updated when new data arrive. A feasible choice of recursive state equation for Eq. (11.3) is

$$\widehat{\theta}_k = \widehat{\theta}_{k-1} + \frac{1}{k}(y_k - \widehat{\theta}_{k-1}) \tag{11.4}$$

According to Eq. (5.23) one can estimate the variance of the least-squares estimate as

$$p_k = \sigma^2(\sum_{i=1}^{k} \phi_i\phi_i^T)^{-1} \approx \mathcal{E}\{(\widehat{\theta} - \theta)(\widehat{\theta} - \theta)^T\} \tag{11.5}$$

The parameter variance estimate in Eq. (11.5) can be expressed as a state vector with updating in each recursion according to

$$p_k^{-1} = p_{k-1}^{-1} + \frac{1}{\sigma^2} \tag{11.6}$$

or

$$p_k = \frac{\sigma^2 p_{k-1}}{\sigma^2 + p_{k-1}} \tag{11.7}$$

where it can be noticed that $p_k \to 0$ as $k \to \infty$. ■

Derivation of recursive least-squares identification

Recursive least-squares identification according to Eq. (11.1) can be derived from the ordinary least-squares estimate according to the following derivation.

Consider as usual the regressor ϕ_i and the observations y_i collected in the matrices

$$\Phi_k = \begin{pmatrix} \phi_1^T \\ \vdots \\ \phi_k^T \end{pmatrix}, \quad \text{and} \quad \mathcal{Y}_k = \begin{pmatrix} y_1 \\ \vdots \\ y_k \end{pmatrix} \tag{11.8}$$

The least squares criterion based on k samples is

$$V(\widehat{\theta}_k) = \frac{1}{2}(\mathcal{Y}_k - \Phi_k\widehat{\theta}_k)^T(\mathcal{Y}_k - \Phi_k\widehat{\theta}_k) = \frac{1}{2}\varepsilon(\widehat{\theta}_k)^T\varepsilon(\widehat{\theta}_k) \tag{11.9}$$

The ordinary least-squares estimate is

$$\widehat{\theta}_k = (\Phi_k^T\Phi_k)^{-1}\Phi_k^T\mathcal{Y}_k = (\sum_{i=1}^{k} \phi_i\phi_i^T)^{-1}(\sum_{i=1}^{k} \phi_i y_i) \tag{11.10}$$

Introduce the matrix

$$P_k = (\sum_{i=1}^{k} \phi_i\phi_i^T)^{-1} = (\Phi_k^T\Phi_k)^{-1} \tag{11.11}$$

A recursive updating is given by

$$P_k^{-1} = P_{k-1}^{-1} + \phi_k \phi_k^T$$
$$\widehat{\theta}_k = P_k\left(\sum_{i=1}^{k-1} \phi_i y_i + \phi_k y_k\right) = P_k(P_{k-1}^{-1}\widehat{\theta}_{k-1} + \phi_k y_k) = \widehat{\theta}_{k-1} + P_k\phi_k(y_k - \phi_k^T\widehat{\theta}_{k-1}) \tag{11.12}$$

It is also feasible to calculate the matrix P_k instead of its inverse. Notice that

$$P_k = (\Phi_k^T\Phi_k)^{-1} = (\Phi_{k-1}^T\Phi_{k-1} + \phi_k\phi_k^T)^{-1} = (P_{k-1}^{-1} + \phi_k\phi_k^T)^{-1} \tag{11.13}$$

In real-time operation it is usually difficult and computationally expensive to do matrix inversion. It is thus preferable to avoid such operation by means of a suitable reformulation of the problem. Application of the matrix inversion relation (see Appendix A)

$$(A + BC)^{-1} = A^{-1} - A^{-1}B(I + CA^{-1}B)^{-1}CA^{-1} \tag{11.14}$$

to the expression in Eq. (11.13) is straightforward and we obtain

$$P_k = (P_{k-1}^{-1} + \phi_k\phi_k^T)^{-1} = P_{k-1} - P_{k-1}\phi_k(I + \phi_k^T P_{k-1}\phi_k)^{-1}\phi_k^T P_{k-1} \tag{11.15}$$

By collecting these formulae one can verify that the recursive least-squares identification in Eq. (11.1) will result in the same parameter estimate $\widehat{\theta}$ as the least-squares method provided that the initial value P_0 of the recursive equation for P_k is chosen appropriately, *i.e.*, so that $P_0^{-1} = 0$. This requirement can be easily satisfied in algorithms for updating of P_k^{-1} but is impossible for updating procedures similar to Eq. (11.15). A heuristic approach is to choose P_0 as an identity matrix multiplied by some large number. However, such an approach does not actually solve the problem of initial estimates and may also cause large initial transients in the parameter estimates. A systematic solution to this problem is therefore to make an initial estimate by means of ordinary least-squares identification, which then provides initial values both for $\widehat{\theta}_0$ and P_0.

11.2 RECURSIVE LEAST-SQUARES IDENTIFICATION

The recursive identification algorithm is

$$\begin{aligned}
\widehat{\theta}_k &= \widehat{\theta}_{k-1} + P_k\phi_k\varepsilon_k \\
\varepsilon_k &= y_k - \phi_k^T\widehat{\theta}_{k-1} \\
P_k &= P_{k-1} - \frac{P_{k-1}\phi_k\phi_k^T P_{k-1}}{1 + \phi_k^T P_{k-1}\phi_k}, \qquad P_0 \text{ given}
\end{aligned} \tag{11.16}$$

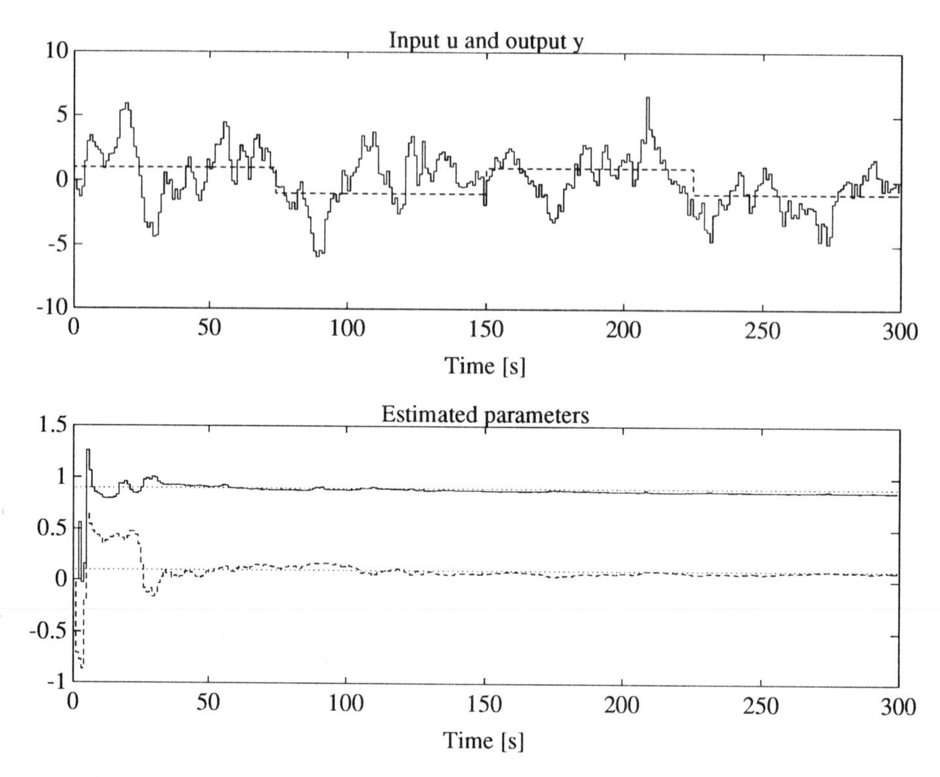

Figure 11.1 A first order system $y_{k+1} = ay_k + bu_k + w_k$ with input u and output y identified by recursive least-squares estimation of a and b . The correct parameter values are $a = 0.9$ and $b = 0.1$ and are indicated by dotted lines.

where $\widehat{\theta}_k$ is the parameter estimate, ε_k is the prediction error. The matrix P_k constitutes, except for a factor σ^2, an estimate of the parameter covariance at recursion number k.

Example 11.2—Recursive least-squares estimation
Consider recursive identification by means of the recursive formulae in Eq. (11.1) when applied to data generated by the system

$$\mathcal{S}: \qquad y_{k+1} = ay_k + bu_k + w_{k+1}, \qquad \text{where} \quad \begin{cases} a = 0.9 \\ b = 0.1 \end{cases} \tag{11.17}$$

where $\{w_k\}$ is a white-noise sequence. Some graphs showing input-output data and parameter convergence are found in Fig. 11.1. ■

Some properties of recursive least-squares estimation

A natural question in the context of recursive identification is how to evaluate parameter accuracy and convergence in the course of recursions. The following convergence analysis provides some relevant partial results.

Consider the following quadratic function of the parameter error

$$Q(\widehat{\theta}_k) = \frac{1}{2}(\widehat{\theta}_k - \theta)^T P_k^{-1}(\widehat{\theta}_k - \theta) = \frac{1}{2}\widetilde{\theta}_k^T P_k^{-1}\widetilde{\theta}_k \tag{11.18}$$

This function develops in each recursion according to

$$\begin{aligned} Q(\widehat{\theta}_k) - Q(\widehat{\theta}_{k-1}) &= \frac{1}{2}\widetilde{\theta}_k^T P_k^{-1}\widetilde{\theta}_k - \frac{1}{2}\widetilde{\theta}_{k-1}^T P_{k-1}^{-1}\widetilde{\theta}_{k-1} \\ &= \frac{1}{2}\widetilde{\theta}_{k-1}^T(P_k^{-1} - P_{k-1}^{-1})\widetilde{\theta}_{k-1} + \widetilde{\theta}_{k-1}^T\phi_k\varepsilon_k + \frac{1}{2}\phi_k^T P_k\phi_k\varepsilon_k^2 \\ &= \frac{1}{2}(\widetilde{\theta}_{k-1}^T\phi_k + \varepsilon_k)^2 + \frac{1}{2}(-1 + \phi_k^T P_k\phi_k)\varepsilon_k^2 \\ &= \frac{1}{2}(\widetilde{\theta}_{k-1}^T\phi_k + \varepsilon_k)^2 - \frac{1}{2}\frac{\phi_k^T P_{k-1}\phi_k}{1 + \phi_k^T P_{k-1}\phi_k}\varepsilon_k^2 \end{aligned} \tag{11.19}$$

Under the linear model assumption $y_k = \phi_k^T\theta + v_k$ so that $\varepsilon_k = -\widetilde{\theta}_{k-1}^T\phi_k + v_k$ one can conclude that

$$Q(\widehat{\theta}_k) - Q(\widehat{\theta}_{k-1}) = \frac{1}{2}v_k^2 - \frac{1}{2}\frac{\phi_k^T P_{k-1}\phi_k}{1 + \phi_k^T P_{k-1}\phi_k}\varepsilon_k^2 \tag{11.20}$$

Clearly, in the noise-free case with $v_k = 0$ for all k it holds that $Q(\widehat{\theta}_k)$ decreases in each recursion step. Moreover, if $Q(\widehat{\theta}_k)$ tends to zero it is implied that $\|\widehat{\theta}_k\|_2$ tends to zero as the sequence of weighting matrices $\{P_k^{-1}\}$ is an increasing sequence of positive definite matrices where $P_k^{-1} \geq P_{k-1}^{-1}$ for all $k > 0$. In such a case parameter convergence follows. It should, however, be borne in mind that it is difficult to verify *a priori* conditions under which $Q(\widehat{\theta}_k)$ tends to zero. This is further complicated in the case of nonzero disturbances for which we give the following result that is valid both for ordinary least-squares identification and for recursive least-squares identification.

Theorem 11.1

The errors of estimated parameters and the prediction error for least-squares estimation have a bound determined by the noise magnitude according to

$$V(\widehat{\theta}_k) + Q(\widehat{\theta}_k) = \frac{1}{2}\varepsilon^T(\widehat{\theta}_k)\varepsilon(\widehat{\theta}_k) + \frac{1}{2}\widetilde{\theta}_k^T P_k^{-1}\widetilde{\theta}_k = \frac{1}{2}v^T v \tag{11.21}$$

(The proof is called for in Exercise 11.1.) ■

Theorem 11.1 thus states that a certain weighted sum of squared prediction errors and the squared parameter errors equal the sum of squared noise components. As the sequence of weighting matrices $\{P_k^{-1}\}$ is an increasing sequence, one finds for recursive least-squares identification by means of the recursion formula for P_k^{-1} that

$$\|\tilde{\theta}_k\|_{P_0}^2 = \frac{1}{2}v^T v - \frac{1}{2}\varepsilon^T(\hat{\theta}_k)\varepsilon(\hat{\theta}_k) - \frac{1}{2}\tilde{\theta}_k^T \Phi_k^T \Phi_k \tilde{\theta}_k \tag{11.22}$$

Under conditions of a stationary stochastic process $\{v_k\}$ with

$$\Phi_k^T \Phi_k > kcI_{p\times p}, \qquad c \text{ constant} \tag{11.23}$$

one can, then, conclude parameter convergence. However, conditions become more complicated in the case when Eq. (11.23) is not valid, which may appear, for instance, in cases of a poor parameterization or for inputs of insufficient excitation. Poor convergence properties have been demonstrated in cases of large disturbances and a rank-deficient $\Phi_k^T \Phi_k$−matrix.

Hence, several important algorithm properties can be derived from the behavior of the P_k−matrix. For instance, the matrix P_k is a positive definite and symmetric matrix ($P_k = P_k^T > 0$) such that $P_k \to 0$ as $k \to \infty$. The matrix P_k is asymptotically proportional to the parameter estimate covariance provided that a correct model structure has been used. It is for this reason it is often called the "covariance matrix."

The result obtained from recursive least-squares estimation is the same as that of ordinary least-squares identification if the initial values P_0 and $\hat{\theta}_0$ can be chosen to be compatible with the results of the ordinary least-squares method. Another approach is to make an initial least-squares estimate by solving the normal equations for some block of initial data. Recursive identification then provides a procedure for updating the parameter estimates.

Modification for time-varying parameters

The recursive least-squares estimation in Eq. (11.16) gives equal weighting to old data and new data. It is, however, natural to pay less attention to old data in many applications where time-varying parameters should be tracked. This can be achieved by introducing the "forgetting factor" λ and the modified performance criterion of least-squares estimation

$$J(\hat{\theta}_k) = \sum_{i=1}^{k} \lambda^{k-i}(y_i - \phi_i^T \hat{\theta}_k)^2, \qquad 0 < \lambda \leq 1 \tag{11.24}$$

This results in the recursive algorithm

$$\begin{aligned} \widehat{\theta}_k &= \widehat{\theta}_{k-1} + P_k \phi_k \varepsilon_k \\ \varepsilon_k &= y_k - \phi_k^T \widehat{\theta}_{k-1} \\ P_k &= \frac{1}{\lambda}\left(P_{k-1} - \frac{P_{k-1}\phi_k \phi_k^T P_{k-1}}{\lambda + \phi_k^T P_{k-1}\phi_k}\right) \end{aligned} \tag{11.25}$$

This algorithm emphasizes the fitting of recent data and reduces the influence of old data. There are, however, some undesirable secondary effects of the algorithm in Eq. (11.25) with problems with a noise sensitivity that becomes more prominent as λ decreases. Another problem is that the P_k−matrix may increase as k grows if the input is such that the magnitude of $P_{k-1}\phi_k$ is small ("P−matrix explosion" or "covariance matrix explosion").

The choice of the forgetting factor λ is determined by a trade-off between the required ability to track a time-varying parameter (*i.e.*, a small value of λ) and the noise sensititivity allowed. A low value of λ results in a system with a good ability to track time-varying parameters but may also give variations in the estimated parameters caused by disturbances. A value of λ close to 1 is less sensitive to disturbances but compromises the ability to track rapid variations in the parameters. A default choice of λ is in the range $0.97 \le \lambda \le 0.995$ although the appropriate choice depends, of course, both of the characteristics of the identified process and the sampling frequency (see Fig. 11.2). The number of data points kept in "memory" can roughly be calculated as

$$\frac{1}{1-\lambda} \tag{11.26}$$

which corresponds to the time constant associated with λ.

Example 11.3—Choice of forgetting factor

Consider once again the process of Example 11.1 and assume that the parameter θ changes abruptly from $\theta = 1$ to the new value $\theta = 2$ at some unknown time. The effect of the forgetting factors $\lambda = 0.99, 0.98, 0.95$ on the tracking of the time-varying parameter can be seen in Fig. 11.2. A lower value of the forgetting factor gives a more rapid response in tracking the new parameter value. A value of λ closer to 1 means that a longer data record is kept so that old data have a considerable impact on the parameter estimate. However, small values of the forgetting factor λ result in a parameter estimate that becomes susceptible to noise. The forgetting factor therefore needs to be chosen as a trade-off between noise sensitivity and the parameter tracking capabilities.

■

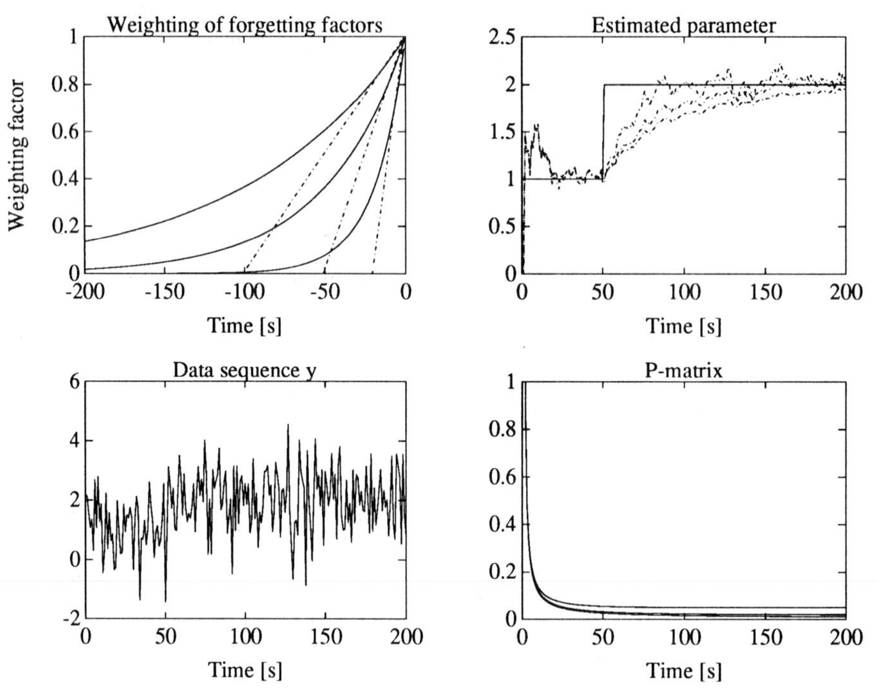

Figure 11.2 Demonstration of the influence of the choice of forgetting factor. The number of data points in memory according to Eq. (11.26) is indicated (*dashed line*) for various forgetting factors.

Alternative approaches in maintaining the least-squares up-to-date consist of keeping a block of recent data and to discard old data according to some criterion of age. Such methods appear to be applications of ordinary identfication methods to which one can refer to other chapters of this book.

Another important issue is how to improve numerical properties for small sampling intervals relative to the time constants of the process. One important approach is to reformulate the recursive algorithms in terms of the δ–operator

$$\delta = \frac{z-1}{h}, \qquad \text{or} \quad z = 1 + h\delta \tag{11.27}$$

with the state-space system representation

$$\begin{cases} x_{k+1} = \Phi x_k + \Gamma u_k \\ y_k = C x_k \end{cases} \Leftrightarrow \begin{cases} \delta x_k = \Phi' x_k + \Gamma' u_k = \frac{1}{h}(\Phi - I)x_k + \frac{1}{h}\Gamma u_k \\ y = C x_k \end{cases} \tag{11.28}$$

This reformulation makes the state-space realization and the corresponding system identification less error-prone due to favorable numerical scaling properties of the $\Phi'-$ and $\Gamma'-$matrices as compared to the ordinary z-transform based algebra; see Middleton and Goodwin (1990) for details.

A Kalman filter interpretation

Assume that the time-varying system parameter θ may be described by the state-space equation

$$\begin{cases} \theta_{k+1} = \theta_k + v_k, & \mathcal{E}\{v_i\} = 0, \quad \mathcal{E}\{v_i v_j^T\} = R_1 \delta_{ij}, \quad \forall i,j \\ y_k = \phi_k^T \theta_k + e_k, & \mathcal{E}\{e_i\} = 0, \quad \mathcal{E}\{e_i e_j^T\} = R_2 \delta_{ij}, \quad \forall i,j \end{cases} \tag{11.29}$$

where $\{y_k\}$ is interpreted as a sequence of scalar indirect observations of θ_k obtained from the observations y_k. The Kalman filter for estimation of θ_k from observations of y_k (see Appendix D) is

$$\begin{aligned} \widehat{\theta}_k &= \widehat{\theta}_{k-1} + K_k \varepsilon_k \\ K_k &= P_{k-1}\phi_k / (R_2 + \phi_k^T P_{k-1} \phi_k) \\ \varepsilon_k &= y_k - \phi_k^T \widehat{\theta}_{k-1} \\ P_k &= P_{k-1} - \frac{P_{k-1}\phi_k \phi_k^T P_{k-1}}{R_2 + \phi_k^T P_{k-1}\phi_k} + R_1 \end{aligned} \tag{11.30}$$

which corresponds to Eq. (11.25) except for the term R_1 added to P_k and R_2 added to the denominator expression of the recursive equation for the estimated covariance matrix P_k. An important difference is, however, that the term R_1 added to P_k causes a change in the dynamics of P_k from an exponential growth rate to a linear growth rate for $\phi_k = 0$. In addition, notice that P_k of the Kalman filter does not approach zero as $k \to \infty$ for a nonzero sequence $\{\phi_k\}$.

11.3 RECURSIVE INSTRUMENTAL VARIABLE METHODS

The ordinary instrumental variable solution

$$\widehat{\theta}_k^z = (Z^T \Phi)^{-1} Z^T \mathcal{Y} = (\sum_{i=1}^{k} z_i \phi_i^T)^{-1} (\sum_{i=1}^{k} z_i y_i) \tag{11.31}$$

with the instrumental variables z_k collected in the matrix Z may be reformulated as the recursive equation

$$\begin{aligned}
\widehat{\theta}_k &= \widehat{\theta}_{k-1} + K_k \varepsilon_k \\
K_k &= P_{k-1} z_k / (1 + \phi_k^T P_{k-1} z_k) \\
\varepsilon_k &= y_k - \phi_k^T \widehat{\theta}_{k-1} \\
P_k &= P_{k-1} - \frac{P_{k-1} z_k \phi_k^T P_{k-1}}{1 + \phi_k^T P_{k-1} z_k}
\end{aligned} \tag{11.32}$$

where the "instrument" vectors z_k replace the regression vectors ϕ_k. A standard choice of z_k for identification of ARX models is

$$z_k = \begin{pmatrix} -x_{k-1} & \ldots & -x_{k-n_A} & u_{k-1} & \ldots & u_{k-n_B} \end{pmatrix}^T \tag{11.33}$$

for some variables x_k uncorrelated with the noise affecting the investigated system. The variables x_k may be, for instance, the estimated output.

The recursive instrumental variable method has indeed some stability problems associated with the choice of the instrumental variables and the updating of the P_k–matrix.

11.4 PSEUDOLINEAR REGRESSION

Pseudolinear regression is also called *recursive maximum likelihood estimation* or *extended least-squares method*. The regression model is

$$\begin{cases} y_k &= \phi_k^T \theta + v_k \\ \theta &= \begin{pmatrix} a_1 & \ldots & a_{n_A} & b_1 & \ldots & b_{n_B} & c_1 & \ldots & c_{n_C} \end{pmatrix}^T \end{cases} \tag{11.34}$$

The recursive equation is

$$\begin{aligned}
\widehat{\theta}_k &= \widehat{\theta}_{k-1} + K_k \varepsilon_k \\
K_k &= P_{k-1} \phi_k / (1 + \phi_k^T P_{k-1} \phi_k) \\
\varepsilon_k &= y_k - \phi_k^T \widehat{\theta}_{k-1} \\
P_k &= P_{k-1} - \frac{P_{k-1} \phi_k \phi_k^T P_{k-1}}{1 + \phi_k^T P_{k-1} \phi_k} + R_1
\end{aligned} \tag{11.35}$$

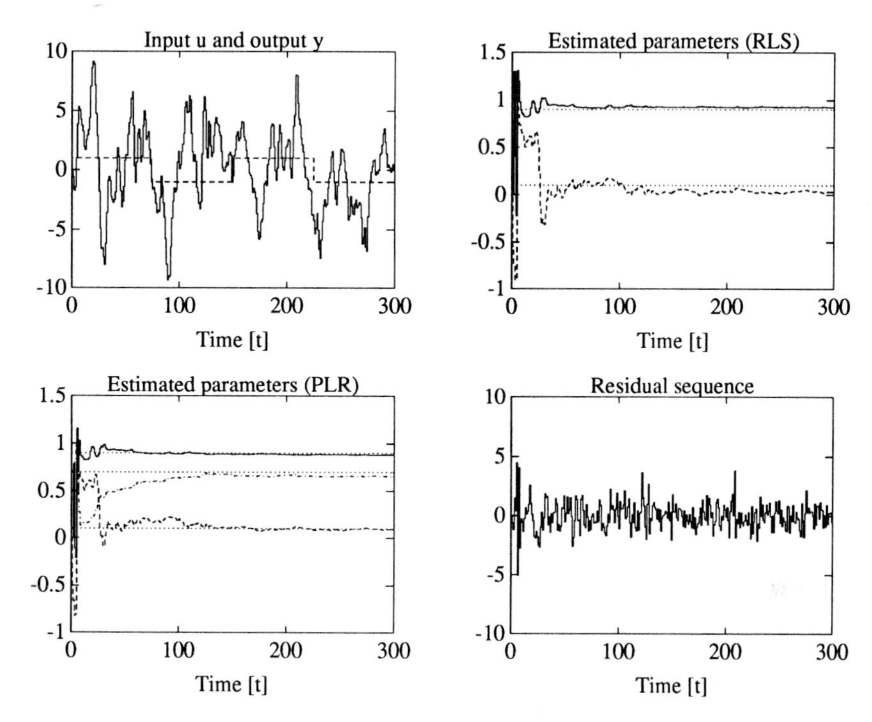

Figure 11.3 A first-order system $y_{k+1} = ay_k + bu_k + w_{k+1} + cw_k$ with input u and output y identified by recursive least-squares estimation of a and b (*upper right*). Notice the bias due to colored noise. Pseudolinear regression (*lower left*) helps to reduce the systematic error by means of estimation of a, b, and c. The correct parameter values are $a = 0.9$, $b = 0.1$, and $c = 0.7$ are indicated by dotted lines.

The regression vector that would be natural to use for linear regression is

$$\begin{pmatrix} -y_{k-1} & \dots & y_{k-n_A} & u_{k-1} & \dots & u_{k-n_B} & v_{k-1} & \dots & v_{k-n_C} \end{pmatrix} \tag{11.36}$$

which, of course, can not be implemented as the disturbance components v_k are not available to measurement. The regression vector ϕ_k–vector has therefore been modified to

$$\phi_k = \begin{pmatrix} -y_{k-1} & \dots & y_{k-n_A} & u_{k-1} & \dots & u_{k-n_B} & \varepsilon_{k-1} & \dots & \varepsilon_{k-n_C} \end{pmatrix}^T \tag{11.37}$$

The name *pseudolinear regression* derives from the fact that disturbance components of the desired regression vector have been replaced by estimated variables. The method is also called *recursive maximum likelihood method* as it tries to use the disturbance components as regressor elements and approximates these unknown elements by the calculated prediction errors $\{\varepsilon_k\}$. This method may also be modified to include a forgetting factor for identification of time-varying systems. The algorithm may also be modified to iterate for the best possible ε within each recursion step.

Example 11.4—Pseudolinear regression
Consider recursive identification of parameters of the system

$$\mathcal{S}: \qquad y_{k+1} = ay_k + bu_k + w_{k+1} + cw_k \tag{11.38}$$

where $\{w_k\}$ is a zero-mean white-noise sequence. A comparison between the results for pseudolinear regression and recursive least-squares identification is shown in Fig. 11.3. ■

Pseudolinear regression is an approximate linear regression method as the unknown regression elements v_k are replaced by the residual ε_k. Such a method is known to work well for ARMAX models with "moderate" noise correlations—*i.e.*, ARMAX models whose $C-$polynomials are close to 1. This vague description can be substantiated by showing that parameter convergence can be expected under stationary conditions for $C-$polynomials satisfying the condition

$$\text{Re}\,\frac{1}{C(z)}\Big|_{z=\exp(i\omega)} \geq \frac{1}{2}, \qquad -\pi < \omega \leq \pi \tag{11.39}$$

for pseudolinear regression based on recursive least-squares identification.

11.5 STOCHASTIC GRADIENT METHODS

This method is sometimes called *stochastic approximation* or *least mean square* (LMS) as advocated by Widrow and the field of digital signal processing. The method is called "steepest descent" as a numerical method. This type of algorithm has been reinvented by several authors. A basic property is that it contains no state to represent collected data, and the estimated parameters constitute a full state-space of the algorithm. Typically, the stochastic gradient methods take on the form

$$\widehat{\theta}_k = \widehat{\theta}_{k-1} + \gamma_k \phi_k \varepsilon_k \tag{11.40}$$

where γ_k is a sequence of step lengths satisfying the properties

$$\sum_{k=0}^{\infty} \gamma_k = \infty, \quad \text{and} \quad \sum_{k=0}^{\infty} \gamma_k^2 < \infty, \quad \text{and} \quad \gamma_k > 0, \quad \forall k \tag{11.41}$$

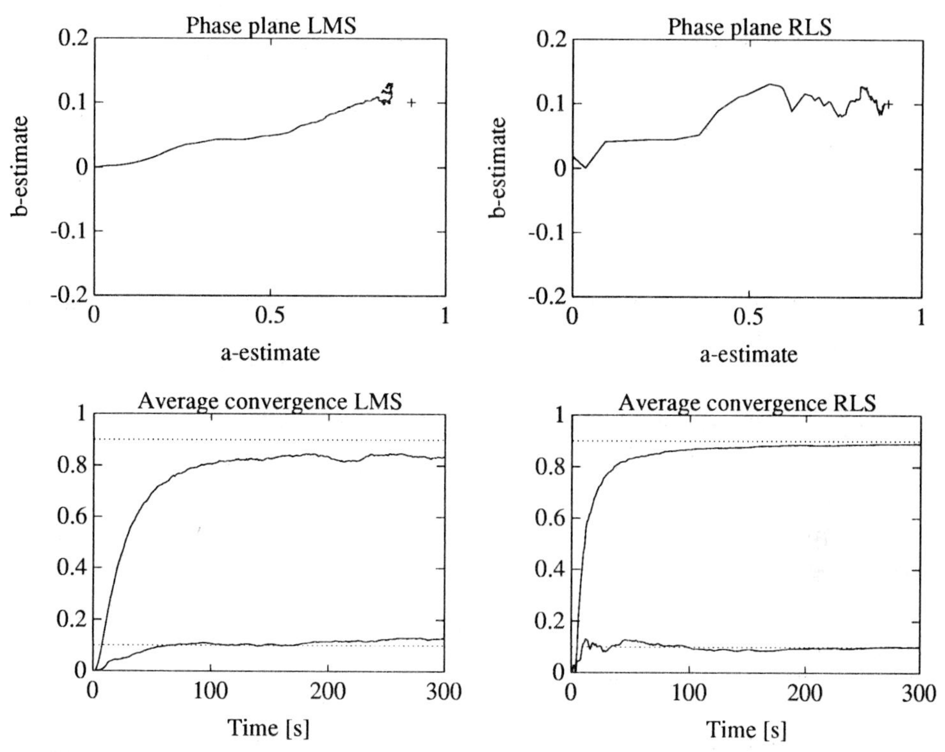

Figure 11.4 Averaged convergence trajectories for recursive least-squares identification (RLS) and least mean-square identification (LMS) for 100 realizations of the data from the system in Example 11.1. The correct parameter values are indicated by + (*upper graphs*) and by dotted lines (*lower graphs*). The RLS identification appears to have faster and more accurate convergence.

A modification to include a time-varying regressor-dependent gain is

$$\begin{aligned}
\widehat{\theta}_k &= \widehat{\theta}_{k-1} + \gamma_k \varepsilon_k \\
\varepsilon_k &= y_k - \phi_k^T \widehat{\theta}_{k-1} \\
\gamma_k &= Q\phi_k / r_k, \qquad Q = Q^T > 0 \\
r_k &= r_{k-1} + \phi_k^T Q^{-1} \phi_k
\end{aligned} \tag{11.42}$$

where Q is some positive definite weighting matrix. Some advantages of this method are that rapid computations are possible as there is no P_k−matrix to evaluate. There is thus little influence on the parameter estimate from past data, which, of course, is often good for detection of time-varying parameters. However, there are several drawbacks such as slow convergence and noise sensitivity as compared to the least-squares-type estimation methods, see Fig 11.4.

Time-varying parameters

The stochastic gradient methods can be applied to time-varying systems without modification. A minor modification of Eq. (11.42) is to include a forgetting factor λ so that

$$r_k = \lambda r_{k-1} + \phi_k^T Q^{-1} \phi_k, \qquad 0 \le \lambda \le 1 \tag{11.43}$$

This algorithm tends to keep the factor r_k at a lower magnitude, which affects the above gain γ_k.

11.6 THE LEVINSON-DURBIN ALGORITHM

Many identification algorithms involve some evaluation of a set of output variables without any known corresponding input variables. Such outputs are often assumed to be filtered white noise, and it is of interest to provide good algorithms for analysis of such data. The Levinson-Durbin algorithm provides an efficient recursive solution to the Yule-Walker equations by using the Toeplitz structure of the correlation matrix in the equation

$$\begin{pmatrix} C_{yy}(0) & C_{yy}(1) & \cdots & C_{yy}(n) \\ C_{yy}(1) & C_{yy}(0) & \cdots & C_{yy}(n-1) \\ \vdots & \vdots & \ddots & \vdots \\ C_{yy}(n) & C_{yy}(n-1) & \cdots & C_{yy}(0) \end{pmatrix} \begin{pmatrix} 1 \\ a_1 \\ \vdots \\ a_n \end{pmatrix} = \begin{pmatrix} \sigma_w^2 \\ 0 \\ \vdots \\ 0 \end{pmatrix} \tag{11.44}$$

and determines the coefficients of the AR-process of order n. The algorithm proceeds recursively in model order and in time k to compute the parameter sets

$$\{a_1^{(1)}, \sigma_1^2\}, \{a_1^{(2)}, a_2^{(2)}, \sigma_2^2\}, \ldots, \{a_1^{(n)}, a_2^{(n)}, \ldots, a_n^{(n)}, \sigma_n^2\} \tag{11.45}$$

where a superscript has been added to the AR-coefficients to denote the model order and where the final set is the desired solution. Expressed in the linear regression terminology we introduce the variables

$$\begin{aligned} \phi_k^{(m)} &= \begin{pmatrix} y_k & y_{k-1} & \cdots & y_{k-m+1} \end{pmatrix}^T \\ a^{(m)} &= \begin{pmatrix} a_1^{(m)} & a_2^{(m)} & \ldots & a_m^{(m)} \end{pmatrix}^T \\ \alpha^{(m)} &= \begin{pmatrix} \alpha_1^{(m)} & \alpha_2^{(m)} & \ldots & \alpha_m^{(m)} \end{pmatrix}^T \\ e_k^{(m)} &= y_k - \widehat{y}_k = y_k - (\phi_{k-1}^{(m)})^T a^{(m)} \\ b_k^{(m)} &= y_{k-m} - \widehat{y}_{k-m} = y_k - (\phi_k^{(m)})^T \alpha^{(m)} \end{aligned} \tag{11.46}$$

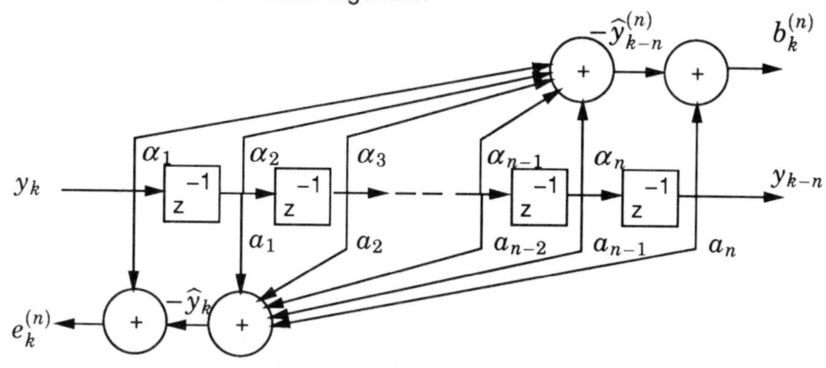

Figure 11.5 A forward/backward prediction error filter according to the Levinson-Durbin algorithm.

where k denotes time index and m model order. The algorithm can be derived by means of an alternative parametrization known as the *lattice structure* according to the relationships

$$
\begin{aligned}
a^{(m)} &= \begin{pmatrix} a^{(m-1)} \\ 0 \end{pmatrix} + K_f^{(m)} \begin{pmatrix} \alpha^{(m-1)} \\ 1 \end{pmatrix} \\
\alpha^{(m)} &= \begin{pmatrix} 0 \\ \alpha^{(m-1)} \end{pmatrix} + K_b^{(m)} \begin{pmatrix} 1 \\ a^{(m-1)} \end{pmatrix}
\end{aligned}
\tag{11.47}
$$

which are sometimes called *Levinson recursions*. The coefficients denoted by $\{K_f^{(1)}, K_f^{(2)}, \ldots, K_f^{(n)}\}$. and $\{K_b^{(1)}, K_b^{(2)}, \ldots, K_b^{(n)}\}$ are known as *reflection coefficients* or *partial correlation coefficients* and exhibit a simple relationship to the autoregressive coefficients $\{a_1^{(1)}, a_2^{(2)}, \ldots, a_n^{(n)}\}$. To visualize the lattice structure we denote the prediction error for an nth order linear predictor of y_k at time k as

$$
\begin{aligned}
e_k^{(n)} &= y_k - \hat{y}_k^{(n)} = y_k + \sum_{j=1}^{n} a_j^{(n)} y_{k-j} \\
&= y_k + \sum_{j=1}^{n-1} (a_j^{(n-1)} + K_f^{(n)} \alpha_j^{(n-1)}) y_{k-j} + K_f^{(n)} y_{k-n} \\
&= e_k^{(n-1)} + K_f^{(n)} b_{k-1}^{(n-1)}
\end{aligned}
\tag{11.48}
$$

where Eq. (11.47) has been used and where we have introduced

$$
b_k^{(n)} = y_{k-n} + \sum_{j=1}^{n} \alpha_j^{(n)} y_{k+1-j} = y_{k-n} - \hat{y}_{k-n}^{(n)} \tag{11.49}
$$

The term $b_k^{(n)}$ is called the *backward prediction error, i.e.*, the error when one attempts to "predict" y_{k-n} on the basis of the sample $y_{k-n+1}, \ldots, y_k$ (see Fig.

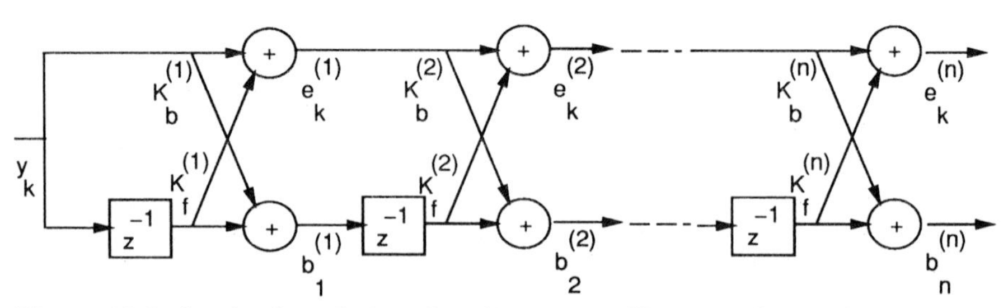

Figure 11.6 Lattice formulation of prediction error filter according to the Levinson--Durbin algorithm. The autoregressive coefficients are reparametrized as *reflection coefficients* or *partial correlation coefficients* $\{K_f^{(i)}\}_{i=1}^n$ and $\{K_b^{(i)}\}_{i=1}^n$.

11.5). By an argument similar to Eq. (11.48) we summarize the recursive equations for model order m and time k as

$$\begin{aligned} e_k^{(m)} &= e_k^{(m-1)} + K_f^{(m)} b_{k-1}^{(m-1)}, &&\text{forward prediction error} \\ b_k^{(m)} &= b_{k-1}^{(m-1)} + K_b^{(m)} e_k^{(m-1)}, &&\text{backward prediction error} \end{aligned} \tag{11.50}$$

where $e_k^{(0)} = b_k^{(0)} = y_k$. The terminology of *forward prediction error* and *backward prediction error* derives from the interpretation of the two variables $e_k^{(m)}$ and $b_k^{(m)}$, of model order m and for time k, as the prediction errors $y_k - \widehat{y}_k^{(m)}$ and $y_{k-m} - \widehat{y}_{k-m}^{(m)}$, respectively. The prediction error equations in Eq. (11.50) have an order recursive structure according to Fig. 11.6 with the transfer function relationships

$$\begin{pmatrix} E^{(m)}(z) \\ B^{(m)}(z) \end{pmatrix} = \begin{pmatrix} 1 & K_f^{(m)} z^{-1} \\ K_b^{(m)} & z^{-1} \end{pmatrix} \begin{pmatrix} E^{(m-1)}(z) \\ B^{(m-1)}(z) \end{pmatrix}$$

where the algorithm is initialized by

$$\begin{pmatrix} E^{(0)}(z) \\ B^{(0)}(z) \end{pmatrix} = \begin{pmatrix} Y(z) \\ Y(z) \end{pmatrix} \tag{11.51}$$

Minimization of $\mathcal{E}\{|e_k^{(m)}|^2\}$ and $\mathcal{E}\{|b_k^{(m)}|^2\}$ with respect to the reflection coefficients $K_f^{(m)}$ and $K_b^{(m)}$ gives the optimal lattice-structure parameters

$$\begin{aligned} K_f^{(m)} &= -\frac{\mathcal{E}\{e_k^{(m-1)} b_{k-1}^{(m-1)}\}}{\mathcal{E}\{|b_{k-1}^{(m-1)}|^2\}}, \qquad m = 1, 2, \ldots, n \\ K_b^{(m)} &= -\frac{\mathcal{E}\{e_k^{(m-1)} b_{k-1}^{(m-1)}\}}{\mathcal{E}\{|e_k^{(m-1)}|^2\}} \end{aligned} \tag{11.52}$$

which follows from the mathematical expectation of the square of Eq. (11.50)– or according to the orthogonality principle. Assuming that the reflection coefficients are chosen according to Eq. (11.52), one can evaluate the following residual covariances

$$p_{be}^{(m)} = \mathcal{E}\{e_k^{(m)} b_{k-1}^{(m)}\} = -\begin{pmatrix} C_{yy}(1) & C_{yy}(2) & \dots & C_{yy}(m+1) \end{pmatrix} \begin{pmatrix} \alpha^{(m)} \\ 1 \end{pmatrix}$$

and

$$\begin{aligned} \sigma_m^2 = p_{ee}^{(m)} &= \mathcal{E}\{|e_k^{(m)}|^2\} = p_{ee}^{(m-1)} + K_f^{(m)} p_{be}^{(m-1)}, \qquad p_{ee}^{(0)} = C_{yy}(0) \\ p_{bb}^{(m)} &= \mathcal{E}\{|b_k^{(m)}|^2\} = p_{bb}^{(m-1)} + K_b^{(m)} p_{be}^{(m-1)}, \qquad p_{bb}^{(0)} = C_{yy}(0) \end{aligned} \tag{11.53}$$

This algorithm is initialized by

$$\begin{aligned} \sigma_0^2 = p_{ee}^{(0)} &= p_{bb}^{(0)} = C_{yy}(0) \\ p_{be}^{(0)} &= -C_{yy}(1) \\ a_1^{(1)} &= \alpha_1^{(1)} = K_f^{(1)} = K_b^{(1)} = -C_{yy}(1)/C_{yy}(0) \end{aligned} \tag{11.54}$$

As the recursive equations for $p_{ee}^{(m)}$ and $p_{bb}^{(m)}$ have the same initial values and take on the same value for all m, it suffices to evaluate σ_m^2 only. Determination of the autoregressive parameters and lattice parameters can, thus, be simplified to the recursive equations

$$\begin{aligned} p_{be}^{(m)} &= -\begin{pmatrix} C_{yy}(1) & C_{yy}(2) & \dots & C_{yy}(m+1) \end{pmatrix} \begin{pmatrix} \alpha^{(m)} \\ 1 \end{pmatrix} \\ a_m^{(m)} &= \alpha_1^{m)} = K_f^{(m)} = K_b^{(m)} = -p_{be}^{(m-1)}/\sigma_{m-1}^2 \\ a_j^{(m)} &= a_j^{(m-1)} + a_m^{(m)} \alpha_j^{(m-1)}, \qquad j = m-1, m-2, \dots, 2, 1 \\ \alpha_{j+1}^{(m)} &= a_j^{(m-1)} + a_m^{(m)} a_j^{(m-1)}, \qquad j = 2, 3, \dots, m-1, m \\ \sigma_m^2 &= (1 - |a_m^{(m)}|^2)\sigma_{m-1}^2 \end{aligned} \tag{11.55}$$

The autoregressive coefficients are, thus, found recursively from correlation data, and the transfer function of the full filter between input $\{y_k\}$ and residuals is

$$\frac{E_n(z)}{Y(z)} = A(z^{-1}) = 1 + a_1^{(n)} z^{-1} + \dots + a_n^{(n)} z^{-n} \tag{11.56}$$

which is the inverse of an AR-process. Hence, one can expect that $\{e_k^{(n)}\}$ restores this input if the filter input is obtained from an nth order AR-process with a white-noise zero-mean input.

As a final remark it should be said that there are several reasons to implement AR-modeling by use of lattice structure calculations as these algorithms are known to combine rapid computation with good numerical properties. In addition, the lattice structure has an interesting relationship to the continued fraction approximation (see Chapter 10), which justifies interpretations in terms of model reduction. Moreover, the model structure is valuable in attempts to identify processes without measurable input variables where it bridges the gap between correlation analysis and state-space realizations. This property makes the lattice model popular and useful for many physical modeling applications—*e.g.*, speech processing and for inverse scattering models in physics.

11.7 SPECTRAL PROPERTIES

Recursive identification methods are very much time-domain oriented and only few connections to frequency-domain methods thus appear. One problem is that the organization of most recursive estimation methods correspond to spectrum estimation methods using rectangular or exponential windows with poor spectral leakage properties. Implementation of sliding windows requires significant block data processing, and spectral estimators of the periodogram or correlogram type are often unattractive for recursive implementation. Another problem is that the extraction of spectral information from ARMAX-type models is not trivial. It has been observed for processes consisting of a sinusoid in noise that the peak location in the autoregressive spectral estimate depends critically on the phase of the sinusoid. Also, the spectral estimate sometimes exhibits two closely spaced peaks falsely indicating a second sinusoid. This phenomenon is known as *spectral line splitting.*

It is for this reason that recursive implementation of spectral estimator often takes the form of block data processing or lattice algorithms.

11.8 BIBLIOGRAPHY AND REFERENCES

The development of recursive identification methods has to a large extent evolved to solve the requirements of implementation, real-time application and adaptive systems. Detailed books that treat recursive identification methods are

- L. LJUNG AND T. SÖDERSTRÖM, *Recursive Identification Methods: Theory and Practice*. Cambridge, MA: MIT Press, 1983.
- G.C. GOODWIN AND K.S. SIN, *Adaptive Filtering, Prediction and Control*. Englewood Cliffs, NJ: Prentice-Hall, 1984.

The δ–operator is described in

- R.H. MIDDLETON AND G.C. GOODWIN, *Digital Control and Estimation: A Unified Approach*, Englewood Cliffs, NJ: Prentice-Hall, 1990.

Recursive identification for applications of detection is surveyed in

- M. BASSEVILLE AND A. BENVENISTE, *Detection of Abrupt Changes in Signals and Dynamical Systems*, Berlin: Springer-Verlag, 1986.

Further references are given in Chapter 15, which treats adaptive systems.

11.9 EXERCISES

11.1 Show that for a linear regression model $\mathcal{Y}_k = \Phi_k\theta + v$ and for $Q(\widehat{\theta}_k)$ according to Eq. (11.18) and for $V(\widehat{\theta}_k) = \sum_{i=1}^{k} \varepsilon_i^T(\widehat{\theta}_k)\varepsilon_i(\widehat{\theta}_k)/2$, it holds for the least-squares solution $\widehat{\theta}_k$ that

$$V(\widehat{\theta}_k) + Q(\widehat{\theta}_k) = \frac{1}{2}v^T v \tag{11.57}$$

Hint: Use the property $\varepsilon^T\Phi_k = 0$.

11.2 Adapt the proof of parameter convergence by means of the function $Q(\widehat{\theta}_k)$ as defined by Eq. (11.18) for the case with a forgetting factor $\lambda < 1$.

11.3 It is clear that the P-matrix in Eq. (11.13) except for a constant factor σ^2 asymptotically represents the parameter covariance in the case of constant parameters and an uncorrelated noise sequence. How should the recursive least-squares algorithm be modified in order to estimate the noise covariance σ^2?

11.4 Show that the function

$$V(\widehat{\theta}_k) = \frac{1}{2}\widetilde{\theta}_k^T Q^{-1}\widetilde{\theta}_k \tag{11.58}$$

decreases in each step of the recursive identification algorithm in Eq. (11.42). Use this result to prove parameter convergence. ■

12

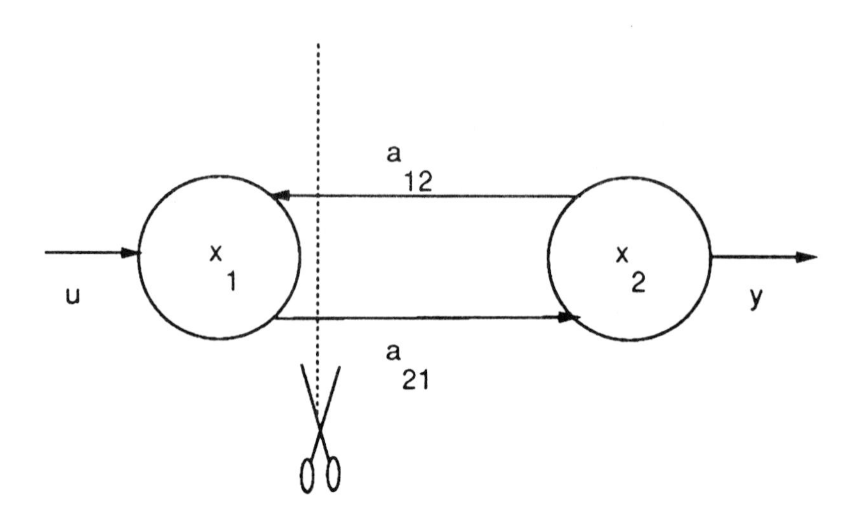

Continuous-time Models

12.1 INTRODUCTION

Accurate knowledge of a continuous-time transfer function is a prerequisite of many methods in physical modeling and control system design. As we have seen in the earlier chapters, system identification is often done by applying time-series analysis to discrete-time transfer function models. A problem with such approaches, however, is that there exists no undisputed algorithm for parameter translation from discrete-time parameters to a continuous-time description. Problems in this context are associated with translation of the system zeros from the discrete-time model to the continuous-time model, whereas the system poles are mapped by means of complex exponentials. As a result, a poor translation tends to affect both the frequency response such as the Bode diagram and transient responses such as the impulse response. One source of error in many existing algorithms is that computation of the system zeros is affected by discrepancy between the assumed and the actual intersample behavior of the control variables.

Another systematic problem associated with the approach of ARMAX-model based parameter estimation is the following simple observation. Assume that a parameter of an ARMAX-type model changes abruptly to a new value. Detection of such a change and convergence of the parameter estimate to the new value would require a time proportional both to the sampling period and to the number of estimated parameters. This delay may be unacceptably long. Not only is it sometimes impossible to improve the response time but a shorter sampling period may be incompatible with good parameter identifiability. Moreover, similar problems arise in adaptive control applications which often are associated with recursive estimation methods. One obvious disadvantage of many discrete-time adaptive control schemes is that the sampling period must be chosen to provide good identifiability rather than good control action.

This chapter deals with the problem of estimating the transfer function of a continuous-time dynamic system in the presence of colored noise. We introduce an operator transformation that allows continuous-time parametrization, whereas the parameter estimation can be made by means of a discrete-time maximum-likelihood algorithm or a recursive algorithm. A comparison is made between the performance of the continuous-time identification method in comparison with a standard identification of an ARMAX model. The method is useful in cases where it is important to not only estimate the coefficients of a continuous-time transfer function but also to maintain a physical interpretation of the transfer function results.

12.2 OUTLINE OF THE METHOD

There is one approach to the identification of continuous-time systems that developed in the 1950s for use on analog computers. The idea is to have "state variable filters" $F_1, \ldots, F_n$ acting on the inputs and outputs of the continuous-

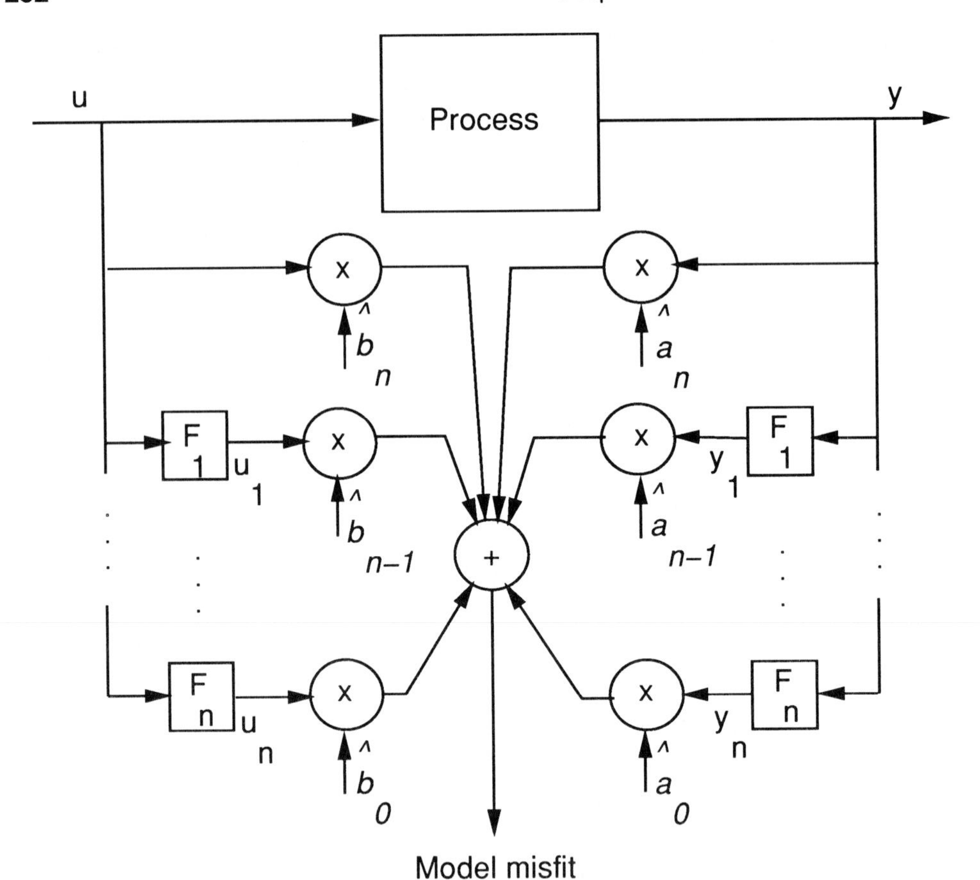

Figure 12.1 Filter arrangement $F_1, \ldots, F_n$ for continuous-time identification and filtered variables $y_1, \ldots, y_n$ and $u_1, \ldots, u_n$.

time process; see Fig. 12.1. One possible alternative is to choose

$$F_1 = \frac{s}{1+s\tau}, \cdots, F_n = (\frac{s}{1+s\tau})^n \tag{12.1}$$

Let all y_i, u_i ; $0 \le i \le n$ be outputs from the filter assembly

$$\begin{aligned} y_i(t) &= F_i\{y(t)\} \\ u_i(t) &= F_i\{u(t)\} \end{aligned} \qquad 0 \le i \le n \tag{12.2}$$

These filter outputs will then give approximative derivatives of the inputs and outputs. If the original input-output model is

$$\mathcal{S}: \quad \frac{d^n y}{dt^n} + a_1 \frac{d^{n-1} y}{dt^{n-1}} + \cdots + a_n y = b_1 \frac{d^{n-1} u}{dt^{n-1}} + \cdots + b_n u \tag{12.3}$$

then we can fit the parameters $\alpha_1, \ldots, \alpha_n, \beta_1 \ldots, \beta_n$ of the linear model

$$\mathcal{M}: \qquad y_n + \alpha_1 y_{n-1} + \cdots + \alpha_n y = \beta_1 u_{n-1} + \cdots + \beta_n u \tag{12.4}$$

by parameter adjustment until y_n and $\widehat{y}_n$ agree

$$\widehat{y}_n = -\widehat{\alpha}_1 y_{n-1} - \cdots - \widehat{\alpha}_n y + \widehat{\beta}_1 u_{n-1} + \cdots + \widehat{\beta}_n u \tag{12.5}$$

and the model misfit $\varepsilon = y_n - \widehat{y}_n \to 0$. For appropriate choices of the state variable filters it holds that

$$\alpha_i \approx a_i, \qquad \beta_i \approx b_i; \qquad 1 \leq i \leq n \tag{12.6}$$

although the filtered signals are only approximatively equal to the true state variables. Methods of this type have therefore been developed as a branch of instrumental variable methods; see Young (1969).

12.3 MODEL TRANSFORMATION

In this section we introduce a modification of this algorithm that is based on an algebraic reformulation of transfer function models and, in addition, we introduce discrete-time noise models. The idea is to find a causal, stable, realizable linear operator that may replace the differential operator without approximations. This must be done in a manner ensuring that we obtain a linear model for estimation of the original transfer function parameters a_i, b_i. Here we shall be considering cases where we obtain a linear model in low pass filter operators. It will be shown that there is always a linear one-to-one transformation which relates the continuous-time parameters and the convergence points for each choice of filter. We then follow investigations on the state-space properties of the introduced filters and the original model. The convergence rate of the parameter estimates is then considered. Finally, two examples are given, one with applications to time invariant systems, the other applicable to time varying systems.

Consider a linear nth order transfer operator formulated with a differential operator $p = d/dt$ and unknown coefficients a_i, b_i.

$$G_0(p) = \frac{b_1 p^{n-1} + \cdots + b_n}{p^n + a_1 p^{n-1} + \cdots + a_n} = \frac{B(p)}{A(p)} \tag{12.7}$$

A and B are assumed to be co-prime. Now introduce the operator

$$\lambda = \frac{a}{p+a} = \frac{1}{1+p\tau}, \qquad \tau = 1/a > 0 \tag{12.8}$$

This allows us to make the following transformation

$$G_0(p) = \frac{B(p)}{A(p)} = \frac{B^*(\lambda)}{A^*(\lambda)} = G_0^*(\lambda) \tag{12.9}$$

with

$$\begin{aligned} A^*(\lambda) &= 1 + \alpha_1\lambda + \alpha_2\lambda^2 + \ldots + \alpha_n\lambda^n \\ B^*(\lambda) &= \beta_1\lambda + \beta_2\lambda^2 + \ldots + \beta_n\lambda^n \end{aligned} \tag{12.10}$$

An input-output model is easily formulated as

$$A^*(\lambda)y(t) = B^*(\lambda)u(t) \tag{12.11}$$

$$y(t) = -\alpha_1[\lambda y](t) - \ldots - \alpha_n[\lambda^n y](t) + \beta_1[\lambda u](t) + \ldots + \beta_n[\lambda^n u](t) \tag{12.12}$$

This is now a linear model of a dynamical system at all points of time. Notice that $[\lambda u]$, $[\lambda y]$, etc., denote filtered inputs and outputs. The parameters α_i, β_i may now be estimated with any method suitable for estimating parameters of a linear model. A reformulation of the model in Eq. (12.12) to a linear regression form is

$$y(t) = \varphi_\tau^T(t)\theta_\tau \tag{12.13}$$

where

$$\theta_\tau = \begin{pmatrix} -\alpha_1 & -\alpha_2 & \ldots & -\alpha_n & \beta_1 & \ldots & \beta_n \end{pmatrix}^T \tag{12.14}$$

and

$$\varphi_\tau(t) = \begin{pmatrix} [\lambda y](t), & [\lambda^2 y](t), & \ldots, & [\lambda u](t), & \ldots & [\lambda^n u](t) \end{pmatrix}^T \tag{12.15}$$

We now have the following continuous-time input-output relations

$$\begin{aligned} y(t) &= G_0(p)u(t) = G_0^*(\lambda)u(t) \\ y(t) &= \varphi_\tau^T(t)\theta_\tau \\ Y(s) &= \Phi_\tau^T(s)\theta_\tau \quad \text{where} \quad \Phi_\tau(s) = \mathcal{L}\{\varphi_\tau(t)\}(s) \end{aligned} \tag{12.16}$$

where $\mathcal{L}\{\cdot\}$ denotes a Laplace-transform. Finally, a Laplace transformation of Eq. (12.16) gives

$$Y(s) = G_0^*(\lambda(s))U(s) \tag{12.17}$$

A particularly attractive feature is the fact that the same linear relation holds not only in both the time domain and the frequency domain, but also without any approximation or selection of data.

Example 12.1—Estimation of two constant parameters
Consider the system with input u, output y, and the transfer operator G_0

$$y(t) = G_0(p)u(t) = \frac{b_1}{p + a_1}u(t) \tag{12.18}$$

which is expressed by means of the differential operator p. Using the operator transformation λ of Eq. (12.8) one obtains

$$\lambda = \frac{1}{1 + p\tau} \tag{12.8}$$

This gives the transformed model

$$G_0^*(\lambda) = \frac{b_1\tau\lambda}{1 + (a_1\tau - 1)\lambda} = \frac{\beta_1\lambda}{1 + \alpha_1\lambda} \tag{12.19}$$

A linear estimation model of the type (12.13) is given by

$$y(t) = -\alpha_1[\lambda y](t) + \beta_1[\lambda u](t) = \varphi_\tau^T(t)\theta_\tau(t) \tag{12.20}$$

with

$$\varphi_\tau(t) = \begin{pmatrix} -[\lambda y](t) \\ [\lambda u](t) \end{pmatrix} \tag{12.21}$$

and the parameter vector

$$\theta_\tau = \begin{pmatrix} \alpha_1 \\ \beta_1 \end{pmatrix} \tag{12.22}$$

The original parameters are found *via* the relationships

$$\begin{pmatrix} a_1 \\ b_1 \end{pmatrix} = \begin{pmatrix} \frac{1}{\tau}(\alpha_1 + 1) \\ \frac{1}{\tau}\beta_1 \end{pmatrix} \tag{12.23}$$

and the corresponding parameter estimates from

$$\begin{pmatrix} \widehat{a}_1 \\ \widehat{b}_1 \end{pmatrix} = \begin{pmatrix} \frac{1}{\tau}(\widehat{\alpha}_1 + 1) \\ \frac{1}{\tau}\widehat{\beta}_1 \end{pmatrix} \tag{12.24}$$

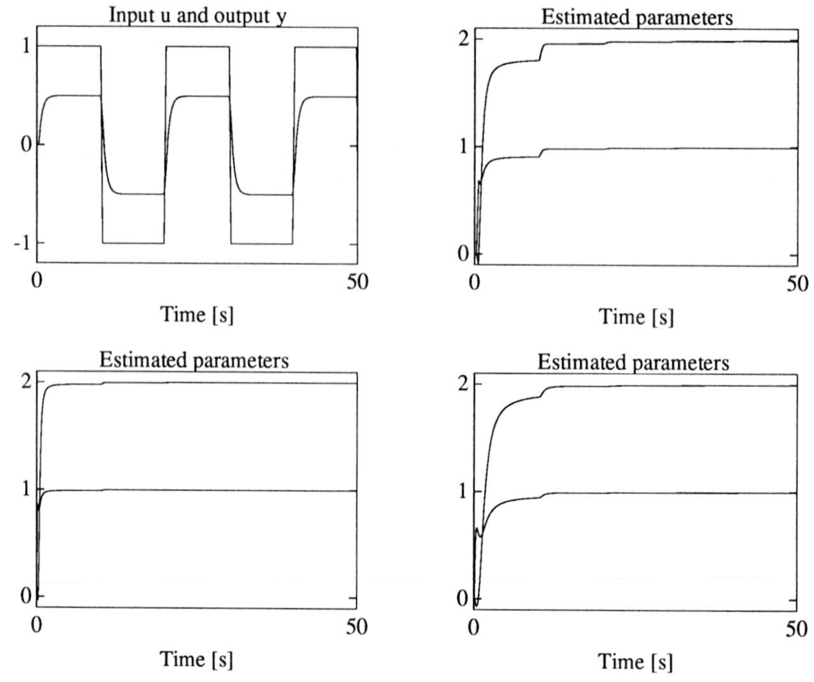

Figure 12.2 Input u and output y of the process $\dot{y} = -ax + bu$ with $a = 2$ and $b = 1$ (*upper left*). Recursively estimated parameters are shown for $\tau = 0.3$, $h = 0.3$ (*upper right*), $\tau = 0.3$, $h = 0.03$ (*lower left*), $\tau = 3.0$, $h = 0.3$ (*lower right*), respectively. Both the sampling rate h and the operator time constant τ affect the convergence properties.

Sampling of all variables in Eq. (12.20) and application of the recursive least-squares estimation algorithm

$$\begin{aligned} \widehat{\theta}_\tau(k) &= \widehat{\theta}_\tau(k-1) + P(k)\varphi_\tau(k)\varepsilon(k) \\ \varepsilon(k) &= y(k) - \varphi_\tau^T(k)\theta_\tau(k-1) \\ P(k) &= P(k-1) - \frac{P(k-1)\varphi_\tau(k)\varphi_\tau^T(k)P(k-1)}{1 + \varphi_\tau^T(k)P(k-1)\varphi_\tau(k)} \end{aligned} \tag{12.25}$$

is obviously possible. The vector $\widehat{\theta}_\tau(k)$ includes the parameter estimates, $\varepsilon(k)$ is the prediction error, and $P(k)$ is the estimate of the covariance at recursion number k. Simulation results for different choices of the filter time constant τ s and the sampling interval h s are based on the input-output data of Fig. 12.2. All simulations have started with initial values at zero for the parameter estimates and the filters. The simulations have been performed with $a_1 = 2$ and $b_1 = 1$ and a moderate excitation by means of a square-wave input. The simulations in Fig. 12.2 indicate that the convergence works satisfactorily

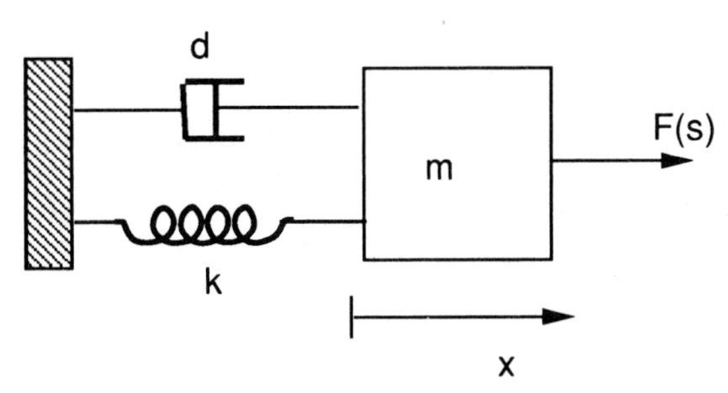

Figure 12.3 A system containing a spring action k, a damping action d, and a mass m. The variable $f(t)$ denotes the external forces on the system and $x(t)$ the resulting position of the mass.

over a large range of values of τ. The estimates are accurate for all the cases of simulation in Fig. 12.2, and recursive estimation performed with the sampling interval $h = 0.03$ s appears to have no limiting effect on the convergence rate. The convergence rate is faster for a shorter time constant τ, but if τ is too short the convergence transient may be violent.

The convergence rate with $h = \tau = 0.3$ s is still good in Fig. 12.2 with a settling time of the same order of magnitude as the process time constant $1/a_1$. It can be seen from Fig. 12.2 that there are acceptable convergence rates over a large range of values of the time constant τ. Notice that the convergence rate is higher for small values of τ, though the parameter transient tends to be more violent.

Example 12.2—Estimation of a time-varying parameter
Assume the spring coefficient k and the mass m to be well known and constant (see Fig. 12.3), whereas the damping coefficient d is unknown and time varying. Assume that the external force f and the position x are measurable. The force f is assumed to be the control input variable. The transfer function from input f to output x is given by

$$\frac{X(s)}{F(s)} = \frac{\frac{1}{m}}{s^2 + \frac{d}{m}s + \frac{k}{m}} = \frac{b_2}{s^2 + a_1 s + a_2} \tag{12.26}$$

The operator translation in Eq. (12.8) gives the transformed transfer operator from force f to position x

$$\frac{X(\lambda)}{F(\lambda)} = \frac{b_2\tau^2\lambda^2}{(1-\lambda)^2 + a_1\tau(1-\lambda)\lambda + a_2\tau^2\lambda^2} \tag{12.27}$$

The unknown coefficient is a_1 for which we find the relation

$$\varphi_\tau(t) \cdot a_1 = y(t) \tag{12.28}$$

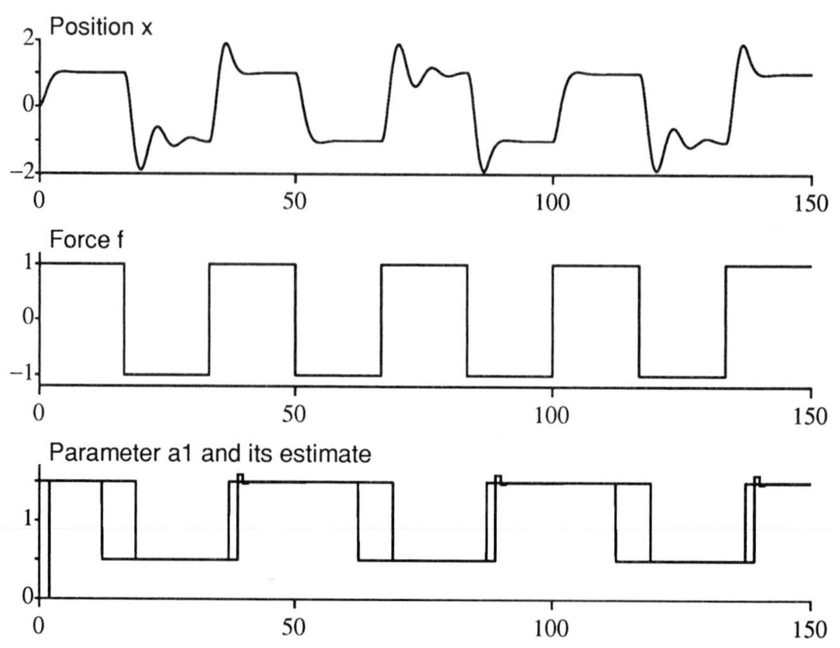

Figure 12.4 Output y (*upper*) and input u (*middle*) for the second order system with time-varying damping. True parameter a_1 and the estimate (*lower*) sampled with 0.5 s. Notice that the parameter estimate can track the true parameter for nonzero velocity only.

where

$$\varphi_\tau(t) = \tau[\lambda(1-\lambda)x](t) \tag{12.29}$$

$$y(t) = -[(1-\lambda)^2 x](t) - a_2\tau^2[\lambda^2 x](t) + \tau^2 b_2[\lambda^2 f](t) \tag{12.30}$$

A simple discrete-time heuristic tracking algorithm for a_1 is the following threshold algorithm

$$\widehat{a}_1(kh) = \begin{cases} \widehat{a}_1(kh-h), & \text{if } |\varphi_\tau(kh)| < 0.1 \\ y(kh)/\varphi_\tau(kh), & \text{if } |\varphi_\tau(kh)| \geq 0.1 \end{cases} ; \qquad k = 1, 2, \ldots \tag{12.31}$$

where the subthreshold updating is chosen in order to avoid division by small numbers. Some simulation studies are presented in Fig. 12.4. As it is possible to utilize the estimated parameter $\widehat{a}_1$ for real-time modifications of the controller, the potential for adaptive control is obvious. ■

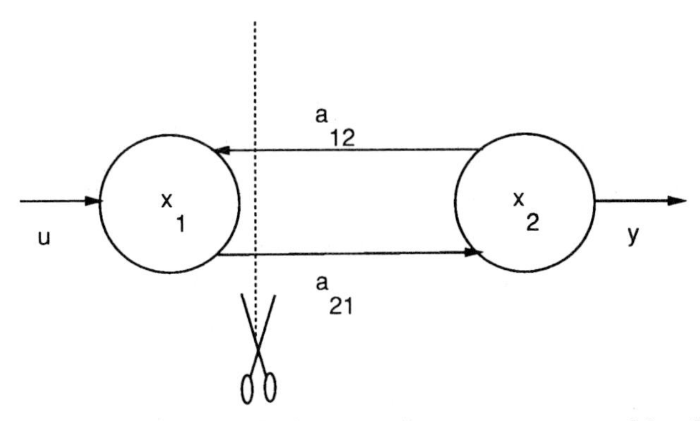

Figure 12.5 A system composed of two components represented by the states x_1 and x_2.

Example 12.3—Systems with interacting subsystems
Consider the composite system of Fig. 12.5 consisting of several subsystems which can be viewed as a system in closed-loop operation. Assume that the system equations are

$$\begin{aligned} \frac{dx_1}{dt} &= a_{11}x_1(t) + a_{12}x_2(t) + u(t) \\ \frac{dx_2}{dt} &= a_{21}x_1(t) \\ y(t) &= x_2(t) \end{aligned} \tag{12.32}$$

with states x_1, x_2 and some coefficients a_{11}, a_{12}, a_{21}. An interesting question is how to estimate the dynamics of one subsystem based on observations from the interacting components x_1 and x_2. What dynamics can be expected of subsystem x_1 if the interaction between x_1 and x_2 ceases? How should the system and the subsystems be identified from observations u and y? The transfer function is

$$Y(s) = \frac{a_{21}}{s^2 - a_{11}s - a_{21}a_{12}} U(s) \tag{12.33}$$

Continuous-time modeling can be performed by introducing the low pass filter λ

$$\lambda = \frac{1}{1 + s\tau} \longleftrightarrow s = \frac{1}{\tau}\frac{1 - \lambda}{\lambda} \tag{12.34}$$

The transformed model where λ has replaced s is

$$\frac{1}{\tau^2}(1 - \lambda)^2 y(t) = a_{11}[\frac{1}{\tau}(\lambda - \lambda^2)y(t)] + a_{21}a_{12}[\lambda^2 y(t)] + a_{21}[\lambda^2 u(t)] \tag{12.35}$$

Implementation of λy, $\lambda^2 y$, $\lambda^2 u$ and subsequent identification by, for example, least-squares identification, is straightforward. Notice that a_{11}, a_{21}, $a_{21}a_{12}$ are identifiable provided that sufficient input excitation is available. Hence, all model parameters a_{11}, a_{21}, a_{12} are identifiable. ■

Parameter transformations

Before proceeding to signal processing aspects, we need to clarify the relationship between the parameters α_i, β_i of Eq. (12.10) and the original parameters a_i, b_i of the transfer function in Eq. (12.7). Let the vector of original parameters be denoted by

$$\theta = \begin{pmatrix} -a_1 & -a_2 & \dots & -a_n & b_1 & \dots & b_n \end{pmatrix}^T \tag{12.36}$$

The relationship between Eq. (12.14) and Eq. (12.36) is then

$$\theta_\tau = F_\tau \theta + G_\tau \tag{12.37}$$

Using the definition of λ in Eq. (12.8) and Eq. (12.10) it can be shown that the $2n \times 2n$–matrix F_τ

$$F_\tau = \begin{pmatrix} M_\tau & 0_{n\times n} \\ 0_{n\times n} & M_\tau \end{pmatrix} \tag{12.38}$$

where

$$M_\tau = \begin{pmatrix} m_{11} & 0 & \cdots & 0 \\ \vdots & \ddots & \ddots & \vdots \\ m_{n1} & \cdots & m_{nn} & \end{pmatrix}; \qquad m_{ij} = (-1)^{i-j}\binom{n-j}{i-j}\tau^j \tag{12.39}$$

Furthermore, the $2n \times 1$–vector G_τ is given by

$$G_\tau = \begin{pmatrix} g_1 & \dots & g_n & 0 & \dots & 0 \end{pmatrix}^T; \qquad g_i = \binom{n}{i}(-1)^i \tag{12.40}$$

The matrix F_τ is invertible when M_τ is invertible, *i.e.*, for all $\tau > 0$. The parameter transformation is then one-to-one and

$$\theta = F_\tau^{-1}(\theta_\tau - G_\tau) \tag{12.41}$$

We may then conclude that the parameters a_i, b_i of the continuous-time transfer function G_0 may be reconstructed from the parameters α_i, β_i of θ_τ by means of basic matrix calculations. Alternatively we may estimate the original parameters a_i, b_i of θ from the linear relationships

$$y(t) = \theta_\tau^T \varphi_\tau(t) = (F_\tau \theta + G_\tau)^T \varphi_\tau(t) \tag{12.42}$$

or

$$y(t) = \varphi_\tau^T(t) F_\tau \theta + \varphi_\tau^T(t) G_\tau \tag{12.43}$$

where F_τ and G_τ are known matrices for each τ.

Hence, the parameter vectors θ and θ_τ are related *via* known and simple linear relationships so that translation between the two parameter vectors can be made without any problems arising. Moreover, identification can be made with respect to either θ or θ_τ.

12.4 A NOISE MODEL

Having treated the case of recursive identification, we now turn our attention to the general identification problem which involves estimation in the presence of colored noise. Consider for modeling purposes the continuous-time system description

$$A(s)Y(s) = B(s)U(s) + C(s)W(s) \tag{12.44}$$

where $A(s)$, $B(s)$, $C(s)$ are polynomials in the Laplace transform variable s and where it is required that $C^{-1}(s)A(s)$ is a stable transfer function.

Let h denote the sampling period used and assume that $\{w_k\}$ is a zero-mean normally distributed white-noise sequence with covariance

$$\mathcal{E}\{w_i w_j^T\} = R\delta_{ij} \tag{12.45}$$

In addition, we assume that $w(t) = w(kh)$ for $kh \le t < (k+1)h$, *i.e.*, $w(t)$ is assumed constant during the sampling interval. According to standard time series analysis it follows that the spectral density of $\{w_k\}$ is constant in the frequency range $[-\pi/h, \pi/h]$. As the Fourier transform of the sampled signal is periodic with a period equal to the sampling frequency $\omega_s = 2\pi/h$, the noise sequence $\{w_k\}$ might be concluded to have constant spectral density. A reason for modeling noise in this way is that a continuous-time white noise representation is obtained without adopting the concept of continuous-time white noise used for analysis of Brownian motion. In principle, this noise model is sufficient to describe any rational spectral density for the sampled noise process in the relevant frequency range $[-\pi/h, \pi/h]$. Application of the operator calculus gives the relationship

$$Y(s) = -\sum_{k=1}^{n} \alpha_k \lambda^k(s) Y(s) + \sum_{k=1}^{n} \beta_k \lambda^k(s) U(s) + \sum_{k=1}^{n} \gamma_k \lambda^k(s) W(s) + W(s) \quad (12.46)$$

which can be separated into the parameter vectors

$$\theta_\tau = \begin{pmatrix} \alpha_1 & \dots & \alpha_n & \beta_1 & \dots & \beta_n & \gamma_1 & \dots & \gamma_n \end{pmatrix}^T \quad (12.47)$$

and

$$\Phi_\tau(s) = \begin{pmatrix} -\lambda(s)Y(s) \\ \vdots \\ -\lambda^n(s)Y(s) \\ \lambda(s)U(s) \\ \vdots \\ \lambda^n(s)U(s) \\ \lambda(s)W(s) \\ \vdots \\ \lambda^n(s)W(s) \end{pmatrix} \quad (12.48)$$

so that

$$Y(s) = \Phi_\tau(s)\theta_\tau + W(s) \quad (12.49)$$

After translation to the time domain and subsequent sampling and application of the z-transform to the corresponding discrete-time variable, we have

$$Y(z) = \Phi_\tau(z)\theta_\tau + W(z) \quad (12.50)$$

As the noise is assumed constant during the sampling interval, it follows that each operator $\lambda^k(s)$ corresponds to the discrete-time transfer function

$$\lambda_k(z) = \frac{Q'_k(z)}{P_k(z)} = \frac{Q_k(z)}{P_n(z)} \quad (12.51)$$

where $Q'_k(z)/P_k(z) = Q_k(z)/P_n(z)$ is the zero-order-hold equivalent of the continuous-time system $\lambda^k(s)$ and where the denominator polynomial $P_n(z)$ is an nth order polynomial with all roots at $z = \exp(-h/\tau)$. The zero-order-hold equivalent of $\lambda^k(s)$ is chosen because the disturbance $w(t)$ is assumed to be constant during the sampling intervals. According to Eq. (12.50) we find that the contribution from $W(z)$ to $Y(z)$ is

$$E(z) = C(z)W(z) = W(z) + \sum_{k=1}^{n} \gamma_k \frac{Q_k(z)}{P_n(z)} W(z) \tag{12.52}$$

where

$$C(z) = 1 + \gamma_1 \lambda_1(z) + \cdots + \gamma_n \lambda_n(z) \tag{12.53}$$

This can be expressed in the form of an autoregressive moving-average model

$$P_n(z)E(z) = P_n(z)W(z) + \sum_{k=1}^{n} \gamma_k Q_k(z) W(z) \tag{12.54}$$

12.5 IDENTIFICATION

Let as usual $\widehat{\theta}$ denote an estimate of the parameter vector θ and let $\widetilde{\theta} = \widehat{\theta} - \theta_\tau$ denote the parameter error. An objective of identification is to choose the optimal θ according to some criterion. From Eq. (12.50) we suggest the residual model

$$\varepsilon(\widehat{\theta}, z) = Y(z) - \Phi_\tau(z)\widehat{\theta} \tag{12.55}$$

Let the residual sample covariance matrix be defined as

$$R_N(\widehat{\theta}) = \frac{1}{N} \sum_{k=1}^{N} \varepsilon_k(\widehat{\theta}) \varepsilon_k^T(\widehat{\theta}) \tag{12.56}$$

where N denotes the number of data points and where $\{\varepsilon_k(\widehat{\theta})\}_{k=1}^{N}$ is the residual sequence obtained from the relationships

$$\begin{aligned} \varepsilon(\widehat{\theta}, z) &= Y(z) - \Phi_\tau(z)\widehat{\theta} = -\Phi_\tau(z)\widetilde{\theta} + W(z), && \text{z-domain} \\ \varepsilon_k(\widehat{\theta}) &= y_k - \phi_k^T \widehat{\theta} = -\phi_k^T \widetilde{\theta} + w_k, && \text{time domain} \end{aligned} \tag{12.57}$$

A goal of system identification is to minimize the model misfit according to some optimization criterion such as least-squares or maximum-likelihood optimization. From Eq. (12.57) we conclude that $\{\varepsilon_k(\theta_\tau)\} = \{w_k\}$, and maximum-likelihood optimization (*e.g.*, in the case of normally distributed disturbances) is approached by maximizing the log-likelihood function

$$\log L(\theta, R) = -\frac{N}{2}\text{tr } R_N(\theta)R^{-1} - \frac{N}{2}\log\det R + \text{constant}$$
$$= -\frac{1}{2}\sum_{k=1}^{N}\varepsilon_k^T(\theta)R^{-1}\varepsilon_k(\theta) - \frac{N}{2}\log\det R + \text{constant} \tag{12.58}$$

where we have neglected transient effects due to initial values. By assuming R and θ to be independently parametrized and the covariance matrix R to be unknown, we have the partial derivatives

$$\frac{\partial \log L(\theta, R)}{\partial \theta} = \sum_{k=1}^{N}\varepsilon_k^T(\theta)R^{-1}\frac{\partial \varepsilon_k(\theta)}{\partial \theta}$$
$$\left(\frac{\partial \log L(\theta, R)}{\partial R_{ij}}\right) = \frac{1}{2}\sum_{k=1}^{N}\varepsilon_k^T(\theta)R^{-1}e_ie_j^TR^{-1}\varepsilon_k(\theta) - \frac{N}{2}\frac{1}{\det R}\text{adj}(R)_{ij} \tag{12.59}$$
$$= \frac{N}{2}(R^{-1}R_N(\theta)R^{-1})_{ij} - \frac{N}{2}(R^{-1})_{ij}$$

where e_i is a zero vector except for 1 in the ith position. It is clear from Eq. (12.59) that a stationary point of $\log L$ with partial derivatives equal to zero only appears for $R = R_N(\theta)$. This yields the covariance estimate

$$\widehat{R} = R_N(\theta) \tag{12.60}$$

We thus reduce the optimization problem by substitution of $R = R_N(\theta)$ in Eq. (6.34)

$$\log L(\theta, R_N(\theta)) = -\frac{N}{2}\log\det R_N(\theta) + \text{constant} \tag{12.61}$$

The maximum of Eq. (6.34) is also the minimum of

$$V_N(\theta) = \log\det R_N(\theta) \tag{12.62}$$

which has a unique minimum for $\theta = \widehat{\theta}_N$ in the sense that

$$\min_{\theta} V_N(\theta) = V_N(\widehat{\theta}_N) = \det R_N(\widehat{\theta}_N) \le \log\det(R_N(\widehat{\theta}_N) + \Delta R) \tag{12.63}$$

for any nonnegative definite matrix ΔR. Numerical optimization of Eq. (12.62) can now be approached by means of the Newton-Raphson method (see Appendix C) so that at iteration order i one evaluates

$$\widehat{\theta}_N^{(i+1)} = \widehat{\theta}_N^{(i)} - \rho_k(\nabla^2 V_N(\widehat{\theta}_N^{(i)})^{-1}\nabla V_N(\widehat{\theta}_N^{(i)}) \tag{12.64}$$

where $\{\rho_k\}$ is a sequence of step lengths chosen such that $\rho_k = 1$ or such that $V(\widehat{\theta}_N^{(i+1)})$ is minimized. Gradients in the scalar case (see Appendix 12.2) can be evaluated by means of

$$\frac{\partial \varepsilon_k^T(\theta)}{\partial \widehat{\theta}}\Big|_{\theta=\widehat{\theta}_N^{(i)}} = \psi_k(\widehat{\theta}_N^{(i)}) \approx \begin{pmatrix} \frac{y_k^{\{1\}}}{\widehat{C}(z)} \\ \vdots \\ \frac{y_k^{\{n\}}}{\widehat{C}(z)} \\ \frac{-u_k^{\{1\}}}{\widehat{C}(z)} \\ \vdots \\ \frac{-u_k^{\{n\}}}{\widehat{C}(z)} \\ \frac{-\varepsilon_k^{\{1\}}}{\widehat{C}(z)} \\ \vdots \\ \frac{-\varepsilon_k^{\{n\}}}{\widehat{C}(z)} \end{pmatrix} \tag{12.65}$$

where we designate the low pass filtered and sampled signals as follows

$$\begin{aligned} y_k^{\{i\}} &= [\lambda^i y](t)|_{t=kh}, \\ u_k^{\{i\}} &= [\lambda^i u](t)|_{t=kh}, \qquad \text{for} \quad i = 0, 1, \ldots, n \quad \text{and} \quad k = 1, 2, \ldots, N \\ \varepsilon_k^{\{i\}} &= [\lambda^i \varepsilon](t)|_{t=kh} \end{aligned} \tag{12.66}$$

and we approximate (see Appendix 12.2)

$$\begin{aligned} \nabla V_N(\widehat{\theta}_N^{(i)}) &= \frac{2}{N}\sum_{k=1}^{N} \psi_k(\widehat{\theta}_N^{(i)}) R_N^{-1}(\widehat{\theta}_N^{(i)}) \varepsilon_k(\widehat{\theta}_N^{(i)}) \\ \nabla^2 V_N(\widehat{\theta}_N^{(i)}) &\approx \frac{2}{N}\sum_{k=1}^{N} \psi_k(\widehat{\theta}_N^{(i)}) R_N^{-1}(\widehat{\theta}_N^{(i)}) \psi_k^T(\widehat{\theta}_N^{(i)})^T \end{aligned} \tag{12.67}$$

At some distance from the optimum where $\nabla^2 V_N$ may become poorly conditioned, it is standard practice to replace the Newton-Raphson procedure by some quasi-Newton method — *i.e.*, $\nabla^2 V_N$ in the algorithm in Eq. (12.64) is replaced by some positive definite matrix (see Appendix C).

12.6 CONVERGENCE AND CONSISTENCY

From Eq. (12.57) we conclude that $\varepsilon(\theta_\tau) = w_k$ for all k so that

$$\lim_{N\to\infty} R_N(\theta_\tau) = \mathcal{E}\{\varepsilon_k(\theta_\tau)\varepsilon_k^T(\theta_\tau)\} = \mathcal{E}\{w_k w_k^T\} = R \tag{12.68}$$

It follows that $\widehat{\theta}_N = \theta_\tau$ is a possible estimate as N increases. However, in order to show that there are no other minimizing elements it is necessary to make the following additional assumptions:

- The identified system is appropriately and uniquely parametrized so that no other parameter vector $\theta'_\tau \neq \theta_\tau$ also describes the same input-output relationship.
- The experimental conditions and the input signal are chosen with appropriate excitation properties so that $\nabla^2 V(\theta_\tau)$ is nonsingular
- The input sequence $\{u_k\}$ is uncorrelated with the disturbance sequence $\{w_k\}$.

If these conditions are satisfied we may conclude that $\widehat{\theta}_N$ is a consistent estimate according to the proofs for consistency of maximum-likelihood and prediction error identification. A Taylor series expansion with two terms gives

$$0 = \nabla V_N(\widehat{\theta}_N) \approx \nabla V_N(\theta_\tau) + \nabla^2 V_N(\theta_\tau)(\widehat{\theta}_N - \theta_\tau) \tag{12.69}$$

where higher-order terms can be neglected as it is known from consistency properties that $\widehat{\theta}_N \to \theta_\tau$ as $N \to N$. The covariance of the parameter estimate $\widehat{\theta}_N$ can thus be estimated as

$$\mathcal{E}\{\widetilde{\theta}_N\widetilde{\theta}_N^T\} = \mathcal{E}\{(\nabla^2 V_N(\theta_\tau))^{-1}\nabla V_N(\theta_\tau)(\nabla V_N(\theta_\tau))^T(\nabla^2 V_N(\theta_\tau))^{-1}\} \tag{12.70}$$

which may serve as an estimate of the accuracy of $\widehat{\theta}$. A covariance estimate can be computed as

$$\widehat{\Sigma}_\theta = (\nabla^2 V_N(\widehat{\theta}_N))^{-1}\widehat{\Sigma}_0(\nabla^2 V_N(\widehat{\theta}_N))^{-1} \tag{12.71}$$

where

$$\widehat{\Sigma}_0 = (\frac{4}{N^2}\sum_{j=1}^{N}\sum_{k=1}^{N}\psi_k(\widehat{\theta}_N)R_N^{-1}(\widehat{\theta}_N)\varepsilon_k(\widehat{\theta}_N)\varepsilon_j^T(\widehat{\theta}_N)R_N^{-1}(\widehat{\theta}_N)\psi_k(\widehat{\theta}_N) \tag{12.72}$$

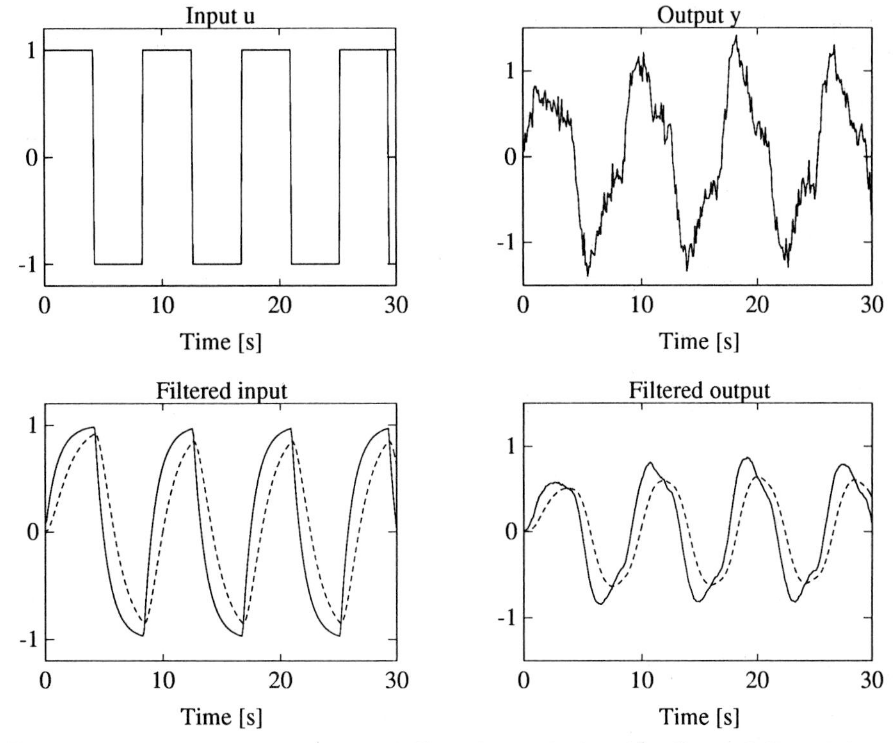

Figure 12.6 Input u, output y, and filtered signals λu, $\lambda^2 u$ (*lower left*) and λy, $\lambda^2 y$ (*lower right*) in Example 12.4. The time record consists of 300 samples and the time constant $\tau = 1$.

The estimate $\widehat{\theta}_N$ thus obtained is asymptotically normally distributed in the sense that

$$\sqrt{N}(\widehat{\theta}_N - \theta_\tau) \tag{12.73}$$

converges in distribution to $\mathcal{N}(0, P)$ where

$$P = (\mathcal{E}\{\psi_k(\theta_\tau)R_N^{-1}(\theta_\tau)\psi_k^T(\theta_\tau)\})^{-1} \tag{12.74}$$

This result, which is similar to Eq. (6.93) and applies to maximum-likelihood methods for ARMAX models, is the Cramér-Rao lower bound (see Appendix 12.1), and we also conclude that $\widehat{\theta}_N$ is asymptotically efficient.

Example 12.4—Continuous-time and discrete-time methods

The data in Fig. 12.6 have been generated from a system with the transfer function relationship

$$\mathcal{S}: \qquad Y(s) = \frac{s+1}{s^2+s+2}U(s) + \frac{2s+3}{s^2+s+2}W(s) \tag{12.75}$$

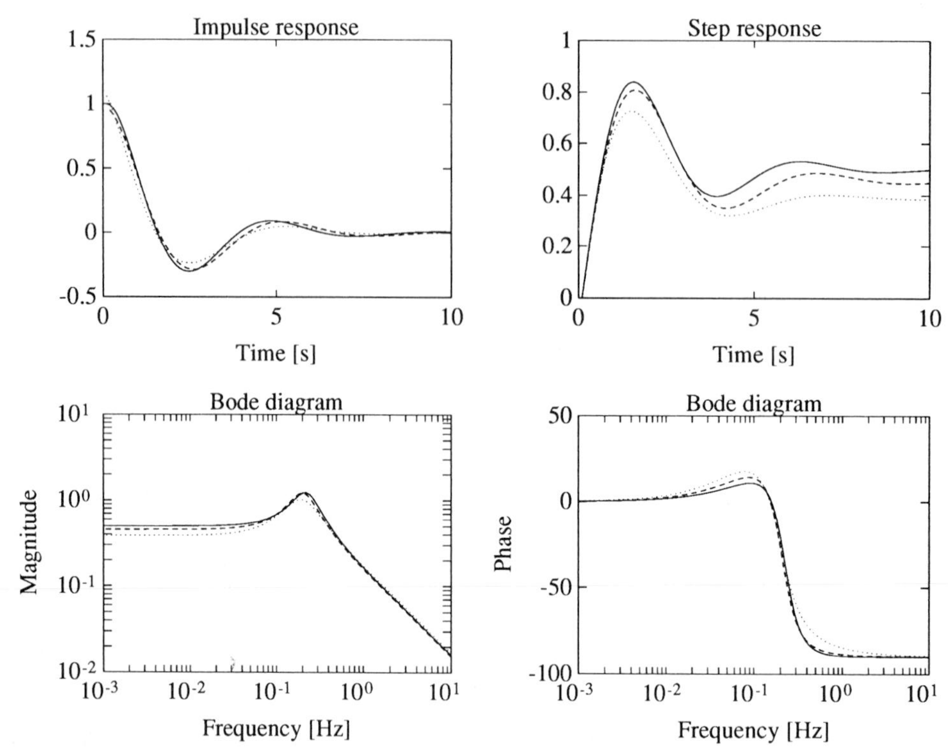

Figure 12.7 The impulse response, step response, and Bode diagram of the system (12.75). The correct value (*solid line*) and the estimates by means of continuous-time identification (*dashed line*) and ARMAX-model identification (*dotted line*).

The simulation in Fig. 12.6 shows time records when zero-mean white-noise with variance $\sigma^2 = 0.01$ is corrupting the output y with a signal-to-noise ratio $S/N = 100$. The simulations have been run with $h = 0.1$ s and $\tau = 1.0$ s, and Fig. 12.6 shows the frequency response, step response, and impulse response for the system and the two identified models. An operator translation $\lambda = 1/(s+1)$ thus gives

$$\mathcal{S}: \qquad Y(s) = \frac{\lambda(s)}{1-\lambda(s)+2\lambda^2(s)}U(s) + \frac{2\lambda(s)+\lambda^2(s)}{1-\lambda(s)+2\lambda^2(s)}W(s) \tag{12.76}$$

Application of the proposed method gives the estimated transfer function relationship

$$\mathcal{M}: \qquad Y(s) = \frac{0.9670s+0.7696}{s^2+0.9386s+1.6941}U(s) + \frac{s^2+2.1154s+1.4033}{s^2+0.9386s+1.6941}W(s) \tag{12.77}$$

A comparison with standard discrete-time identification based on the N=300 sampled values of input u and output y is interesting. Some performance indices are

Method	$R_N(\widehat{\theta})$	$\widetilde{\theta}^T\widetilde{\theta}$	e_{step}	$e_{impulse}$	$\|\|G - G_0\|\|^2$
ARMAX	0.0147	0.359	0.108	0.0227	0.0107
Cont.id.	0.0099	0.152	0.019	0.0076	0.0036

where e_{step} and $e_{impulse}$ are evaluated for u being a unit step input and an impulse, respectively. Errors are evaluated over a time interval of 10 s according to Fig. 12.7 and

$$\begin{aligned}
\widetilde{\theta} &= \begin{pmatrix} \widetilde{a}_1 & \widetilde{a}_2 & \widetilde{b}_1 & \widetilde{b}_2 \end{pmatrix}^T \\
e_{step} &= \int_0^{10} (y(\widehat{\theta}) - y(\theta))^2 dt && \text{step response error} \\
e_{impulse} &= \int_0^{10} (y(\widehat{\theta}) - y(\theta))^2 dt && \text{impulse response error} \\
||G - G_0||^2 &= \int_{0.001}^{10} ||G(\widehat{\theta}_N, i\omega) - G(\theta_\tau, i\omega)||^2 d\omega && \text{Bode diagram error}
\end{aligned} \tag{12.78}$$

Notice that this comparison is favorable for the proposed continuous-time method which provides the more accurate step response, impulse response, and frequency response for a similar value of the loss function of optimization.

The residual sample variance 0.0099 is close to the expected variance 0.01 whereas the ARMAX method does not achieve this bound. Moreover, the residual sample variance approaches this value if the model order is allowed to increase. However, the qualitative difference between the responses remains in such a case. ■

12.7 *STATE-SPACE TRANSFORMATIONS

It is of great importance that no information is lost when doing the operator transformation. This is not obvious from the original approach with state-space filters where a filtered state variable could only approximate the true state variable due to the low pass filter properties. In this section we will show that there is a one-to-one mapping between the state space associated with the original system description and that of the transformed description.

Consider therefore the transfer function

$$G_0(s) = \frac{b_1 s^{n-1} + \cdots + b_n}{s^n + a_1 s^{n-1} + \cdots + a_n} \tag{12.79}$$

The controllable canonical form of Eq. (12.79) with a state vector x and the differential operator p may be written as

$$p \begin{pmatrix} x_1(t) \\ \vdots \\ x_n(t) \end{pmatrix} = \begin{pmatrix} -a_1 & -a_2 & \cdots & -a_{n-1} & -a_n \\ 1 & 0 & \cdots & & 0 \\ 0 & 1 & 0 & \cdots & 0 \\ \vdots & \ddots & \ddots & \ddots & \vdots \\ 0 & \cdots & 0 & 1 & 0 \end{pmatrix} \begin{pmatrix} x_1(t) \\ \vdots \\ x_n(t) \end{pmatrix} + \begin{pmatrix} 1 \\ 0 \\ \vdots \\ 0 \\ 0 \end{pmatrix} u(t)$$

$$y(t) = \begin{pmatrix} b_1 & \ldots & b_n \end{pmatrix} x(t) \tag{12.80}$$

This may be associated with the fractional form

$$\begin{cases} A(p)\xi(t) = u(t) \\ y(t) = B(p)\xi(t) \end{cases} \tag{12.81}$$

with ξ as a scalar internal variable (sometimes called the *partial state*). The components x_i of the state vector x may now be related to ξ *via* the correspondence

$$x_i(t) = p^{n-i}\xi(t); \qquad i = 1, 2, \ldots, n \tag{12.82}$$

The representation in Eq. (12.80) is sufficient to describe the dynamics of the identification object, but the order of the system including both the identification object and the filters is increased by the introduced state variable filters. The filters will thus increase the minimal order of the system. It is possible to find a state space of order $2n$ to describe both the process and the filters although the realization is nonminimal.

$$\begin{cases} A'(p)\xi'(t) = u(t) \\ y(t) = B'(p)\xi'(t) \end{cases} \tag{12.83}$$

with polynomials

$$\begin{aligned} A'(p) &= A(p)(p+a)^n = p^{2n} + a_1' p^{2n-1} + \cdots + a_{2n}' \\ B'(p) &= B(p)(p+a)^n = b_1' p^{2n-1} + \cdots + b_{2n}' \end{aligned} \tag{12.84}$$

A state-space realization is given by

$$\frac{d}{dt}\begin{pmatrix} x_1'(t) \\ \vdots \\ x_{2n}'(t) \end{pmatrix} = \begin{pmatrix} -a_1' & \cdots & -a_{2n-1}' & -a_{2n}' \\ 1 & & & 0 \\ 0 & & & \\ \vdots & \ddots & \ddots & \\ 0 & \cdots & 1 & 0 \end{pmatrix} \begin{pmatrix} x_1'(t) \\ \vdots \\ x_{2n}'(t) \end{pmatrix} + \begin{pmatrix} 1 \\ 0 \\ \vdots \\ 0 \\ 0 \end{pmatrix} u(t) \quad (12.85)$$

with

$$x_i'(t) = p^{2n-i}\xi'(t) \quad (12.86)$$

Each of the components of φ_τ may now be expressed as a linear combination of the state vector components. We have with the arguments of Eqs. (12.8), (12.84), (12.86).

$$\begin{aligned} [\lambda u](t) &= \frac{a}{p+a} A(p)(p+a)^n \xi'(t) \\ &= a(p^{2n-1}\xi'(t)) + \cdots + a^n a_n (p^0 \xi'(t)) = a x_1'(t) + \cdots + a^n a_n x_{2n}'(t) \\ &\vdots \\ [\lambda^n y](t) &= a^n \begin{pmatrix} b_1 & \ldots & b_n & 0 & \ldots & 0 \end{pmatrix} x'(t) = \frac{B(p)}{A(p)}(\frac{a}{p+a})^n u(t) \end{aligned} \quad (12.87)$$

The original state vector x of Eq. (12.82) is related to x' as follows:

$$\begin{aligned} (p+a)^n \xi'(t) &= \xi(t) \\ x_i(t) &= p^{n-i}\xi(t) = p^{n-i}(p+a)^n \xi'(t) \end{aligned} \quad (12.88)$$

From Eq. (12.84) and Eq. (12.88) we find

$$x_i(t) = \sum_{j=1}^{n} \binom{n}{j} a^{n-j} x_{n+i-j}'; \qquad i = 1, 2, \ldots, n \quad (12.89)$$

Consider now the full regression vector

$$\varphi_\tau(t) = \begin{pmatrix} [\lambda y](t) & \ldots & [\lambda^n y](t) & [\lambda u](t) & \ldots & [\lambda^n u](t) \end{pmatrix}^T \quad (12.90)$$

The φ_τ−vector is therefore related to the state vector x' by a linear transformation matrix M' containing coefficients obtained from Eq. (12.89) and Eq. (12.85)

$$\varphi_\tau(t) = M'x'(t) \quad (12.91)$$

Notice that all components of the state space are observable from $[\lambda^n y]$ provided there are no common factors of A and B. This means that the states of $x'(t)$ and $x(t)$ are observable from φ_τ. From the construction of Eq. (12.85) we also see that the state x' is controllable from u provided there are no common factors of A and B. Nor should there be any factor of $(p+a)$ in B. This means that in principle it is possible to determine an input u such that x obtains any direction in the $2n$–dimensional space so that no information is irreversibly lost in the filtering process. The following theorem can be shown.

Theorem 12.1
Let G be a rational function such that

$$G(p) = \frac{B(p)}{A(p)}\left(\frac{p+a}{p+a}\right)^n = \frac{B'(p)}{A'(p)}; \qquad \deg(A) = n; \qquad \deg(B) = m \leq n-1 \tag{12.92}$$

where the polynomial factorization is such that B has no common factor with A or $(p+a)$. Let the following strictly proper transfer operator relationship hold between input u and output y

$$y(t) = \frac{B(p)}{A(p)}u(t) \tag{12.93}$$

Let λ be the operator

$$\lambda = \frac{a}{p+a}$$

Let φ_τ be the vector of filtered inputs and outputs

$$\varphi_\tau(t) = \begin{pmatrix} [\lambda u](t) & \dots & [\lambda^n u](t) & [\lambda y](t) & \dots & [\lambda^n y](t) \end{pmatrix}^T \tag{12.94}$$

and let x' be the state vector of the controllable canonical form of G. Then there exists a linear transformation such that

$$x'(t) = T_\tau \varphi_\tau(t) \tag{12.95}$$

for an invertible matrix T_τ.

Proof: See Appendix 12.3.

Remark: The theorem above has shown that φ_τ is a sufficient state vector for the filter state and the system to be identified. The controllability of x' and φ_τ means that any direction in the $2n$–dimensional space can be reached. Active improvement of identifiability by choice of the input u is also possible.

12.8 SIGNAL PROCESSING FILTERS

In the previous sections we have seen that the transfer operator may be exactly transformed to the linear model

$$y(t) = \varphi_\tau^T(t)\theta_\tau \tag{12.13}$$

with

$$\varphi_\tau(t) = \begin{pmatrix} \lambda y & \lambda^2 y & \dots & \lambda u & \lambda^2 u & \dots \end{pmatrix}^T \tag{12.15}$$

Sometimes it may be desirable to make some data selection. Let us thus denote such a data selection by the filter f in the time domain or F in the frequency domain. Let subscript f denote a signal filtered by f or F. The estimation algorithm will then fit parameters to data from the relationships

$$y_f(t) = \varphi_f^T(t)\theta_\tau \tag{12.96}$$

or

$$Y_f(s) = \Phi_f^T(s)\theta_\tau \tag{12.97}$$

This means that we have the possibility of performing filtering operations in the time domain, or in the frequency domain, or in both. Filtering operations in the frequency domain will entail weightings and selections in certain frequency ranges. Time domain filtering will mean choices of recording times, averaging or sampling.

An interesting possibility is *hybrid identification* which consists of sampling $y(t)$ and all components of $\varphi_\tau(t)$ at certain sequence of time instants $t = t_1, t_2, \dots$ The linear relationship (12.13) then still holds between y and φ_τ. These sampled data may now be used to fit parameters to the continuous time model in Eq. (12.13), Eq. (12.96) by using ordinary discrete-time recursive estimation methods. Notice that there is no discrete-time model involved, although we use sampled data and discrete-time estimation. It is also interesting that such data sampling does not need to be periodic and that "slow sampling" may be used if a lower convergence rate can be accepted. Notice also that in principle the sampling for constant parameters θ may be performed without any anti-aliasing filter. This is due to the fact that θ rather than y is the reconstructed entity. It would, however, be necessary to choose the sampling frequency properly when tracking a time-varying θ.

Choice of the low pass filter λ

A practical issue is to consider the choice of the time constant $\tau = 1/a$ of the low pass filter λ used in the modeling with the input-output relationship

$$Y(s) = G_0(s)U(s) = G_0^*(\lambda(s))U(s) = \Phi_\tau^T(s)\theta_\tau \tag{12.98}$$

The accuracy of a parameter estimate $\widetilde{\theta}_\tau$ can be evaluated, for instance, with a quadratic criterion

$$J_t(\widehat{\theta}_\tau(t)) = \int_0^t (y_f(r) - \varphi_f^T(r)\widehat{\theta}_\tau(t))^2 dr \tag{12.99}$$

It is also possible to use quadratic criteria in the frequency domain. We will make statements for "long" but finite time intervals $[0, t]$ and assume that the Parseval relation holds between the time domain and the frequency domain. A counterpart to Eq. (12.99) in the the frequency domain is then

$$J_\omega(\widehat{\theta}_\tau(t)) = \int_{-\infty}^{+\infty} |Y_f(i\omega) - \Phi_f^T(i\omega)\widehat{\theta}_\tau(t)|^2 d\omega \tag{12.100}$$

By introducing the parameter error vector as

$$\widetilde{\theta}_\tau(t) = \widehat{\theta}_\tau(t) - \theta_\tau \tag{12.101}$$

and the weighting matrix

$$P^{-1}(T) = \int_{-\infty}^{+\infty} \Phi_\tau(-i\omega)\Phi_\tau^T(i\omega)d\omega \tag{12.102}$$

with $\Phi_\tau(s) = \mathcal{L}\{\varphi_\tau(t)\}$ and where T refers to the measurement duration, the optimization criterion Eq. (12.100) can be rewritten as

$$J_\omega = \widetilde{\theta}_\tau^T \left(\int_{-\infty}^{+\infty} \Phi_\tau(-i\omega)\Phi_\tau^T(i\omega)d\omega \right) \widetilde{\theta}_\tau = \widetilde{\theta}_\tau^T P^{-1}(t)\widetilde{\theta}_\tau \tag{12.103}$$

All components of Φ_τ are dependent on the input $U(s)$. From Eqs. (12.13–12.16) it is found that the vector Φ_τ may be reduced to

$$\Phi_\tau(s) = \Gamma_\tau(s)U(s) \tag{12.104}$$

with

$$\Gamma_\tau(s) = \begin{pmatrix} \lambda(s)G_0(s) & \dots & \lambda^n(s)G_0(s) & \lambda(s) & \dots & \lambda^n(s) \end{pmatrix}^T \tag{12.105}$$

In Eq. (12.104) we see that P depends on the spectrum of the input signal u. There is also a dependence on the unknown transfer function G_0. It is therefore difficult to derive any result indicating how to choose τ optimally on the basis of this type of pure quadratic criteria.

Another approach is to demand a certain convergence rate of the parameter estimates which may be achieved with the following modification with a weighted least-squares criterion

$$J_t'(\widehat{\theta}(t)) = \int_0^t e^{2\alpha r}(y_f(r) - \varphi_f^T(r)\widehat{\theta}_\tau(t))^2 dr \qquad (12.106)$$

where $\alpha > 0$ is some constant rate of desired exponential convergence. The weighting matrix Eq. (12.102) modifies to

$$P'^{-1}(t) = \int_0^t e^{2\alpha r}\varphi_f(r)\varphi_f^T(r)dr \qquad (12.107)$$

when evaluated in the time domain. The frequency domain counterpart of Eq. (12.106) is

$$J_\omega'(\widehat{\theta}_\tau(t)) = \frac{1}{2\pi i}\int_{0-i\infty}^{0+i\infty} |Y_f(s-\alpha) - \Phi_f^T(s-\alpha)\widehat{\theta}_\tau(t)|^2 ds \qquad (12.108)$$

By examination of the integrand of Eq. (12.108) we find its convergence properties are related to the properties of $\Gamma(s)$ and $U(s)$, $\Gamma(s)$ being in turn dependent on $\lambda(s)$ and $G_0(s)$. We then find

$$P'^{-1}(t) = \int_{-\infty}^{+\infty} \Gamma_\tau(-\alpha - i\omega)U(-\alpha - i\omega)U^T(-\alpha + i\omega)\Gamma_\tau^T(-\alpha + i\omega)d\omega \quad (12.109)$$

For a nonzero input $U(s)$ we have the following condition for convergence of Eq. (12.109).

- $G_0(s-\alpha)$ stable
- $\lambda(s-\alpha)$ stable $\Rightarrow$ $\tau < 1/\alpha$

This determines the limits of convergence rates for different parameter estimations. It means that we have to require that G_0 is stable and responds rapidly enough to the input u. It is also necessary that the filter time constant τ be smaller than the desired time constant of convergence $1/\alpha$.

Implementation of least-squares estimation

Let us now consider recursive least-squares estimation of parameters of the linear model. A minimization of Eq. (12.99) in the continuous-time domain gives an algorithm of the type

$$
\begin{aligned}
\dot{\widehat{\theta}}_\tau(t) &= P_c(t)\varphi_\tau(t)(y(t) - \varphi_\tau^T(t)\widehat{\theta}_\tau(t)) \\
\dot{P}_c(t) &= -P_c(t)\varphi_\tau(t)\varphi_\tau^T(t)P_c(t)
\end{aligned}
\tag{12.110}
$$

The convergence rate can be evaluated by means of

$$
\frac{d}{dt}\left(\frac{1}{2}\widetilde{\theta}_\tau^T(t)P_c^{-1}(t)\widetilde{\theta}_\tau(t)\right) = -\frac{1}{2}(\widetilde{\theta}_\tau^T(t)\varphi_\tau(t))^2 = -\frac{1}{2}\varepsilon^2 \tag{12.111}
$$

Although it is suboptimal, a discrete-time estimation is normally preferred for reasons of implementation. A discretization of the algorithm in Eq. (12.110) at time-instants $t = 0, h, \ldots, kh$ and a Riemann sum approximation of integration gives the familiar recursive least-squares identification.

$$
\begin{aligned}
\widehat{\theta}_\tau(t) &= \widehat{\theta}_\tau(t-h) + P_s(t)\varphi_f(t)(y_f(t) - \varphi_\tau^T(t)\widehat{\theta}_\tau(t-h)) \\
P_s^{-1}(t) &= P_s^{-1}(t-h) + \varphi_f(t)\varphi_f^T(t), \qquad t = 0, h, \ldots, kh
\end{aligned}
\tag{12.112}
$$

Manipulations of Eq. (12.112) yield a formula to update P_s instead of P_s^{-1}

$$
P_s(t) = P_s(t-h) - \frac{P_s(t-h)\varphi_f(t)\varphi_f(t)^T P_s(t-h)}{1 + \varphi_f(t)^T P_s(t-h)\varphi_f(t)} \tag{12.113}
$$

Of course, more sophisticated numerical integration routines may also be utilized. With trapezoidal interpolation Eq. (12.113) may be replaced by

$$
P_s^{-1}(kh) = P_s^{-1}(kh-h) + \frac{h}{2}(\varphi_f(kh)\varphi_f^T(kh) + \varphi_f(kh-h)\varphi_f^T(kh-h)) \tag{12.114}
$$

to obtain a better approximation of Eq. (12.110).

12.9 CONCLUDING REMARKS

We have formulated an identification method for continuous-time transfer function models and equivalent to ARMAX models for discrete-time systems. The continuous-time method differs from traditional approaches to ARMAX-model identification due to the reformulation of the disturbance model and the new parametrization of the continuous-time transfer function, the parameter estimation method (*i.e.*, maximum-likelihood estimation) being the same as that used in ARMAX-model identification. Nevertheless, as the continuous-time identification algorithm can be implemented as a discrete-time method, its advantages are attributable to the new parametrization. The methodology is of particular relevance for control systems analysis and physical modeling, where it is desirable to avoid discretization of system dynamics.

A relevant question is, of course, why there is no analogue to ARMAX models for continuous-time systems. One reason is that transfer function polynomials in the differential operator can not be immediately used for identification owing to the implementation problems associated with differentiation. The successful ARMAX-models correspond to transfer function polynomials in the forward or the backward shift operators with advantages for modeling and signal processing, respectively, and translation between these two representations is simple. In contrast, there is no commonly used operator in parameter estimation that corresponds to the backward shift operator of discrete-time systems although λ is a suitable candidate for the purpose.

Unfortunately, there are more circumstances that hamper the effective application of the large body of discrete-time identification to problems involving discrete-time transfer functions. First, the model representation formulated in the differential operator must be translated to a shift operator formulation, and there are several ways to do this. However, as exact parameter translation typically requires some matrix exponentiations, a given continuous-time parameter will have a nonlinearly distributed effect on several discrete-time parameters. Accordingly, it becomes very difficult to focus attention upon a particular continuous-time parameter. In order to monitor a particular continuous-time parameter, it is generally necessary to estimate the full order discrete-time parameter vector. In other words, as it is difficult to separate known parameters from unknown ones, partitioning is difficult. Moreover, the discrete-time parameters become abstract with a dependence on the sampling interval. This is a disadvantage as the discrete-time parameters often have little physical meaning.

A related problem is how to identify accurate continuous-time transfer functions from data and, in particular, how to obtain good estimates of the zeros of a continuous-time transfer function. The difficulties in converting a discrete-time transfer function to a continuous-time transfer function are well known and related to the mapping $f(z) = (\log z)/h$. Clearly, a poor parameter translation affects both spectral properties (such as the frequency response) and time-domain properties (such as step and impulse responses).

Another aspect is that, to avoid interference from high-frequency dynamics corrupting the estimation, discrete-time identification requires anti-aliasing filters of the input-output data before sampling. A good frequency cut-off property of a sampling filter would require noncausal operations which cannot be implemented. A causal filter with sufficiently good damping of high frequencies may, on the other hand, eliminate too much of the useful low-frequency contents or introduce a delay. Efficient elimination of the high-frequency components is therefore difficult in the case of on-line identification. Moreover, the sampling filter will be incorporated as part of a discrete-time process model, and it is difficult to separate filter parameters from process parameters of physical significance. This situation is sometimes a dilemma where the shortcomings of discrete-time ARMAX-type approaches to estimation become obvious.

The methodology presented requires implementation of filters $\lambda^1, \dots, \lambda^n$ which operate on input-output data. Maximum-likelihood identification based on the parametric model results in consistent and asymptotically efficient estimates provided that noise is normally distributed. In addition to the maximum-likelihood properties, there is evidence from simulation studies that the method has favorable properties in reproducing transient reponses and frequency responses which are relevant aspects of physical modeling. Finally, the potential for real-time application and adaptive control is very interesting, as the sampling period for the regulator may be chosen independently from that of the identification. Unlike ARMAX-model based discrete-time adaptive regulators, there is no obligation to choose a certain sampling period to satisfy the needs of both control and identication.

12.10 BIBLIOGRAPHY AND REFERENCES

Early treatments of low pass filter methods and extensions deriving from the use of analog computers are described in:

- A.B. Clymer, "Direct system synthesis by means of computers." *Trans. AIEE*, 77, part I, 1959, pp. 798–806.
- H. Unbehauen and G.P. Rao, *Identification of Continuous-Time Systems*. Amsterdam: North-Holland, 1987.

The state variable filter method has been described as an instrumental variable method in:

– P.C. YOUNG, "Applying parameter estimation to dynamic systems." *Control Engineering*, Vol. 16, Oct. 1969, Oct. pp. 119–125, Nov. pp. 118-124.

– P.C. YOUNG, "Parameter estimation for continuous-time models: a survey." *Automatica*, Vol. 17, 1981, pp. 23-29.

The present chapter was based on an algebraic formulation originally described in:

– R. JOHANSSON, "Identification of continuous-time dynamic systems." *Proc. 25th IEEE Conference on Decision and Control*, Athens, Greece, 1986, pp. 1653–1658.

A important transfer function algebra applicable to the operator transformation is described in

– L.PERNEBO, "An algebraic theory for the design of controllers of linear multivariable systems. Part I: Structure matrix and feedforward design; Part II: Feedback realizations and feedback design." *IEEE Trans. Automatic Control*, AC-26, 1981, pp. 171–193.

APPENDIX 12.1 — THE CRAMÉR-RAO LOWER BOUND

The log-likelihood function for Gaussian distributed disturbances is given by

$$\log L(\theta, R) = -\frac{1}{2}\sum_{k=1}^{N} \varepsilon_k^T(\theta) R^{-1} \varepsilon_k(\theta) - \frac{N}{2}\log\det R + \text{constant} \tag{12.115}$$

Assuming R and θ to be independently parametrized and R to be unknown we find

$$\frac{\partial \log L(\theta, R)}{\partial \theta} = -\sum_{k=1}^{N} \frac{\partial \varepsilon_k^T(\theta)}{\partial \theta} R^{-1} \varepsilon_k(\theta)$$

and

$$\begin{aligned}\frac{\partial \log L(\theta, R)}{\partial \theta} &= \frac{1}{2}\sum_{k=1}^{N} \varepsilon_k^T(\theta) R^{-1} e_i e_j^T R^{-1} \varepsilon_k(\theta) - \frac{N}{2}\frac{1}{\det R}\operatorname{adj}(R)_{ij} \\ &= \frac{N}{2}(R^{-1} R_N(\theta) R^{-1})_{ij} - \frac{N}{2}(R^{-1})_{ij}\end{aligned} \tag{12.116}$$

where e_i is a zero vector except for 1 in the ith position.

The Fisher information matrix gives the Cramér-Rao lower bound for an unbiased estimate

$$J = \mathcal{E}\{ \begin{pmatrix} (\frac{\partial \log L(\theta,R)}{\partial \theta})^T \\ (\frac{\partial \log L(\theta,R)}{\partial R_{ij}})^T \end{pmatrix} \begin{pmatrix} \frac{\partial \log L(\theta,R)}{\partial \theta} & \frac{\partial \log L(\theta,R)}{\partial R_{ij}} \end{pmatrix} \} \tag{12.117}$$

evaluated at the correct parameters θ_τ and R for which $\varepsilon_k(\theta) = w_k$. It follows that all terms $\mathcal{E}\{(\partial \log L/\partial \theta_k)\partial \log L/\partial R_{ij}\} = 0$ under the assumption of uncorrelated disturbances w_k whereas

$$\begin{aligned} \mathcal{E}\{(\frac{\partial \log L(\theta, R)}{\partial \theta})^T \frac{\partial \log L(\theta, R)}{\partial \theta}\} &= \mathcal{E}\{\sum_{i=1}^{N}\sum_{j=1}^{N} \psi_i(\theta) R^{-1} w_i w_j^T R^{-1} \psi_j^T(\theta)\} \\ &= N\mathcal{E}\{\psi_j(\theta) R^{-1} \psi_j^T(\theta)\} \end{aligned}$$

We conclude that, assuming noise to be normally distributed, the estimate $\widehat{\theta}$ is asymptotically statistically efficient because $R_N(\widehat{\theta}_N)$ converges to the limit $\lim_{N\to\infty} R_N(\theta_\tau) = R$ as $N \to \infty$. ■

APPENDIX 12.2 — THE HESSIAN MATRIX

Using Eq. (12.66) it is straightforward to express the residual model as

$$\begin{aligned} \varepsilon_k = {} & y_k + \alpha_1 y_k^{\{1\}} + \cdots + \alpha_n y_k^{\{n\}} \\ & - \beta_1 u_k^{\{1\}} - \cdots - \beta_n u_k^{\{n\}} - \\ & - \gamma_1 \varepsilon_k^{\{1\}} - \cdots - \gamma_n \varepsilon_k^{\{n\}} \end{aligned} \tag{12.118}$$

or

$$C(z)\varepsilon(\theta_\tau) = y_k + \alpha_1 y_k^{\{1\}} + \cdots + \alpha_n y_k^{\{n\}} - \beta_1 u_k^{\{1\}} - \cdots - \beta_n u_k^{\{n\}} \tag{12.119}$$

From Eq. (12.119) we derive the partial derivatives

$$\begin{aligned} \frac{\partial \varepsilon_k}{\partial \alpha_j} &= \frac{1}{C(z)} y_k^{\{j\}} \\ \frac{\partial \varepsilon_k}{\partial \beta_j} &= -\frac{1}{C(z)} u_k^{\{j\}} \\ \frac{\partial \varepsilon_k}{\partial \gamma_j} &= -\frac{1}{C(z)} \varepsilon_k^{\{j\}} \end{aligned} \tag{12.120}$$

As $C(z)$ is not exactly known, we approximate it by an estimate $\widehat{C}(z)$ based on an available estimate θ and use the evaluation

$$\begin{aligned} \frac{\partial \varepsilon_k}{\partial \alpha_j} &\approx \frac{1}{\widehat{C}(z)} y_k^{\{j\}} \\ \frac{\partial \varepsilon_k}{\partial \beta_j} &\approx -\frac{1}{\widehat{C}(z)} u_k^{\{j\}} \\ \frac{\partial \varepsilon_k}{\partial \alpha_j} &\approx -\frac{1}{\widehat{C}(z)} \varepsilon_k^{\{j\}} \end{aligned} \tag{12.121}$$

The matrix second-order derivative of the loss function $V_N(\theta)$ is

$$\begin{aligned} \nabla^2 V_N(\theta) &= \frac{2}{N} \sum_{k=1}^{N} \psi_k(\theta) R_N^{-1}(\theta) \psi_k^T(\theta) \\ &+ \frac{2}{N} \sum_{k=1}^{N} \psi_k(\theta) \frac{\partial R_N(\theta)}{\partial \theta} \varepsilon_k(\theta) + \frac{2}{N} \sum_{k=1}^{N} \frac{\partial \psi_k(\theta)}{\partial \theta} R_N^{-1}(\theta) \varepsilon_k(\theta) \end{aligned} \tag{12.122}$$

According to assumptions made concerning uncorrelated disturbances with $\{\varepsilon_k(\theta_\tau)\} = \{w_k\}$, it is possible to neglect the second and third terms close to the stationary point $\widehat{\theta}_N = \theta_\tau$. ■

APPENDIX 12.3 — PROOF OF THEOREM 12.1

Let y_i be the output of i serial operators λ operating on y

$$y_i(t) = [\lambda^i y](t) \tag{12.123}$$

The transfer operator from the input u to the output y_n is then

$$y_n(t) = \frac{B'(p)}{A'(p)} u(t) \tag{12.124}$$

with

$$\begin{aligned} A'(p) &= A(p)(p+a)^n = p^{2n} + a_1' p^{2n-1} + \ldots + a_{2n}' \\ B'(p) &= B(p)(p+a)^n = b_1' p^{2n-1} + \ldots + b_{2n}' \end{aligned} \tag{12.84}$$

A fractional form for Eq. (12.123) is

$$\begin{cases} A'(p)\xi'(t) = u(t) \\ y(t) = B'(p)\xi'(t) \end{cases}$$

The state-space realization on the controllable canonical form is given by

$$\frac{d}{dt}\begin{pmatrix} x_1'(t) \\ \vdots \\ x_{2n}'(t) \end{pmatrix} = \begin{pmatrix} -a_1' & & \cdots & & -a_{2n}' \\ 1 & 0 & \cdots & & 0 \\ 0 & 1 & & & \vdots \\ \vdots & \ddots & \ddots & \ddots & \\ 0 & \cdots & 0 & 1 & 0 \end{pmatrix} \begin{pmatrix} x_1'(t) \\ \vdots \\ x_{2n}'(t) \end{pmatrix} + \begin{pmatrix} 1 \\ 0 \\ \vdots \\ 0 \end{pmatrix} u(t)$$
$$y_n(t) = \begin{pmatrix} b_1' & \ldots & b_{2n}' \end{pmatrix} x'(t) \tag{12.85}$$

where the state vector components are given by

$$x_i'(t) = p^{2n-i}\xi'(t) \tag{12.86}$$

Consider now the fractional form relating u and y

$$\begin{cases} A(p)\xi(t) = u(t) \\ y(t) = B(p)\xi(t) \end{cases}$$

with

$$(p+a)^n\xi'(t) = \xi(t) \tag{12.88}$$

With this state representation it holds that

$$\begin{aligned}
[\lambda u](t) &= \frac{a}{p+a}A(p)(p+a)^n\xi'(t) \\
&= a(p^{2n-1}\xi'(t)) + \ldots + a^n a_n(p^0\xi'(t)) = ax_1'(t) + \ldots + a^n a_n x_{2n}'(t) \\
&\vdots \\
[\lambda^i u](t) &= A(p)a^i(p+a)^{n-i}\xi'(t) \\
[\lambda^i y(t)](t) &= B(p)a^i(p+a)^{n-i}\xi'(t) \\
[\lambda^n y](t) &= a^n \begin{pmatrix} b_1 & \ldots & b_0 & 0 & \ldots & 0 \end{pmatrix} x'(t) = \frac{B(p)}{A(p)}\cdot(\frac{a}{p+a})^n u(t)
\end{aligned} \tag{12.87}$$

This means that all components of φ_τ may be expressed as linear combinations of the components of x'. Hence

$$\varphi_\tau(t) = M'x'(t) \tag{12.125}$$

The next step is to show that x' may be expressed as a linear transformation of φ_τ based on the fractional form expressed in the operator λ.

$$\begin{cases} A^*(\lambda)\xi_\lambda(t) = u(t) \\ y(t) = B^*(\lambda)\xi_\lambda(t) \end{cases} \tag{12.126}$$

with co-prime A^* and B^* and with

$$\xi_\lambda(t) = (p + a)^n \xi(t) \tag{12.127}$$

Recall that

$$(p + a)^n \xi'(t) = \xi(t) \tag{12.88}$$

This gives that

$$\xi'(t) = (\frac{1}{p + a})^{2n} \xi_\lambda(t) \tag{12.128}$$

From Eq. (12.86) and Eq. (12.128) it is found that

$$x_i'(t) = p^{2n-i}(\frac{1}{p + a})^{2n} \xi_\lambda(t) = P_i(\lambda)\xi_\lambda(t) \tag{12.129}$$

where P_i for $1 \leq i \leq 2n$ are polynomials in the operator λ.

$$P_i(\lambda) = \frac{((p + a) - a)^{2n-i}}{(p + a)^{2n}} = a^{-i}(\frac{a}{p + a})^i(1 - \frac{a}{p + a})^{2n-i}$$

It can be seen from the following relationship that all P_i's contain powers of λ from 1 to $2n$.

$$P_i(\lambda) = a^{-i}\lambda^i(1 - \lambda)^{2n-i} = \sum_{j=0}^{2n-i} \binom{2n - i}{j} a^{-i}(-\lambda)^{2n-i} \tag{12.130}$$

The factorization polynomials $A^*(\lambda)$ and $B^*(\lambda)$ are co-prime. The ring of polynomials is an integral domain, and the Diophantine equations

$$A^*(\lambda)R_i^*(\lambda) + B^*(\lambda)S_i^*(\lambda) = P_i^*(\lambda); \qquad i = 1, \ldots, 2n \tag{12.131}$$

therefore have solutions for all i in the given interval. The solutions are such that there are solutions with

$$\begin{aligned} R_i^*(\lambda) &= r_{i1}\lambda + \ldots + r_{in}\lambda^n \\ S_i^*(\lambda) &= s_{i1}\lambda + \ldots + s_{in}\lambda^n \end{aligned} \tag{12.132}$$

From Eq. (12.126) and Eq. (12.131) it is found for $i = 1, \dots, 2n$ that

$$R_i^*(\lambda)u(t) + S_i^*(\lambda)y(t) = P_i^*(\lambda)\xi_\lambda(t) = x_i'(t) \tag{12.133}$$

The constraints on the polynomial degrees are of the form

$$x_i'(t) = \begin{pmatrix} r_{i1} & \dots & r_{in} & s_{i1} & \dots & s_{in} \end{pmatrix} \varphi_\tau(t) \tag{12.134}$$

with

$$\varphi_\tau(t) = \begin{pmatrix} [\lambda y](t) & \dots & [\lambda^n y](t) & [\lambda u](t) & \dots & [\lambda^n u](t) \end{pmatrix}$$

Let the matrix T_τ be

$$T_\tau = \begin{pmatrix} r_{11} & \cdots & r_{1n} & s_{11} & \cdots & s_{1n} \\ \vdots & & \vdots & \vdots & & \vdots \\ r_{(2n)1} & \cdots & r_{(2n)n} & s_{(2n)n} & \cdots & s_{(2n)n} \end{pmatrix} \tag{12.135}$$

Then it holds that

$$x'(t) = T_\tau \varphi_\tau(t) \tag{12.136}$$

It can be concluded from Eq. (12.125) and Eq. (12.136) that

$$T_\tau^{-1} = M' \tag{12.137}$$

Hence, T_τ is an invertible matrix relating x' and φ_τ. ■

12.11 EXERCISES

12.1 Parametrization

Assume that a DC-motor may be described by a transfer function $G_0(s)$ between input u (= voltage) and output y (= angular velocity)

$$G_0(s) = \frac{Y(s)}{U(s)} = \frac{K}{Js + D} \tag{12.138}$$

The parameters K, J, D denote gain, moment of inertia, and damping, respectively.

a. Determine a continuous-time parametrization for identification of K, J, and D. What parameters are identifiable from input-output data? What extra information might be needed for full identifiability?

b. Formulate a recursive estimation algorithm to find the parameters.

12.2 Consider the set-up of Exercise 8.1. Assume that there is a known sinusoidal measurement disturbance at the frequency ω_0 [rad/s].

a. Formulate a regressor filtering that effectively reduces the estimation error.

b. What implementation aspects may be important in the context of regressor filtering?

12.3 Devise a software procedure to plot the frequency response for the DC-motor identification. Also include error bounds of the estimated Bode diagram based on the Parseval relation

$$\begin{aligned}\varepsilon(t) &= y(t) - \widehat{\theta}^T \varphi(t)\\ \varepsilon(s) &= \mathcal{L}\{\varepsilon(t)\}(s) = Y(s) - \widehat{\theta}^T \Phi(s) = (G_0(s) - \widehat{G}_0(s))U(s) \qquad (12.139)\\ &= \{\varepsilon(s), \varepsilon(s)\} = \|G_0(s) - \widehat{G}_0(s)\|^2 \|U(s)\|^2\end{aligned}$$

A Nyquist or Nichols diagram may substitute for the Bode diagram representation.

12.4 Show that the operator λ can be used to formulate a parametric model for estimation of physical coefficients of the robot equations $M(q)\ddot{q} + C(q,\dot{q})\dot{q} + G(q) = \tau$ when q, $\dot{q}$ but not $\ddot{q}$ are available for measurement.

Hint: Show that filtering of the Euler-Lagrange equations (7.5) by λ gives the relationship

$$\frac{1}{\tau_0}\frac{\partial L}{\partial \dot{q}} = \frac{1}{\tau_0}\lambda\{\frac{\partial L}{\partial \dot{q}}\} + \lambda\{\frac{\partial L}{\partial q}\} + \lambda\{\tau\}, \quad \text{where} \quad \lambda = \frac{1}{1 + s\tau_0} \qquad (12.140)$$

and where $L = (1/2)\dot{q}^T M(q)\dot{q} - U(q)$ and τ are the applied external forces. ■

13 Multidimensional Identification

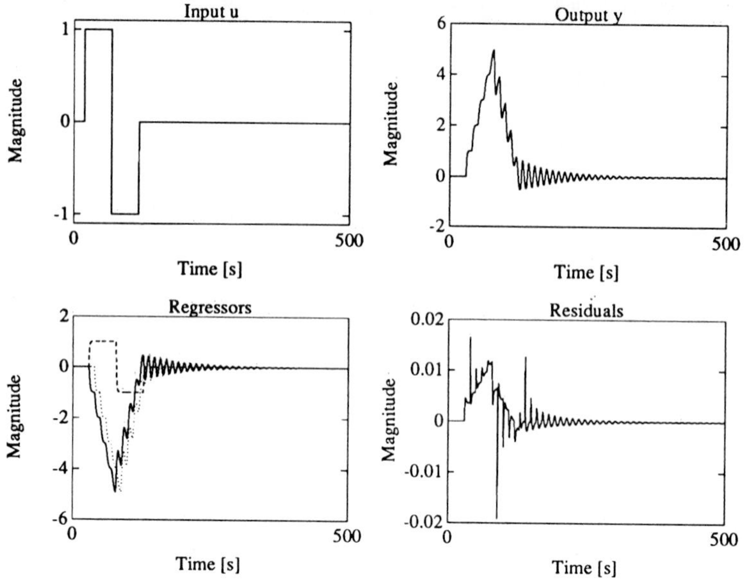

13.1 INTRODUCTION

The problems presented in previous chapters all had time as the independent coordinate variable. There are, however, physical coordinates which in some cases can be regarded as independent variables. Several standard problems deal with both temporal and spatial coordinates as independent variables, and the corresponding physical modeling gives rise to partial differential equations. Measurement devices that are adapted to this kind of modeling are often found in the form of sensor arrays in one or several spatial dimen-

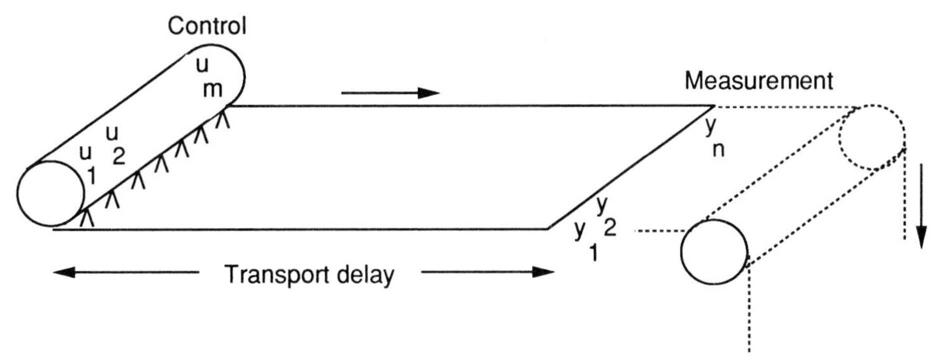

Figure 13.1 Industrial materials handling with control actions $u_1, \dots, u_m$ affecting the material distribution and product quality measured in the variables $y_1, \dots, y_n$ after some time delay τ.

sions. Another example is geometric modeling and image processing in two dimensions (2D) and three dimensions (3D) which usually are represented by discretized models.

In the case of discrete variables there is actually no quite clear-cut distinction between multivariate, multidimensional, and multi-input multi-output systems. Before considering the many difficult problems in multidimensional identification, it is worth mentioning that many problems that appear to be multidimensional at first sight may sometimes be reformulated to standard least-squares problems.

Example 13.1—Industrial materials handling

Consider the industrial materials handling problem in Fig. 13.1 where the control actions $u_1, \dots, u_m$ affect the material distribution through a set of nozzles and where the product quality is measured in the variables $y_1, \dots, y_n$ after some time delay τ. Let the result of the input variables on each one of the output variables be described by a matrix Θ and let the vector-valued measurements and control actions from time $t = t_1, \dots, t_N$ be organized as

$$\begin{pmatrix} y_1(t_1+\tau) & \dots & y_1(t_N+\tau) \\ \vdots & & \vdots \\ y_n(t_1+\tau) & \dots & y_n(t_N+\tau) \end{pmatrix} = \Theta \begin{pmatrix} u_1(t_1) & \dots & u_1(t_N) \\ \vdots & & \vdots \\ u_m(t_1) & \dots & u_m(t_N) \end{pmatrix} \tag{13.1}$$

or

$$\mathcal{Y} = \Theta \mathcal{U}, \qquad \Theta \in R^{n \times m} \tag{13.2}$$

The least-squares estimate of Θ is

$$\hat{\Theta} = \mathcal{Y}\mathcal{U}^T(\mathcal{U}\mathcal{U}^T)^{-1} \tag{13.3}$$

and we conclude that the problem of finding the interaction matrix Θ can be solved with well-known methods. ■

In order to formulate methods for algebraic and spectral analysis applicable to more complicated systems we need the multidimensional Laplace and Fourier transforms. We start this chapter by giving a short overview of the multi-dimensional Laplace transform and the associated transfer function algebra. We also give some attention to the multidimensional fast Fourier transform and alternatives such as the Walsh/Hadamard transform.

13.2 TWO-DIMENSIONAL TRANSFORMS

Two-dimensional Laplace transform is defined *via*

$$\mathcal{L}_2\{x(t_1,t_2)\} = X(s_1,s_2) = \int_{-\infty}^{+\infty}\int_{-\infty}^{+\infty} x(t_1,t_2)\exp(-s_1t_1 - s_2t_2)dt_1dt_2 \tag{13.4}$$

for a variable $x(t_1,t_2)$ that depends on two coordinates t_1,t_2. The first quadrant transform

$$\mathcal{L}_2\{x(t_1,t_2)\} = X(s_1,s_2) = \int_{0}^{+\infty}\int_{0}^{+\infty} x(t_1,t_2)\exp(-s_1t_1 - s_2t_2)dt_1dt_2 \tag{13.5}$$

is the Laplace transform restricted to integration of positive t_1,t_2 and hence corresponds to the one-sided Laplace transform for one-dimensional signals.

The dependency of the output of a linear causal system is characterized by the two-dimensional convolution equation

$$y(t) = \int_0^{\infty}\int_0^{\infty} g(t_1,t_2)u(t-t_1,t-t_2)dt_1dt_2 \tag{13.6}$$

and the transfer function

$$\frac{Y(s_1,s_2)}{U(s_1,s_2)} = G(s_1,s_2) = \mathcal{L}_2\{g(t_1,t_2)\} \tag{13.7}$$

Note that the transform algebra is particularly straightforward when s_1,s_2 are separable variables, *i.e.*, a function whose two-dimensional Laplace transform

$\mathcal{L}_2\{f(t_1,t_2)\} = F(s_1,s_2) = F_1(s_1)F_2(s_2)$ for some functions F_1 and F_2 that each one only depend on one variable.

If we turn our attention to discrete models we have the two-dimensional z-transform of a signal $x(t_1,t_2)$ discretized as the array $\{x_{ij}\}$ defined as

$$X_z(z_1,z_2) = \mathcal{Z}_2\{x\} = \sum_{i=-\infty}^{\infty}\sum_{j=-\infty}^{\infty} x_{ij}z_1^{-i}z_2^{-j} \tag{13.8}$$

Let us consider a two-dimensional weighting function $h(t_1,t_2)$ discretized as the array $\{h_{ij}\}$. The discrete transfer function in two dimensions is defined as

$$H(z_1,z_2) = \mathcal{Z}_2\{h\} = \sum_{i=-\infty}^{\infty}\sum_{j=-\infty}^{\infty} h_{ij}z_1^{-i}z_2^{-j} \tag{13.9}$$

where $h(t_1,t_2)$ is the weighting function and $\{h_{ij}\}$ are its discretized values according to some sampling principle. An important case is the *first quadrant transfer function* where $h(t_1,t_2) = 0$ for $t_1 < 0$ and $t_2 < 0$ which corresponds to a causal system in the one-dimensional case. The first quadrant discrete transfer function in two dimensions is

$$H(z_1,z_2) = \mathcal{Z}_2\{h\} = \sum_{i=0}^{\infty}\sum_{j=0}^{\infty} h_{ij}z_1^{-i}z_2^{-j} \tag{13.10}$$

where $h(t_1,t_2)$ is the weighting function.

13.3 TWO-DIMENSIONAL SYSTEM ANALYSIS

First quadrant transfer functions are easy to implement because the output value y_{mn} at a given point depends only on those points u_{ij} of the input sequence for which $i \leq m, j \leq n$; see Fig. 13.2. A first quadrant transfer function with denominator polynomial $A(z_1,z_2)$ and a numerator polynomial $B(z_1,z_2)$ allows a reformulation of the transfer function polynomials to the denominator polynomial $A^*(z_1^{-1},z_2^{-1})$ and numerator polynomial $B^*(z_1^{-1},z_2^{-1})$ so that

$$H(z_1,z_2) = \frac{Y(z_1,z_2)}{U(z_1,z_2)} = \frac{B(z_1,z_2)}{A(z_1,z_2)} = \frac{B^*(z_1^{-1},z_2^{-1})}{A^*(z_1^{-1},z_2^{-1})}, \quad \text{with} \quad A^*(0,0) = 1 \tag{13.11}$$

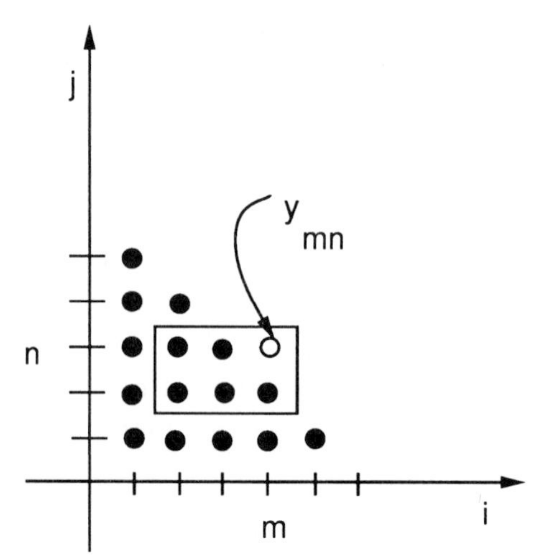

Figure 13.2 The computation of y_{mn} in a first quadrant system only requires values of y_{ij} and u_{ij} for $i \leq m$ and $j \leq n$.

Realization of the transfer function in Eq. (13.11) as a filter is straightforward

$$y(t_1,t_2) = (1 - A^*(z_1^{-1}, z_2^{-1}))y(t_1,t_2) + B^*(z_1^{-1}, z_2^{-1})u(t_1,t_2) \tag{13.12}$$

where z_1^{-1} and z_2^{-1} are interpreted as causal shift operators, *i.e.*, the autoregressive dependency goes for each shift operator in one direction only; see Fig. 13.2.

13.4 STABILITY

Let us consider first quadrant transfer functions on the form of a rational function

$$H(z_1,z_2) = \frac{B(z_1,z_2)}{A(z_1,z_2)} \tag{13.13}$$

where $A(z_1,z_2)$ and $B(z_1,z_2)$ are mutually prime (*i.e*, A and B have no common factors). This transfer function can be expanded as

$$H(z_1,z_2) = \sum_{i=0}^{\infty}\sum_{j=0}^{\infty} h_{ij} z_1^{-i} z_2^{-j} \tag{13.14}$$

A filter of the first quadrant is stable if and only if

$$\sum_{i=0}^{\infty}\sum_{j=0}^{\infty} |h_{ij}| < \infty \tag{13.15}$$

There are also algebraic stability criteria similar to pole-zero analysis for one-dimensional systems. The transfer function H is said to have a *nonessential singularity of the first kind* at the points

$$\{(z_1, z_2): \quad A(z_1, z_2) = 0; \quad B(z_1, z_2) \neq 0\} \tag{13.16}$$

which corresponds to the poles of a one-dimensional system. In addition, there are singularities reminescent of pole-zero cancellations but without any real counterpart for one-dimensional systems. The *nonessential singularity of the second kind* is defined as

$$\{(z_1, z_2): \quad A(z_1, z_2) = 0; \quad B(z_1, z_2) = 0\} \tag{13.17}$$

Note that a nonessential singularity of the second kind may occur even though A and B are mutually prime. The reason is obviously that the pole surface $A(z_1, z_2)$ and the zero surface $B(z_1, z_2)$ intersect in C^2. A sufficient but not necessary condition of stability for discrete systems in two variables is

$$A(z_1, z_2) \neq 0 \quad \forall (z_1, z_2) \in \{(z_1, z_2): |z_1| \geq 1, |z_2| \geq 1\} \tag{13.18}$$

The following example is a counterexample to show that Eq. (13.18) is not a necessary condition of stability.

Example 13.2—Stability of two-dimensional transfer functions
Consider the two transfer functions

$$\begin{aligned} H_1(z_1, z_2) &= \frac{(1 - z_1)^8 (1 - z_2)^8}{2 - z_1 - z_2} \\ H_2(z_1, z_2) &= \frac{(1 - z_1)(1 - z_2)}{2 - z_1 - z_2} \end{aligned} \tag{13.19}$$

The transfer functions $H_1(z_1, z_2)$ and $H_2(z_1, z_2)$ both have nonessential singularities of the second kind at $z_1 = 1$ and $z_2 = 1$. However, a closer investigation of expansions of H_1 and H_2 according to Eq. (13.15) shows that the transfer function H_1 is stable whereas H_2 is unstable. Thus, there is an effect of the numerator on the system stability so that the condition in Eq. (13.18) is sufficient but not necessary for stability. ∎

Let us now consider continuous-time dynamical systems. A condition for stability of a causal two-dimensional dynamical system

$$G(s_1, s_2) = \frac{B(s_1, s_2)}{A(s_1, s_2)} \tag{13.20}$$

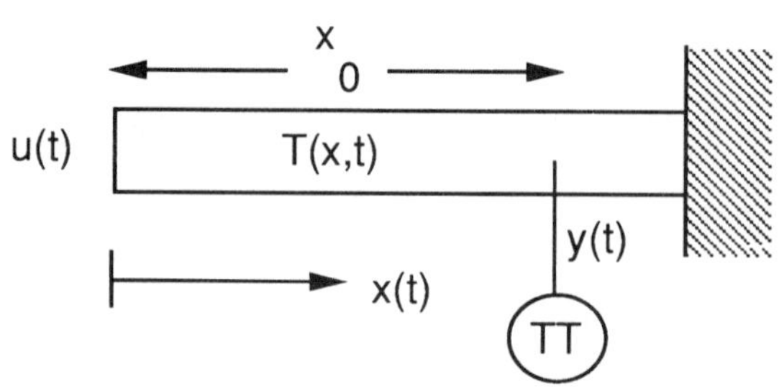

Figure 13.3 One-dimensional heat flow along the spatial coordinate x with control action $u(t)$ and measurement $y(t)$ obtained as a temperature $T(x_0, t)$ at a distance x_0 from the control input.

is that $A(s_1, s_2) \neq 0$ for Re $(s_1) \geq 0$ and Re $(s_2) \geq 0$.

Example 13.3—A partial differential equation for heat flow analysis Consider the one-dimensional heat flow problem shown in Fig. 13.3 where the temperature $T = T(x, t)$ is controlled by means of heating at the boundary at $x = 0$. This control variable is denoted $u(t)$ and is balanced by the fact that the temperature at $x = 1$ is zero. Assume that the temperature measurement at the point $x = x_0$ constitutes the observed variable

$$y(t) = T(x_0, t), \qquad 0 < x_0 < 1 \tag{13.21}$$

The one-dimensional heat flow equation is

$$\rho \frac{\partial T(x,t)}{\partial t} = \frac{\partial^2 T(x,t)}{\partial x^2} \tag{13.22}$$

where $1/\rho$ is a thermal diffusion coefficient. A typical tempearture pulse response in spatial and temporal coordinates is shown in Fig. 13.4. The relevant boundary conditions and initial conditions are

$$\begin{aligned} T(0,t) &= u(t) \\ T(1,t) &= 0 \\ T(x,0) &= T_0(x) \end{aligned} \tag{13.23}$$

Suppose that we forget about initial conditions (*i.e.*, put $T_0(x) = 0$) and look for the transfer function from the input u to the temperature $y(t) = T(x_0, t)$

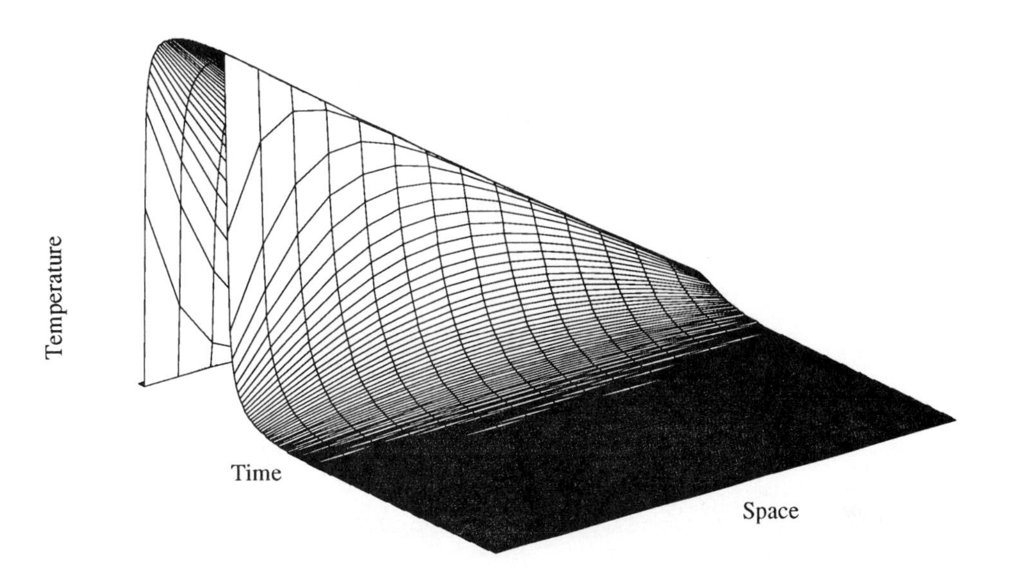

Figure 13.4 Spatial and temporal response of temperature $T(x,t)$ to a rectangular-formed pulse input $u(t)$.

at some given point x_0 along the x-axis. The Laplace transform of Eq. (13.22) with respect to time t is then

$$\begin{aligned} \frac{1}{\rho}\frac{d^2T_t(x,s)}{dx^2} - sT_t(x,s) &= 0 \\ T_t(0,s) &= U(s) \\ T_t(1,s) &= 0 \end{aligned} \tag{13.24}$$

Solving the second-order differential equation for $T_t(x,s)$ and using the boundary conditions yields the transfer function

$$G(s,\rho) = \frac{Y(s)}{U(s)} = \frac{T_t(x_0,s)}{U(s)} = \frac{\sinh(1-x_0)\sqrt{\rho s}}{\sinh\sqrt{\rho s}} \tag{13.25}$$

The Nyquist diagram for some different measurement points x_0 is shown in Fig. 13.5. This transfer function exhibits a complicated dependence on the thermal diffusion coefficient ρ with poles at $s = -k^2\pi^2/\rho$ for $k = 0, 1, 2, \ldots$

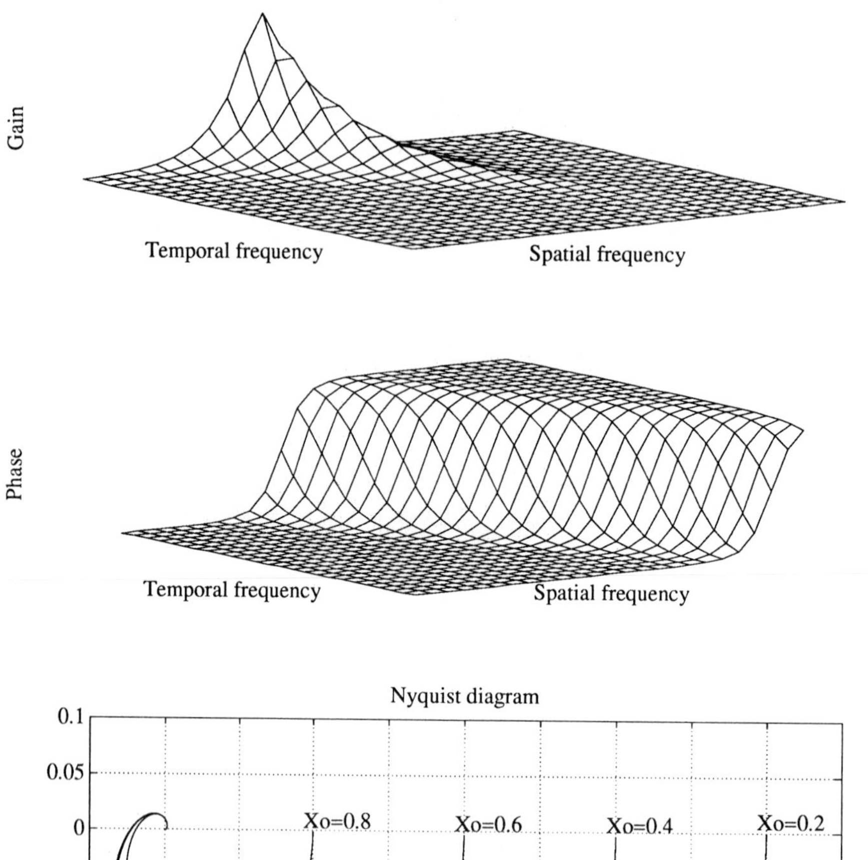

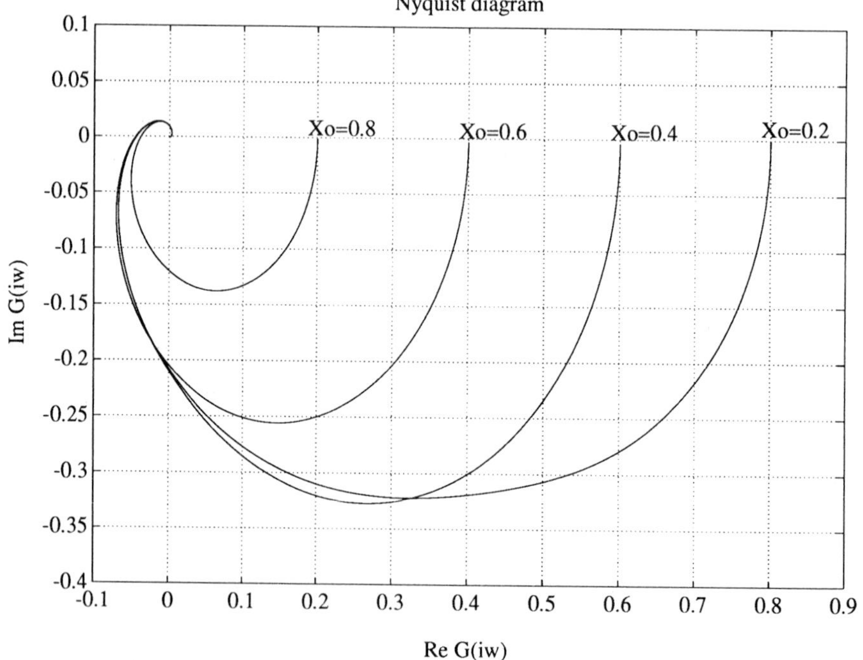

Figure 13.5 Two-dimensional Bode diagram (*upper*) and Nyquist diagram (*lower*) for the system in Example 13.3.

and also an infinite number of zeros. The parameter ρ appears in both the numerator and the denominator of the transfer function, and this complicated dependence motivates the term *distributed parameter*, and systems described by such parameters are accordingly called *distributed parameter systems*.

Now consider the Laplace transform with respect to both x and t and denote the transformed complex frequency variables by s_x and s_t, respectively. The two-dimensional Laplace transform of Eq. (13.22) in temporal frequency s_t and spatial frequency s_x is

$$s_x^2 T_{x,t}(s_x, s_t) - s_x T_t(0, s_t) - \rho s_t T_{x,t}(s_x, s_t) = 0 \tag{13.26}$$

with the boundary conditions

$$\begin{cases} T_t(0, s_t) = U(s_t) \\ T_t(1, s_t) = 0 \end{cases} \tag{13.27}$$

By substituting the boundary condition we obtain

$$s_x^2 T_{x,t}(s_x, s_t) - \rho s_t T_{x,t}(s_x, s_t) = s_x U(s_t) \tag{13.28}$$

The two-dimensional transfer function relationship is

$$T_{x,t}(s_x, s_t) = \frac{s_x}{s_x^2 - \rho s_t} U(s_t) \tag{13.29}$$

The transfer function exhibits a nonessential singularity of the first kind at

$$s_x = \sqrt{\rho s_t} \tag{13.30}$$

This type of singularity corresponds to poles in a one-dimensional system whereas the nonessential singularity of the second kind is $s_x = s_t = 0$. Notice that the expression $\sqrt{\rho s_t}$ for the nonessential singularity of the first kind appears in the transfer function of Eq. (13.25).

As the thermal diffusion takes place in both directions there is no simple one-directional dependence along the x-axis. The system is for this reason not a first quadrant system, and it is more difficult to design a filtering solution to the identification problem. Identification can, however, be made by minimizing the functional

$$\widehat{\rho} = \arg\min J(\rho) = \arg\min \sum_{k=0}^{N-1} |G(i\omega_k, \rho) - \frac{Y_N(e^{i\omega_k h})}{U_N(e^{i\omega_k h})}|^2, \qquad \omega_k = \frac{2\pi}{Nh} k \tag{13.31}$$

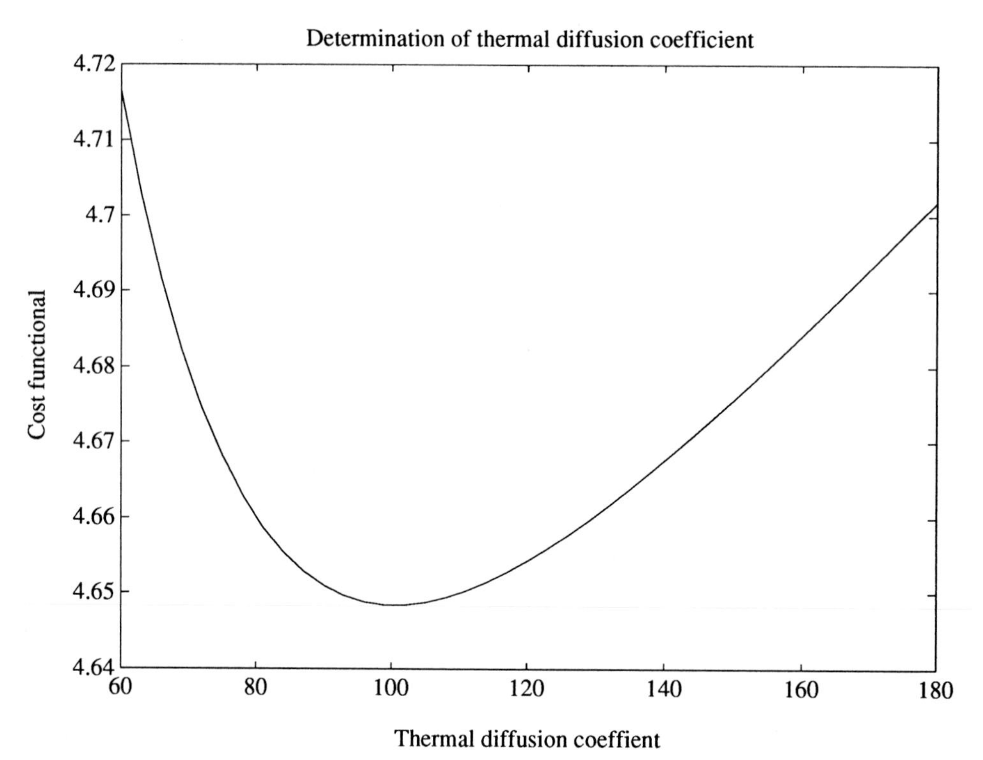

Figure 13.6 Determination of the thermal diffusion coefficient ρ by explicit minimization of the functional in Eq. (13.31).

where the transfer function G in Eq. (13.25) is fitted to the ratio of the discrete Fourier transforms Y_N and U_N. The cost functional is shown in Fig. 13.6 for the case $\rho = 100$ and this explicit optimization obviously solves the identification problem. ■

A pragmatic approach to this identification problem is, of course, to fit a low order rational transfer function to data.

13.5 DELAY-DIFFERENTIAL SYSTEMS

There are several modeling problems arising for instance in physics and technology where the evolution of the system depends not only on the present state but also on past data. Identification problems of this type are obtained for processes with reflections, echoes, and material flows with transport delays and recirculation dynamics. Such temporal dependencies give rise to differential equations including both differential operators and time-delay operators, and systems are for this reason called differential-difference systems or delay-differential systems.

Example 13.4—Identification of a delay-differential system
A typical problem is to estimate the parameters a_1, a_2, b of the model

$$\begin{aligned} \dot{x}(t) &= -a_1 x(t) + a_2 x(t-\tau) + bu(t-\tau) \\ y(t) &= x(t) \end{aligned} \tag{13.32}$$

where the time delay τ is supposed to be known. Notice that the model in Eq. (13.32) is a linear system with the transfer function

$$G(s) = \frac{Y(s)}{U(s)} = \frac{be^{-s\tau}}{s + a_1 - a_2 e^{-s\tau}} \tag{13.33}$$

which is not a rational transfer function because of the denominator term proportional to $e^{-s\tau}$. It is clear from Fig. 13.7 that the impulse response, the Bode diagram, and the Nyquist diagram all exhibit complicated characteristics despite the apparent simplicity of the parametric model in Eq. (13.33) or Eq. (13.32). ■

A transfer function algebra similar to that in Chapter 12 can be proposed in order to solve this problem. The operators may be, for instance, the differential operator p and some time-delay operator. Another choice is the following causal operators

$$\begin{aligned} \lambda_1 &= \frac{1}{1 + p\tau_1} \\ \lambda_2 &= e^{-p\tau_2} \end{aligned} \tag{13.34}$$

for some positive time-constants τ_1, τ_2 which constitute a set of design parameters. These operators are a good choice because they are causal, stable operators with finite gain, and the operators commute in the context of transfer functions. The operator λ_1 is a low pass filter operator similar to that in Chapter 12 and λ_2 represents a time delay τ_2. Notice also that the transformation results in new parameters which are linear with respect to the original parameters. The transfer function polynomials are then

$$A(\lambda_1, \lambda_2) y(t) = B(\lambda_1, \lambda_2) u(t) \tag{13.35}$$

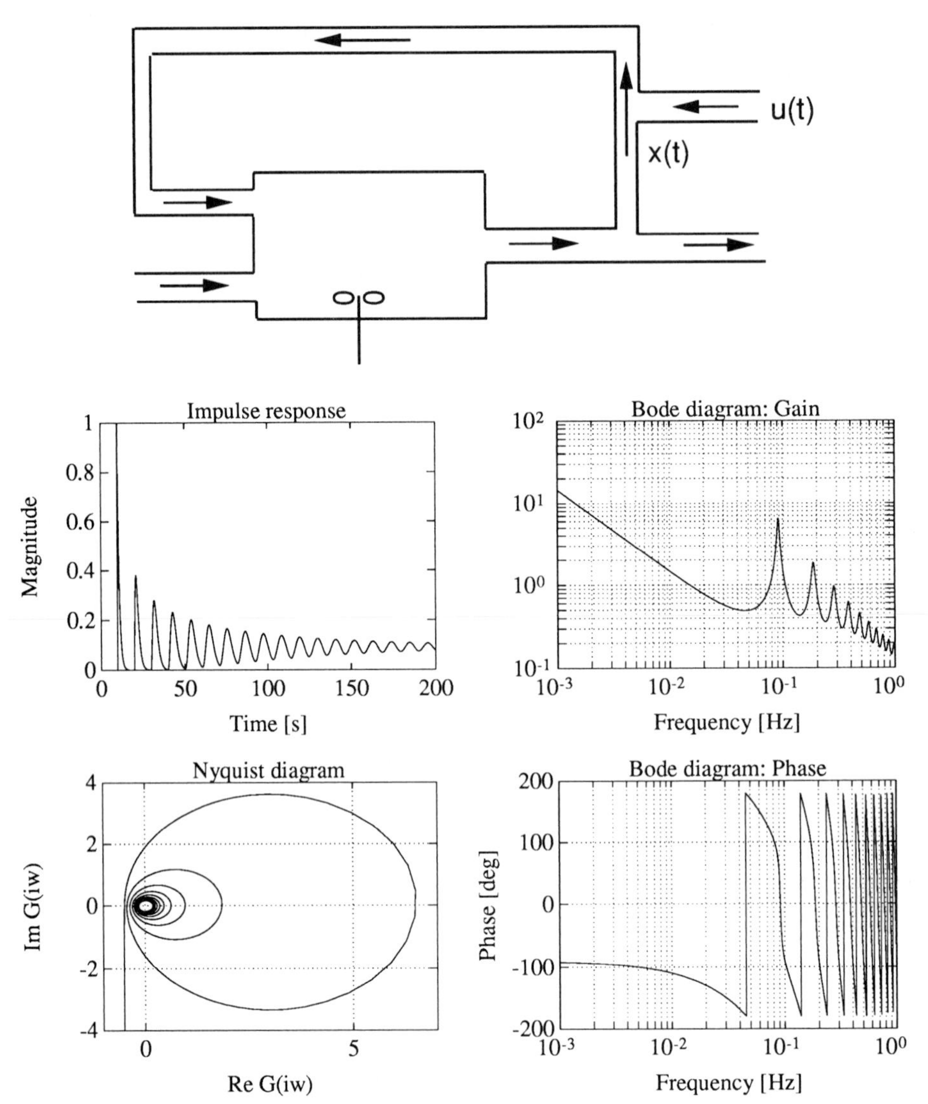

Figure 13.7 Impulse response, Bode diagram, and Nyquist diagram for a delay-differential system with a transfer function in Eq. (13.33) and parameters $a_1 = 1$, $a_2 = 1$, $b = 1$, and $\tau = 10$.

for polynomials in two indeterminates λ_1 and λ_2, *i.e.*,

$$\begin{aligned} A(\lambda_1, \lambda_2) &= \sum_{i=0}^{n_{1A}} \sum_{i=0}^{n_{2A}} a_{ij} \lambda_1^i \lambda_2^j, \qquad a_{00} = 1 \\ B(\lambda_1, \lambda_2) &= \sum_{i=0}^{n_{1B}} \sum_{i=0}^{n_{2B}} b_{ij} \lambda_1^i \lambda_2^j, \end{aligned} \tag{13.36}$$

for some finite orders $n_{1A}, n_{2A}, n_{1B}, n_{2B}$. It is clear from Eq. (13.36) that the system is a first quadrant system which facilitates a filtering solution to identification. The condition $a_{00} = 1$ is imposed for the purpose of normalization and is possible if the transfer function is both causal and proper

$$y(t) = -\sum_{i=0}^{n_{1A}}\sum_{j=0}^{n_{2A}} a'_{ij}\lambda_1^i\lambda_2^j y(t) + \sum_{i=0}^{n_{1B}}\sum_{j=0}^{n_{2B}} b_{ij}\lambda_1^i\lambda_2^j u(t), \quad a'_{ij} = \begin{cases} 0, & i = 0, j = 0 \\ a_{ij}, & \text{otherwise} \end{cases} \tag{13.37}$$

and we formulate in the standard fashion

$$y(t) = \begin{pmatrix} -\lambda_1 y & \dots & -\lambda_1^{n_{1A}}\lambda_2^{n_{2A}} y & \lambda_1 u & \dots & \lambda_1^{n_{1B}}\lambda_2^{n_{2B}} u \end{pmatrix} \begin{pmatrix} a_{10} \\ \vdots \\ a_{n_{1A}n_{2A}} \\ b_{10} \\ \vdots \\ b_{n_{1B}n_{2B}} \end{pmatrix} \tag{13.38}$$

which is in linear regression form $y(t) = \phi^T(t)\theta$. The constant parameters θ may now be estimated by any suitable method for identification of a linear model. A natural choice is to sample $y(t)$ and $\phi(t)$. Clearly, the bandwidth of λ_1 and the sampling period are design parameters which can be modified in order to improve the estimation.

Example 13.5—Identification of a delay-differential system (cont'd.)
Consider the following delay-differential equation for the system

$$\begin{aligned} \dot{x}(t) &= -a_1 x(t) + a_2 x(t-\tau) + bu(t-\tau) \\ y(t) &= x(t) \end{aligned} \tag{13.39}$$

with the transfer operator

$$\frac{Y(\lambda_1, \lambda_2)}{U(\lambda_1, \lambda_2)} = \frac{b\lambda_1\lambda_2}{1 + (a_1 - 1)\lambda_1 - a_2\lambda_1\lambda_2} \tag{13.40}$$

where

$$\lambda_1 = \frac{1}{s+1}, \qquad \lambda_2 = e^{-s\tau} \tag{13.41}$$

Identification of the system based on the data shown in Fig. 13.8 and with $\tau = 10$, $a_1 = 1$, $a_2 = 1$, $b = 1$ provides the estimates

$$\begin{pmatrix} \widehat{a}_1 \\ \widehat{a}_2 \\ \widehat{b} \end{pmatrix} = \begin{pmatrix} 1.0056 \\ 0.9923 \\ 0.9909 \end{pmatrix} \tag{13.42}$$

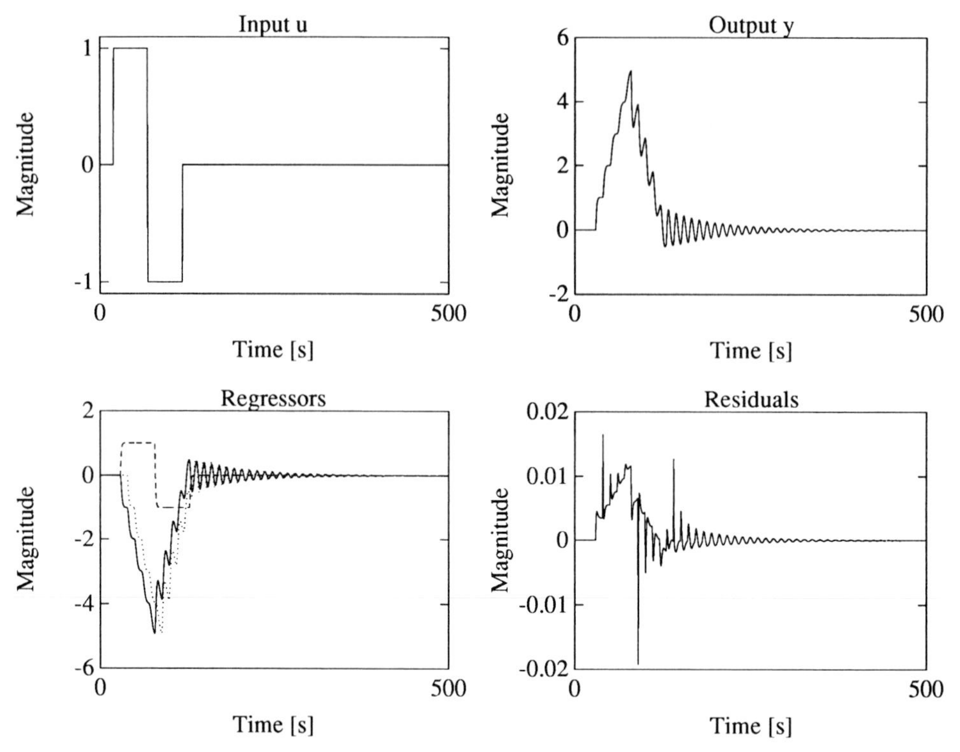

Figure 13.8 A delay-differential system described by the system equation (13.39) with parameters $a_1 = 1$, $a_2 = 1$, and $b = 1$. The least-squares solution provides the estimates $\widehat{a}_1 = 1.0056$, $\widehat{a}_2 = 0.9923$, and $\widehat{b} = 0.9909$.

which shows that method works as expected. ■

The method lends itself to recursive identification, and a similar approach to adaptive control has been reported.

13.6 TWO-DIMENSIONAL SPECTRA

Let $X = \{x_{ij}\}$ denote an array of data and consider the transform Y with components

$$y_{ij} = \frac{1}{\sqrt{MN}} \sum_{m=0}^{M-1} \sum_{n=0}^{N-1} x_{mn} w_{ijmn} \tag{13.43}$$

where w_{ijmn} is some weighting function. In the special case when w_{ijmn} is separable on the two axes, *i.e.*, when $w_{ijmn} = b_{mi} a_{nj}$ for some matrix components $\{a_{ij}\}$ and $\{b_{ij}\}$, it is possible to write the relationship between the original data X and the transformed variable Y as the matrix product

$$Y = \frac{1}{\sqrt{MN}} B^T X A \tag{13.44}$$

where A, B are the matrices of components a_{ij} and b_{ij}, respectively. A standard choice is, of course, to use the matrices in Eq. (5.83) used in the discrete Fourier transform.

$$A = B = \Phi = \begin{pmatrix} 1 & 1 & \cdots & 1 \\ e^{i\omega_0 h} & e^{i\omega_1 h} & \cdots & e^{i\omega_{N-1} h} \\ \vdots & \vdots & & \vdots \\ e^{i\omega_0 (N-1)h} & e^{i\omega_1 (N-1)h} & \cdots & e^{i\omega_{N-1}(N-1)h} \end{pmatrix} \tag{13.45}$$

for discrete frequency points $\omega_k = k(2\pi/Nh)$.

The Walsh/Hadamard transform

Two-dimensional and multidimensional transforms obviously require much computation, and it is natural to search for alternatives and simplifications of the calculation of spectra. One such transform based on binary functions is known as the Walsh/Hadamard transform, which is analogous to the Fourier transform in the sense that its result is a form of spectrum analysis. The binary functions known as Walsh functions form a complete, orthogonal basis on the interval $[-0.5, 0.5]$ of real numbers. A function $f(x)$ defined on the interval $[-0.5, 0.5]$ can thus be expanded as

$$f(x) = a_0 \mathrm{wal}(0, x) + \sum_{n=1}^{\infty} a_n \mathrm{cal}(n, x) + b_n \mathrm{sal}(n, x) \tag{13.46}$$

where

$$\mathrm{wal}(0, x) = \begin{cases} 1, & -1/2 \le x < 1/2 \\ 0, & x < -1/2, \quad \text{or} \quad x \ge 1/2 \end{cases}$$

and where $\mathrm{cal}(n, x)$ and $\mathrm{sal}(n, x)$ correspond to cosine and sine functions, respectively. The first few functions $\mathrm{cal}(n, x)$ and $\mathrm{sal}(n, x)$ can be obtained as the signum function of the corresponding sinusoidal functions (see Fig. 13.9), and the higher-order Walsh functions are easily generated by means of the Hadamard matrices, as shown below.

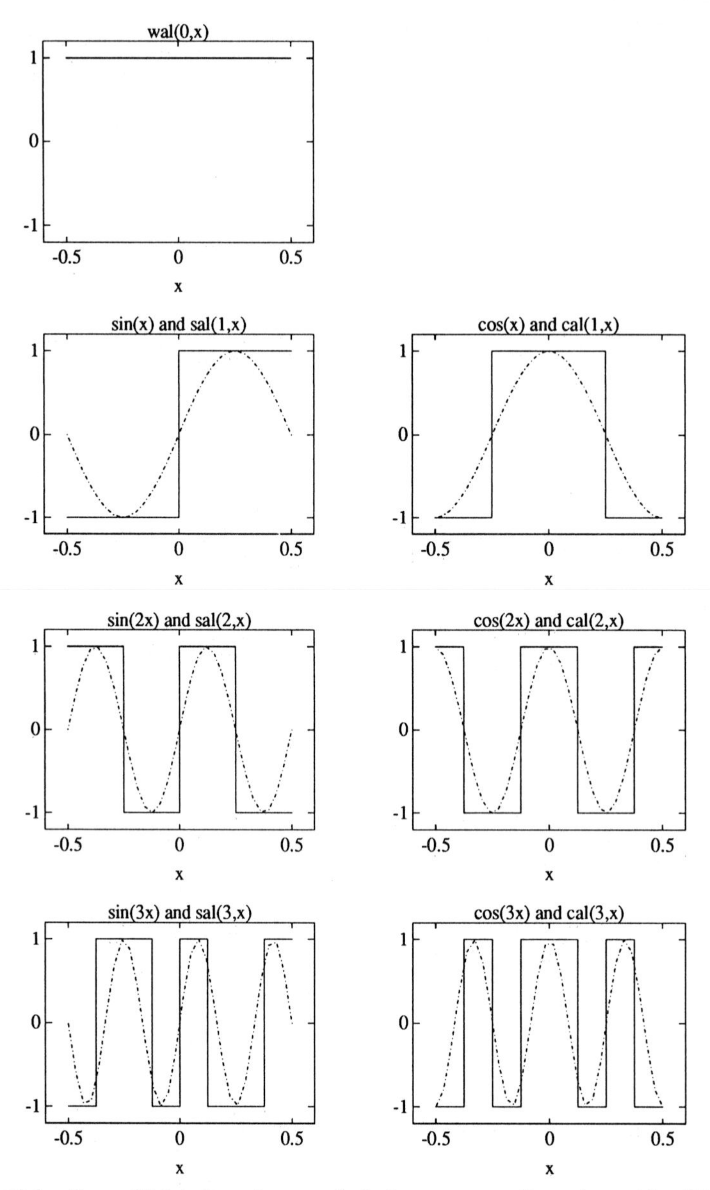

Figure 13.9 Some Walsh functions and their corresponding sinusoids. Note that only the first few Walsh functions are the signum function of their sinusoidal counterpart.

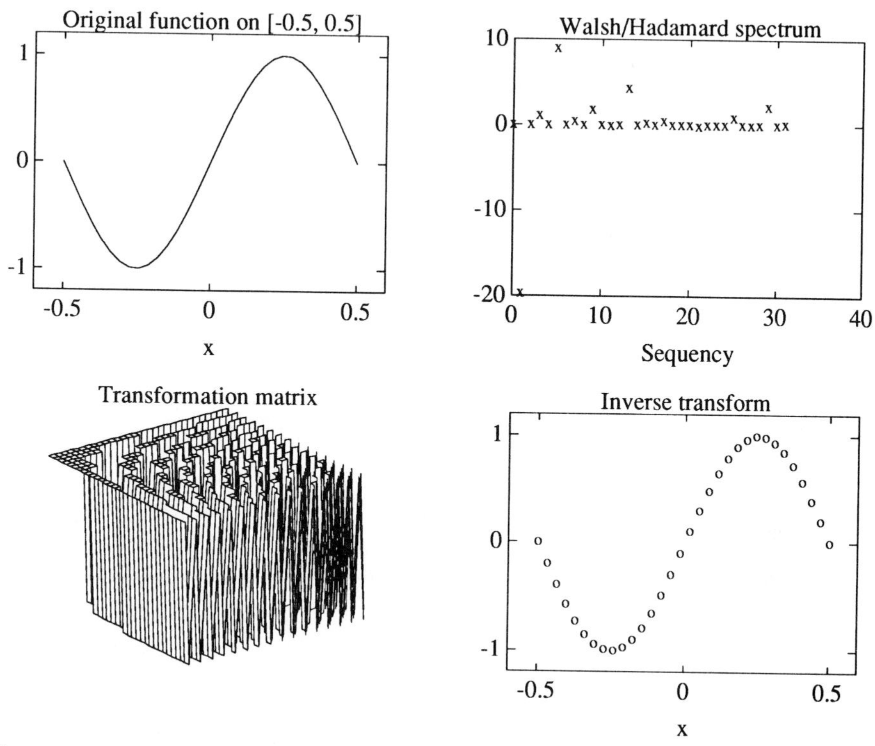

Figure 13.10 A Walsh/Hadamard spectrum of a sinusoid and the inverse transform with sequency in increasing order. All coefficients $\{a_n\}$ are zero.

The coefficients of the Walsh series expansions are obtained as

$$
\begin{aligned}
a_0 &= \int_{-1/2}^{1/2} f(x)\text{wal}(0, x)dx \\
a_n &= \int_{-1/2}^{1/2} f(x)\text{cal}(n, x)dx \\
b_n &= \int_{-1/2}^{1/2} f(x)\text{sal}(n, x)dx
\end{aligned}
\tag{13.47}
$$

The number n determines the number of zero crossings of the functions $\text{cal}(n, x)$ and $\text{sal}(n, x)$, and it is customary to define this number as the *sequency*. This number is analogous to frequency for periodic functions (see Fig. 13.10.)

Example 13.6—A Walsh/Hadamard series expansion
The discrete Walsh/Hadamard spectrum of a sinusoid is shown in Fig. 13.10 together with the inverse transform. The Walsh spectrum is shown in a diagram with b_n versus sequency n (another approach is to represent some other

function such as $\sqrt{a_n^2 + b_n^2}$ versus sequency n). The transform is computationally efficient and gives a spectrum that is analogous to that of traditional Fourier spectra. ■

In particular, the Hadamard matrices H_n with $N = 2^n$ rows and columns for $n = 1, 2, 3, \ldots$ are easy to generate recursively. An $N \times N$–matrix can be generated by the following simple matrix composition rules

$$\begin{aligned} H_1 &= \begin{pmatrix} 1 & 1 \\ 1 & -1 \end{pmatrix} \\ H_{i+1} &= \begin{pmatrix} H_i & H_i \\ H_i & -H_i \end{pmatrix} \end{aligned} \tag{13.48}$$

It is straightforward to verify that

$$HH^T = N \cdot I_{N \times N} \quad \text{for} \quad H \in R^{N \times N} \tag{13.49}$$

An example of a matrix generated this way is

$$H_3 = \begin{pmatrix} 1 & 1 & 1 & 1 & 1 & 1 & 1 & 1 \\ 1 & -1 & 1 & -1 & 1 & -1 & 1 & -1 \\ 1 & 1 & -1 & -1 & 1 & 1 & -1 & -1 \\ 1 & -1 & -1 & 1 & 1 & -1 & -1 & 1 \\ 1 & 1 & 1 & 1 & -1 & -1 & -1 & -1 \\ 1 & -1 & 1 & -1 & -1 & 1 & -1 & 1 \\ 1 & 1 & -1 & -1 & -1 & -1 & 1 & 1 \\ 1 & -1 & -1 & 1 & -1 & 1 & 1 & -1 \end{pmatrix} \tag{13.50}$$

A minor inconvenience is that the Walsh function generated in this manner from the columns or rows of this matrix is not sorted with respect to sequency.

The Walsh/Hadamard two-dimensional transform of X is accordingly

$$Y = \frac{1}{N} H X H^T \tag{13.51}$$

and the inverse transform is

$$X = \frac{1}{N} H^T Y H \tag{13.52}$$

Both the transform and the inverse transform are thus computed with a minimum of computational complexity with respect to multiplications and additions, which is $(N \log_2 N)^2$.

13.7 BIBLIOGRAPHY AND REFERENCES

Example 13.2 derives from

- D. GOODMAN, "Some stability properties of two-dimensional shift invariant digital filters." *IEEE Trans. Circuits and Systems*, CAS-24, 1977, pp. 201–208.
- E.I.JURY, "Stability of multidimensional scalar and matrix polynomials." *Proc. IEEE*, Vol. 66, No.9, 1980, pp. 1018–1047.

Fine sources on two-dimensional signal processing and systems are

- J.S. LIM, *Two-Dimensional Signal and Image Processing*, Englewood Cliffs, NJ: Prentice-Hall, 1990.
- A.K. JAIN, *Fundamentals of Image Processing*, Englewood Cliffs, NJ: Prentice-Hall, 1989.

The section on delay-differential systems is based on the paper

- R. JOHANSSON, "Estimation of direct adaptive control of delay-differential systems." *Automatica*, Vol. 22, No.5, 1986, pp. 555–560.

■

14

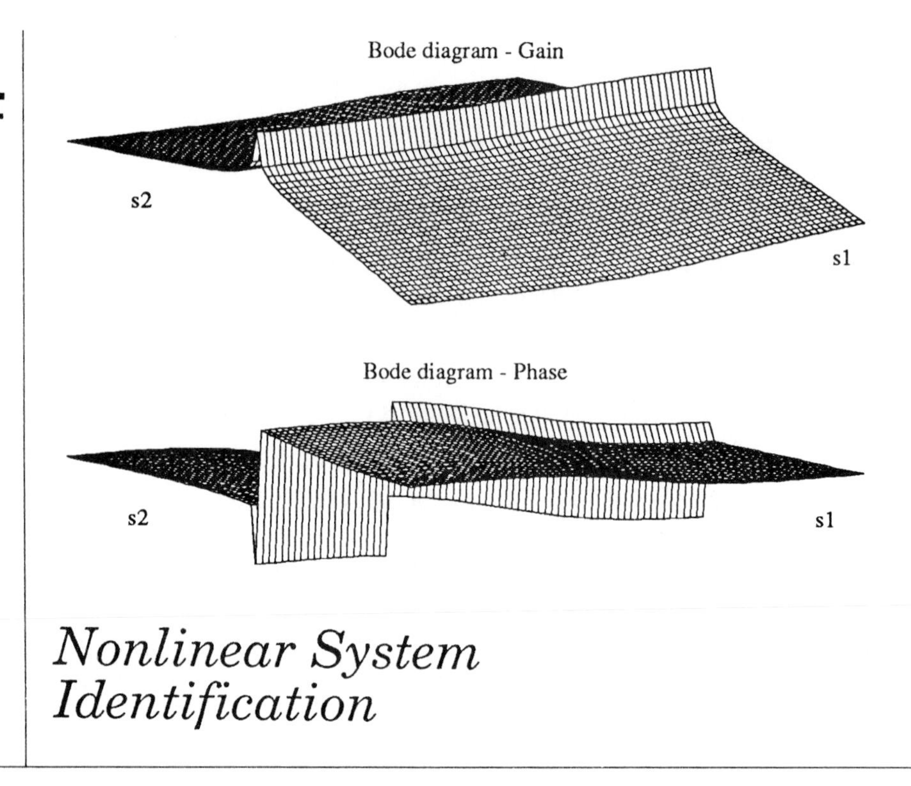

Nonlinear System Identification

Wer immer strebend sich bemüht
den können wir erlösen
—J.W. GOETHE

14.1 INTRODUCTION

The identification of mathematical models in the form of single-input single-output linear systems is well developed both with respect to parameter estimation and structure determination. However, it is considered difficult to apply linear systems identification to nonlinear systems. A pessimistic view is to

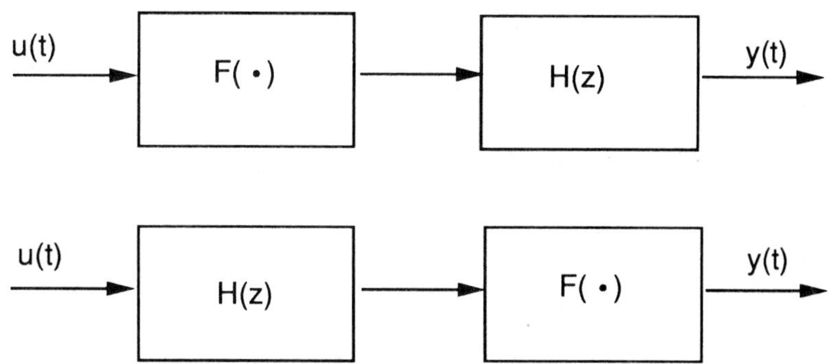

Figure 14.1 Cascaded block structure models with nonlinearities $F(\cdot)$ and transfer functions $H(z)$. These model structures are known as Hammerstein models (*upper*) and Wiener models (*lower*).

regard nonlinear systems as the complement to the class of systems for which systematic identification methods exist. A similar general criticism of linear systems identification in applications is that *a priori* structural knowledge is difficult to include in any useful way. It can be concluded that nonlinear systems identification in general is complicated because there are problems both of parameter estimation and of structure determination (see Mehra, 1979; Billings and Leontaritis, 1982).

Nonlinear systems are often represented by means of differential equations of the form

$$\begin{aligned} \dot{x}(t) &= f(x(t)) + g(x(t), u(t)) + v(t) \\ y(t) &= h(x(t), u(t)) \end{aligned} \tag{14.1}$$

where f and g are some functions and where x, u, v are a state vector, system input, and disturbance, respectively. It is sometimes restrictive to assume the form in Eq. (14.1), and in such cases it might be preferable to use the relationship

$$F(\dot{x}(t), x(t), u(t), v(t)) = 0 \tag{14.2}$$

where F is some nonlinear function. Discrete-time dynamics is often represented in the form

$$x_{k+1} = f(x_k, u_k) \tag{14.3}$$

although such systems are not as much used as the differential equation models in Eq. (14.1). A restricted class of discrete-time models are the cascaded block structure models shown in Fig. 14.1. These models consist of ARMAX-type models with input or output nonlinearities such as

$$y_k = \frac{B(z^{-1})}{A(z^{-1})} F(u_k) \tag{14.4}$$

or

$$y_k = F(\frac{B(z^{-1})}{A(z^{-1})}u_k) \tag{14.5}$$

Models with input nonlinearities F of the type in Eq. (14.4) are known as *Hammerstein models* whereas models with output nonlinearities Eq. (14.5) are a subset of what is known as *Wiener models*. The systems represented by Eq. (14.5) can often be identified using straightforward extensions of linear estimation models with, for instance, a polynomial expansion to represent the nonlinearity F or its inverse (Wigren, 1990).

A nonlinear systems representation with a potential for time domain as well as frequency domain calculations is provided by the Volterra kernel representation which leads to multilinear extensions of the transfer function notions used in linear systems. Identification of such multilinear transfer functions traditionally relies on frequency domain methods, correlation analysis (Barker, 1982; Barrett, 1982), or orthogonal expansions. The use of orthogonal polynomials is motivated in this context because terms can be added to the series and new parameters be evaluated without changing the previously obtained estimates.

A natural requirement is to have methods suitable both for ordinary identification and for recursive methods. It is also preferable to have methods that allow operations both in the frequency domain and in the time domain. The identification methods should allow estimation of unknown parameters of a system with a known structure described by ordinary differential equations. In particular, the real-time identification of a small number of varying parameters is often of interest for detection and adaptation. Such problems often appear in models derived from equations of classical mechanics such as rigid body motion of vehicles, robot manipulators, and in biomechanics.

14.2 WIENER MODELS

The most systematic outline of nonlinear system identification is the Wiener (1958) approach which involves Laguerre and Hermite series expansions to model the dynamic and the nonlinear aspects, respectively. The model outputs are formed as an infinite series of products of Hermite polynomials in the Laguerre coefficients of the past of the input, see Fig. 14.2. The Laguerre operators used can be viewed as a sequence of filters

$$L_k(s) = \frac{1}{1+s\tau}(\frac{1-s\tau}{1+s\tau})^k \tag{14.6}$$

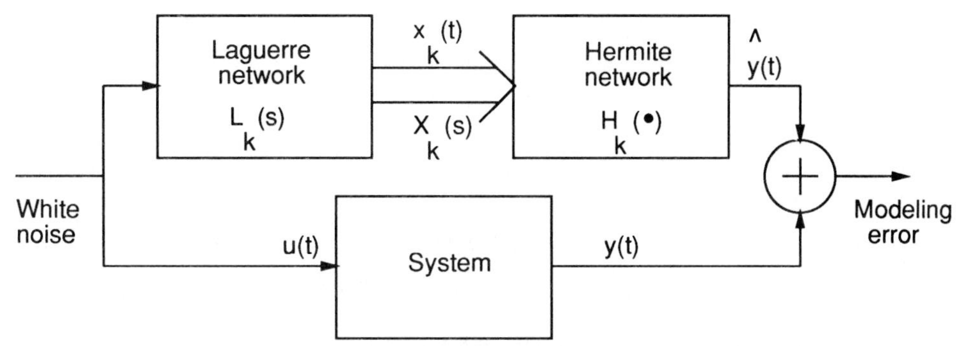

Figure 14.2 The Wiener network for identification of a nonlinear system.

where each transfer function element

$$\frac{1 - s\tau}{1 + s\tau} \tag{14.7}$$

has a constant gain for all $s = i\omega$ and a phase delay $-2\arctan\omega\tau$. That is, each such link represents an all-pass filter which make $L_k(s)$ reminescent of the delay line used for ARMAX models. Consequently, it is natural to regard the filtered inputs $x_k(t) = \mathcal{L}^{-1}\{X_k(s)\}$ as a kind of state-space representation of the system input.

The Hermite polynomial used in the series expansion may be obtained from

$$H_k(x) = (-1)^k e^{x^2} \frac{d^k}{dx^k}(e^{-x^2}), \qquad k = 0, 1, 2, \ldots \tag{14.8}$$

A few examples are

$$\begin{aligned} H_0(x) &= 1 \\ H_1(x) &= 2x \\ H_2(x) &= 4x^2 - 2 \\ H_3(x) &= 8x^3 - 12x \end{aligned} \tag{14.9}$$

Other important relationships are that the Hermite polynomials $H_n(x)$ satisfy the recursive equation

$$H_{k+1}(x) = 2xH_k(x) - 2kH_{k-1}(x) \tag{14.10}$$

and the differential equation

$$H_k''(x) - 2xH_k'(x) + 2kH_k(x) = 0 \tag{14.11}$$

The Hermite polynomials are formed from the Laguerre network outputs $\{x_k(t)\}$ and the model output is formed as

$$\hat{y}(t) = \sum_{k_1=0}^{\infty}\sum_{k_2=0}^{\infty}\cdots\sum_{k_n=0}^{\infty} c_{k_1k_2\ldots k_n} H_{k_1}(x_1(t))H_{k_2}(x_2(t))\ldots H_{k_n}(x_n(t)) \qquad (14.12)$$

where $\{c_{k_1k_2\ldots k_n}\}$ is the set of coefficients to be estimated. In the case of a white noise input it is, for instance, possible to use the following correlation technique to solve for the coefficients; *i.e.*,

$$\hat{c}_{k_1k_2\ldots k_n} = \frac{1}{T}\int_0^T \sum_{k_1=0}^{\infty}\sum_{k_2=0}^{\infty}\cdots\sum_{k_n=0}^{\infty} y(t)H_{k_1}(x_1(t))\ldots H_{k_n}(x_n(t))dt \qquad (14.13)$$

which follows from the orthogonal properties of the Hermite polynomials $H_k(x)$ provided that the measurement duration T is large enough and that the input is white noise. A problem in the context of identification is the overabundant number of coefficients $\{c_{k_1k_2\ldots k_n}\}$ that need to be estimated. Actually, this is such an important restriction that the method is rarely implemented. The Wiener approach, however, has had an important historical impact on identification, and several specialized methods owe their debt to this idea.

14.3 VOLTERRA-WIENER MODELS

The multidimensional Laplace and Fourier transforms introduced in Chapter 13 have an important application also for nonlinear systems.

$$\begin{aligned} F(s_1,\ldots,s_n) &= \mathcal{L}_n\{f(t_1,\ldots,t_n)\} \\ &= \int_{t_1=0}^{\infty}\cdots\int_{t_n=0}^{\infty} e^{-s_1t_1-\ldots-s_nt_n} f(x(t_1,\ldots,t_n))dt_1,\ldots,t_n \end{aligned} \qquad (14.14)$$

The multidimensional Laplace transform with the vector $s_1,\ldots,s_n$ of Laplace variables and the time variables

$$\left(\, t_1t_2\ldots t_n \,\right)^T \qquad (14.15)$$

has an immediate application for frequency domain descriptions of nonlinear systems. Let us consider an nth order convolution that defines the input-output map of a nonlinear system described by an n-dimensional weighting function or *Volterra kernel* g_n. More generally, a nonlinear system output y from the system equation

$$\begin{aligned} \dot{x}(t) &= f(x(t)) + g^T(x(t))u(t) \\ y(t) &= h(x(t)) \end{aligned} \tag{14.16}$$

may be represented as the *Volterra series*

$$y(t) = g_0(t) + \sum_{k=1}^{n} \int_0^{\infty} \dots \int_0^{\infty} g_k(\tau_1, \dots, \tau_k) u(t-\tau_1) \dots u(t-\tau_k) d\tau_1 \dots d\tau_k \tag{14.17}$$

where $g_0(t)$ denotes the dependency on initial values. Notice that all causal Volterra kernels $g_k(t_1, \dots, t_k)$ satisfy the relationship $g_k(t_1, \dots, t_k) = 0$ for any $t_i < 0$ and $i = 1, \dots, k$.

The Volterra series expansion in Eq. (14.17) is clearly a natural extension of the convolution relationships in linear dynamical systems. The multidimensional Laplace transform of g_k results in a transfer function

$$G_k(s_1, \dots, s_k) = \int_0^{+\infty} \dots \int_0^{+\infty} g_k(t_1, \dots, t_k) \exp(-\sum_{i=1}^{k} s_i t_i) dt_1 \dots dt_k \tag{14.18}$$

or

$$\mathcal{L}_k\{g_k(t_1, \dots, t_k)\} = G_k(s_1, \dots, s_k) \tag{14.19}$$

The frequency domain description of the convolution Eq. (14.17) is then for each component

$$Y_k(s_1, \dots, s_k) = G_k(s_1, \dots, s_k) U(s_1) \dots U(s_k) \tag{14.20}$$

so that

$$Y(s_1, \dots, s_n) = G_0(s_1) + \sum_{k=1}^{n} G_k(s_1, \dots, s_k) U(s_1) \dots U(s_k) \tag{14.21}$$

Consider, for instance, the nonlinearly interconnected transfer function of the block diagram in Fig. 14.3. A convolution relationship between input and output variables is

$$\begin{aligned} z(t) &= \int_0^{\infty} g_1(\tau) u(t-\tau) d\tau \int_0^{\infty} g_2(\tau) u(t-\tau) d\tau \\ &= \int_0^{\infty} \int_0^{\infty} g_1(\tau_1) g_2(\tau_2) u(t-\tau_1) u(t-\tau_2) d\tau_1 d\tau_2 \end{aligned} \tag{14.22}$$

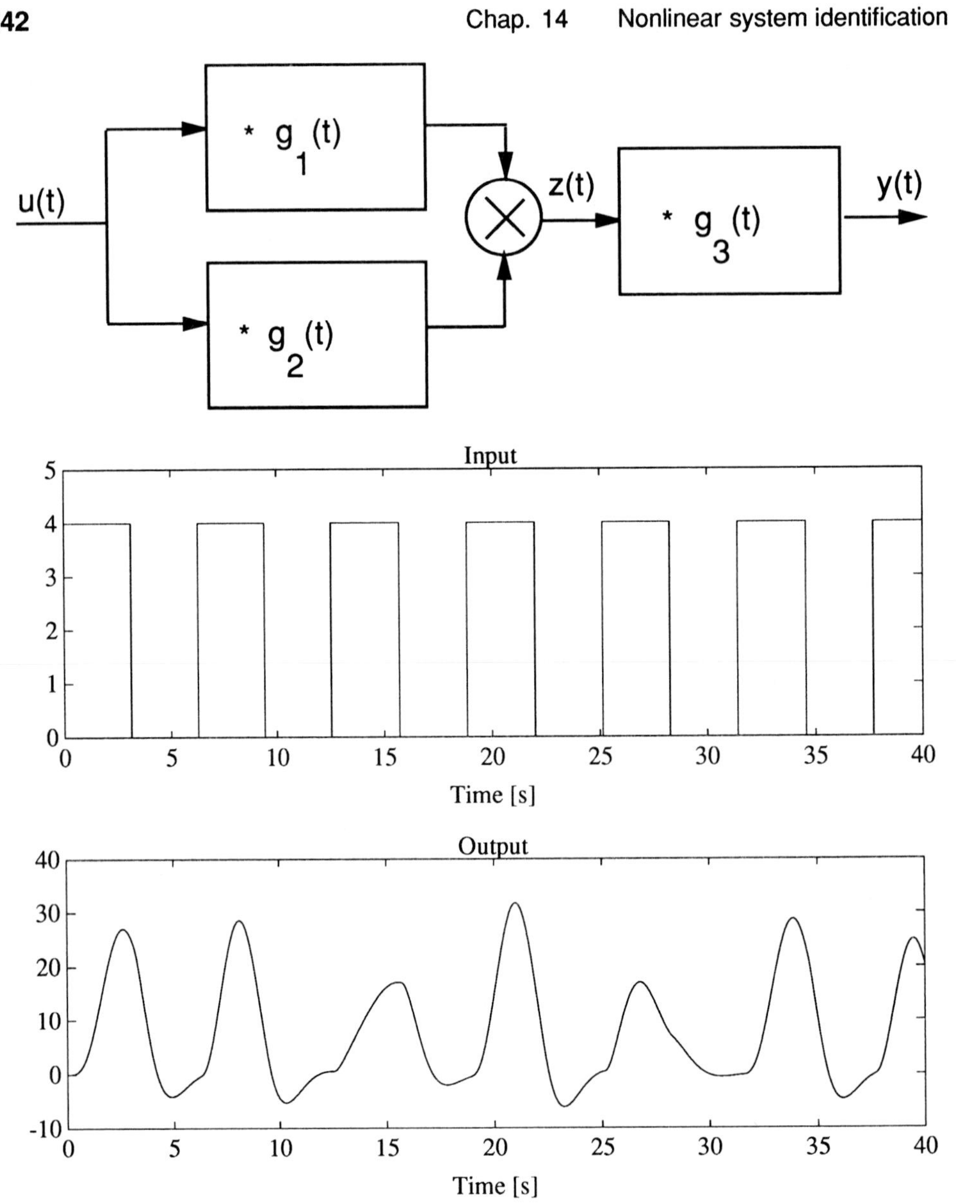

Figure 14.3 A block diagram containing convolution relationships with weighting functions g_1, g_2, g_3 interconnected in a nonlinear manner. Examples of input u and output y are also shown.

and

$$y(t) = \int_0^\infty g_3(\tau_3) z(t - \tau_3) d\tau_3$$
$$= \int_0^\infty \int_0^\infty \int_0^\infty g(\tau_1, \tau_2, \tau_3) u(t - \tau_3 - \tau_1) u(t - \tau_3 - \tau_2) d\tau_1 d\tau_2 d\tau_3 \tag{14.23}$$

where the weighting function of the convolution is

$$g(\tau_1, \tau_2, \tau_3) = g_1(\tau_1) g_2(\tau_2) g_3(\tau_3) \tag{14.24}$$

Let us instead consider the output variable defined for two time variables t_1 and t_2 so that

$$z = z(t_1, t_2) = \int_0^\infty g_1(\tau_1) u(t_1 - \tau_1) d\tau_1 \int_0^\infty g_2(\tau_2) u(t_2 - \tau_2) d\tau_2 \tag{14.25}$$

If we try to find the input-output relationship and thus forget about the initial conditions, then it is straightforward to apply the two-dimensional Laplace transform in the complex frequency variables s_1 and s_2 so that

$$Z(s_1, s_2) = \mathcal{L}_2\{z(t_1, t_2)\} = G_1(s_1) G_2(s_2) U(s_1) U(s_2) \tag{14.26}$$

where the transfer function G_1, G_2 are the ordinary one-dimensional Laplace transforms of the weighting functions g_1 and g_2, respectively.

We now consider the output variable

$$y(t_1, t_2) = (\frac{1}{2\pi i})^2 \int_{\sigma_1 - i\infty}^{\sigma_1 + i\infty} \int_{\sigma_2 - i\infty}^{\sigma_2 + i\infty} Y(s_1, s_2) e^{s_1 t_1 + s_2 t_2} ds_1 ds_2 \tag{14.27}$$

and introduce $t = t_1 = t_2$ and $s = s_1 + s_2$ so that

$$y(t) = y(t_1, t_2) = \frac{1}{2\pi i} \int_{\sigma_1 - i\infty}^{\sigma_1 + i\infty} Y_1(s) e^{st} ds \tag{14.28}$$

where $Y_1(s)$ is the ordinary one-dimensional Laplace transform of $y(t)$. By equating Eq. (14.27) with Eq. (14.28) we find that $Y_1(s)$ must satisfy the relation

$$Y_1(s) = \frac{1}{2\pi i} \int_{\sigma_2 - i\infty}^{\sigma_2 + i\infty} Y(s - s_2, s_2) ds_2 \tag{14.29}$$

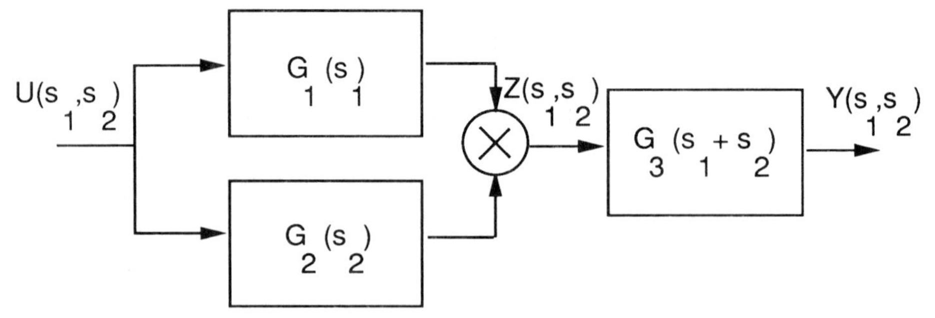

Figure 14.4 A block diagram containing transfer functions G_1, G_2, G_3 interconnected in a nonlinear manner.

Moreover, the one-dimensional Laplace transforms $Y_1(s)$ and $Z_1(s) = \mathcal{L}^{-1}\{z(t)\}$ satisfy the transfer function relationships

$$\begin{aligned} Y_1(s) &= G_3(s)Z_1(s) = G_3(s_1 + s_2)Z_1(s) \\ Y(s_1, s_2) &= G_3(s_1 + s_2)Z(s_1, s_2) \end{aligned} \tag{14.30}$$

A transform algebra relevant for nonlinear systems can now be suggested as follows

$$G_1(s_1)G_2(s_2) = \mathcal{L}_2\{g_1(\tau_1)g_2(\tau_2)\} \tag{14.31}$$

and

$$\begin{aligned} Z(s_1, s_2) &= \mathcal{L}_2\{z(t_1, t_2)\} = G_1(s_1)G_2(s_2)U(s_1)U(s_2) \\ Y(s_1, s_2) &= G_1(s_1)G_2(s_2)G_3(s_1 + s_2)U(s_1)U(s_2) \end{aligned} \tag{14.32}$$

Thus, the multiplicative interaction gives rise to frequency domain models where for some constant s_0 a complex exponential $e^{s_0 t}$ applied on the system input does not propagate through the system independent of other input components.

Example 14.1—A transfer function of a nonlinear system
Consider the transfer functions

$$\begin{aligned} G_1(s) &= \frac{1}{s+1} \\ G_2(s) &= \frac{2}{s^2 + 0.01s + 2} \\ G_3(s) &= \frac{3}{s+3} \end{aligned} \tag{14.33}$$

interconnected according to Fig. 14.4. The resulting transfer function is

$$G(s_1, s_2) = \frac{1}{s_1 + 1} \cdot \frac{2}{s_2^2 + 0.01s_2 + 2} \cdot \frac{3}{s_1 + s_2 + 3} \tag{14.34}$$

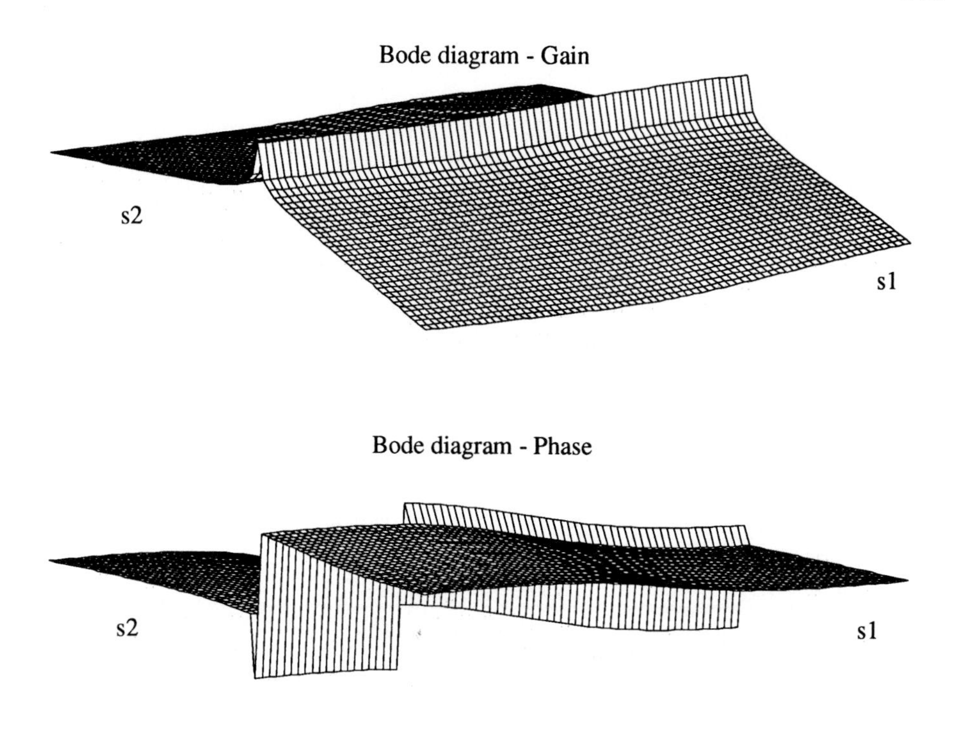

Figure 14.5 A Bode diagram for the transfer function in Example 14.1 in two variables s_1 and s_2 evaluated along the imaginary axes.

It is clear from the example that nonlinear systems also composed of cascaded linear systems with multiplicative interconnections can be described by transfer functions. The corresponding Bode diagram is shown in Fig. 14.1. ■
The method is, however, difficult to extend to systems with feedback, and the method is for this reason of limited value. A drawback in the context of identification is that the two-dimensional representation $y(t_1, t_2)$ is rarely known. It is for this reason difficult to apply methods of spectrum analysis to estimation of transfer functions.

There are, of course, several reasons why the frequency domain approach has limited applicability. One reason for the success of complex exponentials applied to the input of a linear system is that each component propagates through the system independent of other components and is only affected in gain and phase delay. This is not the case for nonlinear systems, and it is hard to find other classes of function with similar properties. Hence, for nonlinear

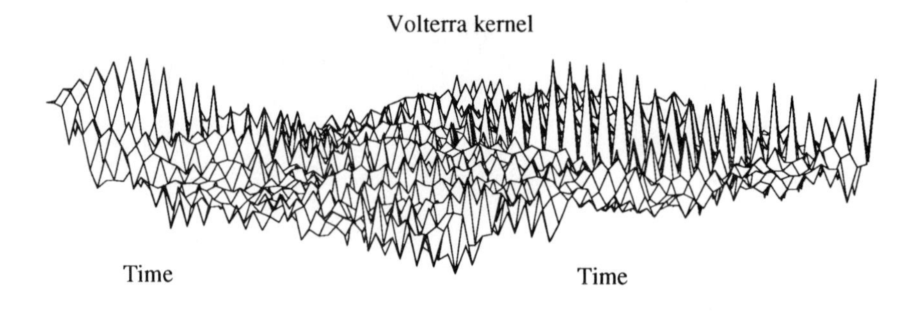

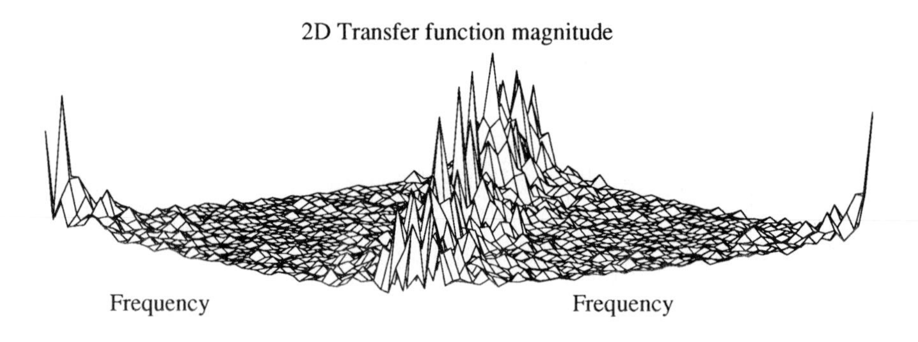

Figure 14.6 Volterra kernel between input u and variable z in Example 14.1 estimated by means of correlation method in the case of a white-noise input.

systems there is in general little point in giving special attention to inputs in the form of complex exponentials. Another point of criticism is the difficulty of relating the estimated Volterra kernels to *a priori* information (see Fig. 14.6).

14.4 POWER SERIES EXPANSIONS

Consider the nonlinear system

$$\dot{x}(t) = \sum_{i=1}^{k} a_i x^i(t) + \sum_{i=0}^{k} \sum_{j=1}^{m} b_{ij} x^i(t) u^j(t) + v(t) \tag{14.35}$$

where the identification problem consists in finding the coefficients $\{a_i\}$ and $\{b_{ij}\}$ given observations of x and u. The system equations (14.35) are clearly of the form

$$\dot{x}(t) = f(x(t)) + g(x(t), u(t)) + v(t) \tag{14.36}$$

As we consider x and u to be the observed variables we can formulate basic time-domain and frequency-domain methods by means of linear regression methods and their extensions. We illustrate these methods by means of the following examples.

Example 14.2—Time-domain identification of nonlinear system
Consider the nonlinear system equation

$$\dot{x} = -a_1 x - b_{11} x u + b_{01} u \tag{14.37}$$

Let us now evaluate the following intgrals over a sequence of intervals determined by the points $\{t_k\}_{k=0}^{N}$

$$\int_{t_k}^{t_{k+1}} \dot{x} dt = -a_1 \int_{t_k}^{t_{k+1}} x(t) dt - b_{11} \int_{t_k}^{t_{k+1}} x(t) u(t) dt + b_{01} \int_{t_k}^{t_{k+1}} u(t) dt \tag{14.38}$$

where the unknown parameters are

$$\theta = \begin{pmatrix} a_1 & b_{11} & b_{01} \end{pmatrix}^T \tag{14.39}$$

These integral relationships derived from Eq. 14.38 provide the equation

$$\begin{pmatrix} x(t_1) - x(t_0) \\ \vdots \\ x(t_N) - x(t_{N-1}) \end{pmatrix} = \begin{pmatrix} -\int_{t_0}^{t_1} x dt & -\int_{t_0}^{t_1} x u dt & \int_{t_0}^{t_1} u dt \\ \vdots & \vdots & \vdots \\ -\int_{t_{N-1}}^{t_N} x dt & -\int_{t_{N-1}}^{t_N} x u dt & \int_{t_{N-1}}^{t_N} u dt \end{pmatrix} \begin{pmatrix} a_1 \\ b_{11} \\ b_{01} \end{pmatrix} \tag{14.40}$$

By formulating Eq. (14.40) on the form $\mathcal{Y} = \Phi\theta$ it is straightforward to solve the estimation problem as a least-squares problem, *i.e.*, $\widehat{\theta} = (\Phi^T\Phi)^{-1}\Phi^T\mathcal{Y}$, or as some extension of a linear estimation problem. ■

Example 14.3—Frequency-domain method of nonlinear system
Consider the nonlinear system equation and its Laplace transform

$$\begin{aligned} \dot{x} &= -a_1 x - b_{11} x u + b_{01} u \\ sX(s) &= -a_1 X(s) - b_{11} \mathcal{L}\{x(t)u(t)\} + b_{01} U(s) \end{aligned} \tag{14.41}$$

By evaluating

$$\begin{aligned} y_k &= i\omega_k X(i\omega_k) \\ \phi_k^T &= \begin{pmatrix} -X(i\omega_k) & -\mathcal{L}\{x(t)u(t)\}|_{s=i\omega_k} & U(i\omega_k) \end{pmatrix} \end{aligned} \tag{14.42}$$

we can arrange this problem as an ordinary least-squares problem with

$$\mathcal{Y} = \begin{pmatrix} y_1 \\ \vdots \\ y_N \end{pmatrix}, \quad \text{and} \quad \Phi = \begin{pmatrix} \phi_1^T \\ \vdots \\ \phi_N^T \end{pmatrix} \tag{14.43}$$

with the solution $\widehat{\theta} = (\Phi^T\Phi)^{-1}\Phi^T\mathcal{Y}$. ■

An attractive property of the methods presented in the examples is that standard statistical validation tests are applicable without extensive modifications.

A standard problem for the frequency-domain methods based on the discrete Fourier transform is that the frequency points chosen have equal spacing in frequency. This implies that the high-frequency fit is favored if no special methods are applied.

We illustrate these methods by means of a differential equation that has received much interest recently.

Example 14.4—Identification of the Lorenz model
Consider the Lorenz equations

$$\begin{aligned} \dot{x} &= \sigma(y - x) \\ \dot{y} &= \rho x - y - xz, \\ \dot{z} &= -\beta z + xy \end{aligned} \quad \text{with} \quad \theta = \begin{pmatrix} \sigma \\ \rho \\ \beta \end{pmatrix} = \begin{pmatrix} 10 \\ 28 \\ 8/3 \end{pmatrix} \tag{14.44}$$

with trajectories according to Fig. 14.7.

$$\begin{pmatrix} x(T) - x(0) \\ y(T) - y(0) + \int_0^T y + xzdt \\ z(T) - z(0) - \int_0^T xydt \end{pmatrix} = \begin{pmatrix} \int_0^T (y - x)dt & 0 & 0 \\ 0 & \int_0^T xdt & 0 \\ 0 & 0 & -\int_0^T dt \end{pmatrix} \theta \tag{14.45}$$

or

$$\begin{pmatrix} 0.0205 \\ -2.6199 \\ -1.6375 \end{pmatrix} \cdot 10^3 = \begin{pmatrix} 2.0535 & 0 & 0 \\ 0 & -93.5693 & 0 \\ 0 & 0 & -614.0740 \end{pmatrix} \theta \tag{14.46}$$

which provide a solution $\widehat{\theta} = \begin{pmatrix} 10.00 & 27.99 & 2.667 \end{pmatrix}^T$ with good accuracy.

■

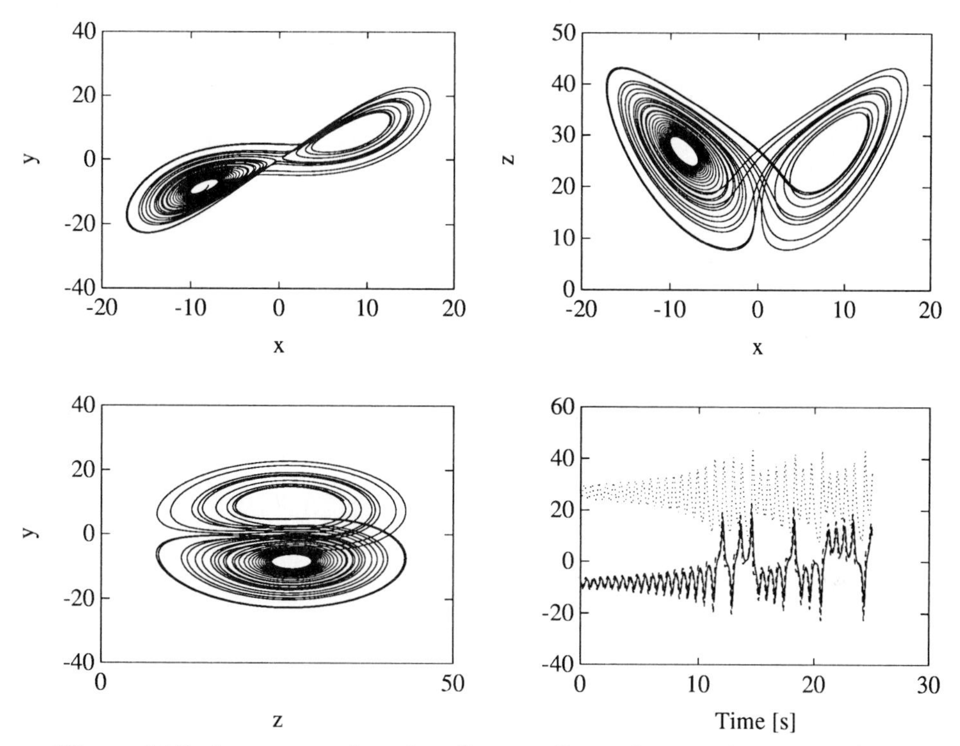

Figure 14.7 Lorenz equations describe a nonlinear dynamical system with three states x, y, z and with three parameters σ, ρ, and β.

Modulating functions

Let us consider the differential equation model in Example 14.2 and introduce the *modulating functions* $\{p_k(t)\}_{k=1}^{N}$. Multiplication of all terms of the differential equation model to be identified and subsequent integration gives the relationship

$$\int_0^T p_k^T \dot{x} dt = -a_1 \int_0^T p_k^T x dt - b_{11} \int_0^T p_k^T x u dt + b_{01} \int_0^T p_k^T u dt \qquad (14.47)$$

for each function $p_k(t)$. The modulating functions can be chosen as any set of differentiable functions, *i.e.*, polynomials in data such as $p_k(t) = x^k(t)$ or trigonometric functions $p_k(t) = \sin(k\omega t)$ or some set of orthogonal functions over an interval $[0, T]$.

A relevant problem is how to evaluate the left-hand integral in Eq. (14.47) since $\dot{x}$ might not be available to measurement. This problem is eliminated if the modulating functions $\{p_k\}$ are chosen such that their time derivatives

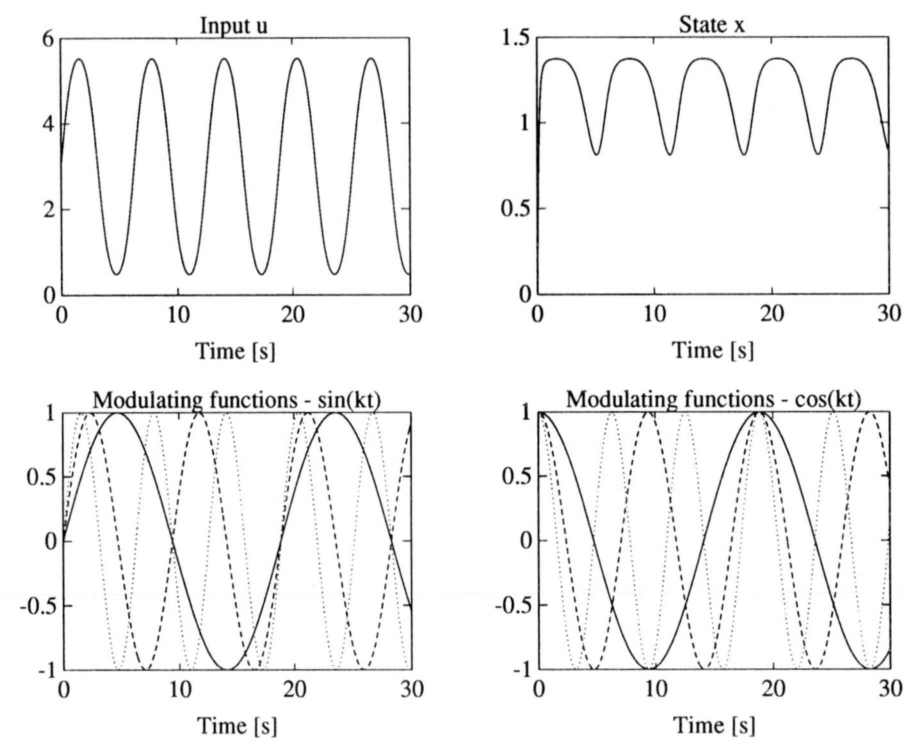

Figure 14.8 Input and output to a system described by the equation $\dot{x} = -a_1 x - b_{11} x u + b_{01} u$. Modulating functions chosen as sinusoids are shown in the lower graphs.

$\{\dot{p}_k\}$ exist. Integrating by parts we have

$$\int_0^T p_k(t)\dot{x}(t)dt = [p_k(t)x(t)]_0^T - \int_0^T \dot{p}_k(t)x(t)dt \tag{14.48}$$

Computation of the integral of Eq. (14.47) for N different modulating functions $\{p_k(t)\}_{k=1}^N$ yields

$$\begin{pmatrix} \int_0^T p_1 \dot{x} dt \\ \vdots \\ \int_0^T p_N \dot{x} dt \end{pmatrix} = \begin{pmatrix} -\int_0^T p_1^T x dt & -\int_0^T p_1^T x u dt & \int_0^T p_1^T u dt \\ \vdots & & \vdots \\ -\int_0^T p_N^T x dt & -\int_0^T p_N^T x u dt & \int_0^T p_N^T u dt \end{pmatrix} \begin{pmatrix} a_1 \\ b_{11} \\ b_{01} \end{pmatrix} \tag{14.49}$$

By formulating Eq. (14.49) on the form $\mathcal{Y} = \Phi\theta$ it is straightforward to solve the estimation problem as a linear estimation problem.

Example 14.5—Identification by means of modulating functions
Consider the input-output data in Fig. 14.8, which are observations $u(t)$ and

$x(t)$ generated by the system

$$\dot{x}(t) = -a_1 x(t) - b_{11} x(t) u(t) + b_{01} u(t), \qquad \theta = \begin{pmatrix} a_1 \\ b_{11} \\ b_{01} \end{pmatrix} = \begin{pmatrix} 1 \\ 2 \\ 3 \end{pmatrix} \tag{14.50}$$

and choose the modulating functions

$$\begin{pmatrix} p_1(t) \\ p_2(t) \\ p_3(t) \\ p_4(t) \\ p_5(t) \\ p_6(t) \end{pmatrix} = \begin{pmatrix} \sin(t/3) \\ \sin(2t/3) \\ \sin(t) \\ \cos(t/3) \\ \cos(2t/3) \\ \cos(t) \end{pmatrix} \tag{14.51}$$

This gives the result

$$\begin{pmatrix} 0.3509 \\ -0.3130 \\ 1.3795 \\ 10.0171 \\ 13.3350 \\ 13.3350 \end{pmatrix} = \begin{pmatrix} -6.0424 & -22.3855 & 17.0548 \\ -1.3142 & -5.4896 & 3.9935 \\ -6.6571 & -61.8896 & 43.9386 \\ -1.0062 & -7.3220 & 8.5557 \\ -1.6966 & -13.3086 & 13.8829 \\ -1.6966 & -13.3086 & 13.8829 \end{pmatrix} \begin{pmatrix} a_1 \\ b_{11} \\ b_{01} \end{pmatrix} \tag{14.52}$$

and we obtain the least-squares estimate $\begin{pmatrix} \hat{a}_1 & \hat{b}_{11} & \hat{b}_{01} \end{pmatrix} = \begin{pmatrix} 1 & 2 & 3 \end{pmatrix}$. ■

Equation error methods

Prediction error methods are for several reasons popular in the domain of discrete-time identification, but they are somewhat difficult to apply to continuous-time systems. One approach in formulating such methods in the context of nonlinear system identification is the following.

Define the error

$$e(t, \theta) = \dot{x} - f(x) - g(x) u(t) \tag{14.53}$$

and introduce the error functional

$$J(\theta) = \frac{1}{2} \int_0^T e^T(t, \theta) e(t, \theta) dt$$

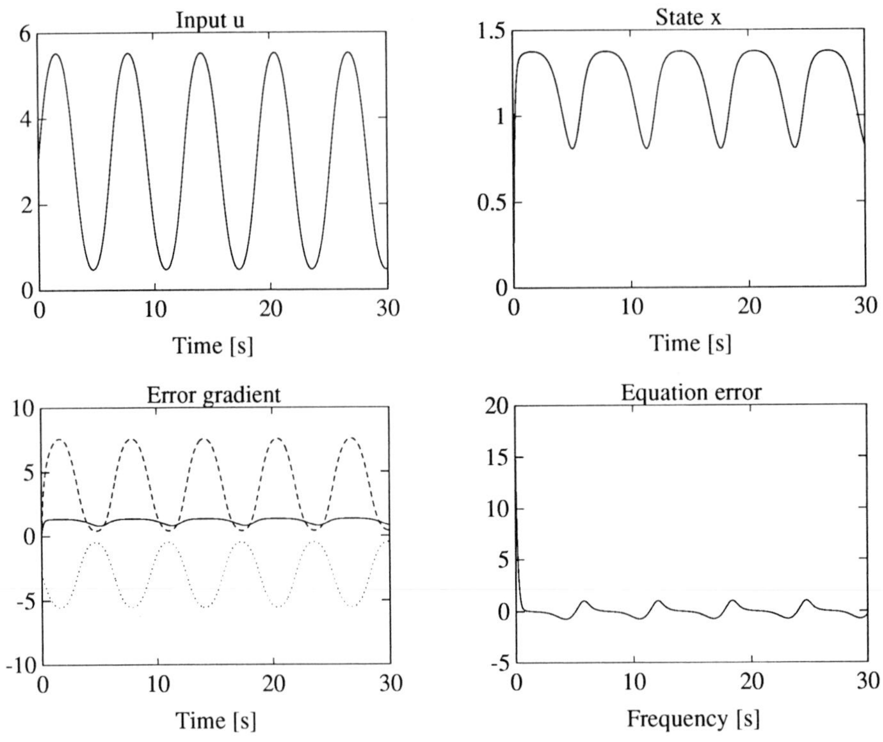

Figure 14.9 Input and output to a system described by the equation $\dot{x} = -a_1 x - b_{11} x u + b_{01} u$.

with the gradient

$$\nabla_\theta J(\theta) = \int_0^T e^T(t,\theta)\nabla_\theta e(t,\theta)dt \tag{14.54}$$

The equation error estimate $\widehat{\theta}$ is taken as the solution to the equation $\nabla_\theta J(\widehat{\theta}) = 0$, *i.e.*, where the gradient of the loss function is zero.

Example 14.6—Identification by means of equation error estimation
Consider the equation

$$\dot{x} = -a_1 x - b_{11} x u + b_{01} u \tag{14.55}$$

$$e = \dot{x} + a_1 x + b_{11} x u - b_{01} u = \dot{x} + \begin{pmatrix} x & xu & -u \end{pmatrix} \begin{pmatrix} a_1 \\ b_{11} \\ b_{01} \end{pmatrix} \tag{14.56}$$

$$\nabla_\theta e = \begin{pmatrix} x \\ xu \\ -u \end{pmatrix} \tag{14.57}$$

By solving the equation

$$0 = \nabla_\theta J = \int_0^T e^T \nabla_\theta e dt = \int_0^T \dot{x} \nabla_\theta e dt + (\int_0^T \nabla_\theta e (\nabla_\theta e)^T dt)\theta \tag{14.58}$$

or

$$0 = \int_0^T \begin{pmatrix} x \\ xu \\ -u \end{pmatrix} \dot{x} dt + (\int_0^T \begin{pmatrix} x \\ xu \\ -u \end{pmatrix} \begin{pmatrix} x & xu & -u \end{pmatrix} dt)\theta \tag{14.59}$$

it is thus straightforward to estimate θ by means of linear estimation methods (see Fig. 14.9). A most restrictive requirement is, of course, that u is differentiable or that $\dot{x}$ is measurable. ∎

Operator transformation methods

The method presented in Chapter 12 can be modified to a class of nonlinear systems, and we show this by means of an example.

Example 14.7—Identification of a nonlinear system

Assume that the objective is to find estimates based on observations of u, y of the unknown parameters a, b in the differential equation

$$\dot{y}(t) = ay(t) + by^2(t) + u(t), \qquad \theta = \begin{pmatrix} a \\ b \end{pmatrix} = \begin{pmatrix} -1 \\ -2 \end{pmatrix} \tag{14.60}$$

A parametric model needs to be developed. A Laplace transformation gives

$$sY(s) = aY(s) + b\mathcal{L}\{y^2(t)\} + U(s) \tag{14.61}$$

with the Laplace variable s. Initial conditions have been ignored. An operator translation Eq. (12.8) with inverse Laplace transformation then gives the parametric model

$$\frac{1}{\tau}(y(t) - [\lambda y(t)]) = a[\lambda y(t)] + b[\lambda(y^2(t))] + [\lambda u(t)] \tag{14.62}$$

Implementation of the low pass filtered signals λy, $\lambda(y^2)$, and λu gives the estimation model

$$\frac{1}{\tau}(y - \lambda y) - \lambda u = \begin{pmatrix} \lambda y & \lambda y^2 \end{pmatrix} \begin{pmatrix} a \\ b \end{pmatrix}, \qquad \theta = \begin{pmatrix} a \\ b \end{pmatrix} \tag{14.63}$$

Sampling of the filtered signals of both hand sides of Eq. (14.63) provides data for identification, and a standard least-squares solution with $\widehat{\theta} = (-1.000 -$

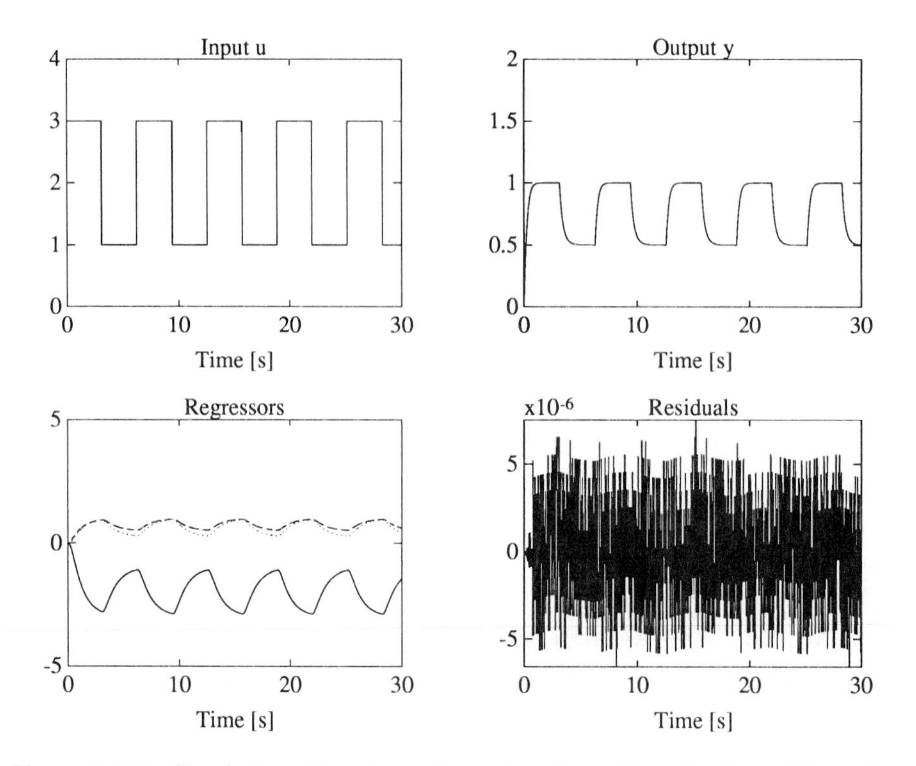

Figure 14.10 Simulation of input u, output y together with estimation of the coefficients $a = 1$, $b = 2$ of Eq. (14.60). The input has been chosen to demonstrate the nonlinear characteristics. All graphs versus time [s].

$2.000)^T$ gives good accuracy (see Fig. 14.9). Real-time identification based on Eq. (14.63) is equally relevant, and sampling of the terms in Eq. (14.63) with subsequent recursive least-squares identification of θ based on the data in Fig. 14.10 gives the result according to Fig. 14.11. It is seen that the method provides good accuracy and convergence rates. The estimation model is similar to that obtained with a heuristic filtering of data appearing at both hand sides of Eq. (14.63). ■

Example 14.8—Identification of a physical parameter in a robot

Adaptive control of robot manipulator motion attracts a considerable interest in the robotics literature. The methods presented for identification are very relevant to the solution of such problems, as will be described in this section.

We consider the two-link example in Fig. 14.12 with point masses m_1, m_2 [kg], lengths l_1, l_2 [m], angles q_1, q_2 [rad], and τ as the vector of joint torques τ_1, τ_2 [Nm]. The end-effector load m_2 is assumed to vary over a certain range.

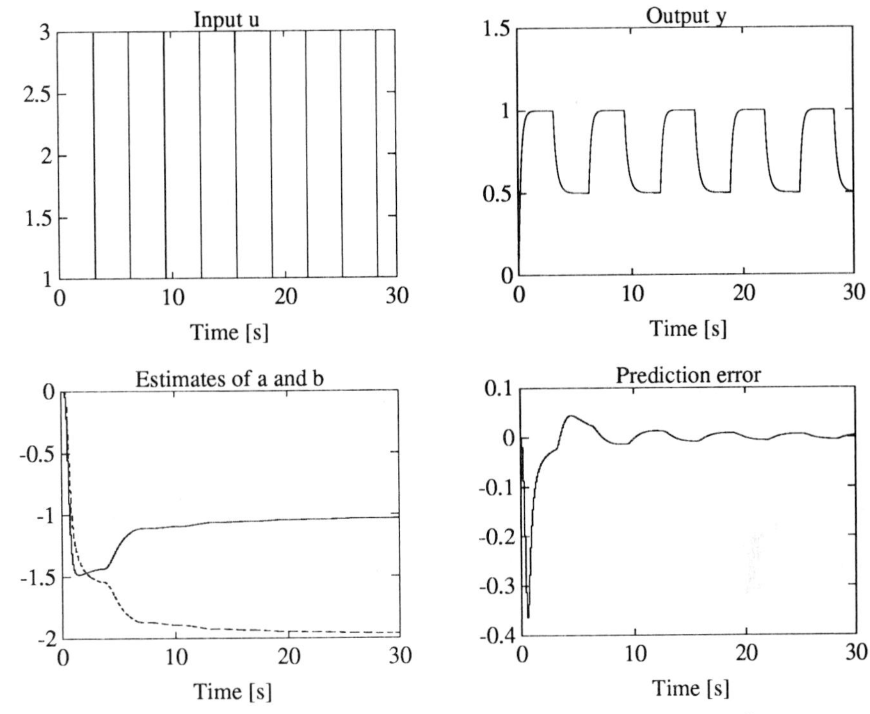

Figure 14.11 Input u and output y of Eq. (14.63) with the estimated parameters a, b by means of recursive identification.

The equations of rigid-body mechanics are:

$$M(q)\ddot{q} + C(q,\dot{q})\dot{q} + G(q) = \tau; \qquad \theta = m_2 \tag{14.64}$$

with the inertia matrix

$$M(q) = \begin{pmatrix} (m_1 + m_2)l_1^2 & m_2 l_1 l_2 (c_1 c_2 + s_1 s_2) \\ m_2 l_1 l_2 (c_1 c_2 + s_1 s_2) & m_2 l_2^2 \end{pmatrix} \tag{14.65}$$

where c_2 is a short notation for $\cos(q_2)$. The Coriolis and centripetal torques are described by the matrix

$$C(q,\dot{q})\dot{q} = m_2 l_1 l_2 (c_1 s_2 - s_1 c_2) \begin{pmatrix} -\dot{q}_2^2 + \dot{q}_1 \dot{q}_2 \\ \dot{q}_1^2 - \dot{q}_1 \dot{q}_2 \end{pmatrix} \tag{14.66}$$

and the gravitation by

$$G(q) = g \begin{pmatrix} -(m_1 + m_2) l_1 s_1 \\ -m_2 l_2 s_2 \end{pmatrix} \tag{14.67}$$

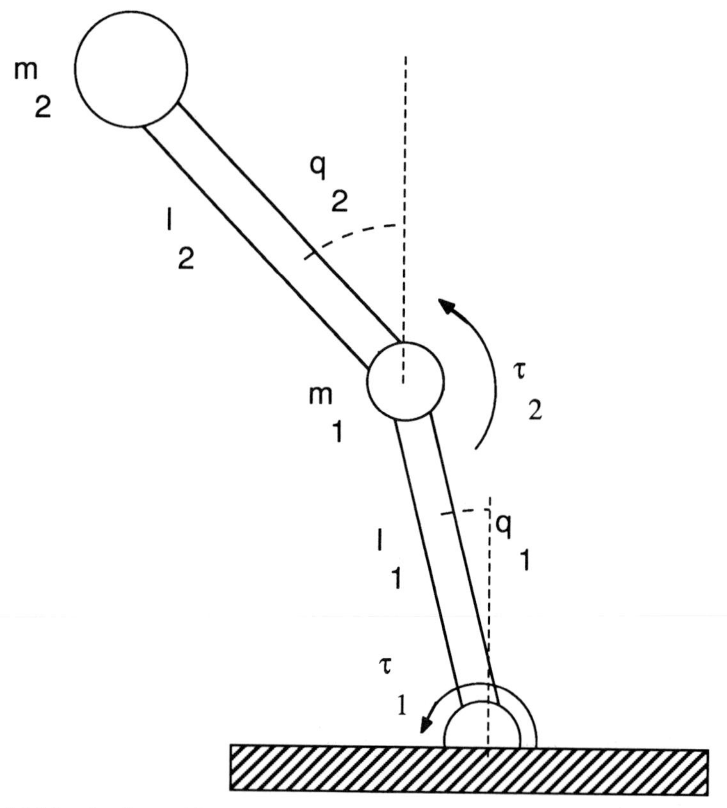

Figure 14.12 A robot with two arm segments of length ℓ_1 and ℓ_2, two point masses m_1 and m_2, and angular coordinates q_1 and q_2 .

These equations are nonlinear in the angular positions q_1, q_2 with polynomials in the derivatives of q_1, q_2. The equations exhibit complicated nonlinear behavior also without any changes in the mass m_2.

$$\begin{aligned}
\int_0^T \dot{q}^T M(q)\ddot{q}dt &= [\frac{1}{2}\dot{q}^T M(q)\dot{q}]_0^T - \frac{1}{2}\int_0^T \dot{q}^T \frac{dM(q)}{dt}\dot{q}dt \\
&= [\frac{1}{2}\dot{q}^T M(q)\dot{q}]_0^T - m_2\int_0^T l_1 l_2(c_1 s_2 - s_1 c_2)(-\dot{q}_1\dot{q}_2^2 + \dot{q}_1^2\dot{q}_2)dt \\
\int_0^T \dot{q}^T C(q,\dot{q})dt &= m_2\int_0^T 2l_1 l_2(c_1 s_2 - s_1 c_2)(-\dot{q}_1\dot{q}_2^2 + \dot{q}_1^2\dot{q}_2)dt \\
\int_0^T \dot{q}^T G(q)dt &= -m_1\int_0^T g l_1 s_1\dot{q}_1 dt - m_2\int_0^T g l_1 s_1\dot{q}_1 + g l_2 s_2\dot{q}_2 dt \\
\int_0^T \dot{q}_1^T \tau dt &= \int_0^T \dot{q}_1^T\tau_1 + \dot{q}_2^T\tau_2 dt
\end{aligned} \tag{14.68}$$

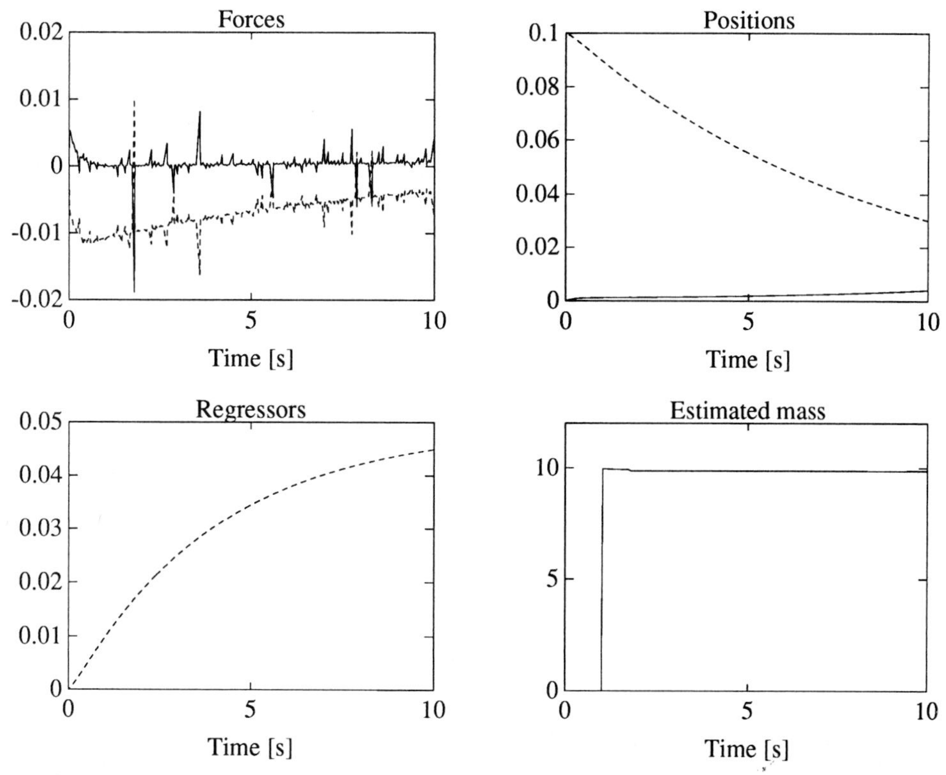

Figure 14.13 Data from robot model Eq. (14.64) with forces τ_1, τ_2 and positions q_1, q_2. The estimated variable $\widehat{m}_2$ and the regressors ϕ_1, and ϕ_2 are also shown. All graphs versus time.

Assume that all terms with m_1 and m_2 as factors are collected in the two terms $m_1\phi_1$ and $m_2\phi_2$, respectively. Then we can summarize Eq. (14.68) and Eq. (14.64) to the linear relationship

$$\int_0^T \dot{q}^T \tau dt = m_1\phi_1(T) + m_2\phi_2(T) \tag{14.69}$$

and we can solve for the unknown m_2 by means of

$$\widehat{m}_2(T) = \frac{1}{\phi_2(T)}\left(\int_0^T \dot{q}^T \tau dt - m_1\phi_1(T)\right) \tag{14.70}$$

Implementation of the above integral is feasible provided that $\dot{q}$ and q and the external forces τ are available for measurement. The unknown parameter m_2 may thus be estimated by linear estimation methods (see Fig. 14.13). ■

14.5 DISCUSSION AND CONCLUSIONS

Hybrid identification involves discrete-time computation although the identification object is formulated as a nonlinear continuous-time model. The discretization usually involves some approximation of the modeling for reasons of computation. Often used trade-off principles are step response equivalence for zero-order hold inputs and bilinear or Euler transformations. The identification objects remain linear in the parameters for the bilinear and Euler transformations, but frequency domain properties may be distorted. The method proposed in this paper is not explicitly affected by such choices of discretization.

As it is sometimes difficult to find transfer functions separated in the variables, it turns out that the frequency domain interpretation of methods such as multidimensional fast Fourier transform is difficult to make. This fact might also complicate the use of identification results for control.

The initial conditions of the filters may have a harmful effect on the identification result. The simplest remedy is to start the filter well before the collecting of data so that transients disappear. A simple rule of thumb is to wait the time of 3τ where τ is a filter time constant associated with λ of Eq. (12.8). The initial conditions of the identification object may also introduce some problems when the influence of initial conditions does not vanish with time.

Several of the methods presented are intuitively very reasonable and appear in most cases as extensions of linear estimation methods. A drawback of several methods is that it is assumed that the state vector is available to measurement. It is in some cases possible to replace nonmeasurable variables by bandwidth limited reconstructions. This is the case for the method of Example 14.6 where the reconstruction bandwidth is determined by the filter time constant τ. The proposed method is in some respects similar to the Laguerre networks, but there is no direct term that complicates the implementation.

14.6 REFERENCES

Some important survey papers on nonlinear system identification are

– S. P. BANKS, *Mathematical Theories of Nonlinear Systems*, London: Prentice-Hall International, 1988, p. 258 ff.

– H.A. BARKER, "Nonlinear system identification by pseudorandom signal testing." *Proc. 6th IFAC Symp. Identification and System Parameter Estimation*, Arlington, VA, 1982, pp. 75–79,

– J.F. BARRETT, "Functional series representation of nonlinear systems—some theoretical comments." *Proc. 6th IFAC Symp. Identification and System Parameter Estimation*, Arlington, VA, 1982, pp. 251–256.

– S.A. BILLINGS, "Identification of nonlinear systems—A survey." *IEE Proc.*, Vol. 127, Pt. D, No. 6, 1980, pp. 272–285.

– S.A. BILLINGS AND I.J. LEONTARITIS, "Parameter estimation techniques for nonlinear systems." *Proc. 6th IFAC Symp. Identification and System Parameter Estimation*, Arlington, VA, 1982, pp. 427–432.

– P. EYKHOFF, *System Identification, Parameter and State Estimation*, London: John Wiley, 1974.

– R. HABER AND H. UNBEHAUEN, "Structure identification of nonlinear dynamic systems—A survey on input/output approaches." *Automatica*, Vol. 26, 1990, pp. 651–677.

– R.K. MEHRA, "Nonlinear system identification selected survey and recent trends", 5th IFAC Symp. Identification and System Parameter Estimation, 1979, pp. 77–83.

Volterra kernels and Wiener models are described in

– P.E. CROUCH, "Dynamical realizations of finite Volterra series." *SIAM J. Contr. Optimiz.*, Vol. 19, 1981, pp. 177–202.

– M.J. KORENBERG, "Parallel cascade identification and kernel estimation for nonlinear systems." *Ann. Biomed. Eng.*, Vol. 19, 1991, pp. 429–455.

– Y.W. LEE AND M. SCHETZEN, "Measurement of the Wiener kernels of a nonlinear system by cross-correlation." *Int. J. Control*, Vol. 2, 1965, pp. 237–254.

– P.Z. MARMARELIS AND V.Z. MARMARELIS, *Analysis of Physiological Systems–The White-Noise Approach.* New York: Plenum Press, 1978.

– G.PALM AND T.POGGIO, "The Volterra representation and the Wiener expansion: Validity and Pitfalls." *Siam J. Appl. Math.*, Vol. 33, No.2, 1977, pp. 195–216.

– W.J. RUGH, *Nonlinear System Theory. The Volterra/Wiener Approach.* Baltimore: Johns Hopkins University Press, 1981.

– M. Schetzen, *The Volterra and Wiener Theories of Nonlinear Systems.* New York: John Wiley, 1980.

– N. Wiener, *Nonlinear Problems in Random Theory.* Cambridge, MA: MIT Press, 1958.

– T.Wigren, *Recursive Identifcation Based on the Nonlinear Wiener Model.* Uppsala, Sweden: Acta Universitatis Upsaliensis, 1990.

A general reference to ordinary differential equations is

– G. Birkhoff and G.C. Rota, *Ordinary Differential Equations.* Lexington, MA: Xerox College Publishing, 1974, p. 144.

14.7 EXERCISES

14.1 Consider the following nonlinear system with the unknown parameters a, b to be estimated from observations of u, y.

$$\ddot{y}(t) + ay(t) \cdot \dot{y}(t) + by(t) = u(t) \tag{14.71}$$

This is a Liénard-type equation that typically arises when modeling mechanical systems with variable friction and damping. The Liénard equation often exhibits limit cycle behavior that depends on the initial conditions (Birkhoff and Rota, 1974).

a. Make a parametric model that allows estimation of the parameters a and b from observations of $\dot{y}$ and y.

b. The initial conditions may have a strong influence on the system trajectories and partly determine the limit cycle behavior. Make a modification of the model so as to compensate for the harmful effect of initial conditions on the estimates. ■

15

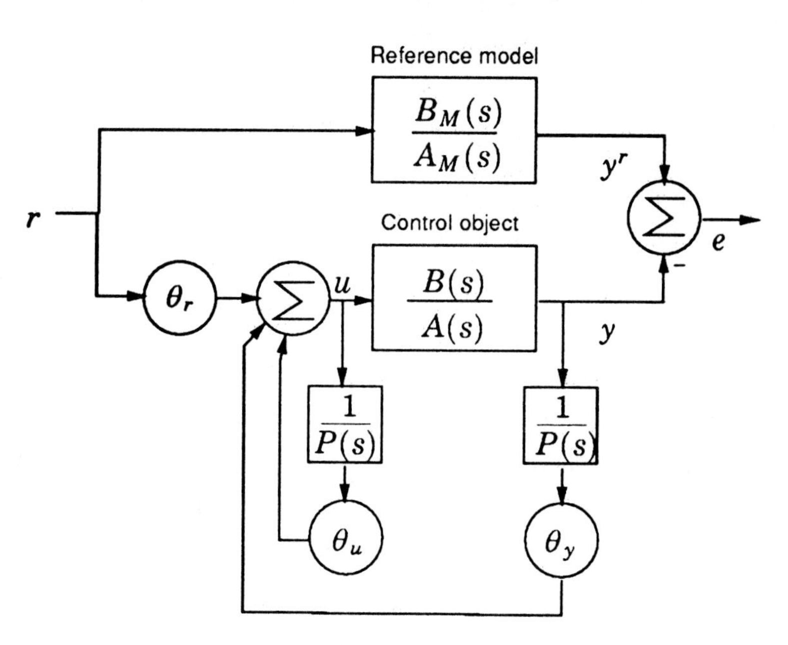

Adaptive Systems

15.1 INTRODUCTION

The ability to adjust behavior with respect to changes in system parameters is called *adaptation*. A typical property is that current information is used to improve operating conditions by eliminating uncertainty so that some optimal state of the system is approached.

There are very important aspects of both identification and control in adaptive systems. Identification methods in real-time application tend to become increasingly important, and such systems are often broadly advertised as *adaptive systems* or *learning systems*. Once an adequate, usually parametric, system representation is available, control actions become meaningful. In a wide sense *adaptive control* may mean modification of the parameters of some control mechanism or control actions in various contexts. The great application potential of adaptive systems is for autonomous systems in which external actions and external perturbations cannot be defined in advance and whose statistical characteristics cannot be determined experimentally in advance. As adaptive systems can be viewed as extensions and applications of identification methodology, we elaborate some aspects in this chapter.

Most adaptive systems, which all require adequate parametric representations, are intended for autonomous operation. In this context there are a number of problems. First, it is necessary to determine the model structure. Second, it is necessary to determine the relevant parameters for a given structure. Third, a suitable control algorithm should be chosen bearing in mind that the result of identification might be inaccurate. There appear to be many difficulties in solving the first and third problem whereas the second problem can be successfully approached by recursive identification and related gradient search methods. A reason for difficulties associated with the first problem is the impact of mismodeled or unmodeled dynamics such as time delays, nonlinearities, or uncertainty in coefficients. As with all model-fitting techniques one can always try to reduce the error by assuming enough degrees of freedom. However, as mathematical modeling involves some idealization of physical properties, there is a source of model uncertainty or incompleteness which usually cannot be described in probabilistic terms.

The third problem is related to parameter uncertainty and the two feedback loops associated with control and adaptation where the control actions supply information to the identification and adaptation procedure which, in turn, generate the control action. The interaction between the adaptation algorithm and the control algorithm is thus much more complex in feedback control operation than for open-loop adaptive systems.

The scientific evolution in adaptive system theory seems to have at least two sources: first, attempts to understand and reproduce biological adaptation and, second, attempts to use optimization methods and parameter estimation in real-time application. Research started with a biomedical interest in neurological adaptation related to behavior, growth, and changes in size of different parts and portions of the body and has, eventually, found important technological application. This historical circumstance has had a clear impact on terminology despite the sometimes inadequate or even misleading biological interpretation.

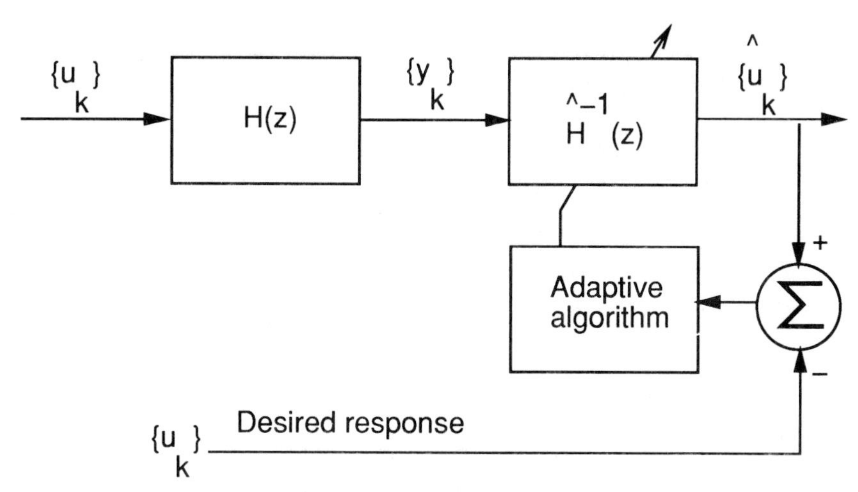

Figure 15.1 Equalization.

First we give an example of adaptive techniques applied to inverse modeling with important application to communication channels.

Example 15.1—Equalization
Consider the following transfer function model for a communication channel

$$y_k = H(z)u_k = \frac{1}{A(z^{-1})}u_k \tag{15.1}$$

where $H(z)$ represents some distortion of a signal $\{u_k\}$ transmitted through the communication channel and where $\{y_k\}$ represents the received signal (see Fig. 15.1). In order to accomplish high-quality communication it is desirable to do restoration of the signal $\{u_k\}$ at the receiver end. Normally there is no other communication of $\{u_k\}$ to the receiver end of the channel, but by means of well-defined test samples sent through the system it is possible to adapt the transfer function $\widehat{H}^{-1}(z)$ so that it implements the inverse of $H(z)$—*i.e.*, to the extent that this is possible. It is noteworthy that the system operates in a feedforward manner except for the feedback involved in the adaptation.

$$\varepsilon_k = \widehat{u}_k - u_k = (\widehat{A}(z^{-1}) - A(z^{-1}))y_k = \phi_k^T\widetilde{\theta} \tag{15.2}$$

where

$$\begin{aligned} \phi_k &= \begin{pmatrix} y_{k-1} & y_{k-2} & \cdots & y_{k-n} \end{pmatrix}^T \\ \widetilde{\theta} &= \begin{pmatrix} \widehat{a}_1 - a_1 & \widehat{a}_2 - a_2 & \cdots & \widehat{a}_n - a_n \end{pmatrix}^T \end{aligned} \tag{15.3}$$

and the parameter estimation problem can be solved by means of standard recursive identification methods (see Chapter 11). ■

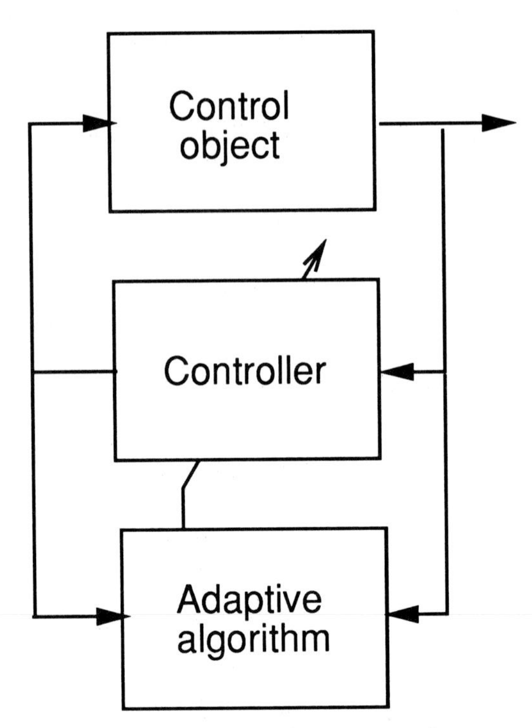

Figure 15.2 Adaptive control.

Equalization as described above is a problem of inverse modeling with deconvolution of $H(z)$ operating on the output $\{y_k\}$. This type of device has proved useful in implementing various deconvolution-type procedures like echo cancellation and channel equalization. The problem is a standard problem in the branch of adaptive systems known as *adaptive signal processing* (see Widrow and Stearns, 1985.)

15.2 HEURISTIC CONTROL METHODS

Several *ad hoc* solutions to practical problems may involve control systems with different characteristics for a range of operating conditions. Control systems designed in such a context may often involve some mechanism for modification of the controller and may be described in a form similar to Fig. 15.2. In its simplest form of *gain scheduling*, the "adaptive algorithm" may consist of look-up tables containing controller parameters for various signal ranges and modes of operation. Such methods are of practical importance and may be used with or without some updating mechanism for the table parameters. However, in order to distinguish adaptive systems from any system with feedback it is desirable that the qualification "adaptive" should be reserved for cases where the look-up-table somehow can be modified from data. In turn, this requires some form of identification procedure.

Indirect adaptive control

A control system coupled to an identification procedure is often called an *adaptive control system*. A more systematic approach involves matching of a system description of a control object obtained by means of identification

$$A(z^{-1})y_k = B(z^{-1})u_k + v_k \tag{15.4}$$

with a feedback control law

$$R(z^{-1})u_k = -S(z^{-1})u_k + T(z^{-1})r_k \tag{15.5}$$

Eliminination of u_k between Eq. (15.4) and Eq. (15.5) yields the closed-loop system

$$(A(z^{-1})R(z^{-1}) + B(z^{-1})S(z^{-1}))y_k = B(z^{-1})T(z^{-1})r_k + R(z^{-1})v_k \tag{15.6}$$

Model matching is a well-known method of control system design that involves specification of the desired system response $Q(z^{-1})/P(z^{-1})$ in terms of a denominator polynomial $P(z^{-1})$ and a numerator polynomial $Q(z^{-1})$. Suitable choices for model matching with the polynomials R, S, and T of Eq. (15.5) are thus found by matching poles and zeros of Eq. (15.6) with those of Q/P. Pole assignment can be accomplished by solving the Diophantine equation

$$A(z)R(z) + B(z)S(z) = P(z) \tag{15.7}$$

The Diophantine equation can be solved by solving the following system of linear equations

$$\begin{pmatrix} 1 & 0 & \cdots & 0 & b_1 & 0 & \cdots & 0 \\ a_1 & 1 & \ddots & \vdots & b_2 & b_1 & \ddots & \vdots \\ a_2 & a_1 & 1 & 0 & \vdots & b_2 & \ddots & 0 \\ \vdots & \vdots & \ddots & 1 & b_{n_B} & \vdots & \ddots & b_1 \\ a_{n_A} & a_{n_A-1} & & a_1 & 0 & b_{n_B} & & b_2 \\ 0 & a_{n_A} & \ddots & a_2 & \vdots & \ddots & \ddots & \vdots \\ \vdots & \ddots & \ddots & \vdots & 0 & \cdots & 0 & b_{n_B} \\ 0 & \cdots & 0 & a_{n_A} & 0 & \cdots & 0 & 0 \end{pmatrix} \begin{pmatrix} r_1 \\ r_2 \\ \vdots \\ r_{n_R} \\ s_0 \\ s_1 \\ \vdots \\ s_{n_S} \end{pmatrix} = \begin{pmatrix} p_1 - a_1 \\ \vdots \\ p_{n_A} - a_{n_A} \\ p_{n_A+1} \\ \vdots \\ p_{n_P} \\ 0 \\ \vdots \\ 0 \end{pmatrix} \tag{15.8}$$

where the matrix structure in Eq. (15.8) is known as a Sylvester matrix. The adaptive control method presented is known as *indirect adaptive control.*

Example 15.2—Indirect adaptive control

Let the system

$$\frac{Y(z)}{U(z)} = H(z) = \frac{b_1 z + b_2}{z^2 + a_1 z + a_2} \tag{15.9}$$

be controlled by means of output feedback control (15.5) designed for pole assignment with the denominator polynomial $P(z) = z^3 + p_1 z^2 + p_2 z + p_3$. The polynomial $T(z^{-1})$ in Eq. (15.5) can, for instance, simply be chosen such that the controlled system has static gain equal to 1—*i.e.*,

$$T(z^{-1}) = \frac{P(1)}{B(1)} = \frac{1 + \sum_{i=1}^{3} p_i}{\sum_{i=1}^{2} b_i} \tag{15.10}$$

These choices suggest the controller

$$u_k = -r_1 u_{k-1} + s_o y_k + s_1 y_{k-1} + t_o r_k \tag{15.11}$$

Equation (15.8) including a Sylvester matrix is then reduced to the linear equation

$$\begin{pmatrix} 1 & b_1 & 0 \\ a_1 & b_2 & b_1 \\ a_2 & 0 & b_2 \end{pmatrix} \begin{pmatrix} r_1 \\ s_0 \\ s_1 \end{pmatrix} = \begin{pmatrix} p_1 - a_1 \\ p_2 - a_2 \\ p_3 \end{pmatrix} \tag{15.12}$$

or

$$A(\theta)\vartheta = b(\theta), \quad \text{with} \quad \vartheta = \begin{pmatrix} r_1 \\ s_0 \\ s_1 \end{pmatrix} \tag{15.13}$$

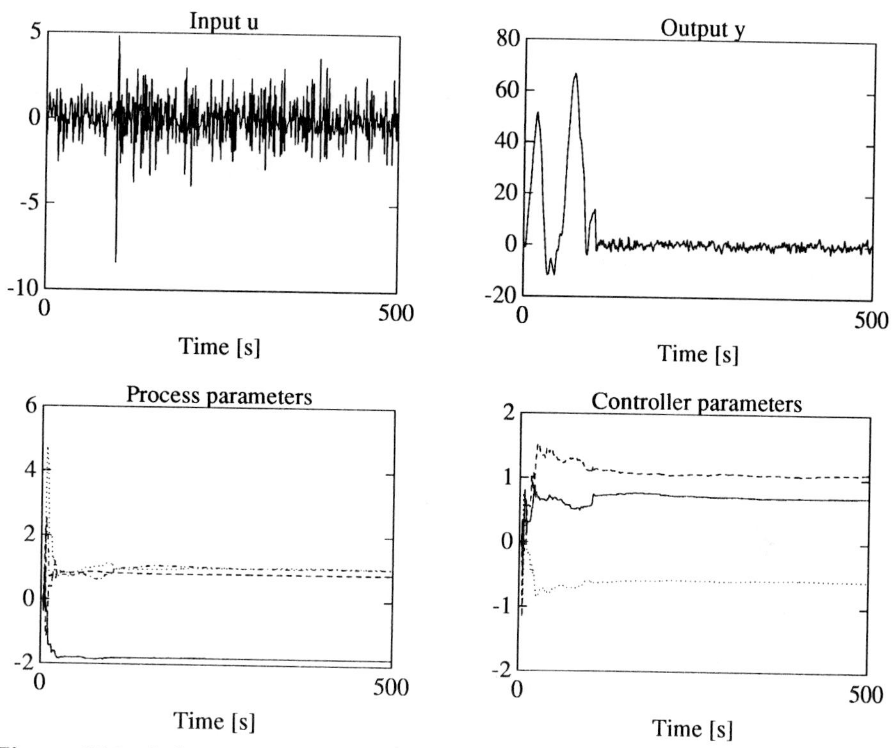

Figure 15.3 Indirect adaptive control starting at time $t = 100$. The second-order process is controlled by an output feedback control law with pole assignment to the origin $z = 0$. The histories of the input u, output y, and the estimated process parameters and the controller parameters are displayed.

A simulation depicting the adaptive control performance is shown in Fig. 15.3. ■

The matrix $A(\theta)$ in Eq. (15.8) can, of course, be singular or close to singular when the model order is overestimated and when the real parameters a_1, a_2, b_1, and b_2 are substituted by their estimated counterparts. Reliable implementations of this type of controller therefore require an equation solver that takes care of the situation with rank deficit (see Appendix A). A systematic solution is, of course, to use the pseudoinverse $A^+(\theta)$ to solve for the controller parameters

$$\vartheta = A^+(\theta)b \tag{15.14}$$

It should be borne in mind that a rank deficient equation of the type (15.13) has an infinite number of solutions and that the solution (15.14) has the smallest 2-norm of all minimizers of $||A\vartheta - b||_2$ (see Appendix A).

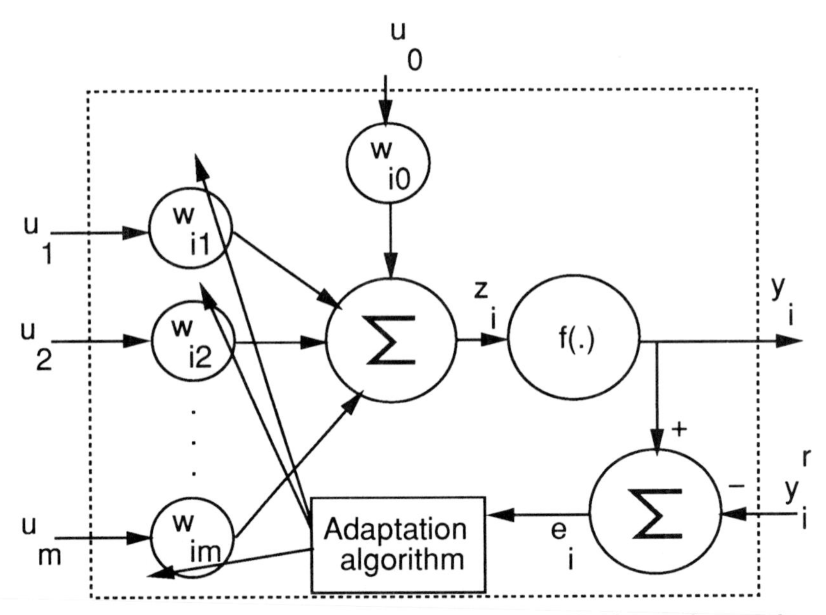

Figure 15.4 Neuron model of the type used in artificial neural networks.

15.3 ASPECTS ON NEURAL NETWORKS

McCulloch and Pitts (1943) introduced a neuron model operating like threshold functions that could generate fairly complicated behavior. Rosenblatt generalized the McCulloch-Pitts model by adding adaptation, calling this model a *perceptron*, which has had an enormous impact on the field.

A neuron model of the type depicted in Fig. 15.4 consists of an input pattern (or input vector) $u = (u_1, u_2, \ldots, u_m)^T$, an output pattern (or output vector) $y = (y_1, y_2, \ldots, y_n)^T$, and gain parameters w_{ij} called *weights*. The input-output relationship is

$$\begin{pmatrix} y_1 \\ y_2 \\ \vdots \\ y_n \end{pmatrix} = \begin{pmatrix} f(z_1) \\ f(z_2) \\ \vdots \\ f(z_n) \end{pmatrix}$$

where

$$\begin{pmatrix} z_1 \\ z_2 \\ \vdots \\ z_n \end{pmatrix} = \begin{pmatrix} w_{10} & w_{11} & \cdots & w_{1m} \\ \vdots & \vdots & & \vdots \\ w_{n0} & w_{n1} & \cdots & w_{nm} \end{pmatrix} \begin{pmatrix} u_0 \\ u_1 \\ u_2 \\ \vdots \\ u_m \end{pmatrix} \tag{15.15}$$

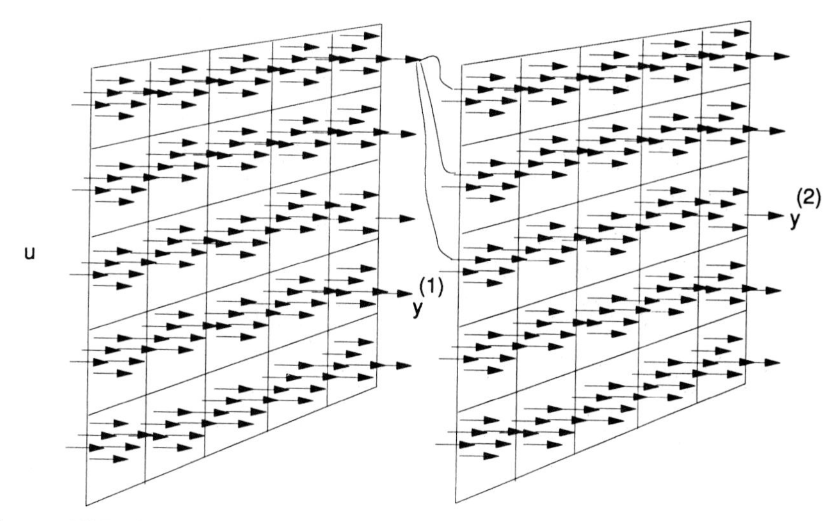

Figure 15.5 Artificial neural networks with two neural cell layers interconnected according to the perceptron principle where the first layer output $y^{(1)}$ provides input to the second layer. Each square element in the figure represents an artificial neuron with many inputs and one output.

where u_0 is an offset. The nonlinear function $f(\cdot)$ should be chosen among the functions that map real numbers onto an interval of the real numbers. Standard choices of $f(\cdot)$ are

$$f(x) = \tanh x, \qquad \text{or} \quad f(x) = \frac{1}{1+e^{-kx}}, \qquad \text{or} \quad f(x) = \frac{x^2}{1+x^2} \tag{15.16}$$

Widrow designed a device called *adaptive linear element* or *adaline* that adjusts its gain parameters (or weights) between the input and the output in response to the error between the computed and desired outputs. The function of an adaline device is closely related to that of a perceptron element (see Figs. 15.4, 15.5) and can be described as follows:

Let y^r be some desired behavior of the artificial neurons. Each error component e_i of the error $e = y - y^r$ is then

$$e_i = y_i - y_i^r = f(z_i) - y_i^r \tag{15.17}$$

By adopting gradient search techniques similar to those in Appendix C, we can approach the minimum of the loss function

$$J(e) = \frac{1}{2}e^T e = \frac{1}{2}\sum(y_i - y_i^r)^2 \tag{15.18}$$

by choosing

$$\dot{w}_{ij} = -\gamma_{ij}\frac{\partial J}{\partial w_{ij}}, \qquad \gamma_{ij} > 0 \tag{15.19}$$

or

$$\dot{w}_{ij} = -\gamma_{ij}(\frac{\partial e_i}{\partial w_{ij}})^T e_i(t) = -\gamma_{ij} f'(z_i) u_j(t) e_i(t), \quad \text{where} \quad f'(z_i) = \frac{df}{dz}|_{z=z_i} \tag{15.20}$$

so that the error e_i converges in the course of time according to

$$\frac{dJ}{dt} = \sum_i \frac{d}{dt} e_i^2(t) = -\sum_{i,j} e_i(t)(\frac{\partial e_i}{\partial w_{ij}})\gamma_{ij}\frac{\partial e_i}{\partial w} e_i(t) \leq 0 \tag{15.21}$$

Much of the effort in the design of artificial neural networks consists of composing large-scale structures of identical elements. Following biological inspiration and early perceptron concepts, it is popular to propose multilayer networks reminiscent of retinal organization in the human eye (see Fig. 15.5). Let $w^{(k)}$ denote the matrix of parameter weightings in layer k, and let y^r denote the prescribed output behavior. Assuming that there are p layers in such a layered network where the output of layer k is connected to the input to layer $(k+1)$ for $k = 1, 2, \ldots, p-1$, the output $y^{(p)}$ of layer p represents the network output. A problem that arises in this context is how to organize the adaptation by using the global output performance $e^{(p)}$ as a means to control the local parameters. In the absence of a "teacher" acting at an intermediate local level in the network, a fruitful idea is to calculate local errors $e^{(k)}$ (*i.e.*, output errors associated to each layer k), based on global error $e^{(p)} = y^{(p)} - y^r$. The local errors can, then, be used for parameter adaptation at each neuron. This approach can be justified by means of optimal control theory according to the following motivations: Consider the performance index

$$J(e^{(p)}) = \frac{1}{2}(e^{(p)})^T e^{(p)} = \frac{1}{2}(e^{(p)})^T e^{(p)} - \sum_{k=1}^{p} \lambda_k^T (y^{(k)} - f(w^{(k)} y^{(k-1)})) \tag{15.22}$$

where the Lagrange multipliers $\{\lambda_k\}_{k=1}^{p}$ have been introduced to "adjoin" the constraint equations imposed by the neuron interconnections in a feedforward neural network organized in p layers. Let the indeterminate Lagrange multipliers be chosen according to

$$\begin{aligned} \lambda_p &= e^{(p)} \\ \lambda_k &= f'(z^{(k)})\lambda_{k+1}, \quad k = 1, \ldots, p-1, \quad \text{where} \quad f'(z^{(k)}) = \frac{df}{dz}|_{z=z^{(k)}} \end{aligned} \tag{15.23}$$

which are the adjoint equations of a $p-$stage optimal control problem. Then it is possible to simplify the gradient of J according to

$$\frac{\partial J}{\partial w_{ij}^{(k)}} = \lambda_{ik} \frac{\partial y^{(k)}}{\partial w_{ij}^{(k)}} \tag{15.24}$$

A gradient search for a minimum value of the loss function of the type (15.19) satisfying the constraint equation is

$$\dot{w}_{ij}^{(k)} = -\lambda_{ik} \frac{\partial y_i^{(k)}}{\partial w_{ij}^{(k)}} = \begin{cases} \lambda_{ik} f'(z_i^{(k)}) y_j^{(k-1)}, & k = 2, \ldots, p \\ \lambda_{ik} f'(z_i^{(k)}) u_j, & k = 1 \end{cases} \tag{15.25}$$

so that

$$\frac{dJ}{dt} = \sum_{i,\,j,\,k} \frac{\partial J}{\partial w_{ij}^{(k)}} \dot{w}_{ij}^{(k)} = - \sum_{i,\,j,\,k} \left(\frac{\partial y_i^{(k)}}{\partial w_{ij}^{(k)}} \lambda_{ik} \right)^2 \le 0 \tag{15.26}$$

The method defined by Eqs. (15.23–15.25), known as *back propagation*, is one such method of computing local errors from the global errors. In turn, the local errors can be used for local adaptation by means of gradient search methods (see Fig. 15.4).

Hence, the back-propagation reformulation enables the transformation of a functional optimization problem to a parametric one. The parametric optimization according to gradient search techniques ensures that the adapting parameters approach at least some local minimum and is thus much related to identification techniques. Other aspects of comparison between neural networks and identification theory, however, seem to be underexploited in this field, and several such aspects are open research issues. For example, artificial neural networks are sometimes characterized by an excessive overparametrization as compared to standard practice in identification methodology. The Akaike final prediction error criterion would suggest that the prediction accuracy of such overparametrized devices is not optimal.

15.4 EXTREMUM CONTROL

As shown in previous chapters, it is standard practice to apply gradient search methods to the minimization of parameter errors. Another possibility, known as *extremum control*, is simultaneous application of parameter estimation and optimization with respect to some control action. This is of particular interest

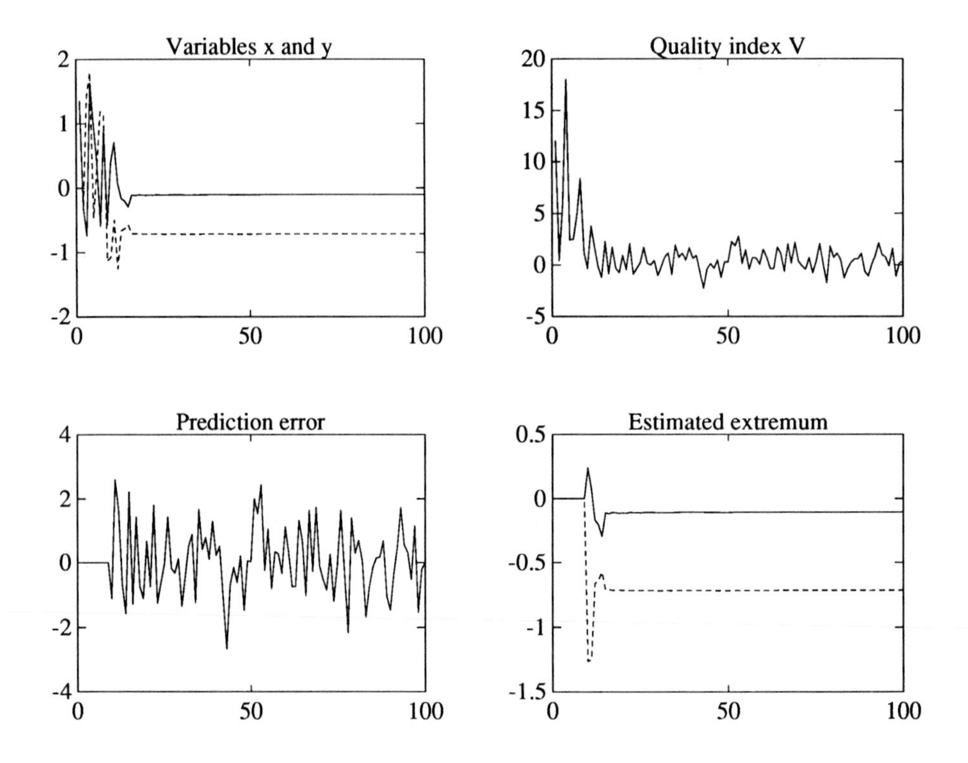

Figure 15.6 Example of extremum control where the estimation result can be used to improve a quality index.

in contexts where control is to be used as a means to improve some quality index. Extremum control is thus relevant in several industrial processes related to energy saving or emission control for environmental protection. Relevant examples of such control variables are the composition of raw materials to some industrial process or the fuel-to-air ratio in combustion engines.

A central problem is to define a suitable quality index or efficiency measure V, which is usually assumed to be a quadratic function of control parameters u and some unknown weighting matrix P_2, a vector P_1, and a scalar P_0.

$$V(u) = \frac{1}{2}u^T P_2 u + u^T P_1 + P_0 \tag{15.27}$$

where the set of control variables u—usually some kind of rate variables—regulates the quality index V toward an extremum. The quadratic function V is a function with its extremum at

$$u^* = -P_2^{-1} P_1 \tag{15.28}$$

and it is necessary to estimate u^* from observations of the control parameter u and the quality measure $V(u)$. A residual model for estimation of P_2, P_1, and P_0 from a control sequence $\{u_k\}$ and observations $\{V(u_k)\}$ is

$$\varepsilon_k = \frac{1}{2}u_k^T \widehat{P}_2 u_k + u_k^T \widehat{P}_1 + \widehat{P}_0 - V(u_k) \tag{15.29}$$

from which parameter estimates $\widehat{P}_0(k), \widehat{P}_1(k), \widehat{P}_2(k)$ may be obtained by methods of previous chapters. A Newton-Raphson gradient search for u^* based on the estimated parameter gives

$$\begin{aligned} \widehat{u}_k &= \widehat{u}_{k-1} - \nabla_u^2 \widehat{V}^{-1}(\widehat{u}_{k-1}, k)\nabla_u \widehat{V}(\widehat{u}_{k-1}, k) = \\ &= \widehat{u}_{k-1} - \widehat{P}_2^{-1}(k)(\widehat{P}_2(k)\widehat{u}_{k-1} + \widehat{P}_1(k)) = -\widehat{P}_2^{-1}(k)\widehat{P}_1(k) \end{aligned} \tag{15.30}$$

Hence, a solution to the extremum control problem involves estimating P_2, P_1, and P_0; computing the extremum u^*; and applying this control parameter to the system.

Example 15.3—Extremum control

Assume that an extremum control problem can be modeled by means of the relationship

$$V(u_k) = \frac{1}{2}u_k^T \begin{pmatrix} 3 & 1 \\ 1 & 3 \end{pmatrix} u_k + u_k^T \begin{pmatrix} 1 \\ 2 \end{pmatrix} + 1 + w_k \tag{15.31}$$

where $\{w_k\}$ is a zero-mean white-noise sequence with $\mathcal{E}\{w_k^2\} = 1$. The extremum of Eq. (15.31) with $V(u^*) = 0.3125$ is at

$$u^* = -P_2^{-1}P_1 = \begin{pmatrix} -0.125 \\ -0.625 \end{pmatrix} \tag{15.32}$$

A simulation is given in Fig. 15.6 where estimation and control starts at $t = 10$, which results in an improvement of the quality index $V(u_k)$. Extremum control based on least-squares estimate of 100 samples gives the estimate

$$\widehat{u} = -\widehat{P}_2^{-1}\widehat{P}_1 = \begin{pmatrix} -0.104 \\ -0.716 \end{pmatrix} \tag{15.33}$$

■

There are a number of problems with this approach. First, it is sometimes difficult to assume a purely quadratic relationship between the control parameter u and the quality measure $V(u)$. Second, the two-step extremum control procedure is numerically sensitive to disturbances and unmodeled dynamics as it involves a two-step optimization where errors in the first step have impact on the second stage. Third, the input-output relationship is assumed to be purely static, and there are few systematic methods to modify the algorithm when system dynamics cannot be neglected.

Extensions to solve these problems require more elaborate nonlinear models such as Hammerstein models including an input nonlinearity

$$A(z^{-1})v_k = P_0 + u_k^T P_1 + \frac{1}{2} u_k^T P_2 u_k \tag{15.34}$$

or Wiener models with output nonlinearity

$$\begin{aligned} v_k &= P_0 + y_k^T P_1 + \frac{1}{2} y_k^T P_2 y_k = \\ &= P_0 + \sum_{i=1}^{n} u_{k-i}^T h_i^T + \frac{1}{2} \sum_{i=1}^{n} \sum_{j=1}^{n} u_{k-i}^T h_i^T P_2 h_j u_{k-j} \end{aligned} \tag{15.35}$$

where $y_k = \sum_{i=1}^{n} h_i u_{k-i}$ is some intermediate signal that represents a filtered input $\{u_k\}$; see Chapter 14.

The estimation accuracy tends to increase at a very slow rate after a rapid initial improvement. Therefore, a successful application is contingent upon a careful implementation of both estimation and control calculations which involve matrix inversions. Active test signals such as pulse sequences or sinusoidals are sometimes proposed in order to improve the adaptation rate although such signals imply suboptimal control, which is applied at a significant cost.

15.5 MODEL-REFERENCE ADAPTIVE CONTROL

It is standard to define an adaptive control system as a control system coupled to an identification procedure. A certain sceptic attitude towards adaptive control is sometimes perceptible with a criticism concerning its unpredictable control actions and stability problems. One reason why adaptive control is more difficult than other problems in adaptive system theory is that two feedback loops interact in a difficult manner—*i.e.*, the feedback control loop and the adaptation. In addition, formulating a control performance that is appropriate for an adaptive control system is not a trivial feat.

A way to simplify the formulation of an adaptive control problem is to request that the controlled system behave similar to some given, well-defined system called reference model representing some prescribed system behavior. In the terminology of neural networks this reference model serves as a teacher that trains the system to have properties similar to its own. Inverse models obtained by feedforward and feedback control are for this reason relevant, and several adaptive control algorithms provide systematic methods of identifying a system inverse so that desired outputs can be transformed into the inputs that generate them. Once an adequate prescribed behavior has been specified, it is straightforward to evaluate performance as well as various design options. One important approach to adaptive control design is the minimization of variance-like or energy-like functions of the system state and output. The need for a rigorous analysis suggests the use of Lyapunov theory and we refer to some background material in Appendix 15.1.

Assume that the control object can be described by the state equation

$$\dot{x} = Ax + Bu = \begin{pmatrix} -a_1 & -a_2 & \cdots & -a_n \\ & I_{(n-1)\times(n-1)} & & 0_{(n-1)\times 1} \end{pmatrix} x + \begin{pmatrix} 1 \\ 0_{(n-1)\times 1} \end{pmatrix} u \quad (15.36)$$

and that

$$u = -\theta^T x \quad (15.37)$$

is an appropriate control law of suitable structure. In the case of a known A it is possible to choose a suitable θ by means of model matching so that $A - B\theta^T = A_m$ for some dynamics matrix A_m representing the prescribed system behavior. This gives the closed-loop system

$$\dot{x} = Ax + Bu = \begin{pmatrix} -a_1 - \theta_1 & -a_2 - \theta_2 & \cdots & -a_n - \theta_n \\ & I_{(n-1)\times(n-1)} & & 0_{(n-1)\times 1} \end{pmatrix} x = A_m x \quad (15.38)$$

Let (15.37) be replaced by the adaptive control law

$$\begin{aligned} \dot{\widehat{\theta}} &= S^{-1} x B^T P x, \qquad S = S^T > 0 \\ u &= -\widehat{\theta}^T x \end{aligned} \quad (15.39)$$

where P solves the Lyapunov equation (see Appendix 15.1)

$$PA_m + A_m^T P = -Q \quad (15.40)$$

The system behavior under adaptive feedback control is

$$\dot{x} = (A - B\theta^T)x - Bx^T\widetilde{\theta} = A_m x - Bx^T\widetilde{\theta} \quad (15.41)$$

In order to prove that the adaptive system defined by Eqs. (15.39-15.41) is stable, we consider the following Lyapunov function candidate (see Sec. 15.8)

$$V(x,\tilde{\theta}) = \frac{1}{2}x^T P x + \frac{1}{2}\tilde{\theta}^T S \tilde{\theta}, \qquad S = S^T > 0 \tag{15.42}$$

with the derivative

$$\begin{aligned}\frac{dV}{dt} &= \frac{1}{2}\dot{x}^T P x + \frac{1}{2}x^T P \dot{x} + \frac{1}{2}\dot{\tilde{\theta}}^T S \tilde{\theta} + \frac{1}{2}\tilde{\theta}^T S \dot{\tilde{\theta}} \\ &= \frac{1}{2}x^T (A_m^T P + P A_m) x - x^T P B x^T \tilde{\theta} + \tilde{\theta}^T S \dot{\tilde{\theta}}\end{aligned} \tag{15.43}$$

If θ is constant then $\dot{\hat{\theta}} = \dot{\tilde{\theta}}$ and

$$\frac{dV}{dt} = -\frac{1}{2}x^T Q x < 0, \qquad \|x\| \neq 0 \tag{15.44}$$

which shows that the system is globally stable in the sense of Lyapunov—*i.e.*, a certain weighted sum of squared control errors and parameter errors is guaranteed to decrease in the course of time. ■

Example 15.4—Adaptive control of a linear system
Consider adaptive stabilization of the system

$$\begin{pmatrix} \dot{x}_1 \\ \dot{x}_2 \\ \dot{x}_3 \end{pmatrix} = \begin{pmatrix} -a_1 & -a_2 & -a_3 \\ 1 & 0 & 0 \\ 0 & 1 & 0 \end{pmatrix} \begin{pmatrix} x_1 \\ x_2 \\ x_3 \end{pmatrix} + \begin{pmatrix} b_1 \\ 0 \\ 0 \end{pmatrix} u \tag{15.45}$$

so that it behaves like the model

$$\dot{x} = \begin{pmatrix} \dot{x}_1 \\ \dot{x}_2 \\ \dot{x}_3 \end{pmatrix} = \begin{pmatrix} -3 & -3 & -1 \\ 1 & 0 & 0 \\ 0 & 1 & 0 \end{pmatrix} \begin{pmatrix} x_1 \\ x_2 \\ x_3 \end{pmatrix} = A_m x \tag{15.46}$$

Application of control algorithm (15.39) for $Q = S = I_{3\times 3}$ and P solving the Lyapunov equation $PA_m + A_m P = -Q$ gives

$$P = \begin{pmatrix} 0.4375 & 0.8125 & 0.5000 \\ 0.8125 & 3.2500 & 1.9375 \\ 0.5000 & 1.9375 & 2.3125 \end{pmatrix} \tag{15.47}$$

A simulation of this adaptive algorithm for $a_1 = a_2 = a_3 = -1$ and $b_1 = 1$ is shown in Fig. 15.7 in which typical transients of control and adaptation are exhibited. ■

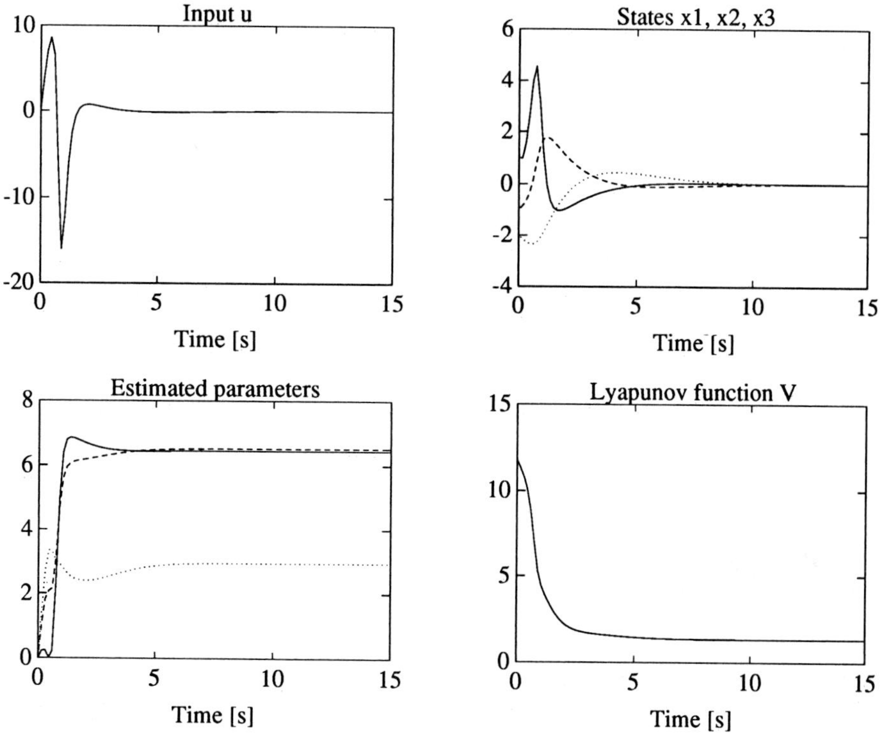

Figure 15.7 Model reference adaptive control of Example 15.4.

Example 15.5—Adaptive control of robot manipulators

The adaptive state feedback methods presented can be extended to certain classes of nonlinear systems. Consider the rigid-body motion of robot manipulators as elaborated in Example 7.3 with nonlinear motion equations

$$M(q)\ddot{q} + C(q,\dot{q})\dot{q} + G(q) = \tau \tag{15.48}$$

expressed in some generalized coordinates $q \in R^n$. Assume that the matrices M, C, G have a known structure but that some parameters might be unknown. An adaptive control objective to follow some given trajectory q_r can be approached by means of energy-like Lyapunov function such as

$$V(\tilde{x},\tilde{\theta},t) = \frac{1}{2}\tilde{x}^T T_0^T \begin{pmatrix} M(q) & 0 \\ 0 & K \end{pmatrix} T_0\tilde{x} + \frac{1}{2}\tilde{\theta}^T S\tilde{\theta}, \quad \tilde{x} = \begin{pmatrix} \dot{\tilde{q}} \\ \tilde{q} \end{pmatrix} = \begin{pmatrix} \dot{\tilde{q}} - \dot{q}_r \\ \tilde{q} - q_r \end{pmatrix} \tag{15.49}$$

Let $Q = Q^T > 0$, $R = R^T > 0$ be some weighting matrices. It can be shown that for certain choices of Q, R it is possible to generate stable adaptive control for $K = K^T > 0$, T_0 solving the algebraic matrix equation

$$\begin{pmatrix} 0 & K \\ K & 0 \end{pmatrix} + Q - T_0^T BR^{-1}B^T T_0 = 0, \tag{15.50}$$

where

$$T_0 = \begin{pmatrix} T_{11} & T_{12} \\ 0 & T_{22} \end{pmatrix}, \quad \text{and} \quad B = \begin{pmatrix} I_{n\times n} \\ 0_{n\times n} \end{pmatrix} \tag{15.51}$$

Let the control law generated by means of V be expressed in terms of unknown parameters $\theta \in R^p$ of M, C, G and the data matrices $\phi \in R^{n\times p}$, $\phi_0 \in R^n$. The vector ϕ_0 contain terms of τ^* that can be computed without reference to unknown or uncertain parameters. In the special case of a diagonal submatrix T_{11} we have

$$\begin{aligned} \tau^* &= M(q)(\ddot{q}_r - T_{11}^{-1}T_{12}\dot{\tilde{q}}) + C(q,\dot{q})(\dot{q} - T_{11}^{-1}B^T T_0\tilde{x}) \\ &+ G(q) - R^{-1}B^T T_0\tilde{x} = \phi\theta + \phi_0 \end{aligned} \tag{15.52}$$

The following adaptation law

$$\dot{\hat{\theta}} = -S^{-1}\phi^T B^T T_0 \tilde{x} \tag{15.53}$$

assures for constant parameters θ that the derivative

$$\dot{V} = -\frac{1}{2}\tilde{x}^T(Q + T_0^T BR^{-1}B^T T_0)\tilde{x} < 0 \tag{15.54}$$

Thus the system is shown to be globally stable in the sense of Lyapunov, and the adaptation eventually optimizes the control system. Simulations of these control principles applied to the robot in Example 7.3 with adaptation with respect to the work load m_2 give results according to Fig. 15.8. This simulation describes a weight-lifting operation from an initial position with all robot arm segments resting on ground, *i.e.* $q_1(0) = q_2(0) = \pi/2$, and the prescribed final position $q_1(\infty) = q_2(\infty) = 0$. ■

Self-tuning minimum variance control

Model-reference adaptive systems are characterized by their gradient search for a set of suitable control parameters. These ideas can be applied also to discrete-time systems, and such direct adaptive control methods can be formulated as alternatives to the indirect adaptive control presented previously. Hence, consider a discrete-time system modelled as

$$\begin{cases} A(z^{-1})y_k = b_0 z^{-1}B(z^{-1})u_k + C(z^{-1})w_k, & \text{Control object} \\ R(z^{-1})u_k = -S(z^{-1})y_k, & \text{Regulator} \end{cases} \tag{15.55}$$

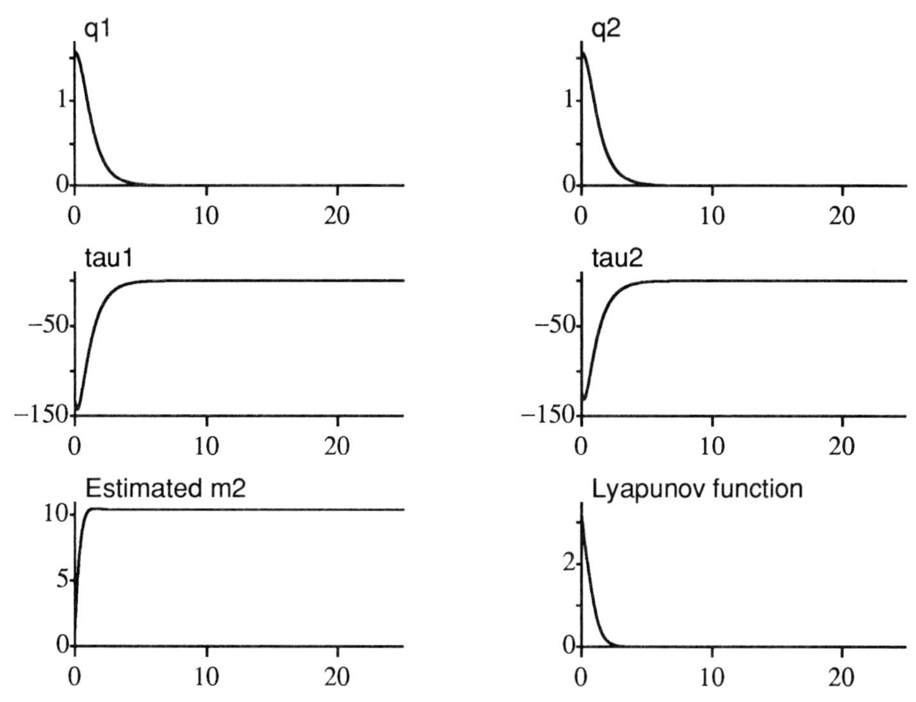

Figure 15.8 Simulation of a robot subject to the self-optimizing adaptive control law. Upper graphs show q_1, q_2, and middle graphs τ_1 and τ_2. The lower left graph shows the estimate $\widehat{\theta}$ of m_2, and the lower right graph the Lyapunov function that decreases everywhere. All graphs versus time [s].

from the input $\{u_k\}$ and the zero-mean normally distributed white-noise process $\{w_k\}$ to the output $\{y_k\}$ with co-prime polynomials

$$\begin{aligned} A(z^{-1}) &= 1 + a_1 z^{-1} + \cdots + a_n z^{-n} \\ B(z^{-1}) &= 1 + b_1 z^{-1} + \cdots + b_{n-1} z^{-n+1} \\ C(z^{-1}) &= 1 + c_1 z^{-1} + \cdots + c_n z^{-n} \end{aligned} \tag{15.56}$$

and where R and S are suitable polynomials in the backward shift operator. The $B-$ and $C-$polynomials are assumed to have no zeros for $|z| > 1$ and the parameter b_0 is a gain factor. The co-primeness of A, B, and C assures that the control object model (15.55) also corresponds to a state-space realization of order n and also the fractional form (see Fig. 15.9)

$$\begin{cases} A(z^{-1})\xi_k = u_k + v_k, \\ y_k = b_0 z^{-1} B(z^{-1})\xi_k + e_k, \end{cases} \quad \text{where} \quad \begin{cases} v_k = \frac{G(z^{-1})}{b_0 B(z^{-1})} w_k \\ e_k = F(z^{-1}) w_k \end{cases} \tag{15.57}$$

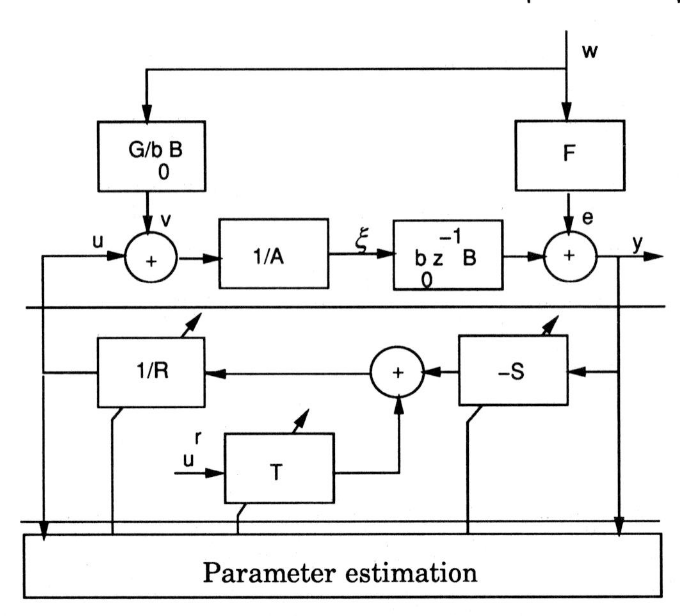

Figure 15.9 Block diagram of the self-tuning regulator with a noise model. Notice that a correctly tuned minimum variance regulator totally decouples ξ from noise interference of w.

with the noise components e_k and v_k for polynomials F and G satisfying the polynomial Diophantine equation

$$AF + z^{-1}G = C \tag{15.58}$$

The use of pole assignment as a design principle suggests that R and S are chosen so that the denominator polynomial $RA + b_0SB$ of closed-loop system matches some prescribed polynomial P, *cf.* Eq. (15.7). Inverse modeling of the output to achieve the pole assignment $RA + b_0SB = P$ can be made by means of the following expansion based on Eq. (15.57) (argument z^{-1} omitted)

$$\begin{aligned} y_{k+1} &= b_0B(z^{-1})\xi_k + e_{k+1} = b_0B\frac{RA + Sb_0z^{-1}B}{P}\xi_k + Fw_{k+1} \\ &= \frac{b_0B}{P}(R(A\xi_k) + S(b_0z^{-1}B\xi_k)) + Fw_{k+1} \\ &= R(\frac{b_0B}{P}u_k) + S(\frac{b_0B}{P}y_k) + \frac{RG - b_0BSF}{P}w_k + Fw_{k+1} \end{aligned} \tag{15.59}$$

This expression is in linear regression form with respect to the feedback parameters of R, S, and the regressors consist of filtered input-output data $\{u_k\}$ and $\{y_k\}$. The corresponding noise model is particularly simple so that the third term disappears for

$$\begin{aligned} P &= BC \\ R &= b_0 BF = b_0 B \\ S &= G = (c_1 - a_1) + \cdots + (c_n - a_n) z^{-n+1} \end{aligned} \tag{15.60}$$

which is the appropriate minimum variance regulator to use in the case of known parameters. Assuming that the input $\{u_k\}$ and the output $\{y_k\}$ are available to measurement, introduce the parameter vector θ and the regressor vector $\{\phi_k\}$ as

$$\begin{aligned} \theta &= \begin{pmatrix} r_1 & \dots & r_{n-1} & s_0 & \dots & s_n \end{pmatrix}^T \\ \phi_k &= \begin{pmatrix} u_{k-1} & \dots & u_{k-n+1} & y_k & \dots & y_{k-n+1} \end{pmatrix}^T \end{aligned} \tag{15.61}$$

Direct minimum variance adaptive control algorithm *ad modum* Åström-Wittenmark (1973) then comprises the following steps.

$$\begin{aligned} \widehat{\theta}_k &= \widehat{\theta}_{k-1} + P_k \phi_{k-1} \varepsilon_k \\ P_k &= P_{k-1} - \frac{P_{k-1}\phi_{k-1}\phi_{k-1}^T P_{k-1}}{1 + \phi_{k-1}^T P_{k-1} \phi_{k-1}} \\ \varepsilon_k &= y_k - \beta u_{k-1} - \widehat{\theta}_{k-1}^T \phi_{k-1} \\ u_k &= -\frac{1}{\beta}(\widehat{\theta}_k^T \phi_k) \end{aligned} \tag{15.62}$$

where the vector of estimated parameters $\widehat{\theta}$ has replaced the parameters θ of the correct desired control law, whereas β_0 is a fixed *a priori* estimate of the gain factor b_0. Convergence analysis of $\widehat{\theta}$ and the output error can be made by means of the function

$$\begin{aligned} V_k &= -\log p(\widetilde{\theta}_k | \mathcal{F}_k) + \log(1 + \frac{1}{\sigma^2} x_k^T Q x_k) \\ &= \frac{n}{2}\log 2\pi + \frac{1}{2}\log\det P_k + \frac{1}{2}\widetilde{\theta}_k^T P_k^{-1} \widetilde{\theta}_k + \frac{1}{2}\log(1 + \frac{1}{\sigma^2} x_k^T Q x_k) \end{aligned} \tag{15.63}$$

The first term of the function V_k consists of the log-likelihood function for the parameter error $\widetilde{\theta}_k$ and the second term is the logarithm of a signal-to-noise ratio with

$$x_k = \begin{pmatrix} \xi_{k-1} & \xi_{k-2} & \dots & \xi_{k-2n+1} \end{pmatrix}^T \tag{15.64}$$

and $\sigma^2 = \mathcal{E}\{w_k^2\}$. In information theory, the mathematical expectation of V_k is known as the system *entropy*, which often is interpreted as a measure of disorder of the system. It can be shown that the entropy function V_k develops as a supermartingale—*i.e.*,

$$V_k \geq \mathcal{E}\{V_{k+1}|\mathcal{F}_k\} \quad \text{a.s.} \tag{15.65}$$

for

$$\frac{1}{C(z^{-1})} - \frac{1}{2} > 0, \quad \text{and} \quad 0 < \frac{b_0}{\beta} < 2 \tag{15.66}$$

As V_k is not positive definite with respect to P_k, it does not qualify as a Lyapunov function. However, the interpretation of V_k as a log-likelihood function admits the conclusion that the parameters converge toward the minimum-variance parameters and that the state vector x_k decreases in magnitude so that $||x_k|| \to 0$.

Example 15.6—Self-tuning minimum variance control
Consider the following ARMAX model with input $\{u_k\}$, output $\{y_k\}$, and the zero-mean white-noise sequence $\{w_k\}$.

$$y_{k+1} = -ay_k + u_k + w_{k+1} + cw_k \tag{15.67}$$

which can be expressed as the following state-space model

$$\begin{aligned} x_{k+1} &= -ax_k + u_k + (c-a)w_k \\ y_k &= x_k + w_k \end{aligned} \tag{15.68}$$

Minimum variance adaptive control applied to this system can be done by means of the algorithm

$$\begin{aligned} \phi_k &= y_k \\ \varepsilon_k &= y_k \\ p_{k+1} &= p_k - \frac{p_k\phi_k\phi_k p_k}{1+\phi_k p_k \phi_k} \\ \widehat{\theta}_{k+1} &= \widehat{\theta}_k + p_k\phi_k\varepsilon_k \\ u_k &= -\widehat{\theta}_k\phi_k \end{aligned} \tag{15.69}$$

The result of a simulation is shown in Fig. 15.10, which shows parameter convergence toward the minimum variance control parameter $\theta = 1.6$ and convergence of $|x_k|$ toward zero. ■

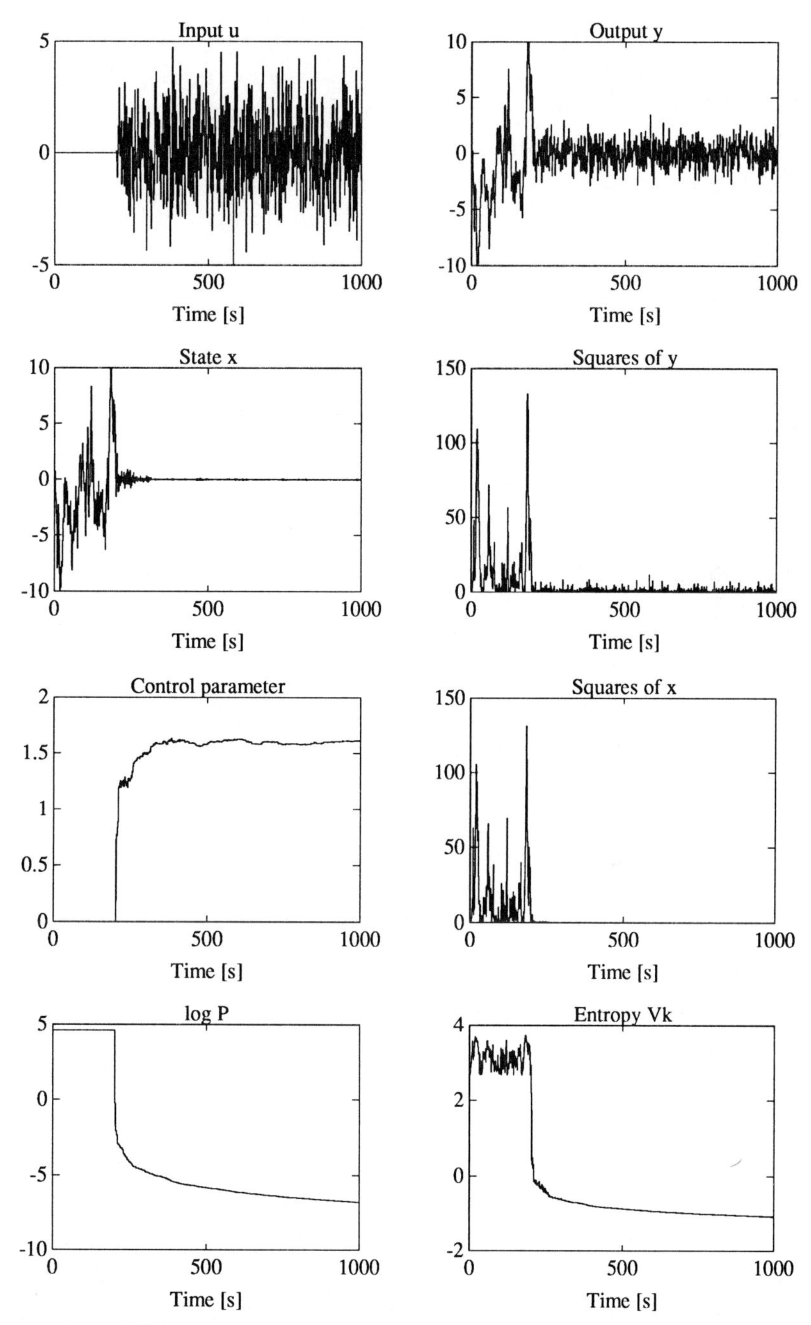

Figure 15.10 Minimum variance adaptive control starting at time $t = 200$ for a first-order ARMAX-model $y_k = 0.9y_{k-1} + u_{k-1} + w_k + 0.7w_{k-1}$. Notice that the state $\{x_k\} = \{y_k\} - \{w_k\}$ eventually becomes noise-free as a result of the adaptive control.

Stability and robustness

Practical application of adaptive system theory requires for reliability reasons some kind of stability proof. Another important aspect is *robustness*, which, roughly speaking, means the ability to tolerate "small" deviations from the model class assumed during the algorithm design. In fact, engineering application of adaptive techniques has been hampered by severe criticism concerning its properties of closed-loop stability and robustness. One aspect of stability that has attracted much attention is that adaptive state feedback control of the type presented in Eq. (15.39) requires measurement of the full state vector for its implementation. A natural question concerns whether measurement of some output variable might suffice. Unfortunately, there is a restrictive answer to this question given in the following theorem and example.

Theorem 15.1—Yakubovich-Kalman
Let the pair (A, b) be controllable. Then there exist positive definite matrices Q and P, a vector w, and a scalar γ solving the equations

$$\begin{aligned} Q + PA + A^T P &= ww^T \\ -c + Pb &= \gamma w \\ \kappa &= \gamma^2 \end{aligned} \tag{15.70}$$

if and only if the transfer function

$$G(s) = \gamma + w^T (sI - A)^{-1} b \tag{15.71}$$

is such that

$$\text{Re} \quad G(i\omega) \geq 0, \qquad \omega \in R \tag{15.72}$$

■

Example 15.7—Adaptive control by means of output feedback
Assume that the complete state vector x is not available to measurement but only some output vector $y = c^T x$.

$$\dot{x} = Ax + Bu = (A - B\theta^T c^T)x - Bx^T c\tilde{\theta} \tag{15.73}$$

Consider the Lyapunov function candidate

$$V(x, \theta) = \frac{1}{2} x^T P x + \frac{1}{2} \tilde{\theta}^T S \tilde{\theta} \tag{15.74}$$

where P solves the Lyapunov function

$$\begin{aligned} Q + PA_m + A_m^T P &= 0 \\ Pb &= c \\ \gamma &= 0 \end{aligned} \tag{15.75}$$

for $Q = Q^T > 0$ and $Pb = c$ as obtained from the Yakubovich-Kalman lemma. A Lyapunov function similar to Eq. (15.42) has the time derivative

$$\begin{aligned} \frac{dV}{dt} &= \frac{1}{2}x^T(A_m^T P + PA_m)x + x^T PBx^T c\tilde{\theta} + \tilde{\theta}^T S\dot{\tilde{\theta}} \\ &= -\frac{1}{2}x^T Qx \end{aligned} \tag{15.76}$$

for the algorithm

$$\begin{aligned} \dot{\hat{\theta}} &= S^{-1}yy^T = S^{-1}yB^T Px \\ u &= -\hat{\theta}^T y \end{aligned} \tag{15.77}$$

This example demonstrates that the Yakubovich-Kalman is a restrictive theorem for the application of output feedback adaptive control as it is required that the $Y(s)/U(s) = G(s) = c^T(sI - A)^{-1}b$ is such that $G(i\omega) \geq 0$ for all $\omega \in R$—*i.e.*, where no other linear combination of x than $y = c^T x$ is available to measurement. ■

As yet there are few effective and systematic means to design adaptive control for systems that do not fulfill the "positive real" condition of the Yakubovich-Kalman theorem. In practice, this means that full state information is required in order to apply continuous-time adaptive feedback control.

Another question is whether the assumptions made on model order are restrictive for adaptive control methods. Unfortunately, there is an affirmative answer also to this question, which is elaborated in the following example adapted from Rohrs† and co-workers.

Example 15.8—Stability problem of adaptive control

Assume that the control object is adequately described by the transfer function

$$Y(s) = \frac{2}{s+1} \cdot \frac{229}{s^2 + 30s + 229} U(s) \tag{15.78}$$

† This example is based, with permission, on the paper of C.E. Rohrs, L. Valavani, M. Athans, and G. Stein. "Robustness of adaptive control algorithms in the presence of unmodeled dynamics." *Proc. IEEE Conf. Decision and Control*, Orlando, Florida, 1982.©1982 IEEE.

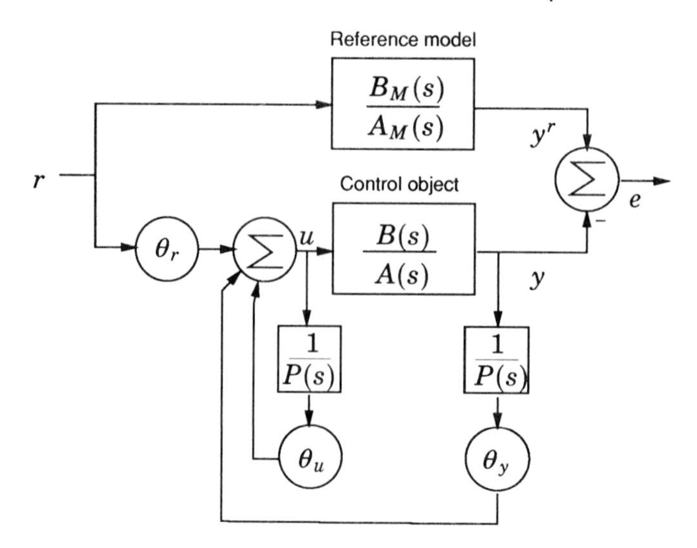

Figure 15.11 An adaptive Rohrs' system exhibiting unstable behavior in response to a sinusoidal reference signal.

whose transient responses are in good agreement with the behavior of a simplified model

$$\frac{Y(s)}{U(s)} \approx G_1(s) = \frac{2}{s+1} \tag{15.79}$$

Furthermore, assume that a suitable reference model would be

$$Y^r(s) = \frac{3}{s+3} R(s) \tag{15.80}$$

A simple proportional control law based on Eq. (15.79) that matches the reference model Eq. (15.80) is

$$u(t) = -\theta_y y(t) + \theta_r r(t) = -y(t) + 1.5r(t) \tag{15.81}$$

whereas it gives the following closed-loop system when applied to the original transfer function Eq. (15.78)

$$\frac{Y(s)}{R(s)} = \frac{687}{s^3 + 31s^2 + 259s + 687} \tag{15.82}$$

The reference signal was chosen as

$$r(t) = 0.3 + 1.85 \sin \omega_o t, \qquad \omega_o = 16.1 \quad [rad/s] \tag{15.83}$$

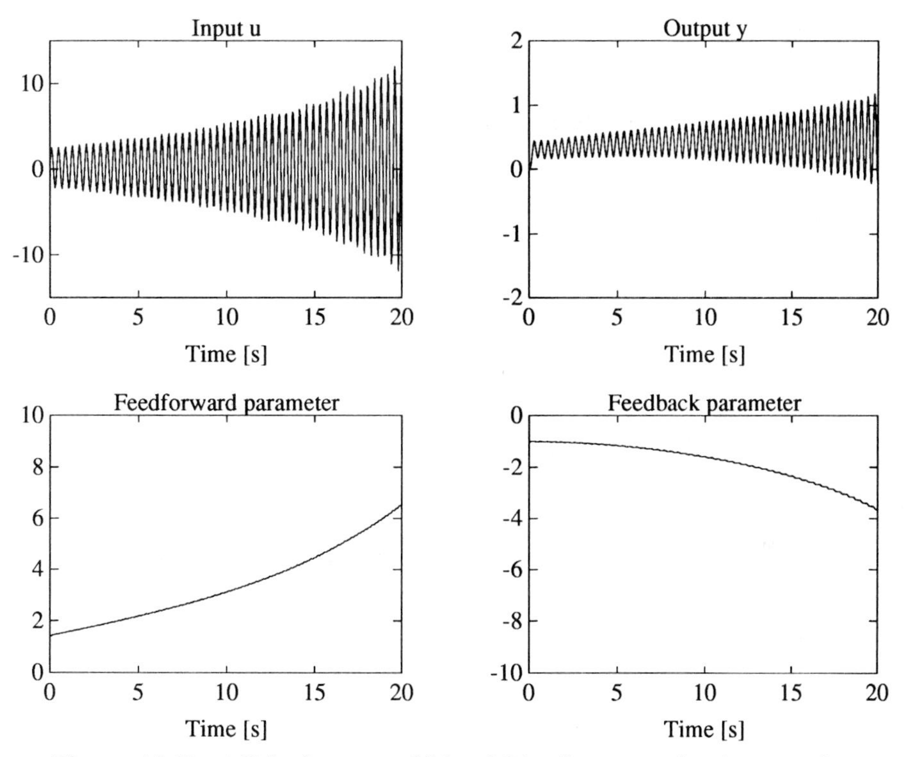

Figure 15.12 A Rohrs' system which exhibits divergent adaptive control.

with ω_o being the frequency of 180° phase lag of the closed-loop system Eq. (15.82). A simple adaptive control system (see Fig. 15.11) based on gradient search and designed for a first-order control object uses the following algorithm

$$
\begin{aligned}
e(t) &= y(t) - y^r(t) && \text{control error} \\
\phi(t) &= \begin{pmatrix} r(t) & y(t) \end{pmatrix}^T && \text{regression vector} \\
\theta(t) &= \begin{pmatrix} \theta_r & \theta_y \end{pmatrix}^T && \text{parameter vector} \\
\dot{\widehat{\theta}} &= \Gamma\phi(t)e(t) && \text{adaptation} \\
u &= \widehat{\theta}^T\phi && \text{control law}
\end{aligned}
$$

It is, unfortunately, obvious from the simulation in Fig. 15.12 that the system has a divergent response to the reference signal. This behavior is reproducible even when starting at the correct initial parameter estimates $\widehat{\theta}_r(0) = 1.5$ and $\widehat{\theta}_y = 1$. ■

The example shows that heuristic model reduction in the context of adaptive control might produce very poor results. It also shows that gradient optimization methods are susceptible to correlated disturbances. This example is, of course, relevant also for neural network theory, a fact which is sometimes overlooked.

15.6 MULTIVARIABLE DIRECT ADAPTIVE CONTROL

The formulation of multi-input, multi-output adaptive control gives rise to some specific mathematical problems. One issue is the system representation problems associated with cross-coupling and system zeros of multivariable systems. A related problem is the parametrization problem. It is, for instance, not trivial to extend notions of gain from single-input, single-output systems to multi-input, multi-output systems.

Consider a time-invariant, finite-dimensional, linear and causal control object with input $\{u_k\}$ and output $\{y_k\}$ and the system equation

$$\begin{aligned} A(z^{-1})\xi_k &= u_k \\ y_k &= B(z^{-1})\xi_k \end{aligned} \quad \text{with} \quad \begin{cases} y_k \in R^n \\ u_k \in R^n \\ \xi_k \in R^n \end{cases}, \quad \text{and} \quad \begin{cases} \text{rank } A(z^{-1}) = n \\ \text{rank } B(z^{-1}) = n \end{cases} \tag{15.84}$$

It is assumed that A and B are right co-prime polynomial matrices, *i.e.*, A and B have no common right factors except for unimodular matrices, and that A and B have no common zeros for $|z| \geq 0$. The transfer operator $H_0(z) = B(z^{-1})A^{-1}(z^{-1})$ from u_k to y_k is assumed to be strictly proper with respect to the forward shift operator z and it is assumed that the control object is stabilizable from the control input u_k.

Consider a linear, causal, and finite-dimensional regulator of the type

$$R_u(z^{-1})u_k = -S_y(z^{-1})y_k + T(z^{-1})r_k \tag{15.85}$$

which generates a control signal u_k by means of feedback of y_k and a reference signal r_k. The closed-loop system obtained from Eq. (15.84) and Eq. (15.85) has the transfer operator relationship

$$y_k = B(z^{-1})\Big(R_u(z^{-1})A(z^{-1}) + S_y(z^{-1})B(z^{-1})\Big)^{-1} T(z^{-1})r_k \tag{15.86}$$

which should be designed to be stable and able to follow the reference signal r_k. The stability requirement implies also internal stability which, in turn, implies that the closed-loop system must not have any uncontrollable or unobservable unstable mode. It is therefore necessary that the polynomial matrices $R_u(z^{-1})$ and $S_y(z^{-1})$ have no common zeros in the unstable region of the z–plane, *i.e.*, for $|z| \geq 1$.

Consider the closed-loop system in Eq. (15.86). It is clear already from the single-input single-output system that attempts to cancel the zeros for $|z| \geq 0$ of the control object will result in unstable modes. An admissible regulator will therefore result in a closed-loop system of the type

$$y_k = B_{LS}(z^{-1})T_R(z^{-1})r_k \tag{15.87}$$

where $B_{LS}(z^{-1})$—the *left structure matrix*— represents the part of the $B(z^{-1})$-matrix which is noninvertible from the right by a control law (15.85) and thus invariant; see (Pernebo, 1981). Choice of the matrix $T_R(z^{-1})$, however, is not further restricted by such stability considerations as long as it is chosen as a stable and causal rational matrix.

The zeros for $|z| \geq 0$ of a system may be described by the Smith form (see Appendix A)

$$B(z^{-1}) = U(z^{-1})\Sigma(z^{-1})V(z^{-1}) \tag{15.88}$$

Here the Σ-matrix contains the invariant polynomials on the diagonal with all nondiagonal elements of $\Sigma(z^{-1})$ being zero whereas the unimodular polynomial matrices U and V have stable and causal inverses. The left structure matrix $B_{LS}(z^{-1})$ can thus be represented by the product of the polynomial matrices $U(z^{-1})$ and $\Sigma(z^{-1})$. The inclusion of $U(z^{-1})$ in the left structure matrix derives from the fact that the regulator identities in (15.85) only operate from the right. The left structure matrix $B_{LS}(z^{-1})$ thus contains information about the cross-couplings and the system zeros for $|z| \geq 1$. Therefore, any reference model to be followed perfectly must contain some representation of this system invariant. Therefore, let the class of reference models be restricted to models of the form

$$\begin{aligned} A_M(z^{-1})r_k &= B_M(z^{-1})u_{c_k} \\ y_k^m &= B_{LS}(z^{-1})T_R(z^{-1})r_k \end{aligned} \tag{15.89}$$

where u_{c_k} is some command signal that generates a filtered reference signal r_k and where y_k^m is the model output to be tracked perfectly. The compensator $T_R(z^{-1})$ can be chosen without restrictions as long as it is stable and the matrices $A_M(z^{-1})$ and $B_M(z^{-1})$ should be left co-prime polynomial matrices for $|z| \geq 1$. The factorization between $A_M(z^{-1})$ and $B_M(z^{-1})$ is usually chosen such that $A_M(0) = I_{nxn}$.

As prior knowledge of cross-couplings and zeros becomes unrealistic to assume in the context of adaptive control, it appears necessary to formulate a method that allows $B_{LS}(z^{-1})$ to be incorporated without actually knowing it. This problem can be solved by certain choices of the compensator $T_R(z^{-1})$ so that $B_{LS}(z^{-1})T_R(z^{-1})$ becomes less complicated than $B_{LS}(z^{-1})$ itself. The dynamic decoupling problem would then include the problem of finding a stable $T_R(z^{-1})$ such that

$$B_{LS}(z^{-1})T_R(z^{-1}) = B_D(z^{-1}) \quad \text{with } B_D(1) = I_{n \times n} \tag{15.90}$$

for some feasible, polynomial matrix $B_D(z^{-1})$. In order to fulfill the desired input-output relationship and to eliminate cross-couplings it would be suitable to choose $T_R(z^{-1})$ in such a way that $B_D(z^{-1})$ becomes a diagonal matrix. A problem formulation in terms of model reference adaptive control would then include the problem of finding $T_R(z^{-1})$ adaptively so that B_{LS} need not be known.

Lemma 15.1
A square $n \times n$ strictly proper transfer operator $H_0(z)$ of full rank and with $\det H_0(1) \neq 0$ may be decomposed into a right co-prime factorization ($A(z^{-1})$, $B(z^{-1})$) for $|z| \geq 1$, *i.e.*, rank $\left(A^T(z^{-1}), B^T(z^{-1})\right)$ for $|z| \geq 1$. The factorization is such that $A(z^{-1})$ contains all the poles of $H_0(z)$ and $B(z^{-1})$ contains all zeros of $H_0(z)$ for $|z| \geq 0$ and

$$H_0(z) = B(z^{-1})A^{-1}(z^{-1}) = B_L(z^{-1})B_S(z^{-1})B_R(z^{-1})A^{-1}(z^{-1}) \tag{15.91}$$

where A is a square, full-rank polynomial matrix and B_L, B_R are polynomial matrices with stable and causal inverses. The diagonal polynomial matrix B_S is satisfying the additional requirements that $A(0) = I_{n \times n}$, $B_S(1) = I_{n \times n}$, and that $B_R(0) = I_{n \times n} + R_0$ is upper right triangular and invertible. The diagonal polynomial matrix $B_S(z^{-1})$ has entries which contain no polynomial factors except for the zeros at infinity and the finite zeros for $|z| \geq 1$ of $H_0(z)$.

Proof: See Johansson (1987). ■

In control design based on pole assignment it is customary to use a polynomial factor $A_0(z^{-1})$ with the interpretation of an observer polynomial. An important property of such a factor is that it can be used in order to modify the control behavior although it does not appear explicitly in the closed-loop transfer function between reference r_k and output u_k. Introduce for this reason a polynomial matrix $A_0(z^{-1})$ with no poles or zeros for $|z| \geq 0$ and with a stable and causal inverse. Pole-assignment design can now be approached by means of the following theorem:

Theorem 15.2
The transfer operator of the closed-loop system (15.86) and the reference model are identical if R_u, S_y, and T are chosen from the polynomial matrix

solutions to the following polynomial equations for given diagonal matrices B_S and B_D.

$$\begin{aligned} T_2 A_0 A_M B_L B_S &= B_S A_0 A_M \\ R_u A + S_y B &= A_0 A_M B_R \\ B_S T &= T_2 A_0 A_M B_D \end{aligned} \tag{15.92}$$

Proof: For proof of existence of these solutions refer to (Johansson 1987). The resulting closed-loop transfer operator from reference r_k to output y_k is then

$$\begin{aligned} H &= B(R_u A + S_y B)^{-1} T = B_L B_S B_R (A_0 A_M B_R)^{-1} T \\ &= (T_2 A_0 A_M)^{-1} B_S (A_0 A_M)(A_0 A_M)^{-1} T = (T_2 A_0 A_M)^{-1} B_S T \\ &= (T_2 A_0 A_M)^{-1} (T_2 A_0 A_M) B_D = B_D \end{aligned} \tag{15.93}$$

which reproduces the requested input-output behavior in Eq. (15.90) for the choice $T_R = (A_0 A_M)^{-1} T$. The calculations involved in Eq. (15.93) are possible as A_0, A_M, and T_2 are stable and causal polynomial matrices with stable and causal inverses. ■

An estimation model

Consider the controlled outputs (argument z omitted)

$$\begin{aligned} y_k &= B(A_0 A_M B_R)^{-1} (R_u A + S_y B) \xi_k \\ &= B_L B_S (A_0 A_M)^{-1} (R_u A + S_y B) \xi_k \end{aligned} \tag{15.94}$$

Application of the equations of Lemma 15.1 gives the relationship

$$T_2 A_0 A_M y_k = B_S (R_u u_k + S_y y_k) \tag{15.95}$$

Now define (argument z^{-1} omitted)

$$e_k^f = A_0 A_M e_k = A_0 A_M (y_k - y_k^m) = A_0 A_M (y_k - B_D r_k) \tag{15.96}$$

Substitution of these relationships into Eq. (15.95) gives the linear estimation model

$$T_2(z^{-1}) e_k^f = B_S(z^{-1}) \Big(u_k + R_0 u_k + R_1(z^{-1}) u_k + S_y(z^{-1}) y_k - T(z^{-1}) r_k \Big) \tag{15.97}$$

in the unknown R_0, $R_1 = R_u - R_u(0)$, S_y, T_2, T and the input-output data e_k^f, u_k, y_k, r_k. Filtering by means of $B_S(z^{-1})$ provides the suitable regressor elements required for linear estimation. Notice that the diagonal matrix

$B_S(z^{-1})$ contains only the time delays and other non-invertible system zeros of the control object.

Example 15.9—Multivariable direct adaptive control
Consider the multi-input, multi-output system

$$Y(z) = B(z^{-1})A(z^{-1})^{-1}U(z) = \begin{pmatrix} b_1 z^{-3} & b_2 z^{-1} \\ 0 & b_3 z^{-2} \end{pmatrix} \begin{pmatrix} 1 & a_3 z^{-1} \\ a_2 z^{-1} & 1 - a_1 z^{-1} \end{pmatrix}^{-1} U(z) \tag{15.98}$$

Some interesting cross-coupling features appear when b_2 is nonzero, in which case it appears difficult to achieve independent control of the two outputs. However, a decomposition of $B(z^{-1})$ according to Lemma 15.1 gives

$$B(z^{-1}) = \begin{pmatrix} 0 & b_2 \\ -\frac{b_1 b_3}{b_2} & b_3 z^{-1} \end{pmatrix} \begin{pmatrix} z^{-4} & 0 \\ 0 & z^{-1} \end{pmatrix} \begin{pmatrix} 1 & 0 \\ \frac{b_1}{b_2} z^{-2} & 1 \end{pmatrix} \tag{15.99}$$

A feasible diagonal matrix $B_D(z^{-1})$ which can be obtained by diagonalization of B from the right is

$$B_D(z^{-1}) = \begin{pmatrix} z^{-3} & 0 \\ 0 & z^{-4} \end{pmatrix} \tag{15.100}$$

Pole assignment to $A_0(z^{-1})A_m(z^{-1}) = I_{2\times 2}$ yields

$$T_2(z^{-1}) = \begin{pmatrix} \frac{1}{b_1} z^{-1} & -\frac{b_2}{b_1 b_2} \\ \frac{1}{b_2} & 0 \end{pmatrix} \tag{15.101}$$

and

$$T(z^{-1}) = \begin{pmatrix} \frac{1}{b_1} & -\frac{b_2}{b_1 b_2} \\ \frac{1}{b_2} z^{-2} & 0 \end{pmatrix} \tag{15.102}$$

A solution of the Diophantine equation gives

$$R_u(z^{-1}) = \begin{pmatrix} 1 + \frac{a_3 b_1}{b_2} z^{-3} & 0 \\ -a_2 z^{-1} + \frac{b_1}{b_2} z^{-2} + \frac{a_1 b_1}{b_2} z^{-3} + \frac{a_2 a_3 b_1}{b_2} z^{-4} + a_3 \frac{b_1^2}{b_2^2} z^{-5} & 1 \end{pmatrix} \tag{15.103}$$

and

$$S_y(z^{-1}) = \begin{pmatrix} -\frac{a_3}{b_2} & -\frac{a_3^2 b_1}{b_2 b_2} z^{-2} \\ \frac{a_1}{b_2} + \frac{a_2 a_3}{b_2} z^{-1} - \frac{b_1 a_3}{b_2^2} z^{-2} & \frac{b_1 a_3}{b_2 b_3} z^{-2} + \frac{b_1 a_2 a_3^2}{b_2 b_3} z^{-3} - \frac{a_3^2 b_1^2}{b_2^2 b_3} z^{-4} \end{pmatrix} \tag{15.104}$$

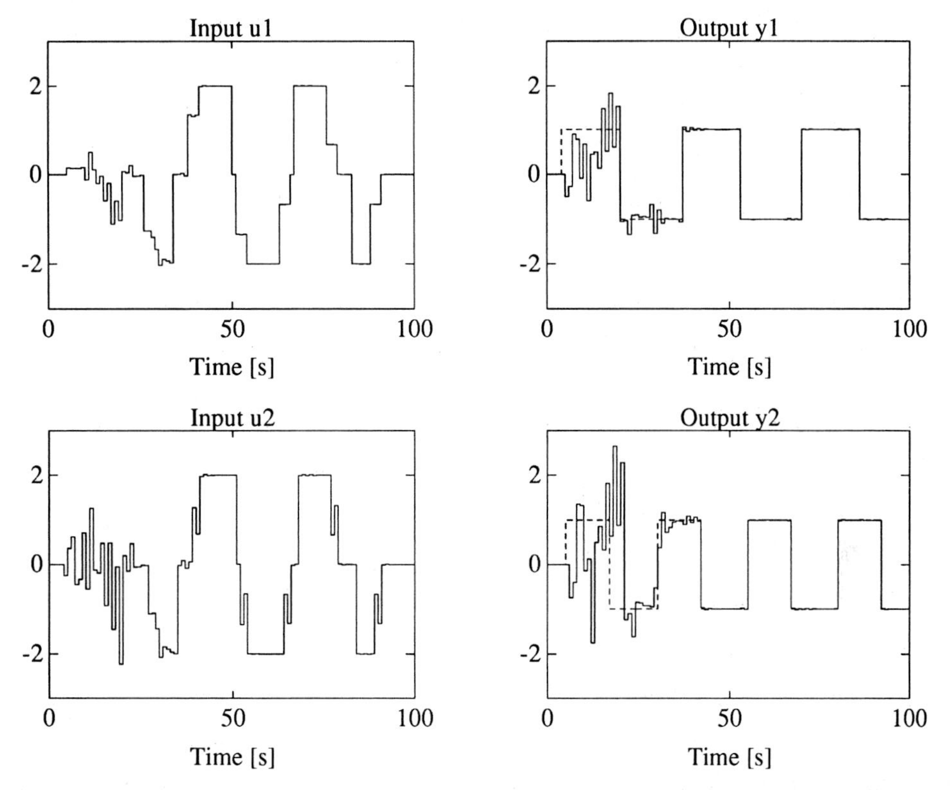

Figure 15.13 Multivariable direct adaptive control of a discrete-time two-input, two-output system with strong cross-coupling properties. The outputs y_1 and y_2 (*solid line*) reproduce the reference signals y_1^m and y_2^m (*dashed line*) after a transient of adaptation. Notice also that the interaction between the outputs disappears after the transient of adaptation. The cross-coupling is visible in the inputs throughout the simulation.

The input estimation model for the inputs u_1 and u_2 applicable at time k is

$$
\begin{aligned}
u_{1_{k-4}} &= \begin{pmatrix} -u_{1_{k-7}} & -y_{1_{k-4}} & -y_{2_{k-6}} & y^m_{1_{k-4}} & y^m_{2_{k-4}} \end{pmatrix} \theta_{11} + \begin{pmatrix} e_{1_{k-1}} & e_{2_k} \end{pmatrix} \theta_{12} \\
u_{2_{k-1}} &= \begin{pmatrix} \bar{\phi}^T_{2u} & \bar{\phi}^T_{2y} & \bar{\phi}^T_{2y^m} \end{pmatrix} \theta_{21} + e_{1_k} \theta_{22}
\end{aligned}
\tag{15.105}
$$

where

$$
\begin{aligned}
\bar{\phi}_{2u} &= \begin{pmatrix} -u_{1_{k-2}} & -u_{1_{k-3}} & -u_{1_{k-4}} & -u_{1_{k-5}} & -u_{1_{k-6}} \end{pmatrix}^T \\
\bar{\phi}_{2y} &= \begin{pmatrix} -y_{1_{k-1}} & -y_{1_{k-2}} & -y_{1_{k-3}} & -y_{2_{k-3}} & -y_{2_{k-4}} & -y_{2_{k-5}} \end{pmatrix}^T \\
\bar{\phi}_{2y^m} &= y^m_{1_{k-3}}
\end{aligned}
\tag{15.106}
$$

and where θ_{11}, θ_{12}, θ_{21}, θ_{22} contains the coefficients of R_u, S_y, T, and T_2. The correct pole-assignment control law would be

$$\begin{aligned} u_{1_k} &= \begin{pmatrix} -u_{1_{k-3}} & -y_{1_k} & -y_{2_{k-2}} & y^m_{1_k} & y^m_{2_k} \end{pmatrix} \theta_{11} \\ u_{2_k} &= \begin{pmatrix} \phi^T_{2u} & \phi^T_{2y} & \phi^T_{2y^m} \end{pmatrix} \theta_{21} \end{aligned} \tag{15.107}$$

where

$$\begin{aligned} \phi_{2u} &= \begin{pmatrix} -u_{1_{k-1}} & -u_{1_{k-2}} & -u_{1_{k-3}} & -u_{1_{k-4}} & -u_{1_{k-5}} \end{pmatrix}^T \\ \phi_{2y} &= \begin{pmatrix} -y_{1_k} & -y_{1_{k-1}} & -y_{1_{k-2}} & -y_{2_{k-2}} & -y_{2_{k-3}} & -y_{2_{k-4}} \end{pmatrix}^T \\ \phi_{2y^m} &= y^m_{1_{k-2}} \end{aligned} \tag{15.108}$$

where parameter estimates $\widehat{\theta}_{11}$ and $\widehat{\theta}_{21}$ enter instead of θ_{11} and θ_{21} in the case of adaptive control. A simulation of adaptive control based on recursive least-squares identification of this system is provided in Fig. 15.13 for parameters

$$\begin{pmatrix} a_1 & a_2 & a_3 & b_1 & b_2 & b_3 \end{pmatrix} = \begin{pmatrix} 2 & 1 & -1 & 1 & 2 & 3 \end{pmatrix} \tag{15.109}$$

The A-parameters result in the characteristic polynomial $\det A(z^{-1}) = 1 - a_1 z^{-1} - a_2 a_3 z^{-2} = (1 - z^{-1})^2$. It is obvious that the multivariable adaptive control achieves the desired closed-loop properties and that the control effectively eliminates cross-coupling between the outputs. ■

15.7 DISCUSSION AND CONCLUSIONS

All the examples of adaptive systems have as a common feature the use of negative feedback as a principle to control or estimate necessary parameters. The principle of negative feedback originating from control systems analysis is used to control the output error signal, *i.e.*, deviation of the system from prescribed operation. The terminology used in the theory of artificial neural network is that *learning* or *supervised learning* or *adaptation* is an adaptive process that incorporates an external “teacher” or some global information with indications of correctness and performance. Supervised learning includes decisions when to turn off adaptation and how to supply *a priori* information. As a constrast there also exist other methods called *self-learning* or *unsupervised learning* or *self-organization*, which have no reference to a feedback principle or a "teacher" but, instead, rely upon internal criteria of correctness.

In the case of adaptive control there are two interacting feedback loops. It has been suggested that a properly organized feedback control loop might eliminate the need for parameter estimation. In fact, it was early observed that the output error and the sensitivity to parameter errors can be drastically reduced by using strong feedback in feedback control loop. Strong feedback can be imposed by increasing the feedback gain parameters ("high gain") or by using certain forms of relay feedback ("sliding modes"). There are, however, two fundamental limitations to the use of strong feedback control. First, the sensitivity to disturbances in the measured output and, second, stability problems caused by the presence of unmodeled time delays (see Section 8.5).

Indeed, there are control methods which systematically try to exploit the reduction of parameter sensitivity associated with high-gain feedback. This approach can be very successful in applications with a maximum transfer function phase shift of 90° over the frequency range. For larger phase shifts this is no possible control principle but—as an extension—it is possible to establish limit-cycle oscillations by introducing a saturating high-gain nonlinearity in the feedback loop—*e.g.*, a relay or a function $f(x) = b \tanh ax$. Limit-cycle oscillations may be characterized in terms of their period and maximum amplitude which, in turn, may provide sufficient information for calculation of gain and phase shift in the associated feedback loops. This idea has been practiced in response to the need for simple control methods and for identification and adaptive control of systems with a large range of gain variation. The resulting class of adaptive systems may exhibit low sensitivity to gain changes, but they may often have a certain sensitivity to external disturbances being amplified.

An important difference between adaptive system theory and identification theory is that stationary stochastic processes are not quite adequate for the analysis of adaptive processes. As assumptions of stationary behavior are irrelevant for adaptive processes, it is often difficult to characterize adaptation. An *ad hoc* method to approach this problem is to apply recursive estimation to the time series and to use the sequence of estimated parameter as a characterization. Additional difficulties arise when the assumptions on the model class are not met in application. As there are no means for interactive modification of model class assumptions or validation in an autonomous system, this should impose several limitations on the use of adaptive techniques.

In experimental work there are often various problems associated with adaptive systems encountered; for example, in biology. As the object of investigation changes owing to adaptation in the course of an experiment, it may be very difficult to establish reproducible experimental conditions. It should be considered as an open problem the question of how to appropriately characterize adaptive processes in experimental work.

15.8 BIBLIOGRAPHY AND REFERENCES

Some early references on adaptive processes include

- W. McCulloch and W. Pitts, "A logical calculus of the ideas immanent in nervous activity." *Bulletin of Mathematical Biophysics*, Vol. 7, 1943, pp. 115–133.
- F. Rosenblatt, "The perceptron—a probabilistic model for information storage and organization in the brain." *Psychological Review*, Vol. 65, 1958, pp. 386–408.
- R.Bellman, *Adaptive Control Processes: A Guided Tour*, Princeton, NJ: Princeton University Press, 1961.
- M.A. Arbib, *The Metaphorical Brain: Neural Networks and Beyond*, 2d ed., New York: John Wiley, 1989.

Back propagation as a means to extend the use of perceptron-like devices is described in

- D.E. Rumelhart and J.L. McClelland, *Parallel Distributed Processing, Vol. 1.* Cambridge, MA: MIT Press, 1986.
- P.J. Werbos, *Beyond Regression: New Tools for Prediction and Analysis in the Behavioral Sciences*, Cambridge, MA: Harvard University Press, 1974.

The relationship between optimal control theory and back propagation in neural networks is described in:

- T. Parisini and R. Zoppoli, "Neural nets for the solution of N-stage optimal control problems." In T. Kohonen *et al.* (Eds.), *Artificial Neural Networks*, New York: Elsevier Science Publishers, 1991.

Self-tuning minimum variance adaptive control was introduced in

- K.J. Åström and B. Wittenmark, "On self-tuning regulators," *Automatica*. Vol. 9, 1973, pp. 185–199.
- D. Clarke and P. Gawthrop, "Self-tuning control." *Proc. IEE, Vol. 122 A*, 1975, pp. 929–934.

Adaptive linear elements, least means square estimation (LMS), and their use in adaptive signal processing are described in

- B. WIDROW AND S.D. STEARNS, *Adaptive Signal Processing.* Englewood Cliffs, NJ: Prentice-Hall, 1985.
- S.CHEN, S.A.BILLINGS, AND P.M.GRANT, "Nonlinear system identification using neural networks." *Int. J. Control*, Vol. 51, 1990, pp. 1191–1214.

The use of limit-cycle oscillations in adaptive processes is described in

- J.H. BLAKELOCK, *Automatic Control of Aircraft and Missiles*, New York: John Wiley, 1965.
- W.W. SOLODOWNIKOW, *Nichtlineare und selbsteinstellende Systeme.* Berlin: VEB Verlag Technik, 1975.

The following book covers a broad range of topics on adaptive systems including filtering and control, pattern recognition, game theory, and operations research.

- YA. Z. TSYPKIN, *Adaptation and Learning in Automatic Systems.* New York: Academic Press, 1971.

Texts describing some of the recent research focus in the adaptive control area are

- K.J. ÅSTRÖM AND B. WITTENMARK, *Adaptive Control.* Reading, MA: Addison-Wesley, 1989.
- G.C. GOODWIN AND K.S. SIN, *Adaptive Filtering, Prediction and Control.* Englewood Cliffs, NJ: Prentice-Hall, 1984.

Example 15.5 on adaptive control of robot manipulators is adapted from

- R. JOHANSSON, "Quadratic optimization of motion coordination and control." *IEEE Trans. Automatic Control*, Vol. 35, 1990, pp. 1197–1208.

Example 15.8 follows the article by

- C.E. ROHRS, L. VALAVANI, M. ATHANS, AND G. STEIN. "Robustness of adaptive control algorithms in the presence of unmodeled dynamics." *Proc. IEEE Conf. Decision and Control*, Orlando, Florida, 1982.

Section 15.6 on multivariable adaptive control is based on the article

- R. JOHANSSON, "Parametric models of linear multivariable systems for adaptive control." *IEEE Trans. Automatic Control*, Vol. AC-32, 1987, pp. 303–313.

Structure matrix concepts for analysis of multi-input, multi-output systems were introduced in

– L. PERNEBO, "An algebraic theory for the design of controllers for linear multivariable systems. Part I: Structure matrix and feedforward design. Part II: Feedback realizations and feedback design." *IEEE Trans. Automatic Control*, Vol. AC-26, 1981, pp. 171–193.

APPENDIX 15.1

This appendix reviews some fundamental results from Lyapunov theory of stability. A useful interpretation of the Lyapunov function $V(x)$ is sometimes that of the energy or variance associated with a variable x.

Theorem 15.3—Lyapunov
Let

$$\dot{x} = f(x,t), \quad x \in \Omega \subseteq R^n \tag{15.110}$$

i. The function $V(x)$ is positive definite—*i.e.*,

$$V(0) = 0,$$
$$V(x) > 0, \quad \|x\| \neq 0$$

ii. The gradient of $V(x)$—*i.e.*,

$$\frac{\partial V}{\partial x} \quad \text{exists for all} \quad x \in \Omega \tag{15.111}$$

iii. The derivative of $V(x)$ for $\dot{x} = f(x,t)$ is negative definite

$$\dot{V} = \frac{\partial V}{\partial x}\dot{x} \leq 0, \quad \forall x \neq 0 \tag{15.112}$$

Theorem 15.4
If there exists a positive definite function $V(x)$ whose derivative for $\dot{x} = f(x,t)$ is negative semidefinite, then the equilibrium is *stable* (in the sense of Lyapunov). ■

Theorem 15.5
If there exists a positive definite function $V(x)$ whose derivative for $\dot{x} = f(x,t)$ is negative definite, then the equilibrium is *asymptotically stable*. ■

Theorem 15.6
If in addition to the assumptions *i–iii* we require that $V(x)$ be *radially unbounded*, *i.e.*,

iv. There is a monotonously increasing function $\phi(\cdot)$ such that

$$V(x) > \phi(\|x\|), \quad \forall x \in R^n \quad \text{and} \quad \lim_{\|x\| \to \infty} \phi(\|x\|) = \infty$$

then the equilibrium is *globally asymptotically stable*—that is, the domain of stability is all of R^n. ■

The condition that $V(x)$ is radially unbounded is necessary to guarantee that the hypersurfaces $V(x) = c$ are closed for each arbitrary c so that for each c there is a scalar δ such that $\|x\| \leq \delta = \phi^{-1}(c)$.

Theorem 15.7—Lyapunov

Let

$$\dot{x} = Ax \tag{15.113}$$

where all eigenvalues of the matrix A have negative real parts. Let the matrix P be the solution to the Lyapunov matrix equation

$$PA + A^T P = -Q \tag{15.114}$$

where Q and P are positive definite matrices. Then

$$V(x) = \frac{1}{2} x^T P x \tag{15.115}$$

is a Lyapunov function for the system $\dot{x} = Ax$. ■

Proof: Direct substitution of

$$P = \int_0^\infty e^{A^T t} Q e^{At} dt \tag{15.116}$$

into Eq. (15.114) shows that P satisfies the Lyapunov equation (15.114).

Quadratic optimal control of a linear system

Optimal control of the state x over a time interval $[t_0, t_f]$ by means of a control variable u of a linear system $\dot{x} = Ax + Bu$ can be formalized as the quadratic optimization problem

$$\text{Minimize } J(u) = \frac{1}{2} x^T(t_f) Q_f x(t_f) + \frac{1}{2} \int_{t_0}^{t_f} x^T(t) Q x(t) + u^T(t) R u(t) dt$$

$$\text{subject to } \dot{x} = Ax + Bu$$

$$\text{for } Q_f = Q_f^T \geq 0, \quad Q = Q^T \geq 0, \quad R = R^T > 0$$

(15.117)

There is a well-known solution to this problem in the theory of optimal control that can be obtained by dynamic programming. The dynamic programming reformulation of the optimization problem is to solve the Hamilton-Jacobi-Bellman equation

$$-\frac{\partial V}{\partial t} = \min_u\{\frac{1}{2}x^T(t)Qx(t) + \frac{1}{2}u^T(t)Ru(t) + \frac{\partial V}{\partial x}(Ax(t) + Bu(t))\} \quad (15.118)$$

By introducing the test function

$$V(x(t)) = \frac{1}{2}x^T(t)P(t)x(t) \quad (15.119)$$

and substituting V into Eq. (15.118) we have (time argument t omitted)

$$-\frac{1}{2}x^T\dot{P}x = \min_u\{\frac{1}{2}x^TQx + \frac{1}{2}u^TRu + x^TP(Ax + Bu)\} \quad (15.120)$$

By completing the squares we have

$$\begin{aligned}-\frac{1}{2}x^T\dot{P}x = \min_u\{&\frac{1}{2}(u + R^{-1}B^TPx)^TR(u + R^{-1}B^TPx)\\ &+ \frac{1}{2}x^TQx + x^TPAx - \frac{1}{2}x^TPBR^{-1}B^TPx\}\end{aligned} \quad (15.121)$$

The right-hand side of Eq. (15.121) has a minimum for the optimal control

$$u = u^* = -R^{-1}B^TPx \quad (15.122)$$

where P solves the Riccati equation

$$-\dot{P} = PA + A^TP + Q - PBR^{-1}B^TP, \quad P(t_f) = Q_f, \quad P = P^T > 0 \quad (15.123)$$

and with the optimal control performance index

$$J(u^*) = V(x(t_0)) \quad (15.124)$$

The value function V is positive definite with respect to x, radially growing and has the negative definite derivative

$$\frac{dV}{dt} = -\frac{1}{2}x^TPBR^{-1}B^TPx - \frac{1}{2}x^TQx < 0, \qquad ||x|| \neq 0 \quad (15.125)$$

Hence, V qualifies as a Lyapunov function for the system, a fact which proves global system stability for the optimal control.

Remark: In the case of adaptive optimal control over a long time horizon $t_f - t_0$ so that P and thus θ are constant, we replace Eq. (15.122) by $u = -\widehat{\theta}\phi = u^* - \widetilde{\theta}\phi$ for $\phi = x$ and introduce the Lyapunov function candidate

$$V(x) = \frac{1}{2}x^T Px + \frac{1}{2}\widetilde{\theta}^T S\widetilde{\theta}, \qquad S = S^T > 0 \tag{15.126}$$

with the derivative

$$\dot{V}(x) = -x^T P^T BR^{-1}B^T Px - x^T Qx - x^T PB\phi^T\widetilde{\theta} + \widetilde{\theta}^T S\dot{\widetilde{\theta}} \tag{15.127}$$

The desirable adaptation law in order to make $\dot{V} < 0$ would be

$$\dot{\widehat{\theta}} = S^{-1}\phi B^T Px = -S^{-1}\phi Ru^* \tag{15.128}$$

This is a possible control law, for example, in learning control when u^* is known but cannot be used in the normal case of adaptive control where u^* is not known in advance. ■

A

$$A = U \Sigma V^*$$

Basic Matrix Algebra

A.1 PRELIMINARIES

We use the following definitions:

Definition A.1

A rectangular array

$$A = \begin{pmatrix} a_{11} & a_{12} & \cdots & a_{1n} \\ a_{21} & a_{22} & \cdots & a_{2n} \\ \vdots & \vdots & \ddots & \vdots \\ a_{m1} & a_{m2} & \cdots & a_{mn} \end{pmatrix} \tag{A.1}$$

with m rows and n columns of elements belonging to a field $\mathcal{F}$ is called a *matrix* over $\mathcal{F}$ of *order* $m \times n$. This is also denoted as $A \in \mathcal{F}^{m \times n}$.

It is standard to use the parallel notations

$$A = (a_{ij}) = A_{m \times n} \tag{A.2}$$

in order to emphasize the elements of A and the order, respectively.

Two $m \times n$ matrices $A = (a_{ij})$ and $B = (b_{ij})$ are equal if $a_{ij} = b_{ij}$ for every i and j.

Definition A.2
A zero matrix $0_{m \times n}$ is an $m \times n$ matrix with all elements equal to zero. A matrix $A_{m \times n}$ is called a *square matrix* of order n if $m = n$. A square matrix of order n

$$I_{n \times n} = (\delta_{ij}), \qquad \delta_{ij} = \begin{cases} 1, & i = j \\ 0, & i \neq j \end{cases} \tag{A.3}$$

is called an *identity matrix*. The identity matrix of order n is sometimes denoted I_n.

Definition A.3—Vectors
An $m \times 1$ matrix x is called a *column vector* whereas a $1 \times n$ matrix y is called a *row vector*.

Definition A.4—Linear independence
A set of vectors $\{x_1, x_2, \ldots, x_n\}$ is said to be *linearly independent* if and only if

$$a_1 x_1 + a_2 x_2 + \cdots + a_n x_n = 0 \tag{A.4}$$

implies that the constants $a_1 = a_2 = \cdots = a_n = 0$. Conversely, the vectors are said to be *linearly dependent* if there are non-zero constants $a_1, \ldots, a_n$ such that (A.4) is satisfied.

Definition A.5
The *transpose* of A, denoted A^T, is the $n \times m$ matrix obtained from A by interchanging rows and columns, *i.e.*, $A^T = (a_{ji})$.

A square matrix A is said to be *symmetrical* if $A = A^T$, *i.e.*, if $a_{ij} = a_{ji}$. Conversely, a square matrix A is said to be *skewsymmetrical* if $A = -A^T$, *i.e.*, if $a_{ij} = -a_{ji}$.

Definition A.6
Let A be a matrix over the complex numbers and let A^* denote the transpose of the complex conjugate of A. A matrix A over the complex numbers is said

to be *Hermitian* if $A = A^*$, *i.e.*, each element a_{ij} is the complex conjugate of a_{ji}.

Definition A.7
The sum of two $m \times n$ matrices A and B is the $m \times n$ matrix $A + B = (a_{ij} + b_{ij})$.

Definition A.8
The product of an $m \times n$ matrix $A = (a_{ij})$ with a scalar α is the matrix $\alpha A = (\alpha a_{ij})$ where α multiplies each matrix element of A.

Definition A.9—Matrix product
Let $A = A_{m \times n} = (a_{ij})$ and $B = B_{n \times p} = (b_{jk})$ be two given matrices with the number of columns of A being equal to the number of rows of B. Then the product of A and B is defined to be the $m \times p$ matrix $C = AB$ with elements

$$c_{ik} = \sum_{j=1}^{n} a_{ij} b_{jk} \tag{A.5}$$

Definition A.10—Determinant
To every square matrix A of order n can be associated a quantity called the *determinant* according to the recursive formula

$$\det A = \begin{cases} A, & n = 1 \\ \sum_{i=1}^{n} a_{ij} A_{ij}, & n > 1 \end{cases} \qquad \text{where} \quad A_{ij} = (-1)^{i+j} \det D_{ij} \tag{A.6}$$

where D_{ij} is the matrix obtained by deleting the ith row and jth column of A. The scalar A_{ij} is called the *algebraic complement* or *cofactor* to the element a_{ij}.

An elementary property of matrix determinants is that for any square matrices A and B it holds that

$$\det(AB) = \det A \cdot \det B \tag{A.7}$$

Definition A.11—Singular matrix
A square matrix A is called *singular* if $\det A = 0$ and *nonsingular* if $\det A \neq 0$.

Definition A.12—Minor
A *minor* (or subdeterminant) of A of order p is defined as

$$\det \begin{pmatrix} a_{i_1 j_1} & a_{i_1 j_2} & \cdots & a_{i_1 j_p} \\ a_{i_2 j_1} & a_{i_2 j_2} & \cdots & a_{i_2 j_p} \\ \vdots & \vdots & \ddots & \vdots \\ a_{i_p j_1} & a_{i_p j_2} & \cdots & a_{i_p j_p} \end{pmatrix} \tag{A.8}$$

provided that $1 \le i_1 < i_2 < \ldots < i_p \le n$ and $1 \le j_1 < j_2 < \ldots < j_p \le n$ where n is the order of A.

Definition A.13—Trace

The sum of the diagonal elements of a matrix A of order n is called the *trace* of A which is denoted $\mathrm{tr}(A)$ so that

$$\mathrm{tr}(A) = \sum_{i=1}^{n} a_{ii} \tag{A.9}$$

Definition A.14—Inverse of a matrix

If for a square matrix of order n there exists another square matrix B such that

$$AB = BA = I_{n \times n} \tag{A.10}$$

then B is called the *inverse* of A and is denoted A^{-1}.

Definition A.15—Orthonormal vectors

A set of vectors $v_1, \ldots, v_n$ is said to be *orthonormal* if

$$v_i^T v_j = \delta_{ij}, \qquad i = 1, \ldots, n; \quad j = 1, \ldots, n \tag{A.11}$$

Definition A.16—Orthogonal matrix

A matrix A with the property $A^T A = I$ is said to be an *orthogonal* matrix.

Result A.1

If A is a square orthogonal matrix, then $A^T = A^{-1}$ and A has orthonormal column vectors as well as orthonormal row vectors.

Definition A.17—Unitary matrix

A *unitary* matrix A is a square matrix of order n such that $A^{-1} = A^*$, *i.e.*, the inverse of A is the same as the transpose of the complex conjugate of A.

Theorem A.1—Cramer's rule

A square matrix A of order n has an inverse if and only if A is nonsingular. The inverse A^{-1} is then unique with elements

$$A^{-1} = \left(\frac{\mathrm{adj}(A)}{\det A}\right) = \frac{1}{\det A}\begin{pmatrix} A_{11} & A_{21} & \cdots & A_{n1} \\ A_{12} & A_{22} & \cdots & A_{n2} \\ \vdots & \vdots & \ddots & \vdots \\ A_{1n} & A_{2n} & \cdots & A_{nn} \end{pmatrix} \tag{A.12}$$

where $\mathrm{adj}(A)$ is the matrix with elements obtained from algebraic complements A_{ij} to the elements of A.

Theorem A.2
Every square matrix A satisfies the relationships

$$\begin{aligned}\sum_{i=1}^{n} a_{ik}A_{ij} &= \delta_{jk}\det A \\ \sum_{i=1}^{n} a_{ki}A_{ji} &= \delta_{jk}\det A\end{aligned} \tag{A.13}$$

Theorem A.3
If A and B are two nonsingular matrices of order n, then the product AB is nonsingular and

$$(AB)^{-1} = B^{-1}A^{-1} \tag{A.14}$$

Definition A.18—Rank of a matrix
The rank of a matrix A of order $m \times n$ is the maximum order of non-zero subdeterminants.

Definition A.19
Given a square matrix A, if there exist a column vector x and a scalar λ such that

$$Ax = \lambda x \tag{A.15}$$

then λ is called an *eigenvalue* of A and x is the *eigenvector* associated with the eigenvalue λ.

The eigenvalues can be obtained by solving the *characteristic equation*

$$\det(A - \lambda I) = 0 \tag{A.16}$$

where $\det(A - \lambda I)$ is called the *characteristic polynomial*. There are n eigenvalues $\lambda_1, \ldots, \lambda_n$ solving Eq. (A.16).

Let Λ denote the following diagonal matrix of eigenvalues

$$\Lambda = \begin{pmatrix} \lambda_1 & 0 & \cdots & 0 \\ 0 & \lambda_2 & \ddots & \vdots \\ \vdots & \ddots & \ddots & 0 \\ 0 & \cdots & 0 & \lambda_n \end{pmatrix} \tag{A.17}$$

Result A.2—Diagonalization
Assume that the $n \times n$ matrix A has n independent eigenvectors x_i organized as the columns of a matrix T. In addition, let Λ be the diagonal matrix of

eigenvalues of A. Then

$$T^{-1}AT = \Lambda \tag{A.18}$$

Proof: Using the fact that the columns of T are eigenvectors of A it is straightforward to establish that

$$AT = A\begin{pmatrix} x_1 & x_2 & \dots & x_n \end{pmatrix} = \begin{pmatrix} \lambda_1 x_1 & \lambda_2 x_2 & \dots & \lambda_n x_n \end{pmatrix} = T\Lambda \tag{A.19}$$

The matrix T is invertible because its columns are linearly independent. Multiplication of Eq. (A.19) from the left by T^{-1} provides the result (A.18).

Result A.3

The eigenvalues of a symmetrical matrix $A \in R^{n \times n}$ are real and there exist orthonormal eigenvectors $x_1, x_2, \dots, x_n$ corresponding to eigenvalues $\lambda_1, \lambda_2, \dots, \lambda_n$ so that

$$A = T\Lambda T^{-1} = T\Lambda T^T \tag{A.20}$$

where the columns of T are eigenvectors of A.

Theorem A.4

Let A and B be any $m \times n$ matrix. Then

$$\text{tr}(AB^T) = \text{tr}(B^T A) \tag{A.21}$$

Proof: Explicit use of the definition of a matrix product gives

$$\text{tr}(AB^T) = \sum_{i=1}^{m}\sum_{j=1}^{n} a_{ij}b_{ij} = \text{tr}(B^T A) \tag{A.22}$$

Result A.4

The n eigenvalues $\lambda_1, \dots, \lambda_n$ solving Eq. (A.16) also satisfy the relations

$$\begin{aligned} \det A &= \lambda_1 \lambda_2 \cdots \lambda_n \\ \text{tr}(A) &= \lambda_1 + \lambda_2 + \cdots + \lambda_n = \text{tr}(\Lambda) \end{aligned} \tag{A.23}$$

Proof: The matrix $A = T\Lambda T^{-1}$ and it follows that $\det A = \det(T\Lambda T^{-1}) = \det T \det \Lambda \det(T^{-1}) = \det \Lambda = \lambda_1 \lambda_2 \cdots \lambda_n$. According to Eq. (A.21) we also find that $\text{tr}(A) = \text{tr}(T\Lambda T^{-1}) = \text{tr}((\Lambda T^{-1})T) = \text{tr}(\Lambda) = \lambda_1 + \lambda_2 + \cdots + \lambda_n$.

Result A.5

Assume that the nonsingular matrix A is partitioned as

$$A = \begin{pmatrix} P & Q \\ R & S \end{pmatrix} \tag{A.24}$$

where P and S are square nonsingular matrices. Then

$$A^{-1} = \begin{pmatrix} (P - QS^{-1}R)^{-1} & -(P - QS^{-1}R)^{-1}QS^{-1} \\ -(S - RP^{-1}Q)^{-1}RP^{-1} & (S - RP^{-1}Q)^{-1} \end{pmatrix} \tag{A.25}$$

Proof: The matrix A can be expressed as

$$\begin{pmatrix} P & Q \\ R & S \end{pmatrix} = \begin{pmatrix} I & 0 \\ RP^{-1} & I \end{pmatrix} \begin{pmatrix} P & 0 \\ 0 & S - RP^{-1}Q \end{pmatrix} \begin{pmatrix} I & P^{-1}Q \\ 0 & I \end{pmatrix} \tag{A.26}$$

or

$$\begin{pmatrix} P & Q \\ R & S \end{pmatrix} = \begin{pmatrix} I & QS^{-1} \\ 0 & I \end{pmatrix} \begin{pmatrix} P - QS^{-1}R & 0 \\ 0 & S \end{pmatrix} \begin{pmatrix} I & 0 \\ S^{-1}R & I \end{pmatrix} \tag{A.27}$$

Straightforward application of Eq. (A.14) on Eq. (A.26) and Eq. (A.27) proves the result (A.25).

Result A.6—Matrix inversion lemma
Let the matrix A of order n be nonsingular and let B and C be $n \times m$ and $m \times n$ matrices, respectively. Then

$$(A + BC)^{-1} = A^{-1} - A^{-1}B(I_{m\times m} + CA^{-1}B)^{-1}CA^{-1} \tag{A.28}$$

Proof: This can be verified by straightforward multiplication of Eq. (A.28) by $(A + BC)$.

Definition A.20—Quadratic forms
A real-valued quadratic form is defined as

$$x^T Ax, \qquad A \in R^{n\times n} \tag{A.29}$$

where A is a symmetric matrix. The quadratic form and the matrix A are said to be *positive definite* if $x^T Ax > 0$ for all non-zero x and all the eigenvalues of a positive definite matrix are positive.

Definition A.21—Projection matrix
A matrix P with the properties $P = P^T$ and $P^2 = P$ is a *projection matrix.*

Theorem A.5—Projection
Let $A \in R^{m\times n}$ be a matrix with column space C. The projection of a vector v into C is Pv where P is the projection matrix

$$P = A(A^T A)^{-1}A^T \tag{A.30}$$

■

A.2 MATRIX NORMS

Let $x \in R^n$ be a vector. For every vector norm $||x||$ it is required that $||x|| \geq 0$ with $||x|| = 0$ only for $x = 0$ and that $||\alpha x|| = \alpha ||x||$ for a scalar α. Moreover, for vectors x, y it is required that

$$\begin{aligned} ||x + y|| &\leq ||x|| + ||y|| \\ ||x^T y|| &\leq ||x|| \cdot ||y|| \end{aligned} \tag{A.31}$$

Some important examples of vector norms are

$$\begin{aligned} ||x||_1 &= |x_1| + |x_2| + \cdots + |x_n| && \text{1-norm} \\ ||x||_2 &= \sqrt{x^T x} = (|x_1|^2 + |x_2|^2 + \cdots + |x_n|^2)^{1/2} && \text{2-norm} \\ ||x||_p &= (|x_1|^p + |x_2|^p + \cdots + |x_n|^p)^{1/p} && \text{p-norm} \\ ||x||_\infty &= \max(|x_1|, |x_2|, \ldots, |x_n|) && \infty\text{-norm} \end{aligned}$$

Let W be some positive definite symmetric matrix. A weighted 2-norm is defined as

$$||x||_W = \sqrt{x^T W x}, \qquad W = W^T > 0 \tag{A.32}$$

Definition A.22—Matrix norms
The matrix norm of a matrix A corresponding to a certain vector norm $|| \cdot ||_p$ is

$$||A||_p = \sup_x \frac{||Ax||_p}{||x||_p} \tag{A.33}$$

Definition A.23—Frobenius norm
The Frobenius norm of a matrix $A \in R^{m \times n}$ is

$$||A||_F = \sqrt{\sum_{i=1}^{m} \sum_{j=1}^{n} |a_{ij}|^2} \tag{A.34}$$

A.3 SINGULAR VALUE DECOMPOSITION

Let A be an $m \times n$ matrix of generally complex-valued elements. Then there exist $m \times m$ and $n \times n$ unitary matrices U and V (*i.e.*, $U^{-1} = U^*$ and $V^{-1} = V^*$) such that

$$A = U \Sigma V^* \tag{A.35}$$

where Σ is an $m \times n$ matrix whose elements are zero except possibly along its main diagonal. These nonnegative diagonal elements are ordered such that

$$\sigma_{11} \geq \sigma_{22} \geq \ldots \geq \sigma_{pp} \geq 0 \quad \text{where} \quad p = \min(m, n) \tag{A.36}$$

The diagonal elements σ_{kk} are known as the *singular values* of the matrix A and the nonzero singular values correspond to the positive square roots of the eigenvalues of the nonnegative Hermitian matrices AA^* and A^*A. The columns of U and V correspond to the orthonormal eigenvectors of the nonnegative Hermitian matrices AA^* and A^*A, respectively. The matrix decomposition (A.35) is known as *singular value decomposition* or SVD.

A useful property of the SVD is that several matrix norms can be represented in terms of the singular values. In particular, one has

$$\begin{aligned} \|A\|_2 &= \sigma_{11} \\ \|A\|_F &= \sqrt{\sigma_{11}^2 + \sigma_{22}^2 + \cdots + \sigma_{pp}^2} \end{aligned} \tag{A.37}$$

Assume that r = rank (A). If we define the matrix $A^+ = V\Sigma^+U^*$ where

$$\Sigma^+ = \begin{pmatrix} 1/\sigma_{11} & & & 0_{r\times(m-r)} \\ & \ddots & & \\ & & 1/\sigma_{rr} & \\ 0_{(n-r)\times r} & & & 0_{(n-r)\times(m-r)} \end{pmatrix} \in R^{n\times m} \tag{A.38}$$

If r = rank $(A) = n$ then $A^+ = (A^TA)^{-1}A^*$ which is called the *pseudo inverse* or (*Moore-Penrose inverse*) of the matrix A. Clearly, if $m = n =$ rank(A), then $A^+ = A^{-1}$. For a real-valued matrix A we define the pseudo inverse $A^+ = V\Sigma^+U^T$.

The singular value decomposition has a close relation to the least-squares problem in that the least-squares solution

$$\hat{x} = \arg\min \|Ax - b\|_2^2 = A^+b \tag{A.39}$$

for a matrix A and a vector b. In the rank deficient case when there are infinitely many solutions that minimize $\|Ax - b\|_2$ then $\hat{x}$ is the solution with the smallest 2-norm. The residual sum at the optimum is

$$\|Ax - b\|_2^2 = \|(AA^+ - I)b\|_2^2 \tag{A.40}$$

The singular value decomposition is valuable to solve the problem of finding the $m \times n$ matrix $A^{(k)}$ of rank k being less or equal to rank(A), which will best approximate A in the 2-norm sense, *i.e.*,

$$\min_{A^{(k)}} ||A - A^{(k)}||_2 \tag{A.41}$$

The unique solution to this problem is

$$A^{(k)} = U\Sigma_k V^* \tag{A.42}$$

where U and V are as in (A.35) while Σ_k is obtained from Σ by setting to zero all but its k largest singular values. The accuracy of this approximation is

$$||A - A^{(k)}||_2 = \sigma_{(k+1)(k+1)}, \quad \text{or} \quad ||A - A^{(k)}||_F = \sqrt{\sum_{j=k+1}^{p} \sigma_{jj}^2} \tag{A.43}$$

Another standard problem that arises in the least-squares normal equations is to find the eigenvalues of AA^T (or A^TA) where A is some given matrix. The singular value decomposition also provides a solution to this problem and the eigenvalues of AA^T are the squared elements of Σ.

A.4 QR-FACTORIZATION

Consider a system of linear equations of the form

$$Ax = b, \qquad A \in R^{m\times n}, \qquad b \in R^m \tag{A.44}$$

Assume that an orthogonal matrix $Q \in R^{m\times m}$ (*i.e.*, $QQ^T = I$) can be computed so that

$$Q^TA = R = \begin{pmatrix} R_1 \\ 0_{(m-n)\times n} \end{pmatrix}, \qquad R_1 \in R^{n\times n} \tag{A.45}$$

with R_1 being upper triangular. Let

$$Q^Tb = \begin{pmatrix} s_1 \\ s_2 \end{pmatrix}, \qquad s_1 \in R^n, \qquad s_2 \in R^{n-m} \tag{A.46}$$

The residual sum of a least-squares problem can then be expressed as

$$||Ax - b||_2^2 = ||Q^TAx - Q^Tb||_2^2 = ||R_1x - s_1||_2^2 + ||s_2||_2^2 \tag{A.47}$$

and the least-squares solution is obtained by solving the upper triangular system

$$R_1\hat{x} = s_1 \tag{A.48}$$

The method described is called *QR-factorization*, and a good numerical method to achieve the triangulation matrix Q is the *Householder method.* As suggested by Householder, such a transformation can be accomplished by means of choosing the triangulation matrix Q as

$$Q = P_n P_{n-1} \dots P_1, \quad \text{where} \quad P_k = \begin{pmatrix} I_{k-1} & 0 \\ 0 & I - 2\frac{v_k v_k^T}{v_k^T v_k} \end{pmatrix} \tag{A.49}$$

where $\{P_k\}_{k=1}^{n}$ is a sequence of unitary matrices with the property $P_k^2 = I$. Let the vector v_1 that defines P_1 be chosen as

$$\begin{aligned} \alpha_1 &= \begin{pmatrix} a_{11} & \dots & a_{m1} \end{pmatrix}^T \\ v_1 &= \alpha_1 - \sqrt{\alpha_1^T \alpha_1} \begin{pmatrix} 1 & 0 & \dots & 0 \end{pmatrix}^T \end{aligned} \tag{A.50}$$

The matrix product $P_1 A$ is then a matrix whose first column is zero except for its first element. This is clearly a first step toward triangulation, and it is suitable to proceed by recursion. Assume that an intermediate result at step $k - 1$ has been obtained as

$$P_{k-1} P_{k-2} \dots P_1 A = \begin{pmatrix} R^{(k-1)} & S^{(k-1)} \\ 0_{(m-k+1)\times(k-1)} & A^{(k-1)} \end{pmatrix} \tag{A.51}$$

where $R^{(k-1)}$ is an upper triangular matrix of dimensions $(k - 1) \times (k - 1)$. Let v_k and thus by extension P_k of Eq. (11.52) be chosen according to the algorithm

$$\begin{aligned} \alpha_k &= \begin{pmatrix} 0_{1\times(k-1)} & a_{11}^{(k-1)} & \dots & a_{(m-k+1)1}^{(k-1)} \end{pmatrix}^T \\ e_k &= \begin{pmatrix} 0_{1\times(k-1)} & 1 & 0_{1\times(m-k)} \end{pmatrix}^T \\ v_k &= \alpha_k - \sqrt{\alpha_k^T \alpha_k} e_k \end{aligned} \tag{A.52}$$

It is then straightforward to verify that Eq. (A.51) is valid also for recursion step k, and by extension to $k = n$ one can verify that the procedure results in the upper triangular form (A.45). (For further details see Golub and Van Loan, 1989.)

A.5 MATRIX DIFFERENTIATION

Let $x \in R^n$ be a vector and let $f(x)$ be a scalar function of x. The partial derivative of f with respect to x is the row vector

$$\frac{\partial f}{\partial x} = \begin{pmatrix} \frac{\partial f}{\partial x_1} & \frac{\partial f}{\partial x_2} & \cdots & \frac{\partial f}{\partial x_n} \end{pmatrix} \tag{A.53}$$

Differentiation of a vector-valued function $f : R^n \to R^m$ with respect to x gives the Jacobian matrix

$$\frac{\partial f}{\partial x} = \begin{pmatrix} \frac{\partial f_1}{\partial x_1} & \frac{\partial f_1}{\partial x_2} & \cdots & \frac{\partial f_1}{\partial x_n} \\ \frac{\partial f_2}{\partial x_1} & \frac{\partial f_2}{\partial x_2} & \cdots & \frac{\partial f_2}{\partial x_n} \\ \vdots & \vdots & & \vdots \\ \frac{\partial f_m}{\partial x_1} & \frac{\partial f_m}{\partial x_2} & \cdots & \frac{\partial f_m}{\partial x_n} \end{pmatrix}, \qquad f(x) \in R^m, \qquad x \in R^n \tag{A.54}$$

Differentiation of a scalar function $f(A)$ with respect to matrix A gives a matrix with elements

$$\frac{\partial f}{\partial A} = \begin{pmatrix} \frac{\partial f}{\partial a_{11}} & \cdots & \frac{\partial f}{\partial a_{1n}} \\ \vdots & & \vdots \\ \frac{\partial f}{\partial a_{n1}} & \cdots & \frac{\partial f}{\partial a_{nn}} \end{pmatrix} \tag{A.55}$$

Example A.1

Let A be a square matrix in $R^{n \times n}$ with elements a_{ij} and let V be the scalar function

$$V(x) = x^T A x = \sum_{i=1}^{n} \sum_{j=1}^{n} a_{ij} x_i x_j, \qquad x \in R^n \tag{A.56}$$

Element-wise differentiation gives

$$\frac{\partial V}{\partial x_k} = \sum_{i=1}^{n} \sum_{j=1}^{n} a_{ij} \delta_{ik} x_j + a_{ij} x_i \delta_{jk} = e_k^T A x + x^T A e_k, \quad \text{where} \quad e_k = \begin{pmatrix} \delta_{1k} \\ \delta_{2k} \\ \vdots \\ \delta_{nk} \end{pmatrix} \tag{A.57}$$

and differentiation with respect to the column vector x yields

$$\frac{\partial V}{\partial x} = Ax + A^T x \tag{A.58}$$

Result A.7

Let A be a square nonsingular matrix of order n. Then

$$\frac{\partial}{\partial A} \log \det A = A^{-T} \tag{A.59}$$

Proof: From Eq. (A.13) it follows that

$$\det A = \sum_{i=1}^{n} a_{ij} A_{ij} \tag{A.60}$$

so that

$$\frac{\partial \log \det A}{\partial a_{ij}} = \frac{A_{ij}}{\sum_{i=1}^{n} a_{ij} A_{ij}} = \frac{A_{ij}}{\det A} \tag{A.61}$$

Identification of terms between Eq. (A.61) and Eq. (A.12) proves the result (A.59).

Result A.8

Let A be a square nonsingular matrix of order n. Then

$$\frac{\partial}{\partial A}(\mathrm{tr}(BA^{-1})) = -(A^{-1}BA^{-1})^{T} \tag{A.62}$$

Proof: From the definition of a matrix inverse it holds that $AA^{-1} = I_{n \times n}$ and differentiation of $I_{n \times n}$ gives

$$\frac{\partial A}{\partial a_{ij}} A^{-1} + A \frac{\partial A^{-1}}{\partial a_{ij}} = 0 \tag{A.63}$$

so that

$$\frac{\partial A^{-1}}{\partial a_{ij}} = -A^{-1} \frac{\partial A}{\partial a_{ij}} A^{-1} \tag{A.64}$$

The matrix $\partial A / \partial a_{ij}$ can be expressed as

$$\frac{\partial A}{\partial a_{ij}} = e_i e_j^T \tag{A.65}$$

where e_i is a zero vector except for 1 in the ith position.

$$\frac{\partial}{\partial a_{ij}} \mathrm{tr}(BA^{-1}) = \mathrm{tr}(B \frac{\partial A^{-1}}{\partial a_{ij}}) = -\mathrm{tr}(BA^{-1} e_i e_j^T A^{-1}) = -e_i^T (A^{-1}BA^{-1})^T e_j \tag{A.66}$$

Hence,

$$\frac{\partial}{\partial A}\mathrm{tr}(BA^{-1}) = -(A^{-1}BA^{-1})^T \tag{A.67}$$

which proves the statement (A.62).

A.6 POLYNOMIALS AND POLYNOMIAL MATRICES

Let $R[x]$ and $R(x)$ denote the polynomials and rational functions in the indeterminate x with coefficients in R, respectively.

A polynomial is said to be *irreducible* if it can not be divided by any other polynomial of degree greater than 0.

Definition A.24—Polynomial matrix
A *polynomial matrix* is a matrix with elements in $R[x]$, and the set of $m \times n$ matrices with elements in $R[x]$ is denoted $R^{m\times n}[x]$.

Definition A.25—Unimodular matrix
A polynomial matrix $A \in R^{n\times n}[x]$ is said to be *unimodular* if and only if $\det A = c$ where $c \in R\backslash\{0\}$.

Definition A.26—Matrix rank
The rank of a polynomial matrix $A \in R^{m\times n}[x]$ is the highest order of any non-zero minor of A.

Definition A.27—Invertibility
The matrix $A \in R^{m\times n}[x]$ is *right invertible* if there is a $B \in R^{n\times m}[x]$ such that $AB = I_{m\times m}$. The matrix $A \in R^{m\times n}[x]$ is *left invertible* if there is a $B \in R^{n\times m}[x]$ such that $BA = I_{n\times n}$. The matrix $A \in R^{n\times n}$ is invertible if it is right invertible and left invertible.

Example A.2—Invertibility
The following unimodular matrix $Q \in R^{2\times 2}[x]$ is invertible

$$Q(x) = \begin{pmatrix} 1 & x + x^2 \\ 0 & 1 \end{pmatrix}, \quad \text{and} \quad Q^{-1}(x) = \begin{pmatrix} 1 & -x - x^2 \\ 0 & 1 \end{pmatrix}, \tag{A.68}$$

where the inverse Q^{-1} is a polynomial matrix.

Definition A.28—Matrix equivalence
The matrices A and B in $R^{m\times n}[x]$ are said to be *equivalent* if there are unimodular matrices $U \in R^{m\times m}[x]$ and $V \in R^{n\times n}[x]$ such that $A = UBV$.

Definition A.29—Left divisors

If $A \in R^{m\times n}[x]$ can be factorized as BC where B and C are polynomial matrices and B has linearly independent columns, then B is a *left divisor* of A.

Definition A.30—Right divisors

If $A \in R^{m\times n}[x]$ can be factorized as BC where B and C are polynomial matrices and C has linearly independent rows, then C is a *right divisor* of A.

Definition A.31—Greatest common left divisor (g.c.l.d.)

Let the polynomial matrices $A \in R^{m\times n}[x]$ and $B \in R^{m\times p}[x]$ and let U belong to $R^{m\times k}[x]$ for some k. If U is a left divisor to both A and B and if every other left divisor to both A and B is also a left divisor to U, then U is called the *greatest common left divisor (g.c.l.d.)* to A and B.

Definition A.32—Greatest common right divisor (g.c.r.d.)

Let the polynomial matrices $A \in R^{m\times n}[x]$ and $B \in R^{p\times n}[x]$ and let V belong to $R^{k\times n}[x]$ for some k. If V is a right divisor to both A and B and if every other right divisor to both A and B is also a right divisor to V, then V is called the *greatest common right divisor (g.c.r.d.)* to A and B.

Result A.9

For every pair of polynomial matrices $A \in R^{m\times n}[x]$ and $B \in R^{m\times p}[x]$ there exists a g.c.l.d.. If rank($[AB]$) $= r$, then any g.c.l.d. U has r columns and can be expressed as

$$U = AX + BY, \qquad U \in R^{m\times r}[x] \tag{A.69}$$

for some $X \in R^{n\times r}[x]$ and $Y \in R^{p\times r}[x]$.

Result A.10

For every pair of polynomial matrices $A \in R^{m\times n}[x]$ and $B \in R^{p\times n}[x]$ there exists a g.c.r.d.. If rank($[A^T B^T]$) $= r$, then any g.c.r.d. V has r rows and can be expressed as

$$V = XA + YB, \qquad V \in R^{r\times n}[x] \tag{A.70}$$

for some $X \in R^{r\times m}[x]$ and $Y \in R^{r\times p}[x]$.

Definition A.33—Smith form

For any polynomial matrix $A(x) \in R^{m\times n}[x]$ of rank r it is possible to find unimodular matrices $U(x) \in R^{m\times m}[x]$ and $V(x) \in R^{n\times n}[x]$ such that

$$A(x) = U(x)\Sigma(x)V(x) \tag{A.71}$$

where

$$\Sigma(x) = \begin{pmatrix} \sigma_1(x) & & & 0_{r\times(n-r)} \\ & \ddots & & \\ & & \sigma_r(x) & \\ 0_{(m-r)\times r} & & & 0_{(m-r)\times(n-r)} \end{pmatrix} \in R^{m\times n}[x] \tag{A.72}$$

and the $\{\sigma_i(x)\}$ are unique polynomials obeying a division property

$$\sigma_i(x)|\sigma_{i+1}(x) \tag{A.73}$$

The matrix $\Sigma(x)$ is called the *Smith form* of $A(x)$ and the $\{\sigma_i(x)\}$ are the *invariant polynomials* of $A(x)$.

Result A.11—Invariant polynomials
Let $\Delta_i(x)$ denote the greatest common divisor of all $i\times i$ minors of a polynomial matrix $A(x)$ and let $\Delta_0(x) = 1$. Then we can obtain

$$\sigma_i(x) = \frac{\Delta_i(x)}{\Delta_{i-1}(x)} \tag{A.74}$$

Definition A.34—Left co-prime matrices
The polynomial matrices $A \in R^{m\times n}[x]$ and $B \in R^{m\times p}[x]$ are said to be *left co-prime* (or *relatively left prime* or *mutually left prime*) if the g.c.l.d. is unimodular.

Definition A.35—Right co-prime matrices
The polynomial matrices $A \in R^{m\times n}[x]$ and $B \in R^{p\times n}[x]$ are said to be *right co-prime* (or *relatively right prime* or *mutually right prime*) if the g.c.r.d. is unimodular.

A.7 BIBLIOGRAPHY AND REFERENCES

Important references are

- F.R. GANTMACHER, *The theory of matrices*, Vols.I–II. New York: Chelsea Publishing Co., New York, 1959.
- G.H. GOLUB AND C.F. VAN LOAN, *Matrix Computations*, 2d ed. Baltimore: The Johns Hopkins University Press, 1989.
- G.H. GOLUB AND W. KAHAN, "Calculating the singular values and pseudo-inverse of a matrix." *J. SIAM Numer. Analysis*, Vol. 2(B), 1965, pp. 205–224.

■

B

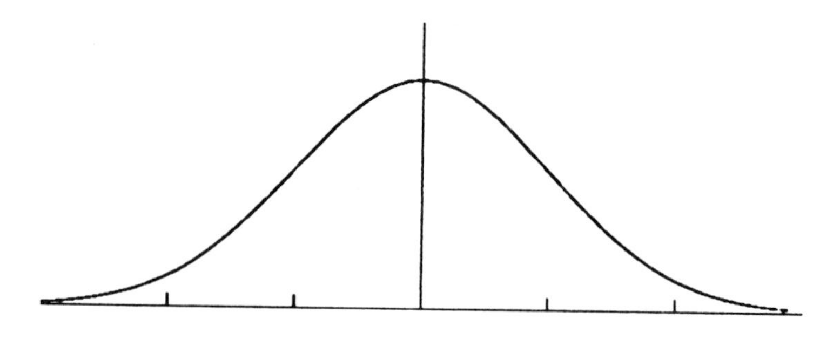

Statistical Inference

B.1 PRELIMINARIES

A *random variable* or a *stochastic variable* has a value which is dependent on chance and which cannot be predicted from a knowledge of the experimental conditions. To describe the outcome of a random variable X it is common practice to introduce the probability distribution function

$$F(x) = \mathcal{P}\{X \leq x\}, \qquad 0 \leq F(x) \leq 1, \quad \forall x \in R \tag{B.1}$$

where $\mathcal{P}\{X \leq x\}$ denotes the probability that $X \leq x$. The derivative $f(x)$ of the distribution function $F(x)$ is called the *probability density function*, and the solution x_α to the equation

$$\mathcal{P}\{X \leq x_\alpha\} = \alpha \tag{B.2}$$

is called the *α-percentile* of the distribution.

In cases when there is no risk of confusion we use x to denote also the random variable. The statistical *mean* μ_y or *expectation* of a variable y, which is a function of a random variable X, is defined as

$$\mu_y = \mathcal{E}\{y\} = \int_{-\infty}^{\infty} y(x)f(x)dx \tag{B.3}$$

and the *mean of the distribution* is

$$\mu_x = \mathcal{E}\{x\} = \int_{-\infty}^{\infty} xf(x)dx \tag{B.4}$$

The variance of a scalar variable $y(x)$ is defined as

$$\mathrm{Var}\{y\} = \mathcal{E}\{(y-\mu_y)^2\} = \int_{-\infty}^{\infty} (y(x)-\mu_y)^2 f(x)dx \tag{B.5}$$

In the case of vector-valued variables it is standard to use the definition of *covariance*

$$\mathrm{Var}\{y\} = \mathrm{Cov}\{y\} = \mathcal{E}\{(y-\mu_y)(y-\mu_y)^T\} = \int_{-\infty}^{\infty} (y(x)-\mu_y)(y(x)-\mu_y)^T f(x)dx \tag{B.6}$$

because it also describes the statistical relations between the components of the vector y. The covariance between two variables x and y is

$$\mathrm{Cov}\{x,y\} = \mathcal{E}\{(x-\mu_x)(y-\mu_y)^T\} = \mathcal{E}\{xy^T\} - \mu_x\mu_y^T \tag{B.7}$$

Definition B.1—Statistically independent variables
Two random variables x and y are said to be *statistically independent* if the probability density function

$$f(x,y) = f_1(x)f_2(y), \qquad \text{for some } f_1, f_2. \tag{B.8}$$

■

Definition B.2—Statistical covariance and correlation

The correlation coefficient between two variables x and y is

$$\rho = \frac{\mathrm{Cov}\{x, y\}}{\sigma_x \sigma_y}, \qquad \text{where} \quad \begin{cases} \sigma_x^2 = \mathrm{Var}\{x\} \\ \sigma_y^2 = \mathrm{Var}\{y\} \end{cases} \tag{B.9}$$

and two variables x, y are *uncorrelated* if

$$\mathrm{Cov}\{x, y\} = 0. \tag{B.10}$$

■

Definition B.3—The pth moment

The pth moment of a probability distribution $F(x)$ is the mathematical expectation

$$\mathcal{E}\{x^p\} = \int_{-\infty}^{\infty} x^p f(x) dx \tag{B.11}$$

■

The *sample mean* of a set of N observed variables x_i for $i = 1, \ldots, N$ is defined as

$$\bar{x} = \frac{1}{N} \sum_{i=1}^{N} x_i \tag{B.12}$$

and the *sample (co)variance* is

$$s^2 = \frac{1}{N} \sum_{i=1}^{N} (x_i - \mu_x)(x_i - \mu_x)^T \tag{B.13}$$

or for unknown μ_x we calculate the sample variance

$$s^2 = \frac{1}{N-1} \sum_{i=1}^{N} (x_i - \bar{x})(x_i - \bar{x})^T \tag{B.14}$$

■

B.2 CONVERGENCE AND CONSISTENCY

It is often desirable to provide an interval $[\theta_L, \theta_U]$ in which the value of a parameter θ would expect to lie with some probability

$$\mathcal{P}\{\theta_L < \theta < \theta_U\} = 1 - \alpha \tag{B.15}$$

In the case $\alpha = 0.05$ this interval $[\theta_L, \theta_U]$ is called a 95% *confidence interval.* More generally, the interval $[\theta_L, \theta_U]$ associated with a probability level $1-\alpha$ is called a $100(1-\alpha)$–percent confidence interval and the statistics θ_L and θ_U are the lower and upper *confidence limits*. In hypothesis testing, α is called the *significance level* of a test.

Definition B.4—Convergence in L^p, $0 < p < \infty$
The sequence $\{x_k\}$ is said to converge in L^p to x if and only if

$$\lim_{k\to\infty} \mathcal{E}\{|x_k - x|^p\} = 0 \tag{B.16}$$

Definition B.5—Convergence almost surely (a.s.)
The sequence $\{x_k\}$ is said to converge *with probability one (w.p.1)* or *almost surely (a.s)* to x if and only if for every $\varepsilon > 0$ we have

$$\lim_{n\to\infty} \mathcal{P}\{|x_k - x| \leq \varepsilon, \quad \forall k \geq n\} = 1 \tag{B.17}$$

Definition B.6—Convergence in probability (in pr.)
The sequence $\{x_k\}$ is said to converge *in probability (in pr.)* to x if and only if for every $\varepsilon > 0$ we have

$$\mathcal{P}\{|x_k - x| > \varepsilon\} = 0 \tag{B.18}$$

Definition B.7—Convergence in distribution (in dist.)
The sequence $\{x_k\}$ is said to converge *in distribution (in dist.)* to F if and only if the sequence $\{F_k\}$ of corresponding distribution functions converge to F. ■

A basic result of probability theory is the following implication relationships between the convergence concepts

$$\begin{array}{ccccc} x_k \overset{L^p}{\rightarrow} x & \overset{\not\Leftarrow}{\Rightarrow} & x_k \overset{\text{prob.}}{\rightarrow} x & \overset{\not\Rightarrow}{\Leftarrow} & x_k \overset{\text{a.s.}}{\rightarrow} F \\ & & \Downarrow \not\Uparrow & & \\ & & x_k \overset{\text{dist.}}{\rightarrow} x & & \end{array} \tag{B.19}$$

Theorem B.1—Central limit theorem
Let $\{x_k\}_{k=1}^{\infty}$ be a sequence of independent random variables with common distribution function F with finite mean μ and variance σ^2. If $S_N = \sum_{k=1}^{N} x_k$, then

$$X_N = \frac{S_N - N\mu}{\sigma\sqrt{N}} \overset{\text{dist}}{\rightarrow} \mathcal{N}(0,1) \tag{B.20}$$

i.e., X_N has a limiting normal distribution with mean 0 and variance 1 as $N \to \infty$. ■

The central limit theorem is commonly used as an approximation theorem in identification theory and is also often used to justify assumptions on normal distribution of variables.

The sample mean $\bar{x}$ and the sample variance s^2 are estimates of the mean μ and the variance σ^2 by means of some functions of data. In general, we use the notion *statistic* (*i.e.*, function of data) in comparing, describing, estimating, and in making decisions on the basis of the results of samples. Of all possible estimates of a parameter or variable some are better than others, and it is important to distinguish good estimates from poor estimates. For this reason it makes sense to introduce the notions of efficiency and consistency.

Definition B.8—Efficient estimate

An estimate $\widehat{\theta}$ is said to be an *efficient* estimate of a parameter θ if

$$\mathcal{E}\{(\widehat{\theta} - \theta)^2\} \leq \mathcal{E}\{(\widehat{\theta}' - \theta)^2\} \tag{B.21}$$

for any other estimate $\widehat{\theta}'$. ■

Definition B.9—Consistent estimate

Let an estimate of a parameter θ based on N samples be denoted $\widehat{\theta}_N$. The estimate $\widehat{\theta}_N$ is said to be a *consistent estimate* (in L^2 or in quadratic mean), of θ if

$$\lim_{N\to\infty} \mathcal{E}\{(\widehat{\theta}_N - \theta)^2\} = 0 \tag{B.22}$$

Consistency (in probability) of an estimate θ_N is defined as the convergence in probability of $\widehat{\theta}_N$ to θ, *i.e.*,

$$\lim_{N\to\infty} \mathcal{P}\{|\widehat{\theta}_N - \theta| > \varepsilon\} = 0, \qquad \text{for any} \quad \varepsilon > 0 \tag{B.23}$$

■

A shorthand way of writing Eq. (B.23) is

$$\operatorname{plim} \widehat{\theta}_N = \theta \tag{B.24}$$

where "plim" is called *probability limit*. An advantage is that there are attractive algebraic properties of the probability limit. For instance, it can be shown that for any continuous function $f(\theta)$ it holds that

$$\operatorname{plim} f(\widehat{\theta}_N) = f(\operatorname{plim} \widehat{\theta}_N) \tag{B.25}$$

For matrices A and B of suitable dimensions it holds that

$$\text{plim}(AB) = \text{plim}(A)\text{plim}(B) \tag{B.26}$$

provided that such probability limits exist (see Wilks, 1962.)

Definition B.10—Unbiased and asymptotically unbiased estimates
An estimate $\widehat{\theta}_N$ of θ based on N data is said to be *unbiased* if $\mathcal{E}\{\widehat{\theta}_N\} = \theta$. The estimate is said to be *asymptotically unbiased* if

$$\lim_{N\to\infty} \mathcal{E}\{\widehat{\theta}_N\} = \theta \tag{B.27}$$

■

Theorem B.2—The CRAMÉR-RAO lower bound
Let $\mathcal{Y}$ be observations of a stochastic variable, the distribution of which depends on an unknown vector θ. Let $L(\mathcal{Y}, \theta)$ denote the likelihood function and let $\bar{\theta} = \bar{\theta}(\mathcal{Y})$ be an arbitrary unbiased estimate of θ. Then

$$\text{Cov}(\bar{\theta}) \geq (\mathcal{E}\{(\frac{\partial \log L}{\partial\theta})^T(\frac{\partial \log L}{\partial\theta})\})^{-1} = -(\mathcal{E}\{\frac{\partial^2 \log L}{\partial\theta\partial\theta^T}\})^{-1} \tag{B.28}$$

Proof: Consider the covariance matrix

$$\mathcal{E}\{\begin{pmatrix} (\frac{\partial \log L}{\partial\theta})^T \\ \tilde{\theta} \end{pmatrix} \begin{pmatrix} \frac{\partial \log L}{\partial\theta} & \tilde{\theta}^T \end{pmatrix}\} = \begin{pmatrix} \mathcal{E}\{(\frac{\partial \log L}{\partial\theta})^T \frac{\partial \log L}{\partial\theta}\} & \mathcal{E}\{\tilde{\theta}\frac{\partial \log L}{\partial\theta}\}^T \\ \mathcal{E}\{\tilde{\theta}\frac{\partial \log L}{\partial\theta}\} & \text{Cov}\{\widehat{\theta}\} \end{pmatrix} \tag{B.29}$$

where the upper-left block contains the Fisher information matrix

$$I_\theta = \mathcal{E}\{(\frac{\partial \log L}{\partial\theta})^T \frac{\partial \log L}{\partial\theta}\} \tag{B.30}$$

and where the lower-right block is the covariance of $\widehat{\theta}$. Using the relationship

$$\frac{\partial p(\mathcal{Y}|\theta)}{\partial\theta} = \frac{\partial \log p(\mathcal{Y}|\theta)}{\partial\theta} p(\mathcal{Y}|\theta) \tag{B.31}$$

and beginning with the off-diagonal block of Eq. (B.29) we find that

$$\begin{aligned} \mathcal{E}\{\tilde{\theta}\frac{\partial \log L}{\partial\theta}\} &= \int \tilde{\theta}\frac{\partial \log p(\mathcal{Y}|\theta)}{\partial\theta} p(\mathcal{Y}|\theta)d\mathcal{Y} \\ &= \int \widehat{\theta}\frac{\partial \log p(\mathcal{Y}|\theta)}{\partial\theta} p(\mathcal{Y}|\theta)d\mathcal{Y} - \theta\int \frac{\partial \log p(\mathcal{Y}|\theta)}{\partial\theta} p(\mathcal{Y}|\theta)d\mathcal{Y} \end{aligned} \tag{B.32}$$

As $\hat{\theta} \in R^p$ is assumed to be an unbiased estimate of θ, we have

$$\mathcal{E}\{\hat{\theta}\} = \int \hat{\theta}(\mathcal{Y})p(\mathcal{Y}|\theta)d\mathcal{Y} = \theta \tag{B.33}$$

so that

$$\frac{\partial}{\partial\theta}\mathcal{E}\{\hat{\theta}\} = \int \hat{\theta}(\mathcal{Y})\frac{\partial p(\mathcal{Y}|\theta)}{\partial\theta}d\mathcal{Y} = I \tag{B.34}$$

provided that differentiation under the integral sign is allowed. In addition, as

$$0 = \frac{\partial}{\partial\theta}\int p(\mathcal{Y}|\theta)d\mathcal{Y} = \int \frac{\partial \log p(\mathcal{Y}|\theta)}{\partial\theta}p(\mathcal{Y}|\theta)d\mathcal{Y} = \mathcal{E}\{\frac{\partial \log p(\mathcal{Y}|\theta)}{\partial\theta}\} \tag{B.35}$$

one can factorize the matrix (B.29) according to the Result A.5 from Appendix A so that

$$\begin{pmatrix} I_\theta & I \\ I & \mathrm{Cov}\{\hat{\theta}\} \end{pmatrix} = \begin{pmatrix} I & 0 \\ I_\theta^{-1} & I \end{pmatrix} \begin{pmatrix} I_\theta & 0 \\ 0 & \mathrm{Cov}\{\hat{\theta}\} - I_\theta^{-1} \end{pmatrix} \begin{pmatrix} I & I_\theta^{-1} \\ 0 & I \end{pmatrix} \tag{B.36}$$

The requirement of a covariance matrix being non-negative definite and symmetric by construction entails that the block diagonal matrix in Eq. (B.36) is non-negative definite or that $\mathrm{Cov}\{\hat{\theta}\} \geq I_\theta^{-1}$ as stated by the theorem. Finally, the equality of Eq. (B.28) is shown by means of Eq. (B.31) and the relationship

$$\begin{aligned} 0 &= \frac{\partial^2}{\partial\theta_i\partial\theta_j}\int p(\mathcal{Y}|\theta)d\mathcal{Y} \\ &= \int \frac{\partial^2 \log p(\mathcal{Y}|\theta)}{\partial\theta_i\partial\theta_j}p(\mathcal{Y}|\theta)d\mathcal{Y} + \int (\frac{\partial \log p(\mathcal{Y}|\theta)}{\partial\theta_i})(\frac{\partial \log p(\mathcal{Y}|\theta)}{\partial\theta_j})p(\mathcal{Y}|\theta)d\mathcal{Y} \end{aligned} \tag{B.37}$$

Mathematical expectation of Eq. (B.37) then proves the theorem. ■

B.3 SOME IMPORTANT PROBABILITY DISTRIBUTIONS

The probability density function of the *normal distribution* is

$$f(x) = \frac{1}{\sqrt{2\pi}\sigma}e^{-(x-\mu)^2/2\sigma^2} \tag{B.38}$$

for a variable x with mean μ and variance σ^2 (see Fig. B.1). Hence, the normal distribution is parametrized by its mean μ and its variance σ^2 and it is

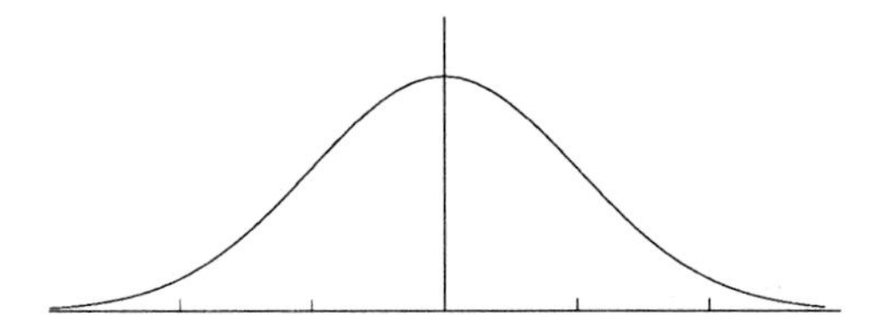

Figure B.1 The probability density function of the normal distribution.

often denoted $\mathcal{N}(\mu, \sigma^2)$ in statistical practice. In consequence, the statement that a variable x is normally distributed is denoted $x \in \mathcal{N}(\mu, \sigma^2)$. The distribution $\mathcal{N}(0, 1)$ is called the *standard normal distribution*, and a stochastic variable $x \in \mathcal{N}(0, 1)$ is called a *standard normal random variable*.

A multivariate normal distribution $\mathcal{N}(\mu, \Sigma)$ with mean $\mu \in R^p$ and covariance matrix $\Sigma \in R^{p \times p}$ has the probability density function

$$f(x) = \frac{1}{(2\pi)^{p/2}\sqrt{\det \Sigma}} \exp\left(-\frac{1}{2}(x - \mu)^T \Sigma^{-1}(x - \mu)\right) \tag{B.39}$$

The probability density function is symmetric around the mean μ and some of the two-tail probabilities, *i.e.*, the probability $\mathcal{P}\{|x - \mu| > \delta\}$ where δ is chosen such that $\mathcal{P}\{|x - \mu| > \delta\} \leq \alpha$. Standard choices of α are shown in Table B.1. The intervals around the mean $[\mu - \delta\sigma, \mu + \delta\sigma]$ define a confidence interval at the probability level $1 - \alpha$. Some of the two-tail probabilities for the standard normal distribution $\mathcal{N}(0, 1)$ are found in Table B.1.

The χ^2–distribution

By the χ^2–distribution we mean the distribution of a sum of squares of the form

$$\chi^2 = x_1^2 + \cdots + x_k^2 \tag{B.40}$$

of k independent, standard normal random variables $\{x_i\}_{i=1}^{k}$ where k is called the *number of degrees of freedom*. The mean of the χ^2–distribution is $\mu = k$ and the variance $\sigma^2 = 2k$ for the χ^2–distribution with k degrees of freedom;

Table B.1 Some confidence intervals at various probability levels for the normal distribution

$\mathcal{P}\{\lvert x\rvert > \delta\}$	δ	Confidence level	Confidence interval
α = 0.001	3.29	99.9%	$[\mu - 3.29\sigma, \mu + 3.29\sigma]$
0.005	3.09	99.5%	$[\mu - 3.09\sigma, \mu + 3.09\sigma]$
0.01	2.58	99%	$[\mu - 2.58\sigma, \mu + 2.58\sigma]$
0.05	1.96	95%	$[\mu - 1.96\sigma, \mu + 1.96\sigma]$

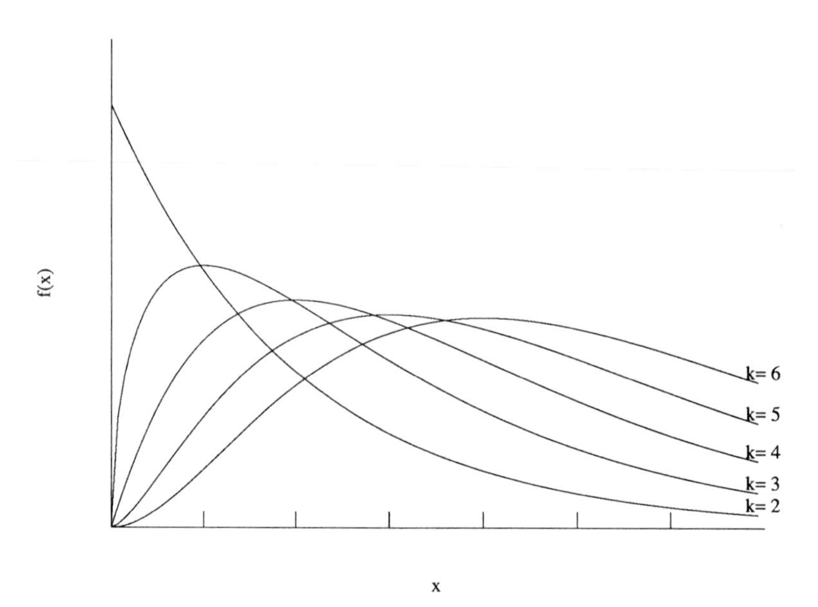

Figure B.2 The probability density function of the χ^2–distribution for some degrees of freedom $k = 2,\ldots,6$.

see Fig. B.2. The probability density function is

$$f(\chi^2) = \frac{1}{2^{k/2}\Gamma(k/2)}(\chi^2)^{(k/2)-1}e^{-\chi^2/2}, \qquad 0 \le \chi^2 < \infty \tag{B.41}$$

where $\Gamma(\cdot)$ denotes the standard Γ–function, which can be found in mathematical software and in statistical tables.

The α–percentile χ^2_α is the value for which

$$\mathcal{P}\{\chi^2 < \chi^2_\alpha\} \le \alpha \tag{B.42}$$

Table B.2 Percentiles χ^2_α of the χ^2–distribution

Degrees of freedom	$\chi^2_{.005}$	$\chi^2_{.01}$	$\chi^2_{.025}$	$\chi^2_{.05}$	$\chi^2_{.95}$	$\chi^2_{.975}$	$\chi^2_{.99}$	$\chi^2_{.995}$
1	0.00	0.00	0.001	0.004	3.84	5.02	6.63	7.88
2	0.010	0.020	0.051	0.103	5.99	7.38	9.21	10.6
3	0.072	0.115	0.216	0.352	7.81	9.35	11.3	12.8
4	0.207	0.297	0.484	0.711	9.49	11.1	13.3	14.9
5	0.412	0.554	0.831	1.15	11.1	12.8	15.1	16.7
6	0.676	0.872	1.24	1.64	12.6	14.4	16.8	18.5
7	0.989	1.24	1.69	2.17	14.1	16.0	18.5	20.3
8	1.34	1.65	2.18	2.73	15.5	17.5	20.1	22.0
9	1.73	2.09	2.70	3.33	16.9	19.0	21.7	23.6
10	2.16	2.56	3.25	3.94	18.3	20.5	23.2	25.2
20	7.43	8.26	9.58	10.9	31.4	34.2	37.6	40.0
30	13.8	15.0	16.8	18.5	43.8	47.0	50.9	53.7
40	20.7	22.1	24.4	26.5	55.8	59.3	63.7	66.8
50	28.0	29.7	32.3	34.8	67.5	71.4	76.2	79.5

A confidence interval at the probability level α around the mean for the χ^2–distribution is $[\chi^2_{\alpha/2}, \chi^2_{1-\alpha/2}]$ (see Table B.2). For degrees of freedom where $k > 30$, it is possible to approximate the percentile as $\chi^2_\alpha = 0.5(z_\alpha + \sqrt{2k-1})^2$ where z_α is the corresponding percentile of the standard normal distribution.

The F-distribution

The F-distribution is defined as the distribution of the ratio of two independent χ^2–variables

$$F = F(n_1, n_2) = \frac{\chi^2_{n_1}/n_1}{\chi^2_{n_2}/n_2} \tag{B.43}$$

with the probability density function

$$f(F) = \frac{\Gamma(\frac{1}{2}(n_1+n_2))}{\Gamma(\frac{1}{2}n_1)\Gamma(\frac{1}{2}n_2)}(\frac{n_1}{n_2})^{\frac{1}{2}n_1}\frac{F^{\frac{1}{2}n_1-1}}{((\frac{n_1}{n_2})F+1)^{\frac{1}{2}(n_1+n_2)}}, \qquad 0 < F < \infty \tag{B.44}$$

where $\Gamma(\cdot)$ is the standard Γ–function. The mean of the F-distribution is

$$\mathcal{E}\{F\} = \frac{n_2}{n_2 - 2}, \qquad n_2 > 2 \tag{B.45}$$

By definition the α−percentile F_α satisfies the relation

$$\alpha = \mathcal{P}\{F < F_\alpha\} = 1 - \mathcal{P}\{F > F_\alpha\} = 1 - \mathcal{P}\{\frac{1}{F} < \frac{1}{F_\alpha}\} \tag{B.46}$$

From the fact that the F-statistic is a ratio and from Eq. (B.46) it can be shown that the percentiles F_α and $F_{1-\alpha}$ are related as follows

$$F_{1-\alpha}(n_1, n_2) = \frac{1}{F_\alpha(n_2, n_1)} \tag{B.47}$$

which in turn helps to determine confidence intervals $[F_{\alpha/2}, F_{1-\alpha/2}]$ for F.

B.4 CONDITIONAL EXPECTATION

When analyzing data the observer is often met with the question of how the outcome of a variable X may be influenced by the outcome of another variable Y. The *conditional probability* is defined as the ratio

$$\mathcal{P}\{X < x | Y < y\} = \frac{\mathcal{P}\{X < x, Y < y\}}{\mathcal{P}\{Y < y\}} \tag{B.48}$$

and the associated *conditional probability density function* is

$$f(x|y) = \frac{f(x,y)}{f(y)} \tag{B.49}$$

Example B.1—Conditional normal distribution
Assume that $x \in R^n$ and $y \in R^m$ are correlated normally distributed variables with expected mean μ_x and μ_y so that

$$z = \begin{pmatrix} x \\ y \end{pmatrix} \in \mathcal{N}(\mu, P) = \mathcal{N}(\begin{pmatrix} \mu_x \\ \mu_y \end{pmatrix}, \begin{pmatrix} P_{xx} & P_{xy} \\ P_{xy}^T & P_{yy} \end{pmatrix}) \tag{B.50}$$

The conditional probability density function is then

$$f(x|y) = \frac{f(x,y)}{f(y)} = \frac{1}{\sqrt{2\pi}^{(n+m)-m} \det P / \det P_{yy}} \frac{\exp(-\frac{1}{2} z^T P^{-1} z)}{\exp \frac{(-1}{2} y^T P_{yy}^{-1} y)} \tag{B.51}$$

According to basic matrix algebra (see Appendix A) this can be simplified to

$$f(x|y) = \frac{1}{\sqrt{2\pi}^n \det Q} \exp(-\frac{1}{2}(x - \mu_{x|y})^T Q^{-1} (x - \mu_{x|y})) \tag{B.52}$$

where the conditional mean $\mu_{x|y}$ and covariance Q of x given the observation y are

$$\begin{aligned} \mu_{x|y} &= \mathcal{E}\{x|y\} &&= \mu_x + P_{xy}P_{yy}^{-1}(y - \mu_y) \\ Q &= \operatorname{cov}\{x|y\} &&= P_{xx} - P_{xy}P_{yy}^{-1}P_{xy}^T \end{aligned} \tag{B.53}$$

■

B.5 STATISTICAL HYPOTHESIS TESTING

Statistical hypothesis testing is a statement about some parameters, *e.g.*, mean or variance, of a probability distribution. In particular, statistical testing of variance properties has become important and is known as *analysis of variance*. Its application is important in contexts where the result and quality of two different experimental methods should be distinguished by means of statistical analysis of data. A relevant observation is that several tests of differences between two experimental methods can be reduced to the question about the equality of variances of the two methods. This can be stated formally as the two alternatives

$$\begin{aligned} H_0 &: \ \sigma_1^2 = \sigma_2^2 \\ H_A &: \ \sigma_1^2 \neq \sigma_2^2 \end{aligned} \tag{B.54}$$

where σ_1^2 and σ_2^2 are the variances of the two methods. The hypothesis H_0, that the two variances are equal, is called the *null hypothesis*, whereas H_A is called the *alternative hypothesis*. A test statistic for testing these hypotheses is the ratio of the two sample variances, *i.e.*,

$$F = \frac{s_1^2}{s_2^2} \tag{B.55}$$

based on the sample sizes n_1 and n_2, respectively. Under the assumption of normal distribution and if H_0 is true, it follows that the statistic F is $F(n_1-1, n_2-1)$−distributed. The null hypothesis H_0 can be accepted according to the *two-sided F-test* if

$$F_{\alpha/2}(n_1 - 1, n_2 - 1) \leq F \leq F_{1-\alpha/2}(n_1 - 1, n_2 - 1) \tag{B.56}$$

Therefore, we should reject the null hypothesis H_0 for F−values that are either too large or too small.

If it is desirable to reject H_0 only if one variance is larger than the other, then the two hypotheses are

$$\begin{aligned} H_0 &: \ \sigma_1^2 = \sigma_2^2 \\ H_A &: \ \sigma_1^2 > \sigma_2^2 \end{aligned} \tag{B.57}$$

The associated *one-sided F-test* consists of accepting the null hypothesis if

$$F \le F_\alpha(n_1 - 1, n_2 - 1) \tag{B.58}$$

Example B.2—A one-sided F-test of variance

In a model order test in identification one wishes to test whether a new proposed method has a larger variance than the old one. The hypothesis to test is

$$\begin{aligned} H_0 &: \ \sigma_1^2 = \sigma_2^2 \\ H_A &: \ \sigma_1^2 > \sigma_2^2 \end{aligned} \tag{B.59}$$

Two samples of sizes $n_1 = 10$ and $n_2 = 20$ are taken, and the sample variances are $s_1^2 = 0.56$ and $s_2^2 = 0.35$ so that the test statistic is $F = s_1^2/s_2^2 = 1.60$. From Table B.3 and Table B.4 we find $F_{0.05}(10, 20) = 2.35$. Clearly $F < F_{0.05}(10, 20)$, which means that the null hypothesis can not be rejected and we have insufficient statistical support for distinguishing between the two methods. ■

Suppose, instead, that it is desirable to test the hypothesis that the variance σ_1^2 is equal to some constant σ^2. The hypotheses are

$$\begin{aligned} H_0 &: \ \sigma_1^2 = \sigma^2 \\ H_A &: \ \sigma_1^2 > \sigma^2 \end{aligned} \tag{B.60}$$

A relevant test statistic is

$$\chi^2 = \frac{(n_1 - 1)s_1^2}{\sigma^2} \tag{B.61}$$

where s_1^2 is the estimate of σ_1^2 based on n_1 data. The test statistic (B.61) is χ^2–distributed under assumptions of normal distribution and under the null hypothesis. In this case we should reject the null hypothesis H_0 for values of χ^2 that are too large, *i.e.*, we reject when

$$\chi^2 > \chi_\alpha^2(n_1 - 1) \tag{B.62}$$

where $\chi_\alpha^2(n - 1)$ is the α-percentile point of the χ^2–distribution with $n - 1$ degrees of freedom.

Table B.3 Percentage points $F_{0.05}$ of the F-distribution

$n_2 \backslash n_1$	1	2	3	4	5	6	8	10	20	30	∞
1	161	200	216	225	230	234	239	242	248	250	254
2	18.5	19.0	19.2	19.2	19.3	19.3	19.4	19.4	19.4	19.5	19.5
3	10.1	9.55	9.28	9.12	9.01	8.94	8.85	8.79	8.66	8.62	8.53
4	7.71	6.94	6.59	6.39	6.26	6.16	6.04	5.96	5.80	5.75	5.63
5	6.61	5.79	5.41	5.19	5.05	4.95	4.82	4.74	4.56	4.50	4.36
6	5.99	5.14	4.76	4.53	4.39	4.28	4.15	4.06	3.87	3.81	3.67
7	5.59	4.74	4.35	4.12	3.97	3.87	3.73	3.64	3.44	3.38	3.23
8	5.32	4.46	4.07	3.84	3.69	3.58	3.44	3.35	3.15	3.08	2.93
9	5.12	4.26	3.86	3.63	3.48	3.37	3.23	3.14	2.94	2.86	2.71
10	4.96	4.10	3.71	3.48	3.33	3.22	3.07	2.98	2.77	2.70	2.54
20	4.35	3.49	3.10	2.87	2.71	2.60	2.45	2.35	2.12	2.04	1.84
30	4.17	3.32	2.92	2.69	2.53	2.42	2.27	2.16	1.93	1.84	1.62
∞	3.84	3.00	2.60	2.37	2.21	2.10	1.94	1.83	1.57	1.46	1.00

A two-sided test is relevant for testing the hypothesis

$$\begin{aligned} H_0 &: \quad \sigma_1^2 = \sigma^2 \\ H_A &: \quad \sigma_1^2 \neq \sigma^2 \end{aligned} \tag{B.63}$$

where the alternative hypothesis should be rejected if

$$\chi^2_{\alpha/2}(n-1) \leq \chi^2 \leq \chi^2_{1-\alpha/2}(n-1) \tag{B.64}$$

This test is relevant to test if two methods are equal in precision.

Example B.3—A two-sided χ^2–test of variance
A sum of squares of 10 random zero-mean variables $x_1, \ldots, x_{10}$ is

$$\chi^2 = \sum_{i=1}^{10} x_i^2 = 18.4 \tag{B.65}$$

The 95% confidence interval for χ^2 is $[\chi^2_{0.025}(9), \chi^2_{0.975}(9)] = [2.70, 19.0]$ under the assumption of normal distribution with $\sigma^2 = 1$. Hence we can not reject the hypothesis that $\sigma^2 = 1$. ■

There is always a risk that a statistical test may lead to false conclusions. Also, good tests may cause rejection of H_0 when it is true or acceptance of H_0

when it is false. It is standard practice to classify these decision errors as

$$\begin{array}{ll} \text{Type I error} & \mathcal{P}\{\text{reject} \quad H_0|H_0 \quad \text{is true}\} = \alpha \\ \text{Type II error} & \mathcal{P}\{\text{accept} \quad H_0|H_0 \quad \text{is false}\} \end{array} \tag{B.66}$$

The type-I error appears by chance with a probability α but it is usually more difficult to quantify the risk of a type-II error. A cautious attitude toward statistical decision and testing is to reject only hypotheses and to avoid accepting an hypothesis that is not rejected. In particular, accepting a very composite alternative hypothesis or a very specific and "narrow" null hypothesis might lead to wrong inferences.

B.6 THE COCHRAN THEOREM

Consider a linear transformation from the vector $U = (U_1, \ldots, U_n)^T$ to $L = (L_1, \ldots, L_m)^T$ where each vector L_i is an $n_i \times n-$vector

$$L = \begin{pmatrix} L_1 \\ \vdots \\ L_m \end{pmatrix} = \begin{pmatrix} A_1 \\ A_2 \\ \vdots \\ A_m \end{pmatrix} = AU \tag{B.67}$$

for some matrices $A_i \in R^{n_i \times n}$ for $i = 1, \ldots, m$. Let the rank of each matrix A_i be denoted r_i and let Q_i denote the sum of squares of the L_i's

$$\begin{aligned} Q_i &= L_i^T L_i = U^T A_i^T A_i U \\ Q &= L^T L = L_1^T L_1 + \cdots + L_m^T L_m = Q_1 + \cdots + Q_m \end{aligned} \tag{B.68}$$

Let $U_1, \ldots, U_n$ be independent, standard normal variables, and suppose that one can write an identity of the form

$$U^T U = \sum_{i=1}^{n} U_i^2 = Q_1 + \cdots + Q_m \tag{B.69}$$

where each Q_i is a sum of squares of linear combinations of the components of U, *i.e.*, $Q_i = U^T A_i^T A_i U$ for matrices A_i of rank r_i. Then, if

$$r_1 + \cdots + r_m = n \tag{B.70}$$

Table B.4 Percentage points $F_{0.01}$ of the F-distribution

$n_2 \backslash n_1$	1	2	3	4	5	6	8	10	20	30	∞
1	4052	5000	5403	5625	5764	5859	5982	6056	6210	6260	6366
2	98.5	99.0	99.2	99.2	99.3	99.4	99.4	99.4	99.4	99.5	99.5
3	34.1	30.8	29.5	28.7	28.2	27.9	27.5	27.3	26.7	26.5	26.1
4	21.2	18.0	16.7	16.0	15.5	15.2	14.8	14.5	14.0	13.8	13.5
5	16.3	13.3	12.1	11.4	11.0	10.7	10.3	10.1	9.55	9.38	9.02
6	13.7	10.9	9.78	9.15	8.75	8.47	8.10	7.87	7.40	7.23	6.88
7	12.2	9.55	8.45	7.85	7.46	7.19	6.84	6.62	6.16	5.99	5.65
8	11.3	8.65	7.59	7.01	6.63	6.37	6.03	5.81	5.36	5.20	4.86
9	10.6	8.02	6.99	6.42	6.06	5.80	5.47	5.26	4.81	4.65	4.31
10	10.0	7.56	6.55	5.99	5.64	5.39	5.06	4.85	4.41	4.25	3.91
20	8.10	5.85	4.94	4.43	4.10	3.87	3.56	3.37	2.94	2.78	2.42
30	7.56	5.39	4.51	4.02	3.70	3.47	3.17	2.98	2.55	2.39	2.01
∞	6.63	4.61	3.78	3.32	3.02	2.80	2.51	2.32	1.88	1.70	1.00

it follows that the variables Q_i have independent χ^2–distributions, each with a number of degrees of freedom given by its rank as a quadratic form. This result is known as the *Cochran theorem* and serves as a theoretical basis for a number of statistical tests.

Example B.4—Application of the Cochran theorem

Let $x_1, \ldots, x_N$ be independent normal variables with mean $\mathcal{E}\{x_i\} = \mu$ and $\mathrm{Cov}\{x_i, x_j\} = \delta_{ij}\sigma^2$

$$\sum_{i=1}^{N}\left(\frac{x_i - \mu}{\sigma}\right)^2 = Q_1 + \cdots + Q_k \tag{B.71}$$

each Q_i being a sum of squares of linear combinations of $x_1, \ldots, x_N$. If we consider the special case

$$Q = \sum\left(\frac{x_i - \mu}{\sigma}\right)^2 = \sum\left(\frac{x_i - \bar{x}}{\sigma}\right)^2 + N\left(\frac{\bar{x} - \mu}{\sigma}\right)^2 = Q_1 + Q_2 \tag{B.72}$$

we notice that Q can be expressed as the sum of two sums of squares with Q_1 related to the sample variance whereas Q_2 is related to the sample mean variance. The variable Q consists of a sum of squares of normal variables and has a χ^2–distribution with N degrees of freedom.

$$L_i = \frac{x_i - \bar{x}}{\sigma} = \frac{x_i - \mu}{\sigma} - \frac{1}{N}\frac{\bar{x} - \mu}{\sigma} \tag{B.73}$$

The sum of these L_i is zero and so the rank of the transformation of Q_1 does not exceed $N - 1$. The distribution of Q is clearly $\chi^2(N)$ and that of Q_2 is $\chi^2(1)$. If Q_1 and Q_2 are found independent then it can be concluded that Q_1 is $\chi^2(N-1)$–distributed. Notice now that

$$L = \begin{pmatrix} L_1 \\ L_2 \\ \vdots \\ L_N \end{pmatrix} = \frac{1}{\sigma N} \begin{pmatrix} N-1 & -1 & \cdots & -1 \\ -1 & N-1 & \ddots & \vdots \\ \vdots & \ddots & \ddots & -1 \\ -1 & \cdots & -1 & N-1 \end{pmatrix} \begin{pmatrix} x_1 \\ x_2 \\ \vdots \\ x_N \end{pmatrix} = A \cdot X \quad \text{(B.74)}$$

where the matrix A that relates L and X is of rank $N - 1$. As the degree of freedom for the sum is N and as the ranks add up as in the condition of the Cochran theorem, the conclusion then follows that Q_1 and Q_2 have independent χ^2–distributions with $N - 1$ and 1 degrees of freedom, respectively. ■

B.7 REFERENCES

Excellent handbooks of statistics are the following monographs

- D.R. COX AND D.V. HINKLEY, *Theoretical Statistics*. London: Chapman and Hall, 1974.
- M.G. KENDALL AND A. STUART, *The Advanced Theory of Statistics, Vol. 1, (3d ed.)*, 1969; *Vol. II, (3d ed.)*, 1973, New York: Hafner Press.
- S.S. WILKS, *Mathematical Statistics*. New York: John Wiley, 1962.

where proofs are found for the theorems included in this summary. ■

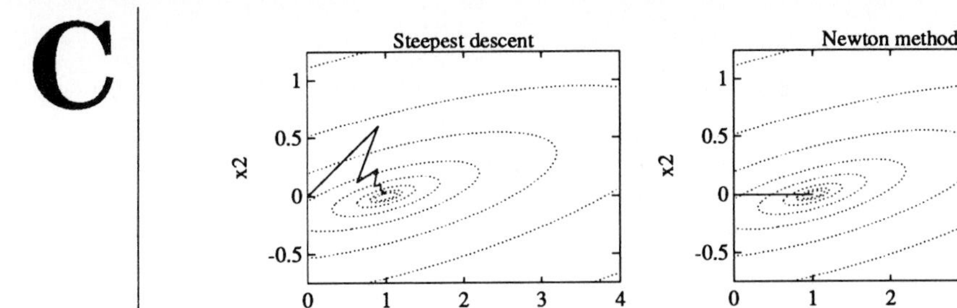

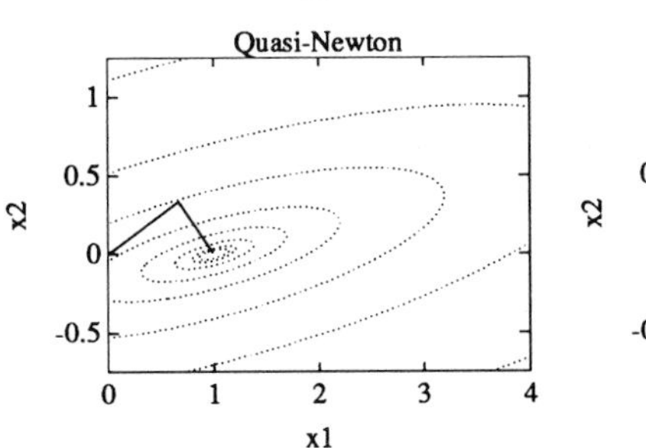

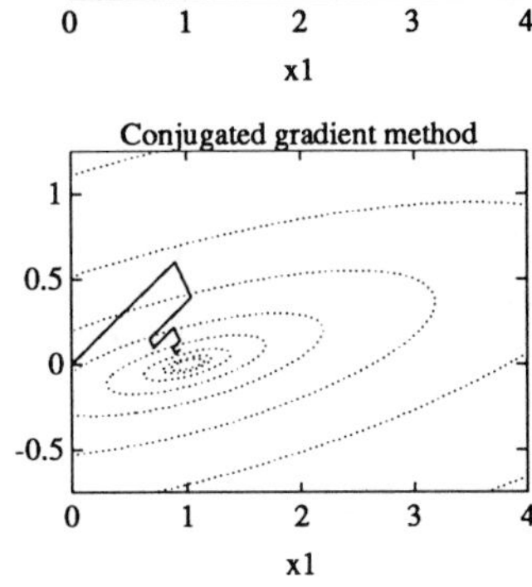

Numerical Optimization

C.1 INTRODUCTION

We will consider the unconstrained optimization problem of

$$\text{Minimize} \qquad f(x), \qquad x = \left(x_1, \ldots, x_n \right)^T \tag{C.1}$$

where x is the vector of free variables to be found. Noniterative methods for finding the optimum $\widehat{x}$ that minimizes $f(x)$ are the *grid search* and *random search*. The grid search consists of constructing a p−dimensional block of points to cover the region where $\widehat{x}$ is known to be, evaluating f at each of its points and choosing the minimum. A random search consists of randomly generating a sequence of points lying within a specified region, evaluating $f(x)$, and choosing the minimum. All these methods are extremely inefficient since they make a large number of unnecessary function evaluations.

C.2 DESCENT METHODS

Several numerical methods for solution of Eq. (C.1) are iterative. Assume that the ith iteration has provided the estimate $x^{(i)}$ and that the minimum lies in the direction d at a distance ρ from $x^{(i)}$. The Taylor series expansion around $x^{(i)}$ is then

$$f(x) = f(x^{(i)} + \rho d) = f(x^{(i)}) + \rho \nabla f(x)|_{x=x^{(i)}}^{T} d + O(\rho^2) \tag{C.2}$$

where the gradient $g = \nabla f(x^{(i)})$ is the gradient of f evaluated at $x^{(i)}$.

In iterative methods we try to improve the estimate $x^{(i)}$ by choosing a value $x^{(i+1)}$ such that $f(x^{(i+1)}) < f(x^{(i)})$ and the search direction d is a *descent direction* at $x^{(i)}$ if and only if $g^T d < 0$. For instance, for all positive definite matrices R we can suggest the descent direction

$$d = -Rg = -R\nabla f(x)|_{x=x^{(i)}} \tag{C.3}$$

as $g^T d = -g^T R g < 0$ for non-zero $g = \nabla f$. An iterative algorithm to improve $x^{(i)}$ is

$$x^{(i+1)} = x^{(i)} - \alpha^{(i)} R g \tag{C.4}$$

so that $f(x^{i+1)}) < f(x^{(i)})$ for suitable choices of the *step length* $\alpha^{(i)}$, beginning with an initial estimate $x^{(0)}$ and proceeding by iterating

$$x^{(i+1)} = x^{(i)} + \alpha^{(i)} d^{(i)} \tag{C.5}$$

One method to choose the step length is, for instance, so that $f(x^{(i)} - \alpha^{(i)} R^{(i)} g^{(i)})$ is minimized.

The choice $R = I$ is called the *method of steepest descent*, and the *steepest descent* search direction is defined by

$$d^{(i)} = -\nabla f(x^{(i)}) \tag{C.6}$$

where $\nabla f(x^{(i)})$ is the gradient of f evaluated at $x^{(i)}$. The steepest descent direction is simple to implement but has the disadvantage of being very slow to converge when the level contours $f(x) = c$ for any constant c are eccentric.

C.3 NEWTON METHODS

Consider the Taylor series expansion of a twice differentiable function $f(x)$ around $x^{(i)}$ with

$$f(x) = f(x^{(i)} + \rho d) = f(x^{(i)}) + \rho \nabla f(x)|_{x=x^{(i)}}^{T} d + \frac{1}{2}\rho^2 d^T \nabla^2 f(x)|_{x=x^{(i)}} d + O(\rho^3) \tag{C.7}$$

If we neglect higher-order terms and look for minimum of f with respect to d, we find that the Newton search direction is

$$d^{(i)} = -(\nabla^2 f(x^{(i)}))^{-1} \nabla f(x^{(i)}) \tag{C.8}$$

where the *Hessian matrix* $\nabla^2 f(x^{(i)})$ is the matrix of second partial derivatives of f evaluated at $x^{(i)}$. The search direction can then be found by solving the linear equation

$$\nabla^2 f(x^{(i)}) d^{(i)} = -\nabla f(x^{(i)}) \tag{C.9}$$

The Newton method has the advantage of very rapid convergence when $x^{(i)}$ is close to the optimum $\widehat{x}$ but has the disadvantage of needing time to calculate the matrix of second partial derivatives. Another problem is that the matrix inverse of second partial derivatives might not exist at some distance from the optimum $\widehat{x}$.

C.4 QUASI-NEWTON METHODS

There are two major classes of optimization techniques, quasi-Newton methods and conjugate gradient methods, that try to avoid some of these disadvantages. Quasi-Newton methods approximate the search direction with

$$d^{(i)} = -H^{(i)} \nabla f(x^{(i)}) \tag{C.10}$$

where $H^{(i)}$ is some positive definite approximation to $(\nabla^2 f(x^{(i)}))^{-1}$ obtained without evaluation of the Hessian matrix. These methods thus avoid problems of the matrix second partial derivatives and can guarantee a descent direction.

C.5 CONJUGATE GRADIENT METHODS

Given a symmetric matrix Q, two vectors v_1 and v_2 are said to be *conjugate* (or Q-orthogonal) with respect to Q if $v_1^T Q v_2 = 0$. Thus if $Q = I$, conjugacy is equivalent to the usual notion of orthogonality. *Conjugate gradient methods* compute the search direction $d^{(i)}$ as a linear combination of the current gradient vector and the previous search direction in the form of a recursive equation

$$d^{(i+1)} = -\nabla f(x^{(i+1)}) + \beta^{(i)} d^{(i)} \tag{C.11}$$

where $\beta^{(i)}$ is a scalar parameter chosen to assure that the sequence of search directions $d^{(i)}$ satisfies the conjugacy condition

$$0 = (d^{(i)})^T Q d^{(i+1)} = -(d^{(i)})^T Q \nabla f(x^{(i+1)}) + \beta^{(i)} (d^{(i)})^T Q d^{(i)} \tag{C.12}$$

A suitable choice of $\beta^{(i)}$ is therefore

$$\beta^{(i)} = \frac{(d^{(i)})^T Q \nabla f(x^{(i+1)})}{(d^{(i)})^T Q d^{(i)}} \tag{C.13}$$

A characteristic property of the search directions used in conjugate gradient methods is that they avoid steps in the same directions as in a few previous steps. This property is accomplished by the condition (C.12). This procedure usually includes some evaluation of the Hessian matrix as a means to define the Q–matrix.

Quadratic functions are of particular importance in nonlinear optimization for several reasons. A general smooth function can be approximated by a quadratic function using the Taylor series expansions. Also, many numerical optimization methods are based on the quadratic optimizations. We illustrate the behavior of some algorithms when applied to a quadratic function.

Example C.1—Comparison of some numerical optimization methods
We apply some of the above-mentioned methods to minimization of the function

$$\begin{aligned} f(x) &= \frac{1}{2} x^T Q_2 x + x^T q_1 + q_0 \\ &= \frac{1}{2} \begin{pmatrix} x_1 & x_2 \end{pmatrix} \begin{pmatrix} 3 & 2 \\ 2 & 3 \end{pmatrix} \begin{pmatrix} x_1 \\ x_2 \end{pmatrix} + \begin{pmatrix} x_1 & x_2 \end{pmatrix} \begin{pmatrix} -3 \\ -2 \end{pmatrix} + 1.5 \end{aligned} \tag{C.14}$$

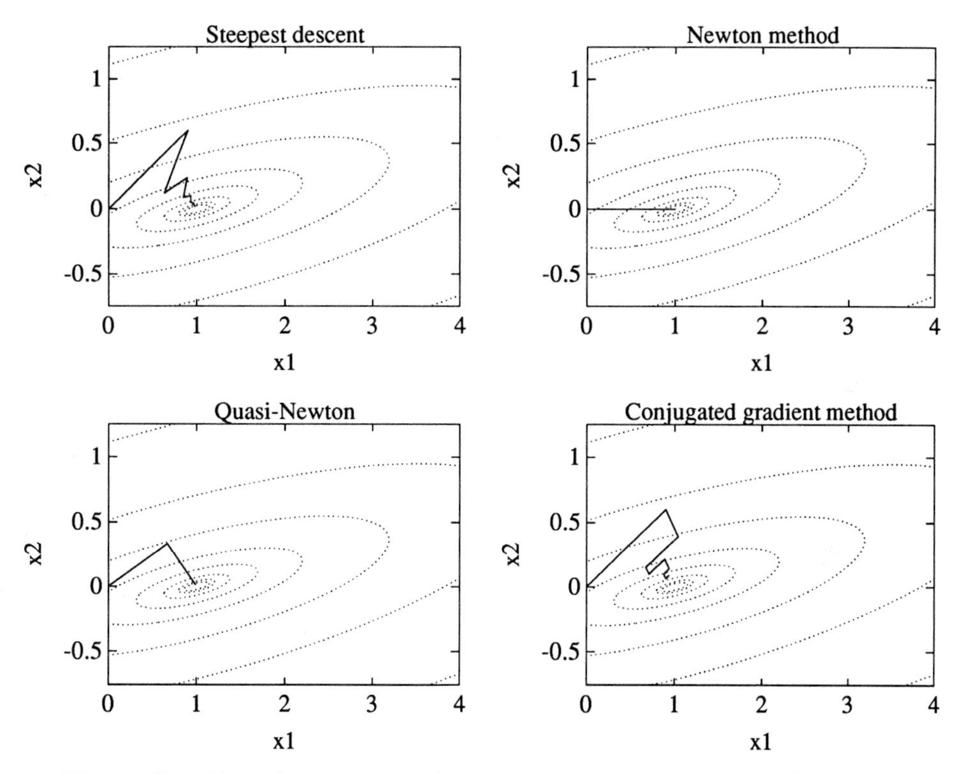

Figure C.1 Typical trajectories of some numerical optimization techniques when finding the minimum at (1, 0) of a quadratic function. Level contours are indicated by dotted lines.

The corresponding gradient is

$$\nabla f(x) = Q_2 x + q_1 \tag{C.15}$$

The steepest descent algorithm was applied with a step-length value $\alpha = 0.3$ and the quasi-Newton method was using a matrix

$$H = \begin{pmatrix} 4 & 1 \\ 1 & 4 \end{pmatrix} \tag{C.16}$$

The numerical solutions may be compared to the analytic solution of the equation

$$\nabla f(\hat{x}) = 0 \quad \Rightarrow \quad \hat{x} = -Q_2^{-1} q_1 = \begin{pmatrix} 1 \\ 0 \end{pmatrix} \tag{C.17}$$

in the present case. The trajectories of the different methods are shown in Fig. C.1 where all trajectories start with the initial value at the origin. ∎

C.6 DIRECT SEARCH METHODS

Consider the optimization problem

$$\text{Minimize} \qquad f(x), \qquad x = \left(x_1, \ldots, x_n \right)^T \tag{C.18}$$

In cases where it is not possible to calculate the partial derivatives and the gradients it might be necessary to use a direct search method. The Nelder-Mead polytope algorithm is one of the most popular and successful direct search methods. In order to estimate $x \in R^n$ it requires $(n+1)$ starting values $x^{(1)}, \ldots, x^{(n+1)}$ ordered so that $f(x^{(1)}) \leq \ldots \leq f(x^{(n+1)})$ and so that $\|x^{(i)} - x^{(j)}\|$ is constant for all $i \neq j$, *i.e.*, so that $x^{(1)}, \ldots, x^{(n+1)}$ form the corners of a regular polytope (triangle, tetrahedron, etc.). By shifting the corner with the highest cost-function value $x^{(n+1)}$ the polytope can be moved in space toward the minimum. This can be accomplished, for instance, by reflecting the worst corner $x^{(n+1)}$ through the centroid

$$c = \frac{1}{n} \sum_{i=1}^{n} x^{(i)} \tag{C.19}$$

The resulting algorithm is

$$x^{(n+2)} = c + \alpha(c - x^{(n+1)}) \tag{C.20}$$

for some step-length α so that the cost function $f(x^{(n+2)})$ shows improvement as compared to the starting values $x^{(i)}$. If the new improved estimate $x^{(n+2)}$ is such that

$$f(x^{(1)}) \leq f(x^{(n+2)}) \leq f(x^{(n+1)}) \tag{C.21}$$

then $x^{(n+2)}$ should replace $x^{(n+1)}$. By reordering all the new $\{x^{(i)}\}_{i=1}^{n+1}$ according to Eq. (C.21) and iterating the procedure, the polytope moves toward the minimum. Notice that this algorithm is usually not competitive with gradient methods in cases where gradients can be calculated.

C.7 PARAMETRIC OPTIMIZATION

We take the approximate maximum-likelihood identification as an example of optimization methods applied to parametric identification.

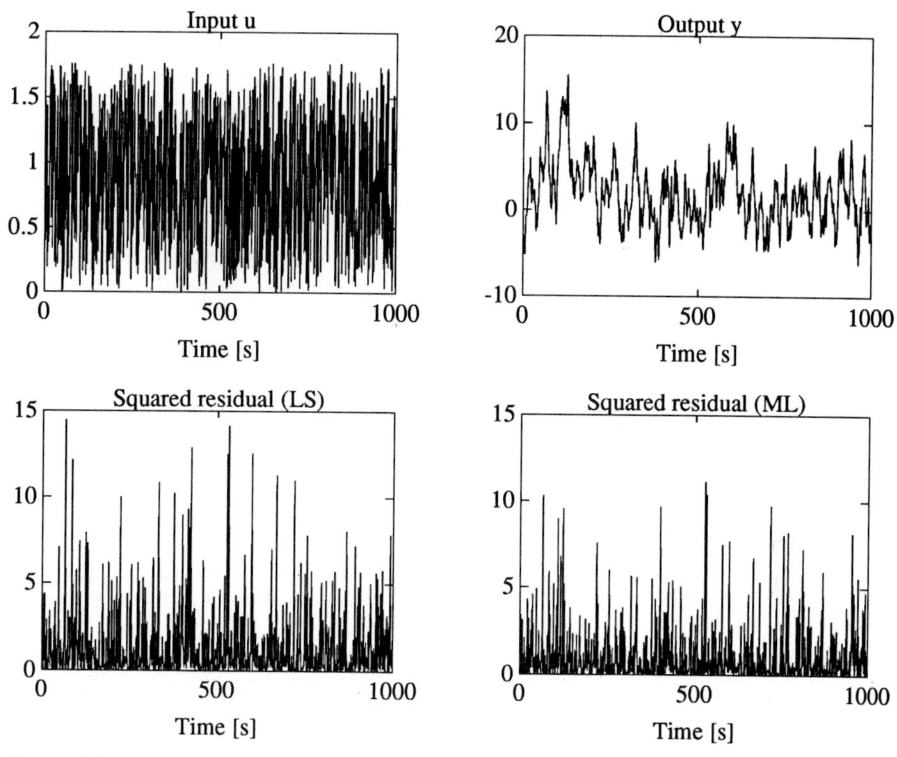

Figure C.2 LS and ML identification of an object described by the ARMAX model $y_k + 0.9y_{k-1} = 0.1u_{k-1} + e_k + 0.7e_{k-1}$. The lower graphs show the squared residuals ε_k^2 for the least squares (LS) and maximum likelihood estimation (ML) methods, respectively.

Example C.2—Approximate ML parameter estimation

Autoregressive moving average models with exogenous input (ARMAX) constitute a model set general enough to describe colored noise that can not be described by the ARX models of the type (6.7). Similar to (6.4) we consider ARMAX models of the type

$$A(z^{-1})y_k = z^{-d}B(z^{-1})u_k + C(z^{-1})v_k \tag{C.22}$$

where the noise covariance matrix $\Sigma_v = \mathcal{E}\{vv^T\}$ is now assumed to be unknown. Formulation of a maximum-likelihood problem involves the formulation of a likelihood function $L(\bar{\theta})$. Considering the case of normally distributed noise we have

$$L(\bar{\theta}) = \frac{1}{(2\pi)^{N/2}(\det \Sigma_v)^{1/2}} \exp(-\frac{1}{2}\varepsilon^T(\bar{\theta})\Sigma_v^{-1}\varepsilon(\bar{\theta})) \tag{C.23}$$

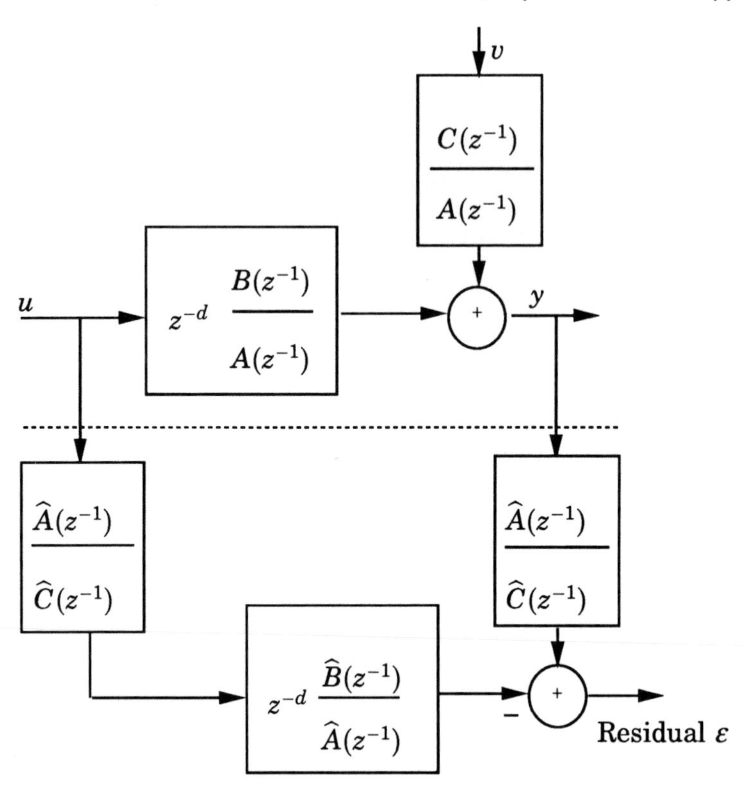

Figure C.3 Block diagram showing the approximate maximum likelihood estimation of an ARMAX model. Minimization of the prediction error is obtained by adjustment of $\widehat{A}$, $\widehat{B}$, and $\widehat{C}$.

or

$$\log L(\bar{\theta}) = -\frac{1}{2}\log(2\pi)^N \det \Sigma_v - \frac{1}{2}\varepsilon^T(\bar{\theta})\Sigma_v^{-1}\varepsilon(\bar{\theta}) \tag{C.24}$$

with ε containing the components $\varepsilon_k = y_k - \phi_k^T\bar{\theta}$ and

$$\begin{aligned} y_k = &- a_1 y_{k-1} - \cdots - a_{n_A} y_{k-n_A} \\ &+ b_1 u_{k-d-1} + \cdots + b_m u_{k-d-n_B} \\ &+ v_k + c_1 v_{k-1} + \cdots + c_{n_C} v_{k-n_C} = \phi_k^T\theta + v_k \end{aligned} \tag{C.25}$$

with

$$\begin{aligned} \phi_k &= \left(-y_{k-1} \ldots - y_{k-n_A} \quad u_{k-d-1} \ldots u_{k-d-n_B} \quad v_{k-1} \ldots v_{k-n_C}\right)^T \\ \theta &= \left(a_1 \ldots a_{n_A} \quad b_1 \ldots b_{n_B} \quad c_1 \ldots c_{n_C}\right)^T \end{aligned} \tag{C.26}$$

where it is a problem that the components $v_{k-1}, \ldots, v_{k-n_c}$ are not known. Another problem is that this optimization criterion is a function of both θ and σ^2 and is thus not known. In the absence of the desired parameters it is therefore only feasible to make approximate solutions by finding successively better estimates of the covariance matrix Σ_v and the parameters θ through some iterative procedure. In the important special case with normally distributed white noise with $\Sigma_v = \sigma^2 I$, we replace Eq. (6.34) by the empirical likelihood function

$$\begin{aligned} \log \widehat{L}(\bar{\theta}, \bar{\sigma}_v^2) &= -\frac{N}{2}\log(2\pi) - \frac{1}{2\bar{\sigma}_v^2}\sum_{k=1}^{N} \varepsilon_k^2(\bar{\theta}) - \frac{N}{2}\log \bar{\sigma}_v^2 \\ &= -\frac{N}{2}\log(2\pi) - \frac{1}{\bar{\sigma}_v^2} V_N(\bar{\theta}) - \frac{N}{2}\log \bar{\sigma}_v^2 \end{aligned} \tag{C.27}$$

where

$$V_N(\bar{\theta}) = \frac{1}{2}\sum_{k=1}^{N} \varepsilon_k^2(\bar{\theta}) = \frac{1}{2}\varepsilon^T(\bar{\theta})\varepsilon(\bar{\theta}) \tag{C.28}$$

The gradient and the second-order derivatives of $\log L(\bar{\theta})$ determine the extrema of $\log L(\bar{\theta})$ as

$$\begin{aligned} 0 &= \frac{\partial}{\partial \bar{\theta}} \log \widehat{L}(\bar{\theta}, \bar{\sigma}_v^2) = -\frac{1}{\bar{\sigma}_v^2}\nabla V_N(\bar{\theta}) \\ 0 &= \frac{\partial}{\partial \bar{\sigma}_v^2} \log \widehat{L}(\bar{\theta}, \bar{\sigma}_v^2) = -\frac{1}{2}\left(-\frac{2}{\bar{\sigma}_v^4} V_N(\bar{\theta}) + \frac{N}{\bar{\sigma}_v^2}\right) \end{aligned} \tag{C.29}$$

with the solution

$$\begin{aligned} \widehat{\sigma}_v^2 &= \frac{2}{N} V_N(\widehat{\theta}) \\ \nabla V_N(\widehat{\theta}) &= 0 \end{aligned} \tag{C.30}$$

A numerical solution to the problem $\nabla V_N(\widehat{\theta}) = 0$ can be obtained as an iterative procedure *via* the Newton(-Raphson) method

$$\theta^{(i+1)} = \theta^{(i)} - \alpha^{(i)}(\nabla^2 V(\theta^{(i)}))^{-1}\nabla V(\theta^{(i)}) \tag{C.31}$$

where $\alpha^{(i)}$ is step length to choose and (i) denotes the iteration order. The elements of this computation can be given the form

$$\begin{aligned}
V_N(\bar{\theta}) &= \frac{1}{2}\sum_{k=1}^{N}\varepsilon_k^2(\bar{\theta}) \\
\psi_k(\bar{\theta}) &= -\nabla\varepsilon_k(\bar{\theta}) \\
\nabla V_N(\bar{\theta}) &= -\sum_{k=1}^{N}\varepsilon_k(\bar{\theta})\psi_k^T(\bar{\theta}) \\
\nabla^2 V_N(\bar{\theta}) &= \sum_{k=1}^{N}\psi_k(\bar{\theta})\psi_k^T(\bar{\theta}) + \sum_{k=1}^{N}\varepsilon_k(\bar{\theta})\nabla^2\varepsilon_k^T(\bar{\theta})
\end{aligned} \tag{C.32}$$

The Newton method is a good numerical procedure with "quadratic" convergence properties. The method must, however, be modified when $\nabla^2 V$ is not invertible, which may constitute a problem at some distance away from the solution. It is therefore a wise idea to start the iterative algorithm with good initial values obtained from some other numerical algorithm or from a least-squares estimate. It is for the same reason difficult to assure global convergence by using the Newton method with arbitrary initial conditions.

Let the expressions in Eq. (C.32) be substituted into

$$\theta^{(i+1)} = \theta^{(i)} - \alpha^{(i)}[\nabla^2 V_N(\theta^{(i)})]^{-1}\nabla V_N(\theta^{(i)}) \tag{C.33}$$

where $\alpha^{(i)}$ determines the step size in the iteration and is nominally equal to 1. An approximate gradient can be calculated as

$$\nabla\varepsilon_k = \frac{\partial\varepsilon_k}{\partial\theta} = \begin{pmatrix} \frac{\partial\varepsilon}{\partial a} \\ \frac{\partial\varepsilon}{\partial b} \\ \frac{\partial\varepsilon}{\partial c} \end{pmatrix} = \begin{pmatrix} \frac{-1}{\widehat{C}(z^{-1})}y_{k-1} \\ \vdots \\ \frac{-1}{\widehat{C}(z^{-1})}y_{k-n_A} \\ \frac{1}{\widehat{C}(z^{-1})}u_{k-1} \\ \vdots \\ \frac{1}{\widehat{C}(z^{-1})}u_{k-n_B} \\ \frac{1}{\widehat{C}(z^{-1})}\varepsilon_{k-1} \\ \vdots \\ \frac{1}{\widehat{C}(z^{-1})}\varepsilon_{k-n_C} \end{pmatrix} \tag{C.34}$$

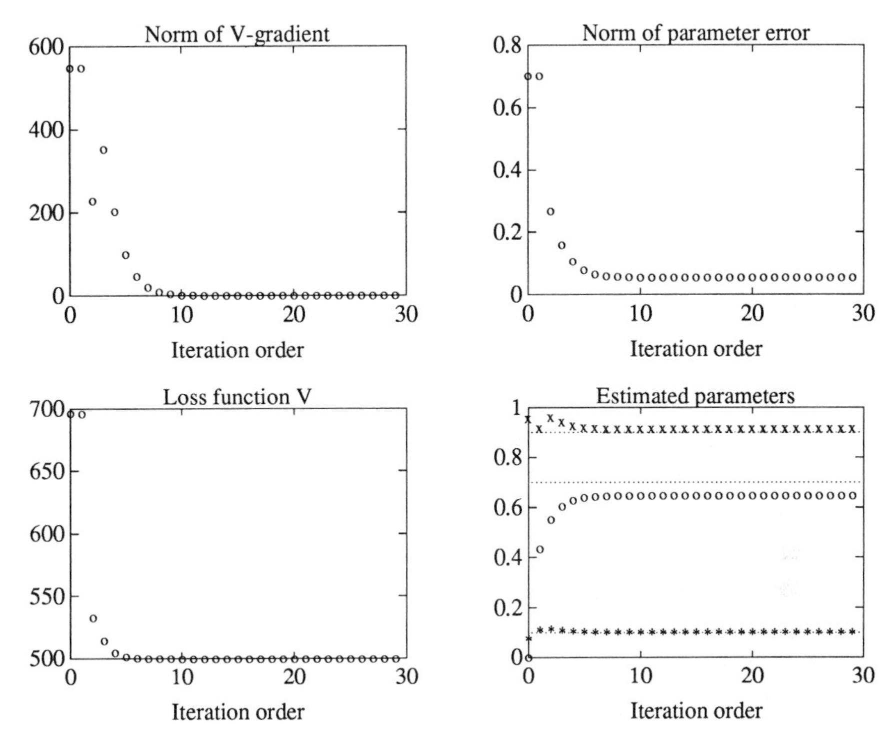

Figure C.4 Typical iterations of Newton-Raphson optimization techniques when finding the parameters at $(a, b, c) = (0.9, 0.1, 0.7)$ of a log-likelihood function. "True" parameters are indicated by dotted lines.

This expression is often expressed in the pseudo-regressor form

$$\psi_k(\theta^{(i)}) = -\nabla \varepsilon_k(\theta^{(i)}) \tag{C.35}$$

By iterating this estimation procedure one obtains a filtering similar to that of Fig. C.3. Some iterations of optimization of the log-likelihood function based on data shown in Fig. C.2. are shown in Fig. C.4 and Fig. C.5. It is clear from these graphs that the Newton methods perform better than the steepest descent search. ■

C.8 BIBLIOGRAPHY AND REFERENCES

Good general references for numerical optimization methods include

[C1] J.E. DENNIS AND R.B. SCHNABEL, *Numerical Methods for Unconstrained Optimization and Nonlinear Equations*. Englewood Cliffs, NJ: Prentice-Hall, 1983.

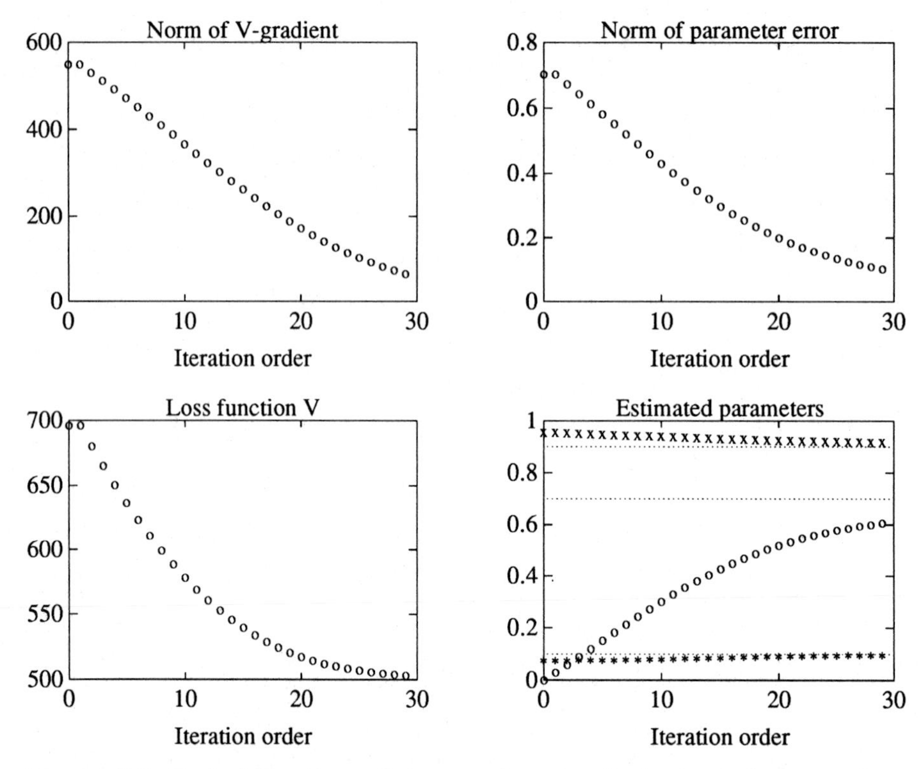

Figure C.5 Typical iterations of steepest descent optimization techniques when finding the parameters at $(a, b, c) = (0.9, 0.1, 0.7)$ of a log-likelihood function. "True" parameters are indicated by dotted lines.

[C2] G.H. Golub and C.F. Van Loan, *Matrix Computations*. Baltimore: The Johns Hopkins University Press, 1983.

[C3] G.H. Golub and W. Kahan, "Calculating the singular values and pseudo-inverse of a matrix." *J. SIAM Numer. Analysis*, Vol. 2(B), 1965, pp. 205–224.

[C4] D.G. Luenberger, *Introduction to Linear and Nonlinear Programming*. Reading, MA: Addison-Wesley, 1974.

The Nelder-Mead direct search algorithm was published in

[C5] J.A. Nelder and R. Mead, "A Simplex Method for Function Minimization." *Comput. J*, Vol. 7, 1965, pp. 308–313. ■

D Statistical Properties of Time Series

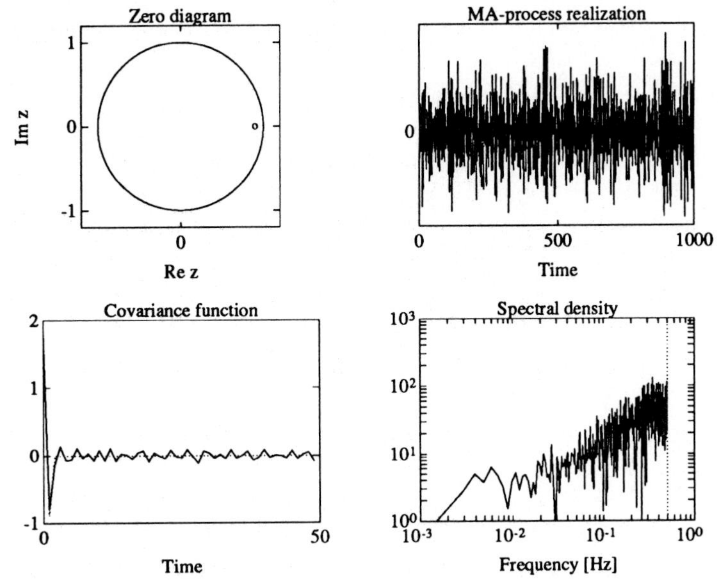

Zero diagram
Im z
Re z
MA-process realization
Time
Covariance function
Time
Spectral density
Frequency [Hz]

Tomorrow will be different

D.1 INTRODUCTION

A *random variable* or a *stochastic variable* has a value which is dependent on chance and which cannot be predicted from a knowledge of the experimental conditions. To describe the outcome of a random variable X it is common practice to introduce the probability distribution function

$$F(x) = \mathcal{P}\{X \leq x\}, \qquad 0 \leq F(x) \leq 1, \quad \forall x \in R \tag{D.1}$$

where $\mathcal{P}\{X \leq x\}$ denotes the probability that $X \leq x$. In addition, $F(x)$ is a monotonically increasing function of x. The derivative $f(x)$ of the distribution function $F(x)$ is called the *probability density function*. In cases when there is no risk of confusion we also use x to denote the random variable.

The statistical *mean* μ_y or *expectation* of a variable y which is a function of a random variable x is defined as

$$\mu_y = \mathcal{E}\{y\} = \int_{-\infty}^{\infty} y(x)f(x)dx \tag{D.2}$$

and the *mean of the distribution* is

$$\mu_x = \mathcal{E}\{x\} = \int_{-\infty}^{\infty} xf(x)dx \tag{D.3}$$

The *variance* of a scalar variable $y(x)$ is defined as

$$\sigma_y^2 = \text{Var}\{y\} = \mathcal{E}\{(y - m_y)^2\} = \int_{-\infty}^{\infty} (y(x) - \mu_y)^2 f(x)dx \tag{D.4}$$

and the *covariance* between two variables x and y is

$$\text{Cov}\{x, y\} = \mathcal{E}\{(x - \mu_x)(y - \mu_y)\} = \mathcal{E}\{xy\} - \mu_x\mu_y \tag{D.5}$$

In the case of a vector-valued variable $y(x)$ it is standard to use the definition of *covariance*

$$\text{Cov}\{y, y\} = \mathcal{E}\{(y - \mu_y)(y - \mu_y)^T\} = \int_{-\infty}^{\infty} (y(x) - \mu_y)(y(x) - \mu_y)^T f(x)dx \tag{D.6}$$

because it also describes the statistical relations between the components of the vector y.

Definition D.1—Statistical covariance and correlation

The correlation coefficient between two variables x and y is

$$\rho_{xy} = \frac{\text{Cov}\{x, y\}}{\sigma_x \sigma_y} \tag{D.7}$$

and two variables x, y are *uncorrelated* if

$$\text{Cov}\{x, y\} = 0 \tag{D.8}$$

■

Example D.1—The normal or Gaussian distribution
Let $x \in R^N$ denote a vector of random variables. A random vector x is called normal (or Gaussian) if its probability density function is

$$f_x(x) = \frac{1}{(2\pi)^{N/2}(\det R)^{1/2}} \exp\{-\frac{1}{2}(x-\mu_x)^T R^{-1}(x-\mu_x)\} \tag{D.9}$$

with the mean μ_x and the variance R. This is denoted $x \in \mathcal{N}(\mu_x, R)$ with

$$\mu_x = \mathcal{E}\{x\} = \begin{pmatrix} \mathcal{E}\{x_1\} \\ \mathcal{E}\{x_2\} \\ \vdots \\ \mathcal{E}\{x_N\} \end{pmatrix}$$

$$R = \mathcal{E}\{(x-\mu_x)(x-\mu_x)^T\} = \begin{pmatrix} \mathrm{Cov}\{x_1,x_1\} & \dots & \mathrm{Cov}\{x_1,x_N\} \\ \mathrm{Cov}\{x_2,x_1\} & \dots & \mathrm{Cov}\{x_2,x_N\} \\ \vdots & & \vdots \\ \mathrm{Cov}\{x_N,x_1\} & \dots & \mathrm{Cov}\{x_N,x_N\} \end{pmatrix} \tag{D.10}$$

If the components of x are uncorrelated, then the covariance matrix (D.10) becomes diagonal. Notice that the covariance matrix is symmetric and positive semidefinite since for any constant vector $a \in R^n$ we have

$$0 \le V\{a^T x\} = \mathrm{Cov}\{a^T x, a^T x\} = a^T V\{x\} a \tag{D.11}$$

A noteworthy special case is when $a^T V\{x\} a = 0$, which means that the components of x are linearly dependent. ■

Linear transformations

Consider the following linear transformation from $x \in R^n$ to $y \in R^m$ by means of a constant $m \times n$-matrix A and a constant m-vector b

$$y = Ax + b \tag{D.12}$$

The mean μ_y and covariance of y are then related to the mean and covariance of x in the following manner

$$\begin{aligned} \mathcal{E}\{y\} &= A\mathcal{E}\{x\} + b = A\mu_x + b \\ V\{y\} &= \mathcal{E}\{((Ax+b)-(A\mu_x+b))((Ax+b)-(A\mu_x+b))^T\} = AV\{x\}A^T \end{aligned} \tag{D.13}$$

D.2 STOCHASTIC PROCESSES

A function $x(t) = x(t,\omega)$ whose values depend on a random variable ω is called a *random* or *stochastic process*. For each value of time t, the function is a function of ω alone and, consequently, it is a random variable. For each fixed value on ω the $x(t,\omega)$ depends only on t and is thus an ordinary function of one real variable, and each such function is called a *realization* of the stochastic process. For each fixed ω the function $x(t,\omega)$ is called a *trajectory* or *realization* or *sample function*. In discrete time we find a stochastic process in the form of an infinite sequence $\{x_k\}_{k=\infty}^{\infty}$ or a sequence $\{x_k\}_{k=0}^{N}$ over some interval of time, *i.e.*, in both cases *time series*.

Each discrete random variable x_k should have some fixed probability distribution, usually assumed to be the normal distribution, with zero mean and variance σ_w^2. A sequence of mutually independent random variables $\{w_k\}_{k=-\infty}^{\infty}$ is called *white noise* in engineering terminology.

Definition D.2—White noise
A sequence of N uncorrelated stochastic variables $\{w_i\}_{i=1}^{N}$ with $\mathcal{E}\{w_i\} = 0$, $\mathcal{E}\{w_i w_j\} = \delta_{ij}\sigma^2$ for all i,j is known as *white noise* in the domain of time-series analysis. ■

Autocovariance and cross covariance

An important special case of time-series analysis is when the dispersion of the inputs to a linear system may be described according to stochastic distributions. The dependent variables—static or dynamic—then also behave as random variables. A major objective in applications is to describe the random process in such a way that predictions (in a probabilistic sense) of future values can be made. As the disturbances affecting a system are not known beforehand, it is important to consider various temporal covariances in order to make accurate predictions possible.

The *cross covariance function* and the *autocovariance function* are defined as

$$\begin{aligned} C_{xy}(t_1,t_2) &= \text{Cov}\{x(t_1), y(t_2)\} \\ C_{yy}(t_1,t_2) &= \text{Cov}\{y(t_1), y(t_2)\} \end{aligned} \qquad \text{(D.14)}$$

and the *cross correlation function* and *autocorrelation function* as

$$\begin{aligned} \rho_{xy}(t_1,t_2) &= \frac{C_{xy}(t_1,t_2)}{\sqrt{C_{xx}(t_1,t_1)}\sqrt{C_{yy}(t_1,t_2)}} \\ \rho_{yy}(t_1,t_2) &= \frac{C_{yy}(t_1,t_2)}{\sqrt{C_{yy}(t_1,t_1)}\sqrt{C_{yy}(t_2,t_2)}} \end{aligned} \qquad \text{(D.15)}$$

The stochastic processes $\{x_k\}$ and $\{y_k\}$ are said to have *stationary correlations* if $C_{xy}(t_1, t_2)$ depends on $\tau = t_1 - t_2$ only. In this case it is customary to denote the autocovariance and cross covariance functions by $C_{yy}(\tau)$ and $C_{xy}(\tau)$, respectively. For stochastic processes with stationary correlations it holds that

$$C_{xy}(\tau) = C_{yx}(-\tau) \tag{D.16}$$

Stationary stochastic processes

Definition D.3—Weakly stationary stochastic processes

A random process $x(t, \omega)$ is called *weakly stationary* if its expectation $\mu_x(t) = \mathcal{E}\{x(t, \omega)\}$ is constant and independent of time t and if the covariance function $C_{xx}(t_1, t_2)$ depends only on the time shift $\tau = t_1 - t_2$. ■

This definition sometimes takes on a slightly different form when applied to discrete stochastic processes. A discrete random process $\{x_k\}_{k=-\infty}^{\infty}$ is called weakly stationary if its expectation $\mu_k = \mathcal{E}\{x_k\}$ and covariance function $\text{Cov}\{(x_k - \mu_k)(x_{k-q} - \mu_{k-q})^T\}$ are independent of k. The covariance function is then denoted by $C_{xx}(\tau) = C_{xx}(qh)$ where $\tau = qh$ and where q is the number of samples corresponding to the time shift τ.

Definition D.4—Strictly stationary stochastic processes

A stochastic process is said to be *strictly stationary* if the joint probability distribution of some set of N observations $x_1, \ldots, x_N$ is the same as that associated with the N observations $x_{1+k}, \ldots, x_{N+k}$ for any k. ■

Definition D.5—Uncorrelated stochastic processes

Two stochastic processes $\{x_k\}$ and $\{y_k\}$ are *uncorrelated* if and only if cross covariance function $C_{xy}(\tau) = 0$ for all τ. ■

Spectra

The power spectrum of a stochastic process x is the Fourier transform of the autocovariance function

$$S_{xx}(i\omega) = \mathcal{F}\{C_{xx}(\tau)\} \tag{D.17}$$

where ω is complex frequency. The power cross spectrum between two stochastic processes x and y is

$$S_{xy}(i\omega) = \mathcal{F}\{C_{xy}(\tau)\} = \begin{cases} \int_{-\infty}^{\infty} C_{xy}(\tau) e^{-i\omega\tau} d\tau, & \text{continuous time} \\ h \sum_{q=-\infty}^{\infty} C_{xy}(qh) e^{-i\omega qh}, & \text{discrete time } (\tau = qh) \end{cases} \tag{D.18}$$

which is sometimes shown in a diagram as $\mathrm{Re}(S_{xy})$ (*amplitude power spectrum*) and $\mathrm{Im}(S_{xy})$ (*phase power spectrum*). The inverse of Eq. (D.18) is

$$C_{xy}(\tau) = \begin{cases} \frac{1}{2\pi i}\int_{\sigma-i\infty}^{\sigma+i\infty} S_{xy}(i\omega)e^{i\omega\tau}d\omega & \text{continuous time} \\ \frac{1}{2\pi}\int_{-\pi/h}^{\pi/h} S_{xy}(i\omega)e^{i\omega\tau}d\omega & \text{discrete time} \end{cases} \tag{D.19}$$

Example D.2—Spectral density of a white-noise process
Assume that the spectral density of a discrete white noise process $\{w_k\}_{k=-\infty}^{\infty}$ is such that

$$\begin{aligned} \mathcal{E}\{w_k\} &= 0 \\ \mathrm{Cov}\{w_i, w_j\} &= \sigma^2\delta_{ij} \end{aligned} \tag{D.20}$$

The spectral density is then

$$S_{ww}(i\omega) = h\sigma^2, \qquad -\frac{\pi}{h} < \omega \le \frac{\pi}{h} \tag{D.21}$$

which is constant over the spectral range. In addition, according to the inversion formula (D.19) it is verified that $C_{xy}(0) = (1/2\pi)\int_{-\pi/h}^{\pi/h} h\sigma^2 d\omega = \sigma^2$. ■
Remark: Notice that the spectral density is often defined in the following slightly different way

$$\Phi_{xy}(\omega) = \frac{1}{2\pi}\int_{-\infty}^{\infty} C_{xy}(\tau)e^{-i\omega\tau}d\tau \tag{D.22}$$

Effectively, the two definitions of S_{xy} and Φ_{xy} differ by a factor of 2π, which can be regarded as a difference in the definition of the Fourier transform. ■

Linear stationary models

Assume that $u(t) = \{u_k\}$ and $v(t) = \{v_k\}$ are uncorrelated weakly stationary stochastic processes and that $y(t)$ is related to u and v *via* the convolution

$$y(t) = h(t) * u(t) + v(t) \tag{D.23}$$

and for discrete-time variables

$$y_k = \sum_{j=0}^{\infty} h_j u_{k-j} + v_k \tag{D.24}$$

where $h(\tau) = \{h_k\}$ is the weighting function. As Eq. (D.24) also can be interpreted as a convolution (D.23) we make no distinction between Eq. (D.23) and Eq. (D.24) in the case of discrete stochastic processes.

Let $\tau = qh$ denote a multiple q of the sampling period h and assume that the autocovariance function of $\{u_k\}$ is $C_{uu}(\tau)$. The mean value of $\{y_k\}$ is then constant and independent of time, and the cross covariance function between the output y and the input u is

$$\begin{aligned} C_{yu}(\tau) = C_{yu}(qh) &= \text{Cov}\{\sum_{j=0}^{\infty} h_j u_{k-j} + v_k, u_{k-q}\} \\ &= \sum_{j=0}^{\infty} h_j \text{Cov}\{u_{k-j}, u_{k-q}\} + \text{Cov}\{v_k, u_{k-q}\} = h(\tau) * C_{uu}(\tau) \end{aligned} \tag{D.25}$$

Hence, it follows that the output $\{y_k\}$ is a weakly stationary random process.

In particular, if the input to a linear system is zero-mean white noise with $C_{uu}(\tau) = C_{uu}(qh) = \sigma^2 \delta_q$ then

$$C_{yu}(\tau) = \sigma^2 h(\tau) \tag{D.26}$$

where $h(\tau)$ is the weighting function.

D.3 DIFFERENCE EQUATIONS

Consider a stochastic process with an output that depends on previous outputs according to the difference equation

$$y_k = -a_1 y_{k-1} - a_2 y_{k-2} - \cdots - a_n y_{k-n} + w_k \tag{D.27}$$

or

$$A(z^{-1}) y_k = w_k, \qquad \text{with} \quad A(z^{-1}) = 1 + a_1 z^{-1} + \cdots + a_n z^{-n} \tag{D.28}$$

where $\{w_k\}$ is a sequence of uncorrelated stochastic variables with $\mathcal{E}\{w_k\} = 0$ for all k. Stochastic models according to Eq. (D.27) and Eq. (D.28) are known as *autoregressive models*. The transfer function $H(z) = 1/A(z^{-1})$ is stable if and only if the poles, *i.e.*, the complex numbers $z_1, \ldots, z_n$ solving the equation $A(z^{-1}) = 0$, are strictly inside the unit circle—*i.e.*, $|z_i| < 1$. The polynomial $A(z^{-1})$ is called the *generating polynomial* of the stochastic process.

The output sequence $\{y_k\}$ is a weakly stationary sequence $\{y_k\}_{k=-\infty}^{\infty}$ with mean

$$\begin{aligned} \mu_y = \mathcal{E}\{y_k\} &= \mathcal{E}\{-a_1 y_{k-1} - a_2 y_{k-2} - \cdots - a_n y_{k-n} + v_k\} \\ &= -\mu_y (a_1 + \cdots + a_n) \end{aligned} \tag{D.29}$$

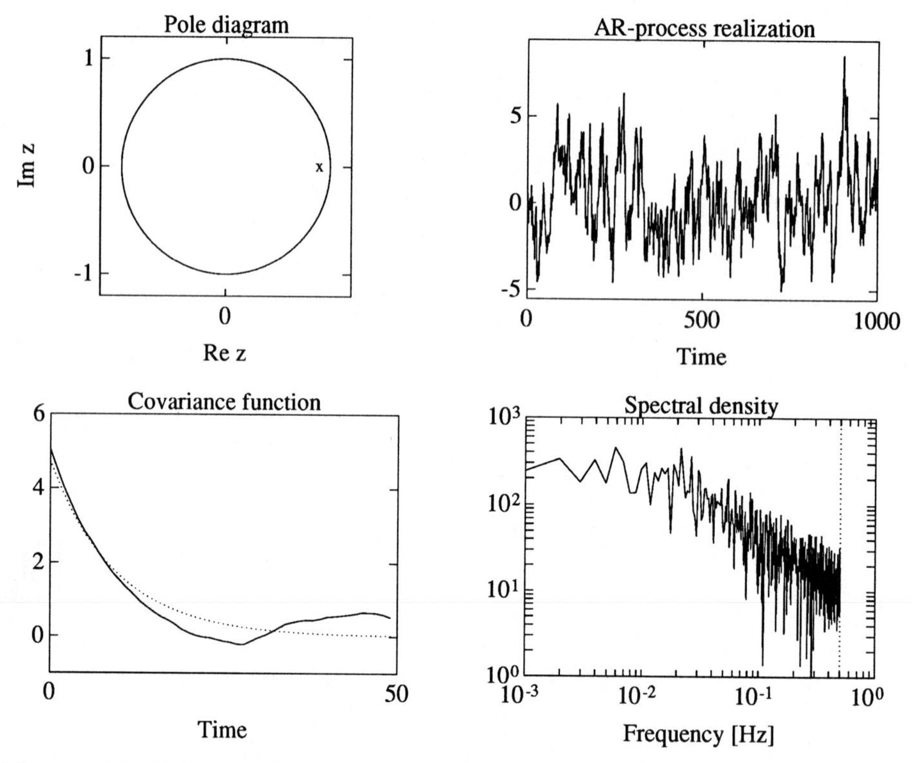

Figure D.1 Autoregressive process $y_k = 0.9y_{k-1} + w_k$ with pole diagram, a realization with $\mathcal{E}\{w_k\} = 0$ and $\mathcal{E}\{w_k^2\} = \sigma^2 = 1$, theoretical (*dotted line*) and empirical (*solid line*) covariance function, and amplitude spectrum.

or

$$\mu_y A(1) = 0 \tag{D.30}$$

so that $\mu_y = 0$ except in the case when $A(1) = 0$ which corresponds to autoregressive dynamics with integral action. Notice that this case is precluded inasmuch as $A(z^{-1})$ belongs to the set of polynomials with all zeros strictly inside the unit circle; see Figs. D.1, D.2, and D.3.

The covariance function for an autoregressive stochastic process satisfies the following difference equation (Yule-Walker equation)

$$C_{yy}(kh) + a_1 C_{yy}((k-1)h) + \cdots + a_n C_{yy}((k-n)h) = 0, \qquad k = 1, 2, \ldots \tag{D.31}$$

with the initial condition

$$C_{yy}(0) + a_1 C_{yy}(h) + \cdots + a_n C_{yy}(nh) = \sigma^2 \tag{D.32}$$

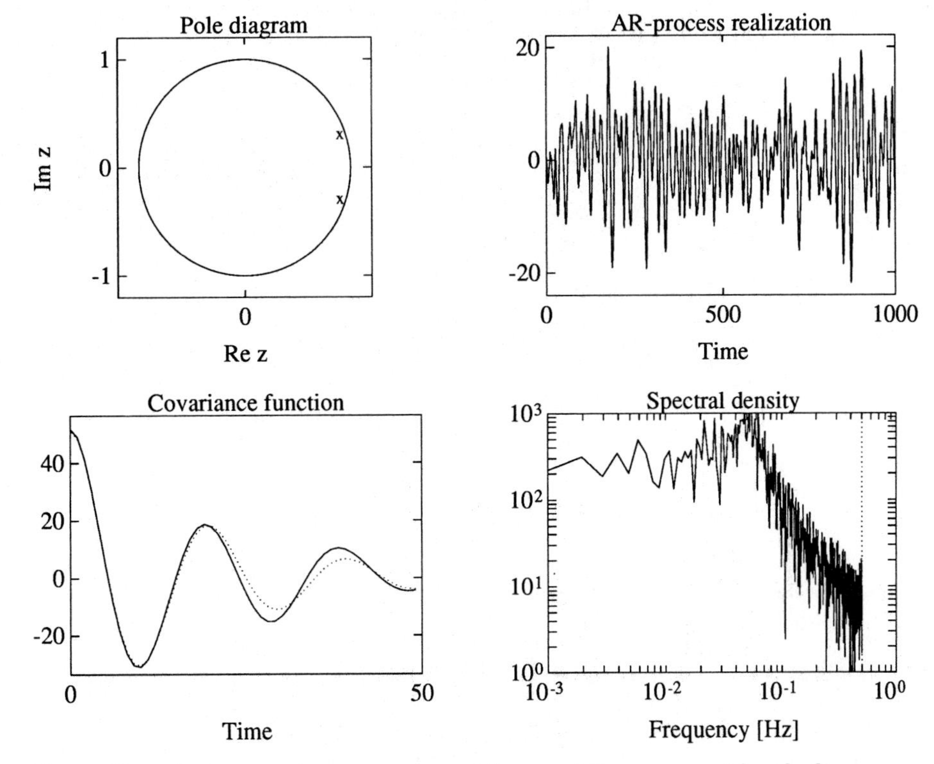

Figure D.2 Autoregressive process $y_k = 1.8y_{k-1} - 0.9y_{k-2} + w_k$ with pole diagram, a realization with $\mathcal{E}\{w_k\} = 0$ and $\mathcal{E}\{w_k^2\} = \sigma^2 = 1$, theoretical (*dotted line*) and empirical (*solid line*) covariance function, and amplitude spectrum.

Proof: By multiplying Eq. (D.27) by y_{k-q} and calculating the expected values

$$\begin{aligned}\mathcal{E}\{w_k y_{k-q}\} &= \mathcal{E}\{y_k y_{k-q} + a_1 y_{k-1} y_{k-q} + \cdots + a_n y_{k-n} y_{k-q}\} \\ &= C_{yy}(qh) + a_1 C_{yy}((q-1)h) + \cdots \\ &+ \cdots a_n C_{yy}((q-n)h)\}\end{aligned} \tag{D.33}$$

where the left-hand side may be simplified to

$$\mathcal{E}\{w_i y_{i-j}\} = \sigma^2 \delta_{ij} \tag{D.34}$$

because w_i and y_{i-j} are uncorrelated for $j = 1, 2, 3, \ldots$. ■

The spectral density for an autoregressive process is

$$S_{yy}(i\omega) = \frac{h}{A(e^{i\omega h})A(e^{-i\omega h})}\sigma^2 = \frac{h}{|A(e^{i\omega h})|^2}\sigma^2 \tag{D.35}$$

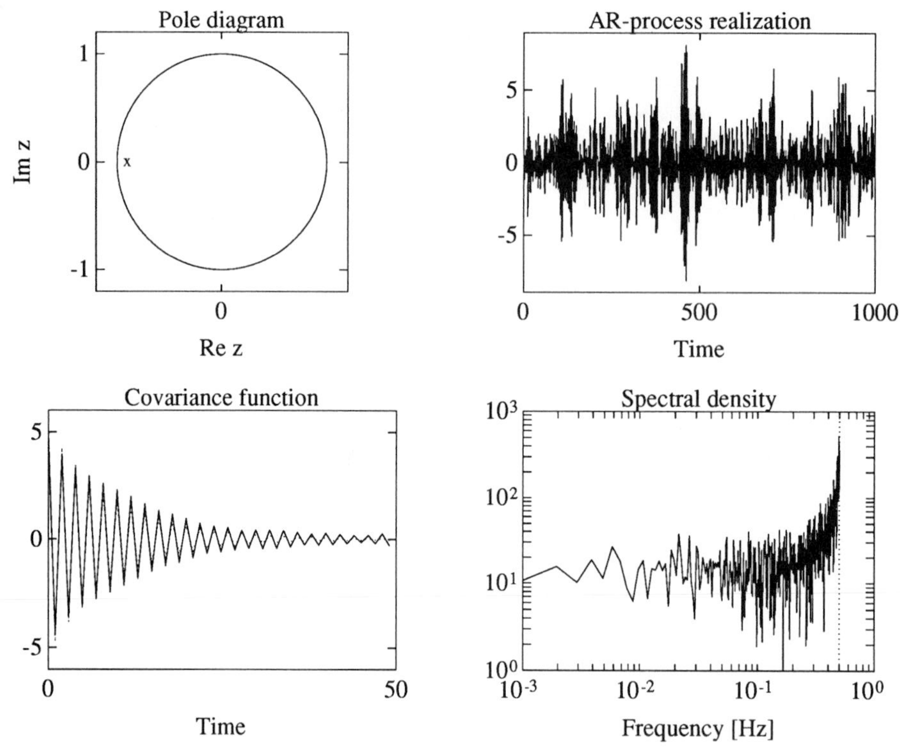

Figure D.3 Autoregressive process $y_k = -0.9y_{k-1} + w_k$ with pole diagram, a realization with $\mathcal{E}\{w_k\} = 0$ and $\mathcal{E}\{w_k^2\} = \sigma^2 = 1$, theoretical (*dotted line*) and empirical (*solid line*) covariance function, and amplitude spectrum.

for $|\omega| < \omega_N = \pi/h$.

Example D.3—Yule-Walker equations for a first-order process

Consider the stochastic process with sampling interval h and with the first-order autoregressive dynamics

$$y_k = 0.9y_{k-1} + w_k; \qquad \text{with} \quad \mathcal{E}\{w_k\} = 0, \quad \text{and } \mathcal{E}\{w_i w_j\} = \sigma^2 \delta_{ij} \tag{D.36}$$

The Yule-Walker equation and the initial condition give

$$\begin{aligned} C_{yy}(0) + a_1 C_{yy}(-h) &= \sigma^2 \\ C_{yy}(h) + a_1 C_{yy}(0) &= 0 \\ &\vdots \\ C_{yy}(kh) + a_1 C_{yy}((k-1)h) &= 0 \end{aligned} \tag{D.37}$$

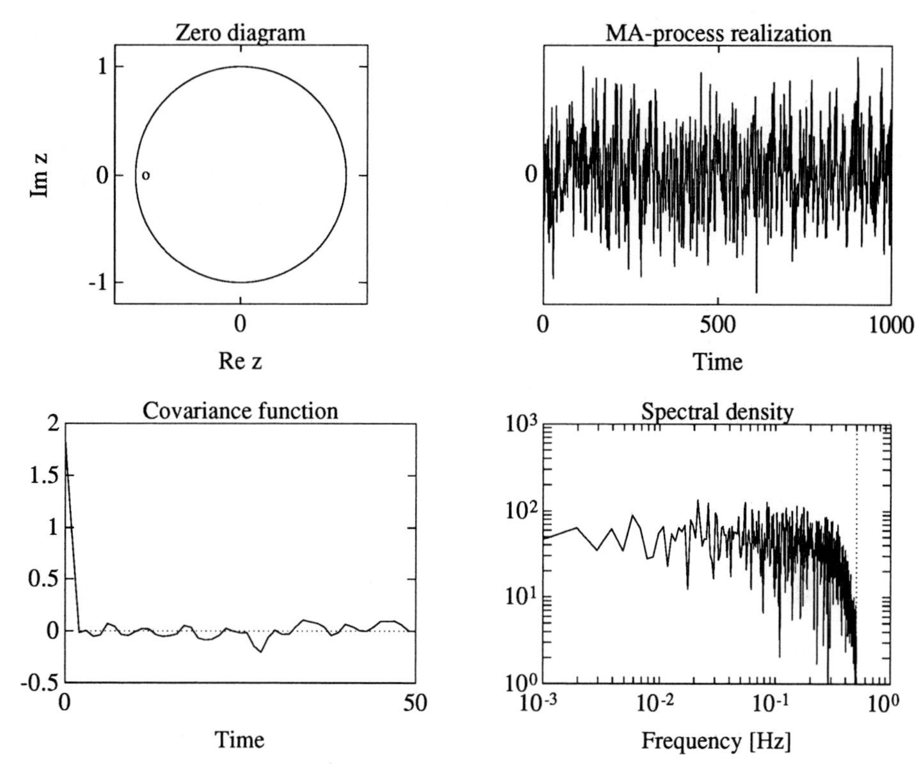

Figure D.4 Moving average process $y_k = w_k + 0.9w_{k-1}$ with zero diagram, a realization with $\mathcal{E}\{w_k\} = 0$ and $\mathcal{E}\{w_k^2\} = \sigma^2 = 1$, theoretical (*dotted line*) and empirical (*solid line*) covariance function, and amplitude spectrum.

The first two equations give the solution

$$\begin{pmatrix} C_{yy}(0) \\ C_{yy}(h) \end{pmatrix} = \begin{pmatrix} 1 & a_1 \\ a_1 & 1 \end{pmatrix}^{-1} \begin{pmatrix} \sigma^2 \\ 0 \end{pmatrix} = \frac{\sigma^2}{1 - a_1^2} \begin{pmatrix} 1 \\ -a_1 \end{pmatrix} \tag{D.38}$$

The difference equation for the autocovariance function then gives the explicit solution

$$C_{yy}(kh) = \frac{\sigma^2}{1 - a_1^2}(-a_1)^k \tag{D.39}$$

and the autospectrum is

$$S_{yy}(i\omega) = \frac{1}{1 + a_1 e^{-i\omega h}} \frac{1}{1 + a_1 e^{i\omega h}} \sigma^2 h = \frac{1}{1 + a_1^2 + 2a_1 \cos \omega h} \sigma^2 h \tag{D.40}$$

■

Moving average models

Stochastic processes of the following type where the weights c_i are zero for

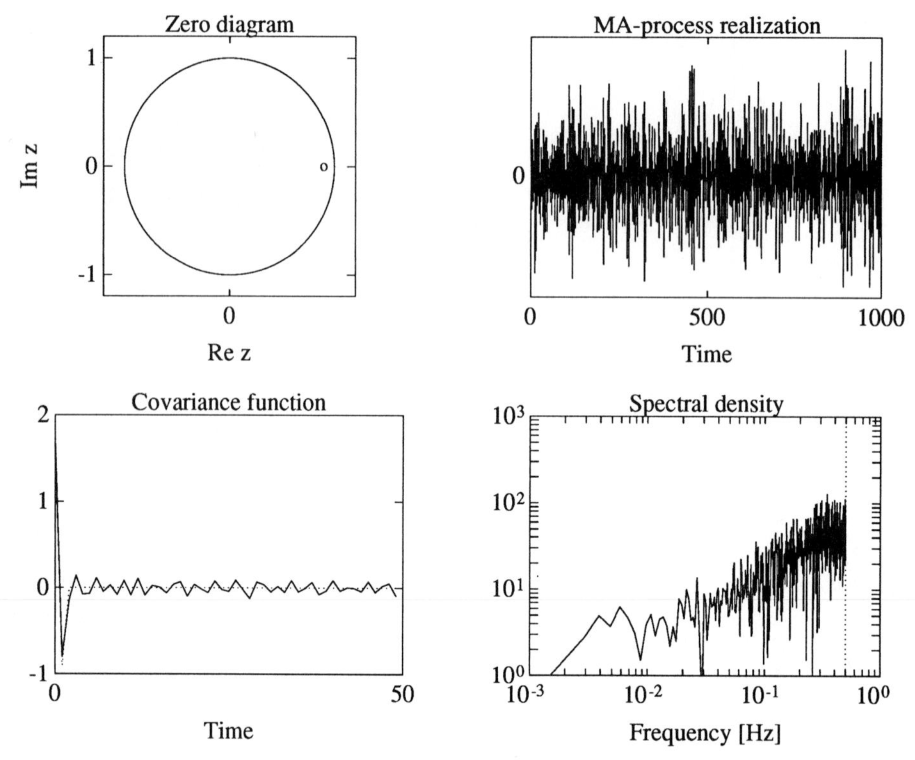

Figure D.5 Moving average process $y_k = w_k - 0.9w_{k-1}$ with zero diagram, a realization with $\mathcal{E}\{w_k\} = 0$ and $\mathcal{E}\{w_k^2\} = \sigma^2 = 1$, theoretical (*dotted line*) and empirical (*solid line*) covariance function, and amplitude spectrum.

$i > m$ are known as *moving average processes*

$$y_k = w_k + c_1 w_{k-1} + \cdots + c_m w_{k-m} \tag{D.41}$$

The transfer function that relates the output sequence $\{y_k\}$ to the input $\{w_k\}$ is

$$H(z) = C(z^{-1}) = c_0 + c_1 z^{-1} + \cdots + c_m z^{-m}; \qquad c_0 = 1 \tag{D.42}$$

Notice that the terminology "moving average" is here somewhat misleading as there is no restriction that the coefficients should add to 1 or that the coefficients are non-negative. An alternative description is *finite impulse response* or *all-zero filter*.

The covariance function associated with Eq. (D.41) is

$$C_{yy}(qh) = \begin{cases} \sigma^2 \sum_{j=0}^{m} c_j c_{j+|q|} & \text{if } |q| \leq m \\ 0 & \text{if } |q| > m \end{cases} \tag{D.43}$$

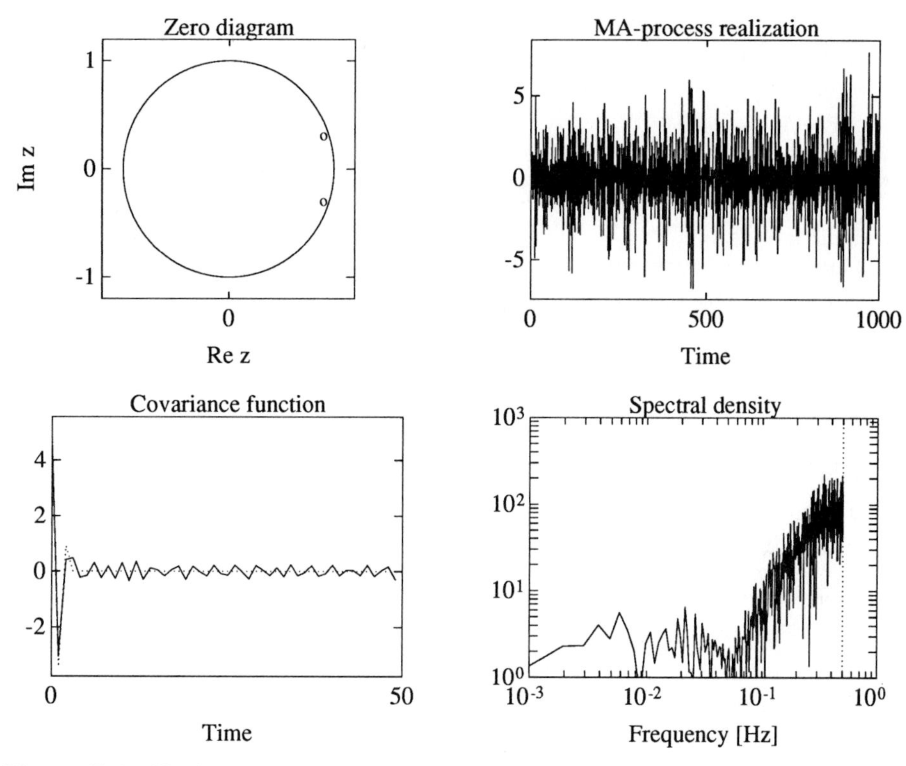

Figure D.6 Moving average process $y_k = w_k - 1.8w_{k-1} + 0.9w_{k-2}$ with zero diagram, a realization with $\mathcal{E}\{w_k\} = 0$ and $\mathcal{E}\{w_k^2\} = \sigma^2 = 1$, theoretical (*dotted line*) and empirical (*solid line*) covariance function, and amplitude spectrum.

The spectral density is

$$S_{yy}(i\omega) = |C(e^{i\omega h})|^2 \sigma^2 h \tag{D.44}$$

Realizations of various MA-processes are shown in Figs. D.4, D.5, and D.6.

Example D.4—Covariance functions for MA-processes
Consider the two MA-processes

$$\begin{aligned} y_k &= w_k + c w_{k-1} \\ x_k &= c w_k + w_{k-1} \end{aligned} \tag{D.45}$$

with the generating polynomials $1 + cz^{-1}$ and $c + z^{-1}$, respectively. The input sequence $\{w_k\}$ is assumed to be zero-mean white noise. The corresponding covariance functions of the processes $\{y_k\}$ and $\{x_k\}$ are

$$C_{yy}(qh) = C_{xx}(qh) = \begin{cases} \sigma^2(1 + c_1^2) & \text{if} \quad q = 0 \\ \sigma^2 c_1 & \text{if} \quad |q| = 1 \\ 0 & \text{if} \quad q \geq 2 \end{cases} \tag{D.46}$$

Notice that the two covariance functions C_{yy} and C_{xx} are equal despite the different generating polynomials. ■

D.4 AUTOREGRESSIVE MOVING AVERAGE MODELS

Consider an autoregressive signal $\{x_k\}$ and noisy observations $\{y_k\}$ represented by the following stochastic process

$$\begin{aligned} x_k &= -a_1 x_{k-1} - \cdots - a_n x_{k-n} + v_k \\ y_k &= x_k + w_k \end{aligned} \tag{D.47}$$

where $\{v_k\}$ and $\{w_k\}$ are uncorrelated white-noise sequences with $\mathcal{E}\{v_k\} = \mathcal{E}\{w_k\} = 0$ and $\mathcal{E}\{v_k^2\} = \sigma_v^2$ and $\mathcal{E}\{w_k^2\} = \sigma_w^2$. The spectral density of the noise-disturbed output y is

$$S_{yy}(i\omega) = \frac{\sigma_v^2 h}{A(e^{i\omega h})A(e^{-i\omega h})} + \sigma_w^2 h = \frac{\sigma_v^2 + \sigma_w^2 A(e^{i\omega h})A(e^{-i\omega h})}{A(e^{i\omega h})A(e^{-i\omega h})} h \tag{D.48}$$

Spectral estimators based on the autoregressive model tend to give very poor results when applied to data generated by a system with observation noise of the type (D.47). A reason for this sensitivity can be sought in the spectral density (D.48), which is clearly characterized by poles as well as zeros. It is obvious that the presence of zeros in Eq. (D.48) is not compatible with the original autoregressive model which motivates an extension of the model set to autoregressive moving average (ARMA) models of the type

$$y_k = -a_1 y_{k-1} - \cdots - a_n y_{k-n} + w_k + c_1 w_{k-1} + \cdots + c_m w_{k-m} \tag{D.49}$$

with the spectral density

$$S_{yy}(i\omega) = \frac{C(e^{i\omega h})C(e^{-i\omega h})}{A(e^{i\omega h})A(e^{-i\omega h})}\sigma^2 h \tag{D.50}$$

Spectral densities on the form Eq. (D.50) are sometimes called *rational spectral densities* (or *rational spectra*) as they derive from the rational function $C(z)/A(z)$.

The covariance function may be calculated from the Yule-Walker equations

$$\begin{aligned} C_{yy}(qh) &= -a_1 C_{yy}((q-1)h) - \cdots - a_n C_{yy}((q-n)h) \\ &\quad + C_{yw}(qh) + c_1 C_{yw}((k-1)h) + \cdots + c_m C_{yw}((q-m)h) \end{aligned} \tag{D.51}$$

where $C_{yw}(qh) = \text{Cov}\{y_k w_{k-q}\}$ satisfies the relationship

$$C_{yw}(qh) = -a_1 C_{yw}((q-1)h) - \cdots - a_n C_{yw}((q-n)h) + c_q \sigma_w^2 \tag{D.52}$$

and where $C_{yw}(qh)$ is equal to zero for $q < 0$.

Example D.5—Yule-Walker equation for an ARMA process
Consider the first-order autoregressive moving average model

$$y_{k+1} = -a_1 y_k + w_{k+1} + c_1 w_k \tag{D.53}$$

The Yule-Walker equation gives

$$\begin{aligned} C_{yy}(0) + a_1 C_{yy}(-h) &= \sigma^2 + c_1(-a_1 + c_1)\sigma^2 \\ C_{yy}(h) + a_1 C_{yy}(0) &= c_1 \sigma^2 \\ &\vdots \\ C_{yy}((k-1)h) + a_1 C_{yy}((k-1)h) &= 0 \end{aligned} \tag{D.54}$$

As $C_{yy}(-h) = C_{yy}(h)$ the first equations above give a solvable system of linear equations

$$\begin{pmatrix} 1 & a_1 \\ a_1 & 1 \end{pmatrix} \begin{pmatrix} C_{yy}(h) \\ C_{yy}(0) \end{pmatrix} = \begin{pmatrix} c_1 \\ 1 - a_1 c_1 + c_1^2 \end{pmatrix} \sigma^2 \tag{D.55}$$

The solution to Eq. (D.55) and the recursive Yule-Walker equation provide the explicit solution as

$$\begin{pmatrix} C_{yy}(0) \\ C_{yy}(h) \\ \vdots \\ C_{yy}(qh) \end{pmatrix} = \begin{pmatrix} 1 + c_1^2 - 2a_1 c_1 \\ (1 - a_1 c_1)(c_1 - a_1) \\ \vdots \\ (1 - a_1 c_1)(c_1 - a_1)(-a_1)^{q-1} \end{pmatrix} \tag{D.56}$$

■

Spectral factorization

It was shown in Eq. (D.48) that an autoregressive signal disturbed by white noise gave rise to an ARMA-type rational spectral density of the type (D.50) with a factor $C(z)/A(z)$ evaluated for $z = e^{-i\omega h}$. In fact, the generalization is valid so that it is possible to find a transfer function factor $H(z)$ for any rational spectral density generated by a state-space representation

$$\begin{aligned} x_{k+1} &= \Phi x_k + v_k \\ y_k &= C x_k + w_k \end{aligned} \tag{D.57}$$

where Φ is stable, (Φ, C) is observable, and where the independent white-noise sequences $\{v_k\}$ and $\{w_k\}$ have zero mean and covariances Σ_v and Σ_w, respectively. This procedure is called *spectral factorization* and consists of solving the Riccati equation

$$P = \Phi P \Phi^T - \Phi P C^T (C P C^T + \Sigma_w)^{-1} C P \Phi^T + \Sigma_v \tag{D.58}$$

and evaluating the matrices

$$\begin{aligned} K &= \Phi P C^T (C P C^T + \Sigma_w)^{-1} \\ \Sigma &= C P C^T + \Sigma_w \end{aligned} \tag{D.59}$$

and the transfer function

$$H(z) = C(zI - \Phi)^{-1} K + I \tag{D.60}$$

The spectral density for $|\omega| \le \omega_N = \pi/h$ is then

$$\begin{aligned} S_{yy}(i\omega) &= (C(e^{i\omega h} - \Phi)^{-1} \Sigma_v (e^{-i\omega h} - \Phi)^{-1} C^T + \Sigma_w) \\ &= H(e^{i\omega h}) \Sigma H^T (e^{-i\omega h}) \end{aligned} \tag{D.61}$$

both for Eq. (D.57) and for a stochastic process

$$\begin{aligned} y_k &= H(z) e_k \\ \mathcal{E}\{e_k\} &= 0 \\ \mathcal{E}\{e_k e_q^T\} &= \Sigma \delta_{kq} \end{aligned} \tag{D.62}$$

where $\{e_k\}$ is a noise sequence filtered by $H(z)$.

D.5 SAMPLE COVARIANCE FUNCTIONS AND SPECTRA

The theoretical covariance functions and hence the corresponding spectra can be calculated, for instance, by means of solving the Yule-Walker equations. Calculations of the empirical counterparts, that is, computation of covariances and spectra from data, sometimes require reformulation in order to consider effects of finite data records.

The sample mean $\bar{x}$ based on N samples is

$$\bar{x} = \frac{1}{N} \sum_{k=1}^{N} x_k \tag{D.63}$$

Two standard covariance estimators based on N samples with the constant sampling interval h are

$$\begin{aligned}\widehat{C}_{xx}(qh) &= \frac{1}{N}\sum_{k=q}^{N-1}(x_k - \bar{x})(x_{k-q} - \bar{x}) \\ \widehat{C}'_{xx}(qh) &= \frac{1}{N-q}\sum_{k=q}^{N-1}(x_k - \bar{x})(x_{k-q} - \bar{x})\end{aligned} \qquad \text{(D.64)}$$

The major difference between the two estimators is the normalization factor which takes on the values $N - q$ and N in the two cases. In both cases it is required to make some correction for low-frequency trends which, at least, entails removal of constant sample mean levels $\bar{x}$.

Notice that these sample covariance functions (D.64) are not derived from analytical considerations and are, indeed, chosen more because of their similarity to the theoretical counterpart than as a result of theoretical motivations.

Error analysis

Assume that $\mathcal{E}\{x_k\} = 0$. The mean values of the two covariance estimates are then

$$\mathcal{E}\{\widehat{C}'_{xx}(qh)\} = \mathcal{E}\{\frac{1}{N-q}\sum_{k=q}^{N-1} x_k x^*_{k-q}\} = \frac{1}{N-q}\sum_{k=q}^{N-1}\mathcal{E}\{x_k x^*_{k-q}\} = C_{xx}(qh) \qquad \text{(D.65)}$$

Thus, $\widehat{C}'_{xx}(qh)$ is an unbiased estimator of C_{xx} whereas $\widehat{C}_{xx}(qh)$ is only asymptotically unbiased as the record length tends to infinity.

$$\begin{aligned}&\mathrm{Cov}\{\widehat{C}_{yy}(qh), \widehat{C}_{yy}(rh)\} \\ &\quad\approx \frac{1}{N}\sum_{k=-\infty}^{\infty} C_{yy}(kh)C_{yy}((k-q+r)h) + C_{yy}((k+r)h)C_{yy}((k-q)h) \\ &\mathrm{Cov}\{\widehat{C}_{xy}(qh), \widehat{C}_{xy}(rh)\} \\ &\quad\approx \frac{1}{N}\sum_{k=-\infty}^{\infty} C_{xx}(kh)C_{yy}((k-q+r)h) + C_{xy}((k+r)h)C_{yx}((k-q)h)\end{aligned} \qquad \text{(D.66)}$$

Notice that the approximation in Eq. (D.66) refers to the case of normally distributed noise. It is obvious that the cross-covariance estimates at different times might be correlated.

It can be concluded that sample autocovariance functions with a normalization factor $(N-q)$ are unbiased, but their variance is larger than that of estimators with the normalization factor N.

D.6 NONSTATIONARY STOCHASTIC MODELS

Time series collected from cases of application often exhibit non-zero mean value and even systematic fluctuations of the mean value in the course of the time series. The nature of the fluctuating, non-zero mean value may be very diverse and may exhibit, for instance, linear trends or periodic behavior. In the theory of stochastic processes this is known as *trends*.

Trends can often be approached analytically if conditions remain stable over a certain period of time. A nonstationary time series $\{y_k\}$ may sometimes be split up into a trend series $\{f_k\}$ and a stationary residual series $\{x_k\}$ according to

$$y_k = f_k + x_k \tag{D.67}$$

where $\mathcal{E}\{x_k\} = 0$. It is often somewhat more difficult to suggest a decomposition for trends in the variance although it is often suggested to take logarithms of the original time series.

The trend can sometimes be represented by a polynomial in time of low degree—for instance, an offset, a linear trend, or a parabolic trend with an increasing trend at the beginning of the time series and a decreasing trend at the end, or *vice versa*. If the trend is periodic, it might be represented by a finite Fourier series. Of course, standard regression techniques can be used to fit such trends.

For analytical reasons it is often desirable to remove low-frequency trends by subtracting some function fitted to data, by eliminating deterministic functions, or by means of filtering. Such trend elimination of constant offsets or linear trends usually presents no statistical or practical problems.

Seasonal or periodic trends can sometimes be eliminated by means of filtering the original data with the filter

$$\nabla_p = 1 - z^{-p}, \qquad p = \text{period of trend} \tag{D.68}$$

This filter effectively eliminates purely periodic trends, but sometimes it gives rise to strange initial and end-point effects (see Fig. D.7). It is noteworthy that

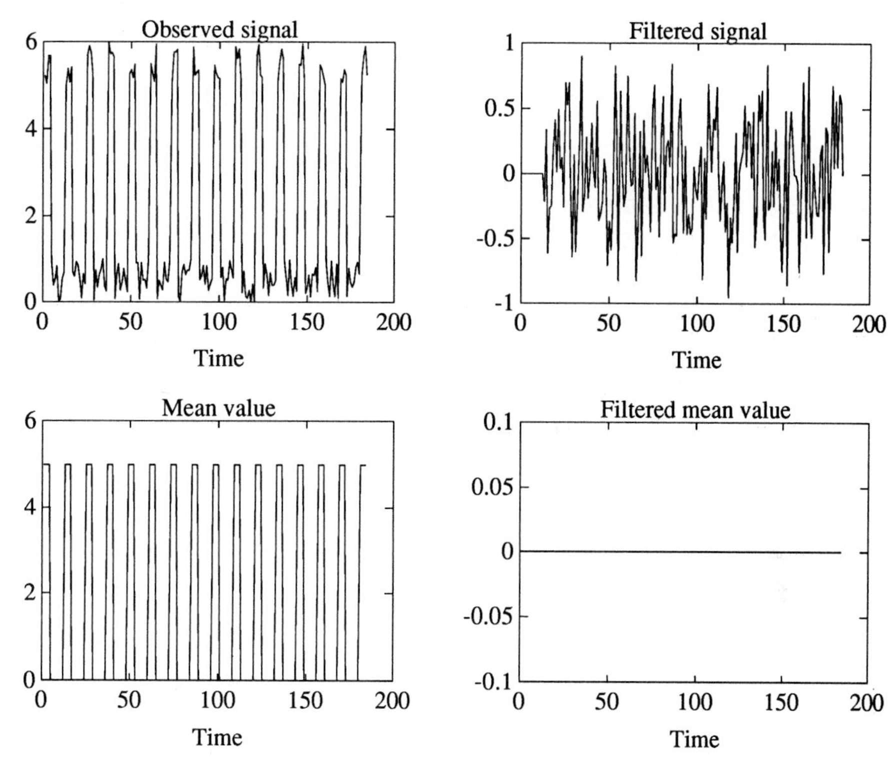

Figure D.7 Filtering of signal ($p = 12$) with periodic mean value extracted from time series by means of a filter $1 - z^{-12}$. The residual time series exhibits stationary characteristics.

it is critical to compensate with the correct period of data as there is otherwise a risk of additional data distortion. It is for the same reason important to choose the sampling period such that the data period p appears as a multiple of the sampling period.

ARMAX models

An important class of nonstationary stochastic processes is where some deterministic response to an external input and a stationary stochastic process are superimposed. This is relevant, for instance, when the external input cannot be effectively described by some probabilistic distribution.

The ARMA model can be extended by adding an external input $\{u_k\}$ which is usually considered to be known

$$\begin{aligned} y_k &= -a_1 y_{k-1} - \cdots - a_{n_A} y_{k-n_A} + \\ &+ b_1 u_{k-1} + \cdots + b_{n_B} u_{k-n_B} + \\ &+ w_k + c_1 w_{k-1} + \cdots + c_{n_C} w_{k-n_C} \end{aligned} \tag{D.69}$$

In polynomial form we can express Eq. (D.69) as

$$A(z^{-1})y_k = B(z^{-1})u_k + C(z^{-1})w_k \tag{D.70}$$

Because of linearity $\{y_k\}$ can be separated into one purely deterministic process $\{x_k\}$ and one purely stochastic process $\{v_k\}$

$$\begin{aligned} A(z^{-1})x_k &= B(z^{-1})u_k \\ A(z^{-1})v_k &= C(z^{-1})w_k \end{aligned} \quad \Rightarrow \quad y_k = x_k + v_k \tag{D.71}$$

The type of decomposition (D.71) which separates the deterministic and stochastic processes is known as the *Wold decomposition.*

D.7 PREDICTION AND RECONSTRUCTION

Consider the problem of predicting the output d steps ahead when the output $\{y_k\}$ is generated by the ARMA model

$$A(z^{-1})y_k = C(z^{-1})w_k \tag{D.72}$$

which is driven by a zero-mean white noise $\{w_k\}$ with covariance $\mathcal{E}\{w_i w_j\} = \sigma_w^2 \delta_{ij}$. In other words, assuming that observations $\{y_k\}$ are available up to the present time, how should the output d steps ahead be predicted optimally?

Assume that the polynomials $A(z^{-1})$ and $C(z^{-1})$ are mutually prime with no zeros for $|z| \geq 1$. Let the C-polynomial be expanded according to the *Diophantine equation*

$$C(z^{-1}) = A(z^{-1})F_d(z^{-1}) + z^{-d}G_d(z^{-1}) \tag{D.73}$$

which is solved by the two polynomials

$$\begin{aligned} F_d(z^{-1}) &= 1 + f_1 z^{-1} + \cdots + f_{n_F} z^{-n_F}, & n_F &= d - 1 \\ G_d(z^{-1}) &= g_0 + g_1 z^{-1} + \cdots + g_{n_G} z^{-n_G}, & n_G &= \max(n_A - 1, n_C - d) \end{aligned} \tag{D.74}$$

Using Eqs. (D.72) and (D.73) we find

$$y_{k+d} = F_d(z^{-1})w_{k+d} + \frac{G(z^{-1})}{C(z^{-1})}y_k \tag{D.75}$$

Let us by $\widehat{y}_{k+d|k}$ denote linear $d-$step predictors of y_{k+d} based upon the measured information available at time k. As the term $F_d(z^{-1})w_{k+d}$ of Eq. (D.75) is unpredictable at time k, it is natural to suggest the following $d-$step predictor

$$\widehat{y}_{k+d|k} = \frac{G(z^{-1})}{C(z^{-1})} y_k \tag{D.76}$$

The prediction error satisfies

$$\begin{aligned} \varepsilon_{k+d} &= (\widehat{y}_{k+d|k} - y_{k+d}) = \frac{G(z^{-1})}{C(z^{-1})} y_k - \frac{A(z^{-1})F(z^{-1}) + z^{-d}G(z^{-1})}{C(z^{-1})} y_{k+d} = \\ &= -F_d(z^{-1})w_{k+d} \end{aligned} \tag{D.77}$$

Let $\mathcal{E}\{\cdot|\mathcal{F}_k\}$ denote the *conditional mathematical expectation* relative to the measured information available at time k. The conditional mathematical expectation and the covariance of the $d-$step prediction relative to availaible information at time k is

$$\begin{aligned} \mathcal{E}\{\widehat{y}_{k+d|k} - y_{k+d}|\mathcal{F}_k\} &= \mathcal{E}\{-F_d(z^{-1})w_{k+d}|\mathcal{F}_k\} = 0 \\ \mathcal{E}\{(\widehat{y}_{k+d|k} - y_{k+d})^2|\mathcal{F}_k\} &= \mathcal{E}\{(F_d(z^{-1})w_{k+d})^2|\mathcal{F}_k\} \\ &= \mathcal{E}\{(w_{k+d} + f_1 w_{k+d-1} + \cdots + f_{d-1}w_{k+1})^2|\mathcal{F}_k\} \\ &= (1 + f_1^2 + \cdots + f_{n_F}^2)\sigma_w^2 = 0 \end{aligned} \tag{D.78}$$

It follows that the predictor (D.76) is unbiased and that the prediction error only depends on future, unpredictable noise components. It is straightforward to show that the predictor (D.76) achieves the lower bound (D.78) and that the predictor (D.76) is optimal in the sense that the prediction error variance is minimized.

Example D.6—An optimal predictor for a first-order model
Consider for the first-order ARMA model

$$y_{k+1} = -a_1 y_k + w_{k+1} + c_1 w_k \tag{D.79}$$

The variance of a one-step-ahead predictor $\widehat{y}_{k+1|k}$ is

$$\begin{aligned} \mathcal{E}\{(\widehat{y}_{k+1|k} - y_{k+1})^2|\mathcal{F}_k\} &= \mathcal{E}\{(\widehat{y}_{k+1|k} + a_1 y_k - c_1 w_k)^2|\mathcal{F}_k\} + \mathcal{E}\{w_{k+1}^2|\mathcal{F}_k\} \\ &= \mathcal{E}\{(\widehat{y}_{k+1|k} + a_1 y_k - c_1 w_k)^2|\mathcal{F}_k\} + \sigma_w^2 \geq \sigma_w^2 \end{aligned} \tag{D.80}$$

The optimal predictor satisfying the lower bound (D.80) is obtained from Eq. (D.80) as

$$\widehat{y}^o_{k+1|k} = -a_1 y_k + c_1 w_k \tag{D.81}$$

which, unfortunately, is not realizable as it stands because w_k is not available to measurement. Therefore, the noise sequence $\{w_k\}$ has to be substituted by some function of the observed variable $\{y_k\}$. A linear predictor chosen according to Eq. (D.76) is

$$\widehat{y}_{k+1|k} = \frac{G_1(z^{-1})}{C(z^{-1})} y_k = \frac{c_1 - a_1}{1 + c_1 z^{-1}} y_k \tag{D.82}$$

Let the difference between the predictors (D.81) and (D.82) be denoted

$$\delta_k = \widehat{y}_{k+1|k} - \widehat{y}^o_{k+1} \tag{D.83}$$

By direct substitution it follows that

$$(1 + c_1 z^{-1})\delta_k = 0, \qquad \text{for all } k \tag{D.84}$$

and so

$$\delta_k = (-c_1)^k \delta_0 \tag{D.85}$$

for some initial value δ_0. Any possible initial difference between the two estimators thus disappears as k grows and as the predictor $\widehat{y}_{k+1|k}$ approaches stationarity. Hence, the linear predictor (D.82) achieves the lower bound (D.80) and is, consequently, optimal. ■

D.8 THE KALMAN FILTER

Consider the linear state-space model

$$\begin{aligned} x_{k+1} &= \Phi x_k + v_k, & x_k &\in R^n \\ y_k &= C x_k + w_k, & y_k &\in R^m \end{aligned} \tag{D.86}$$

where $\{v_k\}$ and $\{w_k\}$ are assumed to be independent zero-mean white-noise processes with covariances Σ_v and Σ_w, respectively. It is assumed that $\{y_k\}$ but not $\{x_k\}$ is available to measurement and that it is desirable to predict $\{x_k\}$ from measurements of $\{y_k\}$.

Introduce the state predictor

$$\begin{aligned} \widehat{x}_{k+1|k} &= \Phi \widehat{x}_{k|k-1} - K_k(\widehat{y}_k - y_k), & \widehat{x}_{k|k-1} &\in R^n \\ \widehat{y}_k &= C\widehat{x}_{k|k-1}, & y_k &\in R^m \end{aligned} \tag{D.87}$$

The predictor (D.87) has the same dynamics matrix Φ as the state-space model (D.86) and, in addition, there is a correction term $K_k(\widehat{y}_k - y_k)$ with a factor K_k to be chosen. The prediction error is

$$\widetilde{x}_{k+1|k} = \widehat{x}_{k+1|k} - x_{k+1} \tag{D.88}$$

The prediction-error dynamics is

$$\widetilde{x}_{k+1} = (\Phi - K_k C)\widetilde{x}_k + v_k - K_k w_k \tag{D.89}$$

The mean prediction error is governed by the recursive equation

$$\mathcal{E}\{\widetilde{x}_{k+1}\} = (\Phi - K_k C)\mathcal{E}\{\widetilde{x}_k\} \tag{D.90}$$

The mean square error of the prediction error is governed by

$$\begin{aligned} \mathcal{E}\{\widetilde{x}_{k+1}\widetilde{x}_{k+1}^T\} &= \mathcal{E}\{((\Phi - K_k C)\widetilde{x}_k + v_k - K_k w_k)((\Phi - K_k C)\widetilde{x}_k + v_k - K_k w_k)^T\} \\ &= (\Phi - K_k C)\mathcal{E}\{\widetilde{x}_k\widetilde{x}_k^T\}(\Phi - K_k C)^T + \Sigma_v + K_k \Sigma_w K_k \end{aligned} \tag{D.91}$$

If we denote

$$\begin{aligned} P_k &= \mathcal{E}\{\widetilde{x}_k\widetilde{x}_k^T\} \\ Q_k &= \Sigma_w + CP_kC^T \end{aligned} \tag{D.92}$$

then Eq. (D.91) is simplified to

$$P_{k+1} = \Phi P_k \Phi^T - K_k C P_k \Phi - \Phi^T P_k C^T K_k^T + \Sigma_v + K_k Q_k K_k^T \tag{D.93}$$

By completing squares of terms containing K_k we find

$$\begin{aligned} P_{k+1} &= \Phi P_k \Phi^T + \Sigma_v - \Phi P_k C^T Q_k^{-1} C P_k \Phi^T \\ &\quad + (K_k - \Phi P_k C^T Q_k^{-1}) Q_k (K_k - \Phi P_k C^T Q_k^{-1})^T \end{aligned} \tag{D.94}$$

where only the last term depends on K_k. Minimization of P_{k+1} can be done by choosing K_k such that the positive semidefinite K_k–dependent term in Eq. (D.94) disappears. Thus P_{k+1} achieves its lower bound for

$$K_k = \Phi P_k C^T (\Sigma_w + C P_k C^T)^{-1} \tag{D.95}$$

and the *Kalman filter* (or *Kalman-Bucy filter*) takes the form

$$\begin{aligned} \widehat{x}_{k+1|k} &= \Phi \widehat{x}_{k|k-1} - K_k(\widehat{y}_k - y_k) \\ \widehat{y}_k &= C\widehat{x}_{k|k-1} \\ K_k &= \Phi P_k C^T (\Sigma_w + C P_k C^T)^{-1} \\ P_{k+1} &= \Phi P_k \Phi^T + \Sigma_v - \Phi P_k C^T (\Sigma_w + C P_k C^T)^{-1} C P_k \Phi^T \end{aligned} \tag{D.96}$$

which is the optimal predictor in the sense that the mean square error (D.91) is minimized in each step.

Example D.7—Kalman filter for a first-order system
Consider the state-space model

$$\begin{aligned} x_{k+1} &= 0.9x_k + v_k \\ y_k &= x_k + w_k \end{aligned} \tag{D.97}$$

where $\{v_k\}$ and $\{w_k\}$ are zero-mean white-noise processes with covariances $\mathcal{E}\{v_k^2\} = 1$ and $\mathcal{E}\{w_k^2\} = 1$, respectively.

The Kalman filter (D.96) takes on the form

$$\begin{aligned} \widehat{x}_{k+1|k} &= 0.9\widehat{x}_{k|k-1} - K_k(\widehat{x}_{k|k-1} - y_k) \\ K_k &= \frac{0.9P_k}{1 + P_k} \\ P_{k+1} &= 0.9^2 P_k + 1 - \frac{0.9^2 P_k^2}{1 + P_k} \end{aligned} \tag{D.98}$$

The result of one such realization is shown in Fig. D.8. ■

D.9 BIBLIOGRAPHY AND REFERENCES

Much of the material on prediction theory was originally developed by

– A.N. Kolmogorov, "Sur l'interpolation et extrapolation des suites stationnaires." *C.R. Acad. Sci.*, Vol. 208, 1939, pp. 2043–2045.
– N. Wiener, *Extrapolation, Interpolation and Smoothing of Stationary Time Series with Engineering Applications*. New York: John Wiley, 1949.

A good general reference on stochastic processes is

– J.L. Doob, *Stochastic Processes*. New York: John Wiley, 1953.

An important textbook on time-series analysis containing several interesting data records is

– T.W. Anderson, *The Statistical Analysis of Time Series*. New York: John Wiley, 1971.

Important textbooks on time-series analysis are

– G.M. Jenkins and D.G. Watts, *Spectral Analysis and Its Applications*. San Franscisco: Holden-Day, 1968.

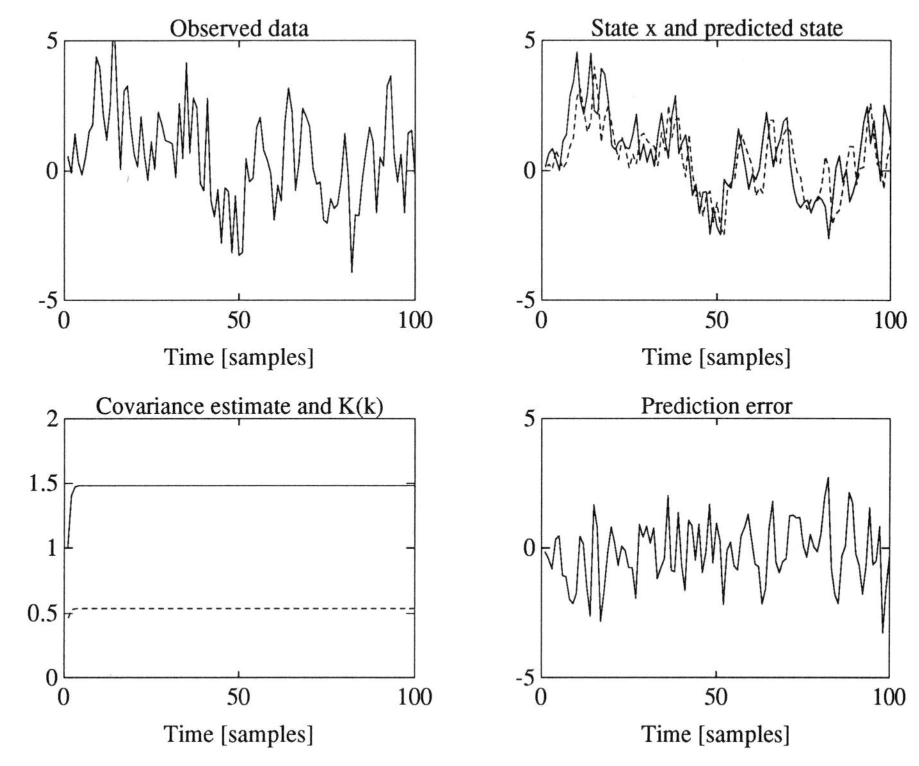

Figure D.8 Kalman filter applied to one-step-ahead prediction of x_{k+1} in Eq. (D.98). The observed variable $\{y_k\}$, the state $\{x_k\}$ and the predicted state $\{\widehat{x}_k\}$, the estimated variance $\{P_k\}$ and $\{K_k\}$, and the prediction error $\{\widetilde{x}_k\}$ are shown in a 100-step realization of the stochastic process.

- G.E.P. Box and G.M. Jenkins, *Time Series Analysis: Forecasting and Control*. San Francisco: Holden-Day, 1970.

The formula in Eq. (D.66) is obtained from

- M.S. Bartlett, *Stochastic Processes*. Cambridge: Cambridge University Press, 1955.

The example in Eq. (D.48) was obtained from

- M. Pagano, "Estimation of models of autoregressive signal plus white noise." *Ann. Statistics*, Vol. 2, 1974, pp. 99–108.

■

E

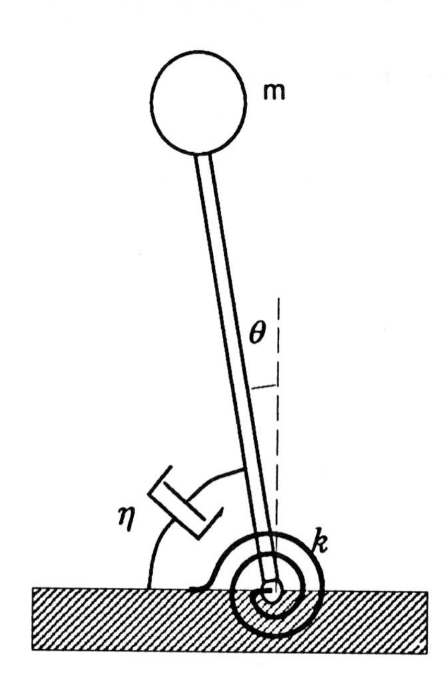

A Case Study

E.1 INTRODUCTION

The study† consists of an analysis of measurements related to human postural dynamics, which was investigated in six healthy subjects by means of a force platform recording body sway induced by vibrators attached to the

† Based on "Identification of Human Postural Dynamics" by R. Johansson, M. Magnusson, M. Åkesson which appeared in *IEEE Transactions Biomedical Engineering* Vol. 35, No. 10, pp. 858–869; October 1988. ©1988 IEEE

calf muscles. The model of body mechanics adopted was that of an inverted pendulum, and posture control was quantified in three variables—swiftness, stiffness, and damping—which were assessed by means of parametric identification of a transfer function representing the stabilized inverted pendulum. The identification fulfills statistical validation criteria, and it is conjectured that the state feedback parameters identified are suitable for use in assessing ability to maintain posture.

E.2 SUMMARY

Human posture control is maintained by proprioceptive, vestibular, and visual feedback, integrated within the central vestibular and locomotor system. Lesions to the sensory feedback system, or to the central nervous system, may impair postural control and equilibrium. It is therefore of interest to assess the ability of postural control by measuring the displacement of the body center of gravity. Recordings of the amplitude and frequency of spontaneous oscillations around the equilibrium position may describe the sway and thus, by extension, the control of posture. Normally, spontaneous oscillation appears in healthy individuals during stance, and the oscillating behavior of the body sway is often irregular or complex. Another problem is to analyze response to an external disturbance in the presence of spontaneous motion. To understand the biological correlates of the posture control variables, it is also desirable to make a model-based analysis of the control system.

For the present study, we developed a model for posture control based on exposure of the subject to erroneous proprioceptive input. The stimulus is produced by vibration of the calf muscles, which results in activation of muscle spindles (see References [E7], [E12]). Vibration is believed to activate the muscle spindles, as occurs during passive muscle stretch, which causes a reflex contraction. In the present experiments, the stimulus used is vibration of the calf muscles. Body sway is measured with a force platform. The model adopted is that of the standing human body as an inverted pendulum, equipped with a servo-mechanism for balance. The model is designed so that the spectral analysis is compatible with a dynamic systems approach, and Laplace transform methods are used for transient input-output analysis. Parametric estimation is done with maximum-likelihood estimation of coefficients in ARMAX models. Model fitness and parameter uncertainty are analyzed statistically. The aim of the present study is to identify feedback parameters useful in evaluating ability to maintain posture control.

E.3 METHODS AND MATERIALS

Tests were done on naive human subjects, three males and three females (mean age 28; range 23–39 years), none of whom had any history of vertigo, central nervous disorder, ear disease, or injury to the lower extremities. At investigation, no subject was on any form of medication or had consumed alcoholic beverages for at least 48 hours.

Equipment and experimental setup

The equipment consisted of a square force platform connected to a computer for data recording and computation. The platform is equipped with strain gauges to measure vertical force at each corner at four symmetrically located points. Measurements obtained from the strain gauges are recorded by the computer, and represent the differential distribution of forces exerted by the feet on the platform. The equipment allows simultaneous recording of body sway both in the sagittal and frontal planes—*i.e.*, longitudinal and lateral motion. The stimulus is produced by vibration of the calf muscles at frequencies of 60 Hz and 100 Hz and of 0.4 mm amplitude. The subject stood with heels together on the platform while staring at a spot on the opposite wall. A small vibrator was attached to the calf muscle in each leg with elastic straps. The subject stood erect but not at attention either with closed or open eyes as instructed, and the recording was started. First, spontaneous sway was recorded. Then, the vibrators were turned on/off and modulated pseudorandomly (PRBS) according to a program executed in the computer while recording continued.

The frequency of the vibrators depended linearly on the input voltage v which had been checked for all vibrators before use. As part of routine laboratory practice, it was verified that there was no interference (aliasing) between the sampling frequency and the vibration frequency. The test sequence took 180 s.

E.4 MODELING OF POSTURE CONTROL

When exposed to a saggittal perturbation a subject may regain equilibrium by two different strategies: "ankle strategy," in which muscular forces rotate

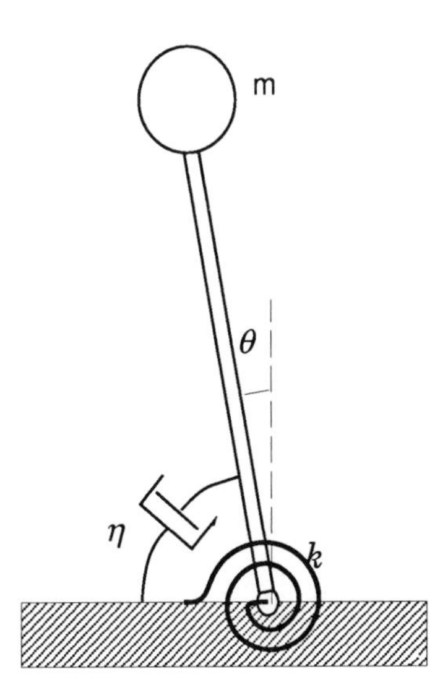

Figure E.1 Inverted pendulum model of human postural dynamics with the balancing torque T_{bal} similar to that achievable with a spring (k) and a dashpot (η).

the body around the ankle joint; or "hip strategy," involving flexion at the hips and knees. The "ankle strategy" is sufficient to counteract minor perturbations that occur during natural stance, and it fits the model of posture control as an inverted pendulum. Hip strategy has to be employed when large correction forces are needed. In hip strategy there is a potential problem with shear forces against the supporting surface. However, the force platform has been constructed so that shear forces do not interfere with the recorded signal. The moment of inertia may also change in pronounced movements because the center of body mass will be lowered. Thus, it is arguable that where gross compensatory movements are concerned (*e.g.*, preventing the subject's imminent fall), the inverted pendulum model may be insufficient. However, in natural stance and in the minor perturbations induced by the vibratory stimulus used here, the inverted pendulum model is fully adequate to account for the corrective movements used to control body posture.

The model is formulated for dynamics in the sagittal plane with the body conceived of as an inverted pendulum. The inverted pendulum has an unstable equilibrium point at $\theta = 0$ (see Fig. E.1) which means that active stabilizing forces must compensate for deviations in position in order to maintain posture. The balancing forces exerted are the result of a complex event invoking all body muscles acting in concert. A model of balance as a servo mechanism need not, however, be more complicated than suffices to describe the resulting behavior as reflected in the measurements.

The model, then, consists of an inverted pendulum to explain the pure body mechanics and a balance control system which acts like the shock absorber of a motor vehicle. The "suspension" is characterized by a spring constant k and a damping η, which keep the body in an upright position and capable of counteracting disturbance. The response to an impulse is determined by the values of k and η, as well as m (body weight) and l (distance of the body center of mass from the platform surface). The following assumptions are made in order to formalize and simplify analysis.

Assumption 1: The body is stiff, and has a mass m [kg].

Assumption 2: The body center of mass is located at distance l [m] from the platform surface.

Assumption 3: There is a dynamic equilibrium between the torque of the foot and the forces acting on the "pendulum."

A person who does not counteract the forces of gravity may be modeled by the force equilibrium of an inverted pendulum. Introduce J as the body moment of inertia around the ankle, and the tangential torque equilibrium for a standing person subject to gravitation g is then

$$J\frac{d^2\theta}{dt^2} = mgl\sin\theta(t), \qquad J = ml^2 \tag{E.1}$$

It is easy to understand both mathematically and intuitively that there is no stable equilibrium at $\theta = 0$. A person who does not counteract the gravitational torque with a stabilizing response will inevitably fall. The following two assumptions are introduced to model balance action and the effect of disturbances from the environment.

Assumption 4: Assume that there is a stabilizing ankle torque, $T_{bal}(t)$.

Assumption 5: Assume that there is a disturbance torque, $T_d(t)$, from the environment.

The torque balance now has the form

$$J\frac{d^2\theta}{dt^2} = mgl\sin\theta(t) + T_{bal}(t) + T_d(t); \qquad J = ml^2 \tag{E.2}$$

We assume that PID-control (proportional, integrating, derivative) *via* the ankle torque T_{bal} is sufficient to represent the nature of the stabilizing control.

Assumption 6: Assume that T_{bal} stabilizes the posture with PID-control with the components P, I, D determined by coefficients k, η, and ρ.

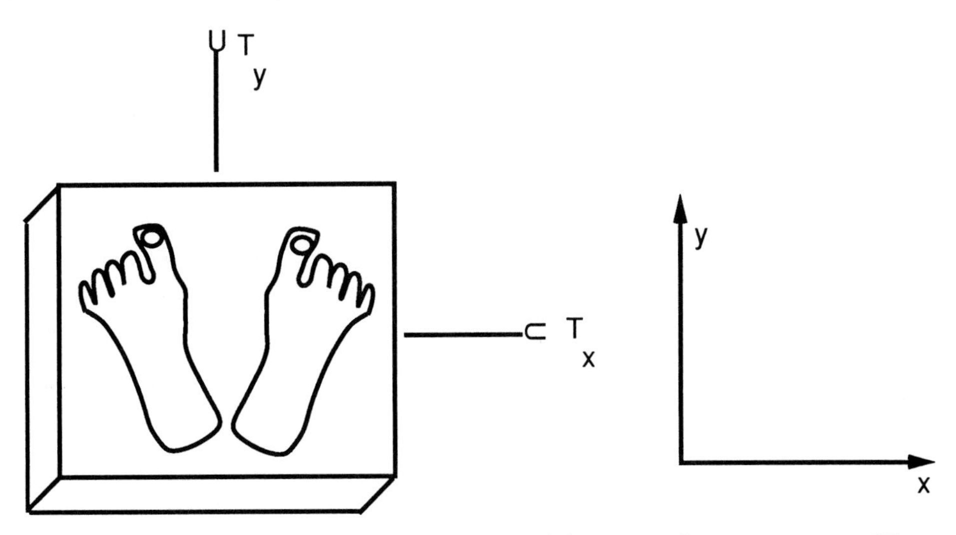

Figure E.2 A sketch of a force platform used for postural measurements. The variable T_x and T_y denote the torques around the $x-$ and $y-$directions, respectively.

P: $-mgl\sin\theta(t) - kJ\theta(t)$

I: $-\rho J\int_{t_0}^{t}\theta(t)dt$

D: $-\eta J\dot{\theta}(t)$

PID-control is chosen here because the proportional, derivative, and integral actions are fundamental modes of control. The components (P,D), and that $k, \eta > 0$ are indispensable for stability according to the Routh criterion of stability. The integral component (I) accounts for (slow) compensation of bias in θ; as (I) is not *a priori* necessary for stability, one of the aims of the experiment is to show its presence. The parameter k may be interpreted as a spring constant, and η might be compared with a viscous damping as obtained with a dashpot. The parameter ρ may be interpreted as a constant for the slow reset action in the control system.

Finally, it is necessary to model the effect of the vibration stimulus.

Assumption 7: The vibration v introduces erroneous input into the stabilizing system, causing misperception of the position θ (stretch) and the angular velocity $\dot{\theta}$ (rate) so that the P,D actions of feedback system are modified to

P: $-mgl\sin\theta(t) - kJ\theta(t) + b_1 v(t)$

D: $-\eta J\dot{\theta}(t) + b_2 v(t)$

where it is assumed that v disturbs both stretch and rate perception but at different proportions, b_1 and b_2, respectively.

A transfer function

The torque equilibrium of Eq. (E.2) and Assumption $A6$ give the two equations

$$\begin{aligned} J\frac{d^2\theta}{dt^2} &= mgl\sin\theta(t) + T_{bal}(t) + T_d(t) \\ T_{bal}(t) &= -mgl\sin\theta(t) - kJ\theta(t) - \eta J\dot{\theta}(t) - \rho J\int_{t_0}^{t}\theta(t)dt \end{aligned} \tag{E.3}$$

According to Eq. (E.3) there are three states that affect motion, namely angular velocity $d\theta/dt$, angular position θ, and the bias compensation. A transfer function from vibration stimulus $V(s) = \{v(t)\}$ and disturbance T_d to the torque T_{bal} is found *via* Eqs. (E.2), (E.3) (see Appendix E.1).

$$T_{bal}(s) = \frac{(b_1+b_2)(s^3 - \frac{g}{l}s)}{s^3+\eta s^2+ks+\rho}V(s) - \frac{\eta s^2 + (k+\frac{g}{l})s+\rho}{s^3+\eta s^2+ks+\rho}T_d(s) \tag{E.4}$$

It is of interest here to estimate the indispensable positive coefficients k and η, and to decide from data whether there is any integral action.

E.5 FORCES ON THE PLATFORM

Before signal processing may proceed, it is necessary to establish the relationship of the measurement signal μ to the angular position θ, and a static force equilibrium argument would go as follows: A signal which represents the center of force on the force platform is measured. With static equilibrium between the force on the platform and the body weight, it follows that the force center also represents the projection on the platform of the body center of gravity; see Fig. E.2.

However, such a model is not entirely satisfactory for the purposes of dynamic analysis, as the force center and the vertical projection of center of body mass do not generally coincide at the same point. The foot may, for example, exert a corrective force on the platform to initiate an angular acceleration of the body. As described in Appendix E.1 it holds that the measurement μ is related to the torque T_{bal} for a certain body mass m so that

$$\mu(t) = \frac{2\gamma}{a+b}T_{bal}(t) + \gamma\frac{b-a}{a+b}mg \tag{E.5}$$

for positions a and b, with a gain factor γ. This means that the measurement μ represents the ankle torque T_{bal} except for a gain factor and a bias term.

It is part of signal processing to compensate for the gain factor and the bias term in the recorded measurements.

E.6 A DYNAMIC RESPONSE CLASSIFICATION

We have given one interpretation of the coefficients in terms of a mechanical model with a spring k and a dashpot effect η. Naturally, a more rapid reflex system requires a balanced increase both of spring action and damping action. It is therefore desirable to quantify mutually independent characteristics of motion. Normalization of the transfer function (E.4) with respect to frequency gives for the stimulus dependence

$$T_{bal}(s) = \frac{(b_1 + b_2)((\frac{s}{\omega_0})^3 - \frac{g}{l\omega_0}(\frac{s}{\omega_0}))}{(\frac{s}{\omega_0})^3 + \frac{\eta}{\omega_0}(\frac{s}{\omega_0})^2 + \frac{k}{\omega_0^2}(\frac{s}{\omega_0}) + 1} V(s), \qquad \omega_0 = \sqrt[3]{\rho} \tag{E.6}$$

A more functional characterization of the motion based on the transfer function properties may therefore be formulated using the concepts

- **Swiftness:** $\omega_0 = \sqrt[3]{\rho} \quad [rad/s]$
- **Stiffness:** k/ω_0^2
- **Damping:** η/ω_0

This classification describes the posture dynamics by one swiftness parameter and two stability parameters. The swiftness parameter is a bandwidth [rad/s] and provides information about the highest angular frequency of the disturbance for which the posture control system gives adequate correction. The stiffness and damping are dimensionless stability parameters, independent of posture control swiftness because the dependence on ω_0 is eliminated. A high value of swiftness means rapid response to disturbance, *i.e.*, rapid compensation for small deviations from equilibrium. A high value of damping means good damping of sway velocity.

E.7 EXPERIMENTS

Experiments were performed with six subjects to evaluate the model and the method. The first experiment tested the difference between performance with open eyes and that with closed eyes. Other experiments were performed to test the difference between two choices of stimulation frequency of the method by using asymmetric stimulation. Time and frequency domain properties of the stimulus are presented in the results section. The following recordings were made with a sampling interval of 0.04 s, *i.e.*, the sampling frequency 25 Hz.

Experiment A: The empty experiment to measure electronic offsets.

Experiment B: A test sequence with a vibration stimulus of 100 Hz that is switched on and off according to pseudorandom binary sequence (PRBS), the subject standing with open eyes; see Fig. E3.

Experiment C: A test sequence with a vibration stimulus of 100 Hz that is switched on and off according to a PRBS, the subject standing with closed eyes; see Fig. E.4.

Experiment D: A test sequence with a vibration stimulus of 60 Hz that is switched on and off according to a pseudorandom binary sequence, the subject standing with closed eyes.

Data analysis was performed in the following order:

- Autospectrum of
  - Stimulus v (vibration)
  - Response μ (force distribution in direction x)
- Cross spectrum between v and μ
- Coherence between v and μ
- Transfer function from v to μ computed from spectra
- Maximum-likelihood identification of an ARMAX model
- Validation by test of residuals
  - Changes of signs (χ^2-test)
  - Autocorrelation (χ^2-test)
  - Cross correlation between v and residuals (χ^2−test)
  - Normal distribution of residuals
- Validation by simulation
- Translation from ARMAX model to continuous-time transfer function

Examples of responses are shown in Figs. E.3–E.7.

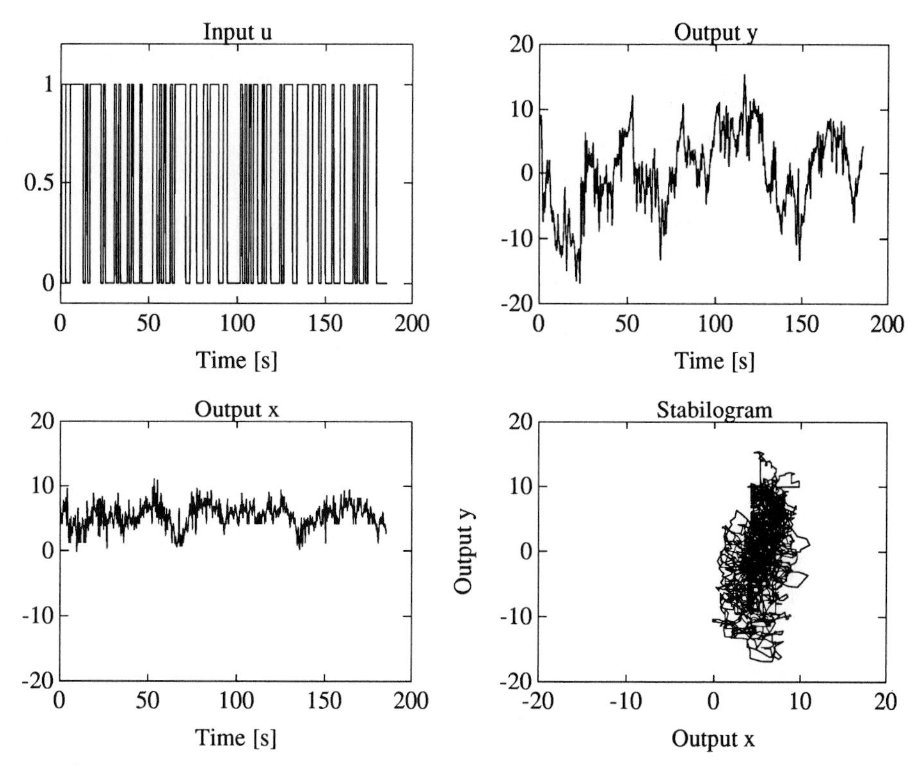

Figure E.3 Experiment B (open eyes, 100 Hz). Input voltage to vibrators versus time. Longitudinal response y and lateral response x versus time.

E.8 RESULTS OF THE EXPERIMENTS

Coherence between stimulus and response was tested for the different experiments. A detailed presentation of calculations and numerical results is given in Appendix E.2. It was found that coherence was lower for all experiments with open eyes (B) than with closed eyes. Response of frontal sway was also shown to be low for all subjects. Thus, computations of transfer functions based on such data are not to be recommended.

The results with closed eyes and symmetric stimulation were quite convincing, with good coherence between vibration stimulus v and body sway in the sagittal plane, *i.e.*, longitudinal motion. This fact indicates that there is a reasonable response to vibration at least in the absence of visual input. The continuous-time pole polynomial (transfer function denominator) of Eq. (E.6)

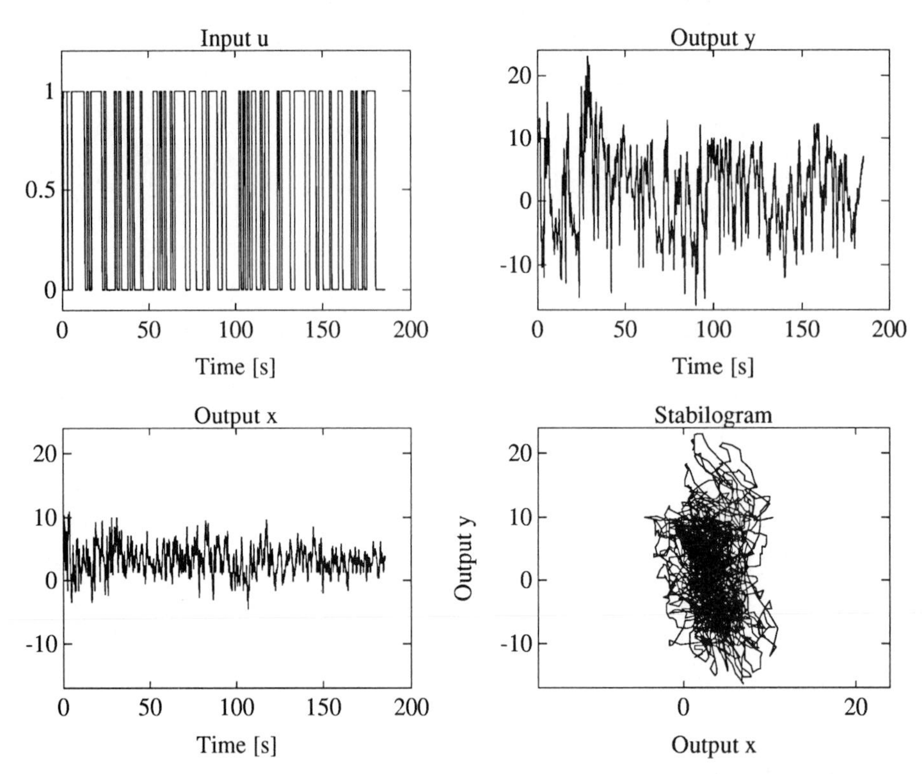

Figure E.4 Experiment C (closed eyes, 100 Hz). Input voltage to vibrators versus time. Longitudinal response y and lateral repsonse x versus time.

was computed. The third-order model pole polynomial

$$A(s) = s^3 + \eta s^2 + ks + \rho \tag{E.7}$$

was fitted with data from Experiments C and D. Results according to Table E.1 were obtained from Experiment C (closed eyes, 100 Hz).

These parameters characterize a very well damped regulation system. The dynamic response classification describes posture dynamics by one swiftness parameter and two stability parameters (see above).

The results of experiments are listed with comments on good (+) or poor (−) properties of the present approach in estimating ability to maintain posture control. The arguments for these conclusions are given in the previous section and in Appendix E.2.

\+ There is acceptably strong coherence in sagittal plane motion with closed eyes. The power of the oscillation increases by a factor of two, which means that there is a reasonable response to the vibration stimulus.

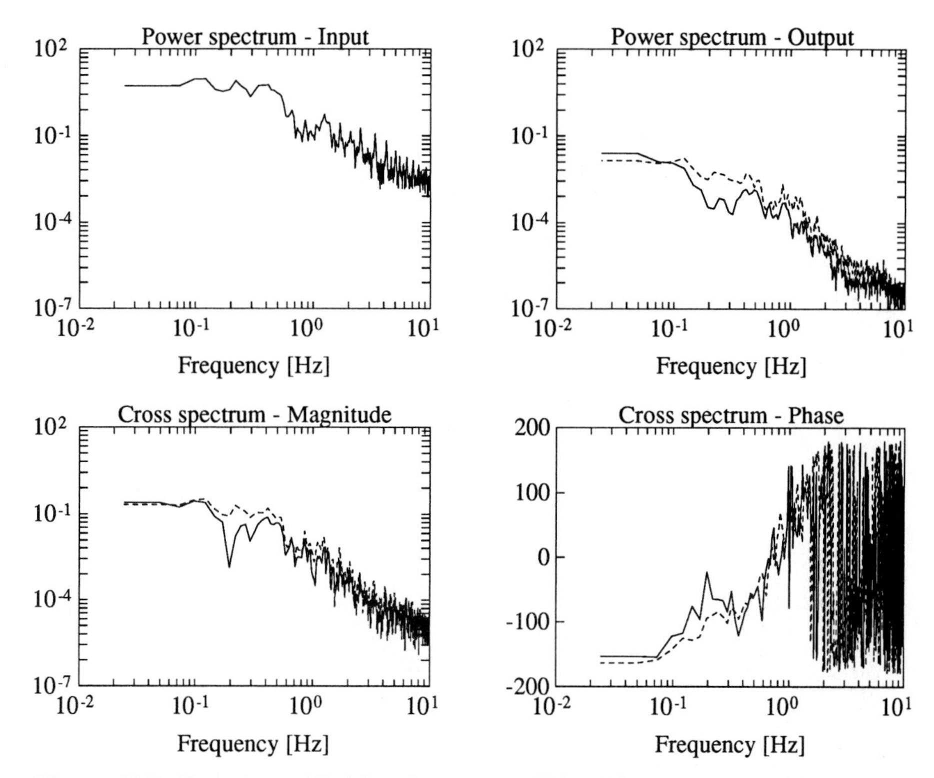

Figure E.5 Experiment *B* (closed eyes, 100 Hz). Vibrator input voltage power spectrum (1) versus frequency [rad/s]. Longitudinal sway power spectrum (2) lateral sway power spectrum (3) versus angular frequency [rad/s].

+ There is weak coherence with open eyes.
+ There is weak coherence to sway in the frontal plane.
+ The data fit very well to a linear model.
+ It is possible to identify the feedback parameters with good accuracy.
+ The residual signal has a small oscillative component of 0.2 – 0.3 Hz which may correspond to breathing.
– The method is sensitive to assymmetry in stimulation.

E.9 DISCUSSION

The identified coefficients k, η, and ρ of Assumption A6, represent different aspects of the posture control system. The amplitude of body sway may become large for a small k, whereas a large k gives good postural control of the angular

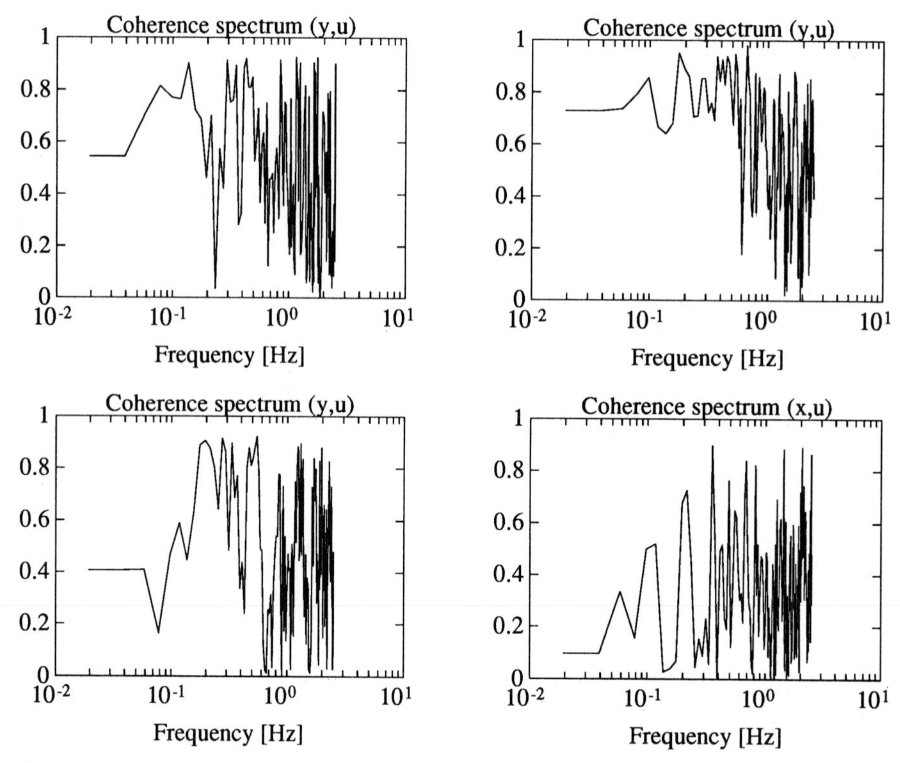

Figure E.6 Coherence spectra between input (100 Hz) and longitudinal body sway in the case of open eyes (*upper left*) and closed eyes (*upper right*). Coherence spectrum between input (60 Hz) and longitudinal body sway (*lower left*) and coherence spectrum between input (100 Hz) and lateral body sway (*lower right*). All spectra versus frequency Hz.

position. The parameter η represents the damping of body sway. Too small an η value means low damping of body sway whereas a large value means rigidity. The parameter ρ represents the automatic reset, *i.e.*, compensative action to eliminate bias in the angular position.

With a combination of the parameters k, η, and ρ, a large variety of body sway patterns can be described. The proportional and derivative actions represented by the parameters k and η are indispensable to maintain stability. The third-order model is statistically validated; it is accurate and explains data well. The strong cross covariance of the estimates of k and ρ constitutes a practical difficulty, however.

We have given one interpretation of the coefficients in terms of a mechanical model with a spring effect k and a dashpot effect η. The integral component of Eq. (E.3) and Assumption A6 is responsible for a slow reset action (see Reference [E2]), an action that is biologically feasible considering the anatomical

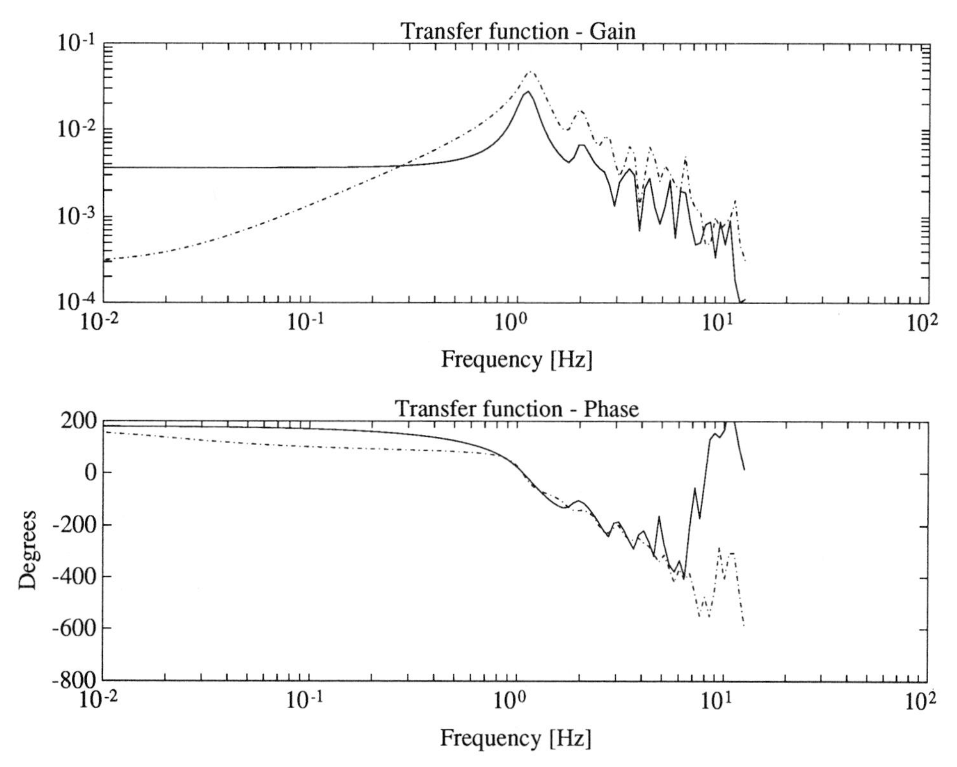

Figure E.7 Transfer function from spectra for experiments B, C with open eyes (*solid line*) and closed eyes (*dashed line*), respectively.

and physiological background. Vestibular, visual, and somatosensory information reaches the spinal motoneurons from the vestibular nucleus *via* several vestibulospinal and reticulospinal tracts with or without modulation in the cerebellum (see Reference [E10]). Spinal motoneurons are also influenced by interneurons with information from antagonistic muscles. An induced perturbation changes the visual, vestibular, and somatosensory inputs which affect the spinal motoneurons at different latencies (see Reference [E10]). Naturally, a more rapid reflex system requires a balanced increase both of spring and damping action. It is therefore desirable to quantify mutually independent characteristics of motion. A more functional characterization of the motion based on the transfer function properties may be formulated *via* normalized parameters by means of the concepts of swiftness, stiffness, and damping. A high value of swiftness means rapid response to disturbances of equilibrium, and a high value of stiffness means small deviations from equilibrium. A high value of damping means good attenuation of sway velocity.

With the model presented here, the effect of vibration on muscle stretch per-

Table E.1 Results of parametric identification with data from experiment C with 100 Hz vibration and closed eyes

Subject	η	k	ρ
1	5.24	44.26	45.64
2	5.18	26.11	14.32
3	1.37	19.26	1.75
4	6.00	33.16	26.15
5	4.56	26.35	17.13
6	6.53	60.53	99.01

Subject	Swiftness	Stiffness	Damping
1	3.57	3.47	1.47
2	2.43	4.43	2.13
3	1.21	13.3	1.14
4	2.97	3.76	2.02
5	2.58	3.97	1.77
6	4.63	2.83	1.41

ception cannot be distinguished from that on rate perception. The use of coherence functions makes it possible to quantify the relative importance of visual feedback *vis-à-vis* vestibular and proprioceptive feedback in different frequency ranges.

The choice of suitable experimental conditions for future clinical development is not self-evident although those applied here have proved reasonably satisfactory. The following aspects deserve further consideration: test duration, vibration amplitude (intensity), vibration frequency, vibration pattern, and the possibility of stimulating other muscle groups. A shorter test duration may be preferable for clinical purposes, as might different choices of vibration amplitude and frequency. The amplitude must be chosen so that the vibration stimulus does not induce any excessive sway or falling reactions. Thus application of the method is naturally limited to patients who are able to tolerate additional loading of their postural control. Other choices of frequency and amplitude may give different but statistically acceptable results for each testing condition, though standardization of test conditions is, of course, necessary to permit comparison of results.

The stimulus in the present experiments is produced by vibration of the calf muscles, sway being recorded in the sagittal and frontal planes for longitudinal and lateral motion, respectively. The vibration caused only insignificant motion in the frontal plane. Vibration to other muscles may provide tests which induce two-dimensional motion. The choice of a pseudo-random vibration pattern with a flat power spectrum is not dictated by methodological considerations, though there is a certain advantage in having a stimulus that is unpredictable by the subject.

A future clinical application of the present approach is in patients where defect postural control may be suspected. A large test material of both normals and subjects with well-defined lesions has to be analyzed to determine the reliability, sensitivity, and discriminatory power of the parameters. However, the coherence function is sufficiently large for good reproducibility to be expected.

Conclusions are only drawn from the coefficient values of the denominator polynomial, which means that attention is focused on effects of recovery from a perturbation, rather than on the onset of perturbation. This is important, because the stimulus intensity may vary, and there may be substantial interindividual variation in the primary effect of perturbations. The parameters of swiftness, stiffness, and damping presented here may therefore prove useful for interindividual comparison both in clinical practice and in research.

E.10 CONCLUSIONS

A postural test involving a force platform has been analyzed quantitatively by means of a new method. The proposed model-oriented transfer function approach also allows angular position θ (or displacement of the body center of gravity) as well as sway velocity to be computed from the measurements recorded with the force platform. Parameters to quantify the body's ability to maintain posture have been proposed, and the following conclusions are made.

- The ankle torque T_{bal} represents the body's feedback control to maintain stability. It is emphasized that the force platform measurement may best be understood as the feedback actuated by the body.

- A quantitative analysis of the feedback properties of posture control is made. The control action is analyzed with classical control concepts. It is shown that there is corrective action with respect to angular position θ, angular velocity $\dot{\theta}$, and a slow reset control of bias in θ.
- The results of computation show that the proposed quantifiers of posture k, η, and ρ may be estimated with good accuracy according to generally accepted statistical validation criteria.
- The model complexity is chosen as a linear system of order 3, which is sufficient to explain the outcome of measurements.
- The method is sensitive to symmetry of stimulation.
- The proposed model is compatible with earlier attempts to represent measurements of the posture dynamics by spectral analysis (see Reference [E5]). Spectral analysis supported by parametric identification is advantageous because it allows quantitative statistical analysis as well as physiological interpretation.
- The approach with parametric identification of a transfer function between stimulus and response can be made with higher confidence than can parametric analysis of spontaneous motion. The coherence function gives a measure of the dependence of the response on variations in the stimulus.

APPENDIX E.1 — TRANSFER FUNCTION

The torque equilibrium of Eq. (E.2) and Assumption $A6$ give the two equations

$$J\frac{d^2\theta}{dt^2} = mgl\sin\theta(t) + T_{bal}(t) + T_d(t)$$

$$T_{bal}(t) = -mgl\sin\theta(t) - kJ\theta(t) - \eta J\dot{\theta}(t) - \rho J\int_{t_0}^{t}\theta(t)dt$$

Elimination of T_{bal} gives

$$J\frac{d^2\theta}{dt^2} = -\eta J\frac{d\theta}{dt} - kJ\theta(t) - \rho J\int_{t_0}^{t}\theta(\tau)d\tau + T_d(t)$$

There are three states that affect motion, namely angular velocity $d\theta/dt$, angular position θ, and the bias compensation. A Laplace transformation and algebraic simplification gives the transfer function

$$\theta(s) = \frac{\frac{1}{J}s}{s^3 + \eta s^2 + ks + \rho} T_d(s) \tag{E.8}$$

With a vibration stimulus $v(t)$, according to Assumption A7 there is one more transfer function, namely that from stimulus v to θ

$$\theta(s) = \frac{\frac{1}{J}(b_1 + b_2)s}{s^3 + \eta s^2 + ks + \rho} V(s) + \frac{\frac{1}{J}s}{s^3 + \eta s^2 + ks + \rho} T_d(s) \tag{E.9}$$

A reduced model without any integrating compensation ($\rho = 0$) gives the simplification

$$\theta(s) = \frac{\frac{1}{J}(b_1 + b_2)}{s^2 + \eta s + k} V(s) + \frac{\frac{1}{J}}{s^2 + \eta s + k} T_d(s)$$

A transfer function from vibration stimulus $V(s) = \mathcal{L}\{v(t)\}$ and disturbance T_d to the torque T_{bal} is found *via* Eqs. (E.2) and (E.8) for linearized motion around the equilibrium $\theta = 0$, where $\sin\theta \approx \theta$ and

$$\begin{aligned} T_{bal}(s) &\approx (Js^2 - mgl)\theta(s) - T_d(s) \\ &= \frac{(b_1 + b_2)(s^3 - \frac{g}{l}s)}{s^3 + \eta s^2 + ks + \rho} V(s) - \frac{\eta s^2 + (k + \frac{g}{l})s + \rho}{s^3 + \eta s^2 + ks + \rho} T_d(s) \end{aligned} \tag{E.10}$$

It is of interest here to estimate the indispensable positive coefficients k and η, and to decide from data if there is any integral action. ■

APPENDIX E.2 — FORCE BALANCES

The distances a and b denote horizontal distances from the ankle point of rotation to each one of the support points at the edges of the force plate. Let P_{foot} denote the pressure of the soles exerted on the force plate, and Ω denote the area of contact between the feet and the force platform (see Figs. E.8 and E.9). The forces F_a and F_b represent the support forces at the edges of the force plate. The measurements μ are force differences given by

$$\mu = \gamma(F_a - F_b) \tag{E.11}$$

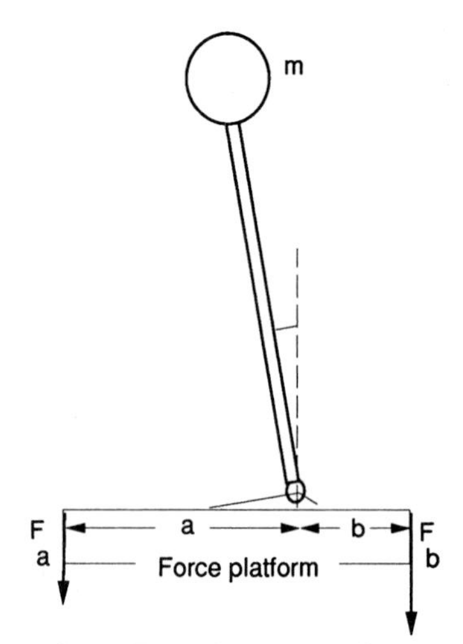

Figure E.8 Anterior force F_a and posterior force F_b on the force plate.

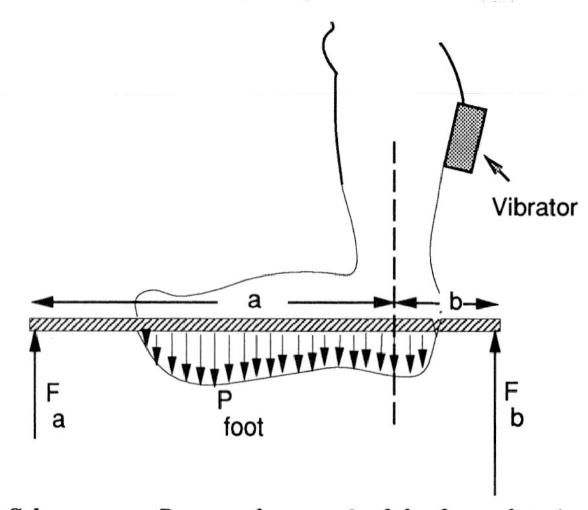

Figure E.9 Sole pressure P_{foot} on the area Ω of the force plate in the xz–plane.

with γ as a gain factor due to strain gauges and the electronics. The force equilibria related to the foot pressure P_{foot} on the support surface are thus

$$\int_{\Omega}\int P_{foot}(x,y)dxdy = mg \tag{E.12}$$

and

$$\int_{\Omega}\int P_{foot}(x,y)dxdy = F_a + F_b \tag{E.13}$$

on the body and on the force plate, respectively. The corresponding torque equilibria are

$$T_{bal} = T_y = \int_{\Omega}\int P_{foot}(x,y)x\,dx\,dy$$
$$T_x = \int_{\Omega}\int P_{foot}(x,y)y\,dx\,dy \tag{E.14}$$

The forces F_a and F_b act on the distances a and b from the origin, with the resulting torque

$$-F_a a + F_b b + T_{bal} = 0 \tag{E.15}$$

where the ankle torque equilibrium results in body sway given by

$$J\frac{d^2}{dt^2}\theta(t) = mgl\sin\theta(t) + T_{bal}(t), \qquad J = ml^2 \tag{E.16}$$

From Eq. (E.11) and Eq. (E.15), we find for body mass m [kg] that

$$F_a + F_b = mg, \qquad \text{and} \qquad \gamma(F_a - F_b) = \mu \tag{E.17}$$

Solving these equations with respect to F_a and F_b gives

$$F_a = \frac{1}{2}mg + \frac{1}{2\gamma}\mu, \qquad \text{and} \qquad F_b = \frac{1}{2}mg - \frac{1}{2\gamma}\mu \tag{E.18}$$

With F_a and F_b it is possible to express the torque T_{bal} as

$$T_{bal} = aF_b - bF_a = \frac{a-b}{2}mg + \frac{a+b}{2\gamma}\mu \tag{E.19}$$

Solving for μ shows that μ represents T_{bal} *via* the linear relation

$$\mu = \frac{2\gamma}{a+b}T_{bal} + \gamma mg\frac{b-a}{a+b} \tag{E.20}$$

Calibration experiments give the values $a+b$ =0.327 [m] and γ =0.044 [V/N].

APPENDIX E.3 — CALCULATIONS AND ANALYSIS

The results of computation are presented in this section together with certain conclusions. The presentation essentially follows the order of computation given in Section E.6 and capital letters (*A–D*) refer to experiments presented

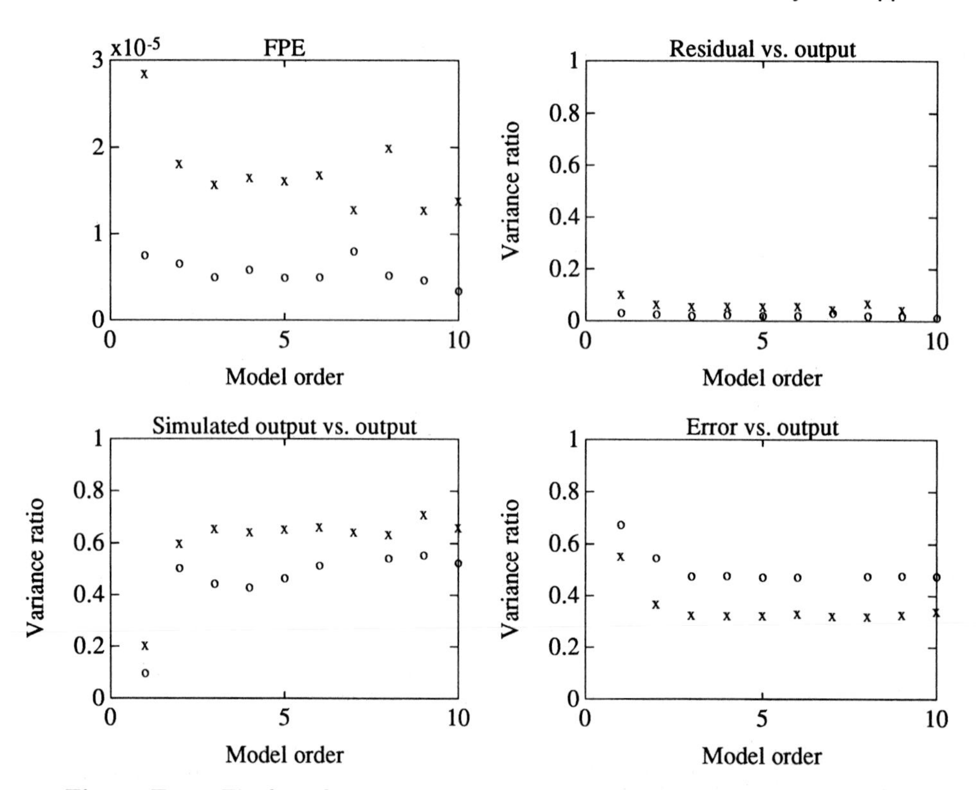

Figure E.10 Final prediction error (FPE) for various model orders (*upper left*) for models describing behavior with closed eyes (x) and open eyes (o). Shown are the variance ratio between residuals and output (*upper right*), simulated output and output (*lower left*), and model error and output (*lower right*). All graphs versus model order.

above. Graphical presentation of experimental results is given in Fig. E.3 and Fig. E.4 for open and closed eyes, respectively. It is noticeable from these graphs that the lateral response to vibration stimulus is much smaller in amplitude than is the longitudinal response.

Spectral analysis

The autospectra (power spectra) show the frequency contents of the signals investigated (see Fig. E.5). Notice that the spectrum should not be confused with the vibration frequency of the stimulus.

A coherence spectrum between input and recorded response variables was made (see Fig. E.6). Recall that a coherence spectrum can be interpreted as a correlation analysis made for each frequency. A large absolute value close to 1 indicates that the input and the output are correlated. A coherence of 0.5 denotes that half of the output variation may be explained by variations

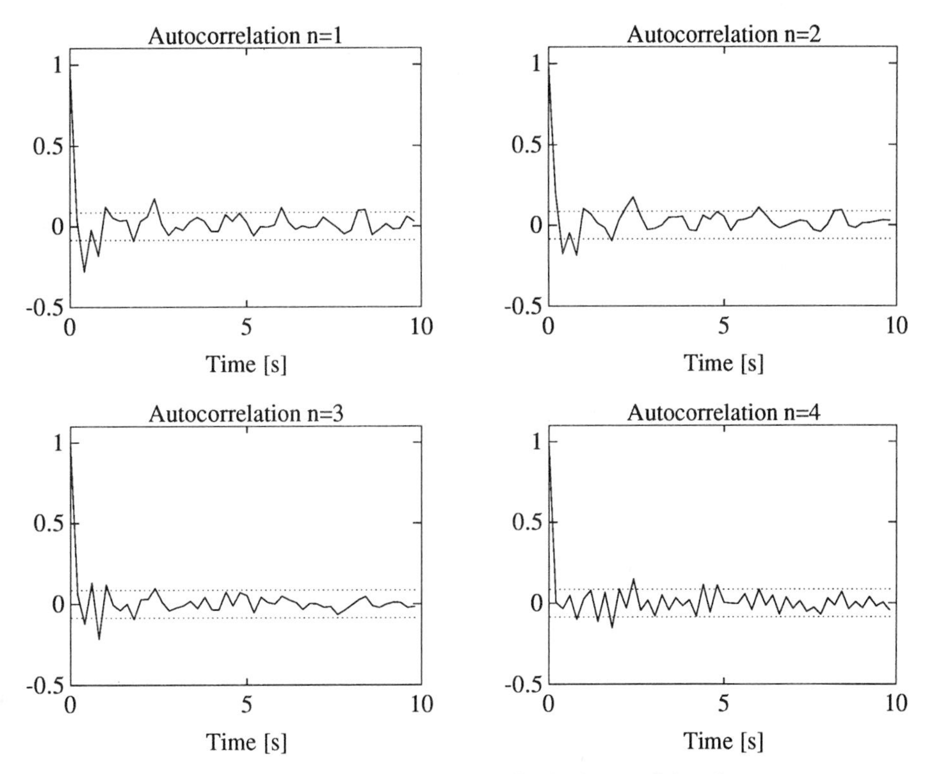

Figure E.11 Test of autocorrelation of residuals for model orders $n = 1, \ldots, 4$. Confidence interval (99%) is displayed.

in the stimulus input. It may be concluded from the coherence spectra that the coherence is quite satisfactory. The coherence is better for longitudinal sway than for lateral sway. This makes sense as the vibrators are mounted on both calves to stimulate longitudinal motion. Also notice in Fig. E.6 that the coherence is low for frequencies below 0.05 Hz. This may indicate that the reflex reaction affected by the vibration stimulus operates with a bandwidth down to 0.05 Hz.

Transfer functions from spectra

Division of the cross spectrum between input v and output x by autospectrum of v gives the transfer function, *i.e.*, gain and phase lag for a range of frequencies (see Fig. E.7).

Estimation of the delay time T_d in the feedback loop is possible by checking the phase lag for high frequencies. A pure delay appears in a transfer function as the factor $\exp(-sT_d)$, which tends to dominate the phase delay in the high frequency range. Hence, for high angular frequencies ω it holds that the phase

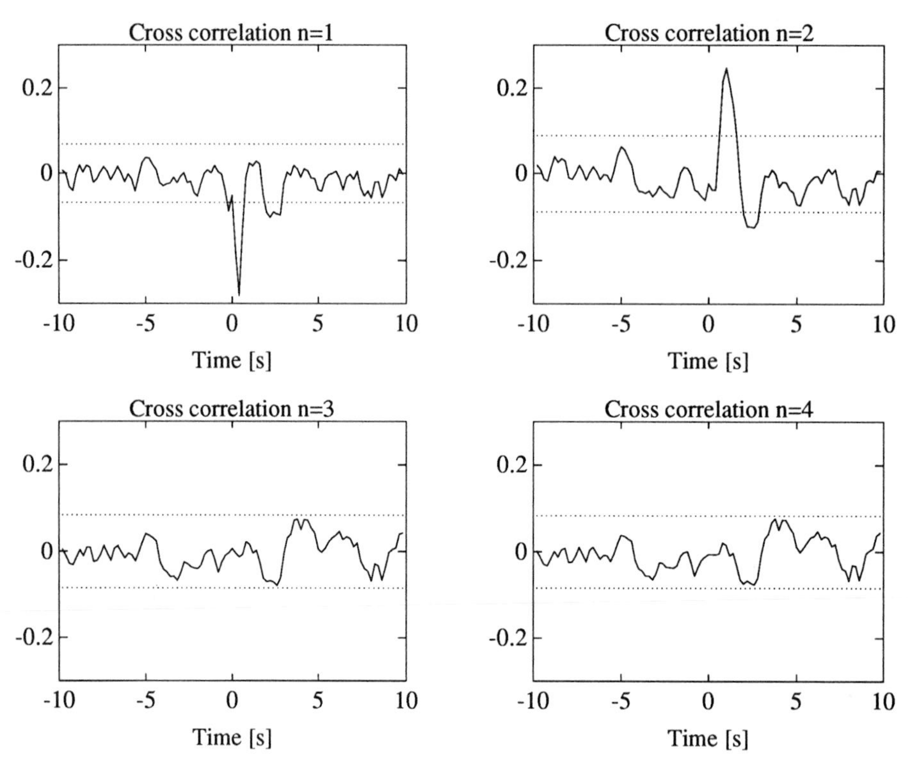

Figure E.12 Test of cross correlation of residuals for model orders $n = 1, \dots, 4$ with support for the choice of model order $n = 3$. Confidence interval (99%) is displayed.

lag is ωT_d, and from a transfer function estimate (see Fig. E.7) we find, for instance, that at the frequency 10 Hz ($\omega = 63$ rad/s) we have a phase lag of 600° so that we can approximate

$$T_d \approx \frac{\pi 600}{180}\frac{1}{63} = 0.17s \tag{E.21}$$

This value should be compared to other measures of the time required for a signal to complete a round-trip in the neurological circuit.

There is some evidence that the muscle spindles react in different ways to different vibration frequencies (see Reference [E9]), though this is not a predominant feature. Different numerical results may, however, be expected.

Maximum-likelihood identification

The time delay was estimated to $T_d \approx 0.17$ s and it is therefore desirable to estimate model parameters at sampling rates of this order of magnitude.

The following ARMAX models all have a sampling interval of 0.20 s which is obtained by extraction of every fifth sample from the original time series.

Parameter identification with estimation of initial values was made for model orders one, two, three, and four. Statistical tests are satisfied for orders $n = 3$ and $n = 4$ but not for the second-order model. The Akaike test criteria AIC and FPE do not change much but do indicate $n = 3$ as the appropriate model order (see Fig. E.10). The model order of choice is therefore a third-order model.

Validation by test of residuals

The purpose of residual tests is to find remaining correlations which indicate whether the model order is adequate. With an adequate model order, the residual noise is white noise only and of sufficiently small magnitude. The ratio between the residual variance and the output variance is shown for model orders $n = 1, \ldots, 10$ in Fig. E.10. The residual χ^2-tests for a third-order model give *significant* (95% confidence) validation with respect to changes of sign, independence of residuals, normality, and independence between residuals and input (see Fig. E.11 and Fig. E.12).

Validation by simulation

Real and simulated data have been compared using the vibration signal as a deterministic input (see Fig. E.13). We studied to what extent the experiment data are explained by the deterministic input-output behavior of the estimated model.

Conversion to continuous-time parameters

For the third-order model, we have estimated an ARMAX pole polynomial

$$A(z) = z^3 + a_1 z^2 + a_2 z + a_3 \tag{E.22}$$

with the following result of parameter values and standard deviations for subject no. 6

$$a_1 = -1.227 \pm 0.106, \qquad a_2 = 0.734 \pm 0.154, \qquad a_3 = -0.370 \pm 0.073 \tag{E.23}$$

where the standard deviations have been calculated from the estimated covariance matrix for the parameter estimates. Conversion to continuous-time parameters requires inverse sampling

$$\frac{1}{h} \log \Phi \tag{E.24}$$

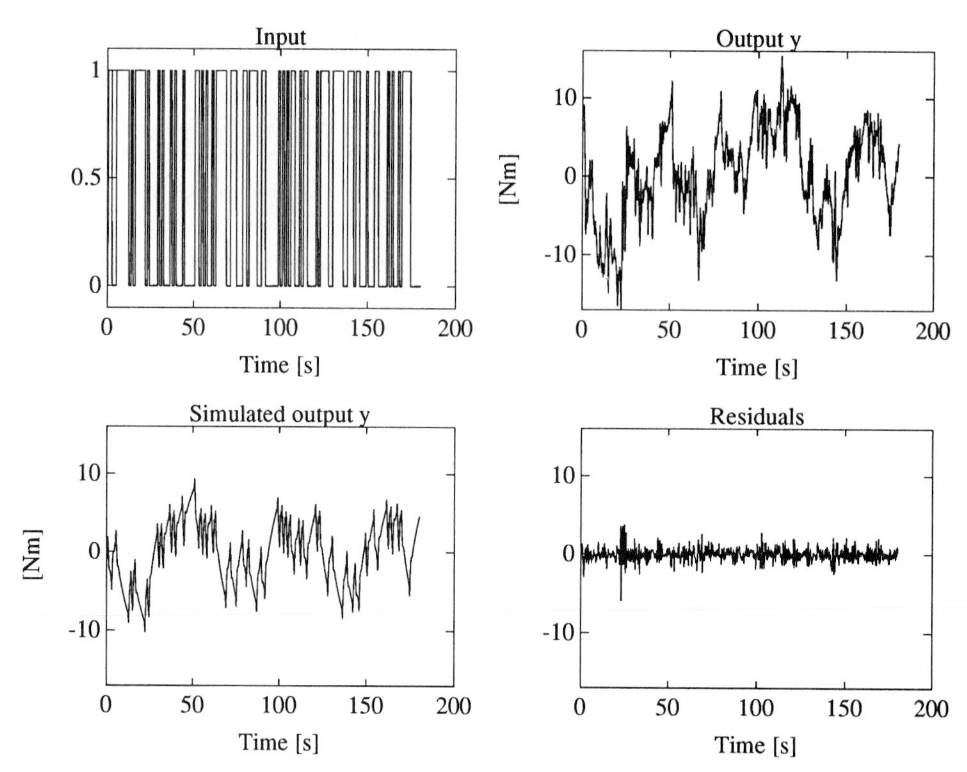

Figure E.13 Longitudinal sway of model with input and output of experiment C (closed eyes, 100 Hz) in upper graphs. Simulated output from third order estimated model (*lower left*) and residual sequence (*lower right*).

with a sampling interval h and a dynamics matrix Φ. Computation of the continuous-time characteristic polynomial gives

$$A(s) = s^3 + 4.97s^2 + 49.4s + 32.0 \tag{E.25}$$

This formulation allows identification of the physiological feedback parameters in terms of the third-order model pole polynomial of Eq. (E.22) where the coefficients of $A(s)$ determine the postural behavior.

$$A(s) = s^3 + \eta s^2 + ks + \rho \tag{E.26}$$

Parameters are already normalized in the model with respect to body weight m and body height l, in terms of the moment of inertia J with results for the test group according to Table E.1 and Table E.2. We have given one interpretation of the coefficients in terms of a mechanical model with a spring (k) and a dashpot component (η). The more functional characterization of the motion in terms of *swiftness* ($\sqrt[3]{\rho}$), *stiffness* ($k/(\sqrt[3]{\rho})^2$) and *damping* $\eta/\sqrt[3]{\rho}$, is based

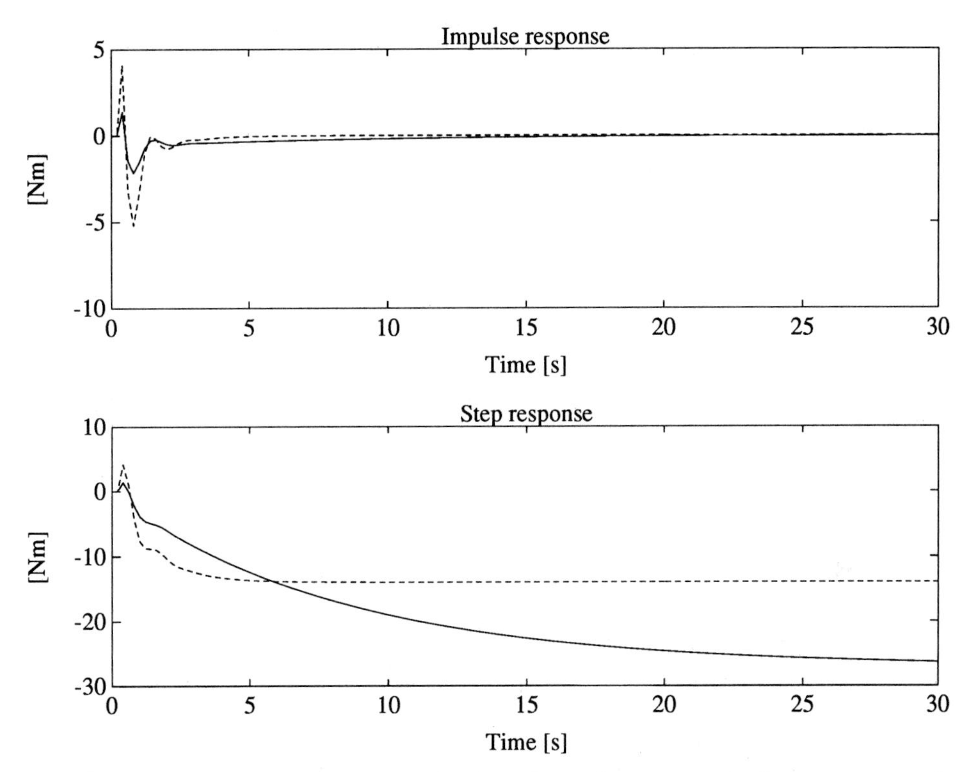

Figure E.14 Longitudinal sway. Simulated impulse response and step response of force on the plate from third order estimated model for open eyes (*solid line*) and closed eyes (*dashed line*), respectively.

on the dynamic response. This classification describes the postural dynamics by one swiftness parameter and two stability parameters. A high value of swiftness means rapid response to disturbance, and a high value of stability means small deviations from equilibrium. The swiftness parameter is a bandwidth [rad/s] and provides information about the highest angular frequency of disturbance for which the posture control system gives adequate correction. Other useful representations are impulse response and step response (see Fig. E.14) and zero-pole diagrams and transfer functions (see Fig. E.15).

BIBLIOGRAPHY AND REFERENCES

The material presented in this appendix is based on the paper by

[E1] R. JOHANSSON, M. MAGNUSSON, AND M. ÅKESSON, "Identification of human postural dynamics." *IEEE Trans. Biomed. Eng.*, Vol. BME-35, 1988, pp. 858–869.

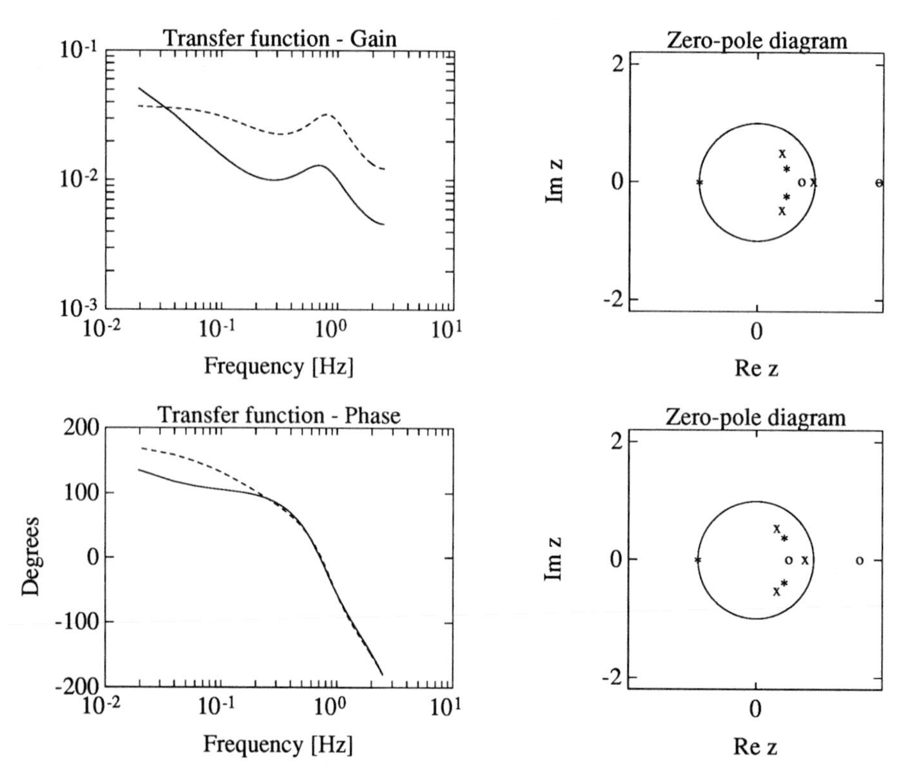

Figure E.15 Transfer functions for stimulus-response relationships for open eyes (*solid line*) and closed eyes (*dashed line*) as obtained from third-order ARMAX models. The zero-pole diagrams for open eyes (*upper*) and closed eyes (*lower*) are shown. The symbols 'x' and 'o' denote poles and zeros whereas '*' denotes zeros of the noise spectrum.

Some important references on dynamic posturography are

[E2] H.C. Diener, J. Dichgans, B. Guschbauer and M. Bacher, "Role of visual and static vestibular influences on dynamic posture control." *Human Neurobiology*, Vol. 5, 1986, pp. 105–113.

[E3] N.G. Henriksson, G. Johansson, L.G. Olsson, and H. Östlund, "Electrical analysis of the Rhomberg test." *Acta Otolaryngol. Suppl.*, Vol. 224, 1967, pp. 272–279.

[E4] A. Ishida and S. Miyazaki, "Maximum likelihood identification of a posture control system." *IEEE Trans. Biomedical Engineering*, Vol. BME-34, 1987, pp. 1–5.

[E5] H. Östlund, (Ed.), "A study of aim and strategy of stability control in quasistationary standing." Report from Dept. of Neurology, Dept. of Research, S:t Lars Sjukhus, Lund, Sweden, 1979.

Table E.2 Results of parametric identification with data from Experiment D with 60 Hz vibration and closed eyes

Subject	η	k	ρ
1	6.09	49.25	18.67
2	4.46	43.99	10.46
3	3.64	32.15	14.85
4	2.90	10.44	4.39
5	6.89	47.79	28.68
6	4.97	49.45	31.99

Subject	**Swiftness**	**Stiffness**	**Damping**
1	2.65	7.00	2.29
2	2.19	9.20	2.04
3	2.46	5.32	1.48
4	1.64	3.90	1.77
5	3.06	5.10	2.25
6	3.17	4.91	1.57

The physiologic background to vibration-induced body sway is

[E6] G. EKLUND, "Some physical properties of muscle vibrators used to elicit tonic proprioceptive reflexes in man." *Acta Soc. Med. Upsal.*, Vol. 76, 1971, pp. 271–280.

[E7] G.M. GOODWIN, D.I. MCCLOSKEY, AND P.B.C. MATTHEWS, "The contribution of muscle afferents to kinesthesia shown by vibration induced illusion of movements and by the effects of paralyzing joint afferents." *Brain*, Vol. 95, 1972, pp. 705–748.

[E8] R. GRANIT, "The functional role of the muscle spindles—Facts and hypotheses." *Brain*, Vol. 98, 1975, pp. 531–556.

[E9] K. TAKANO AND S. HOMMA, "Muscle spindle responses to vibratory stimuli at certain frequencies." *The Japanese Journal of Physiology*, Vol. 18, 1968, pp. 145–156.

[E10] V.J. Wilson and G. Melville-Jones, *Mammalian Vestibular Physiology*. New York: Plenum Press, 1979.

Some references related to modeling of the human posture control system including the notions of "ankle strategy" and "hip strategy" are found in

[E11] L.M. Nashner, "Conceptual and biomechanical models of postural control: Strategies for organization of human posture." In M. Igarashi, O. Black (Eds.), *Vestibular and Visual Control on Posture and Locomotor Equilibrium*, 7th Int. Symp. Int. Soc. Posturography, Houston, Texas, 1983, pp. 1–8.

[E12] I. Pyykkö, G.Å. Hansson, L. Schalén, N.G. Henriksson, C. Wennmo, and M. Magnusson, "Vibration-induced body sway." In C.F. Claussen and M.V. Kirtane (Eds.), *Computers in Neuro-otologic Diagnosis*, Werner Rudat, 1983, pp. 139–155.

■

Index

G

H

I

Y

Z